Theoretical and Experimental Sonochemistry Involving Inorganic Systems

Pankaj • Muthupandian Ashokkumar
Editors

Theoretical and Experimental Sonochemistry Involving Inorganic Systems

Springer

Editors
Prof. Pankaj
Department of Chemistry
Faculty of Science
Dayalbagh Educational Institute
Agra 282 110 Uttar Pradesh
India
pankaj2@sancharnet.in;
executive_editor_jicc@yahoo.com

Prof. Muthupandian Ashokkumar
University of Melbourne
School of Chemistry
3010 Parkville Victoria
Australia
masho@unimelb.edu.au

ISBN 978-90-481-3886-9 e-ISBN 978-90-481-3887-6
DOI 10.1007/978-90-481-3887-6
Springer Dordrecht Heidelberg London New York

Library of Congress Control Number: 2010937992

Cover design: WMXDesign GmbH

Printed on acid-free paper

Springer is part of Springer Science+Business Media (www.springer.com)

Foreword

I began my research into Sonochemistry over 30 years ago now and at that time it was for me an exploration of the unknown. In 1988 with my colleague Phil Lorimer we wrote the first book to carry in its title the word Sonochemistry with a subtitle "Theory applications and uses of ultrasound in chemistry". In recent years, *Sonochemistry* has shown significant growth in a variety of fields no longer limited to chemistry with special attention being paid to materials science, environmental protection, food technology and therapy. Indeed the overall breadth of sonochemistry is expanding to such an extent that it now encompasses hybrid technologies involving combinations of ultrasound with electrochemistry, photochemistry and microwaves. In particular great attention has been focused on the synthesis of functional nano- and micro-particles involving both biological and inorganic materials.

The publication of new text books and monographs reflects the health of a subject and so it is with great pleasure that I write this Foreword for the book, *Theoretical and Experimental Sonochemistry Involving Inorganic Systems*, Edited by Professors Pankaj and Ashokkumar.

Theoretical and Experimental Sonochemistry Involving Inorganic Systems is a unique compilation of theoretical and experimental studies involving water based systems and chemical rather than biological species. This is really where sono-chemistry began and so it is appropriate to have the more recent studies in aqueous systems brought together in one volume. When ultrasound is introduced into such systems the chemistry becomes quite fascinating as a result of the influence of acoustic cavitation both from the points of view of chemical and physical effects. This book contains chapters that deal with various aspects of sonochemical research in aqueous solutions with a particular emphasis on inorganic systems. This will be an important text for all those interested in or directly involved with current sonochemistry research.

September 2010

Timothy J. Mason
Professor of Chemistry
Coventry University, UK

Preface

The themes of several books published in the field of sonochemistry revolve around physical and chemical aspects involving mainly organic chemistry or a combination of physics, chemistry and other areas. The sonochemical studies involving inorganic reactions, although numerous, are scarcely discussed and compiled in the existing literature. This prompted us to editing this book. This was welcomed and has been made successful by many contributors, as can be seen through various chapters of this book. Besides, the availability of a book devoted to inorganic systems in sonochemistry may also help undergraduate students, juvenile workers and senior researchers alike to learn about sonochemistry and publicize the sonochemistry research field to a much broader community.

The book offers a theoretical introduction in the first three chapters, provides recent applications in material science in the next four chapters, describes the effects of ultrasound in aqueous solutions in the following five chapters and finally discusses the most exciting phenomenon of sonoluminescence in aqueous solutions containing inorganic materials in subsequent two chapters, before ending with a few basic introductory experiments of sonochemistry and sonoluminescence in the concluding chapter.

Prof. Yasui discussed the fundamentals of acoustic cavitation and sonochemistry through the splitting of water to generate free radicals as a consequence of exceptionally high temperatures, pressures and mass flow conditions generated during acoustic cavitation in solutions. Dr. Gogate has discussed the design aspects of cavitation reactors and examined the effect of intensity and frequency of ultrasound, geometry of the reactor, physicochemical properties of liquids and the operational temperature on the intensity of cavitation for the maximization of process efficiencies. Later, Dr. Gogate and Prof. Pandit have described the phenomenon of hydrodynamic cavitation for the scale up operation of several physical, chemical and biological processes. Prof. Garcia has discussed the combined effects of electrochemistry and ultrasound for the production of gas, metal deposits and metal oxides, in addition to providing a summary of the fundamental aspects, experimental set-up and different applications of a rather new field of applied sonoelectrochemistry. Prof. Okitsu has illustrated the synthesis of metal nanoparticles and the effects of dissolved gases, rate of reduction and the concentration of organic additives on the size and shapes of nanoparticles. To advance the portrayal further, Assistant Prof. Anandan and Prof. Ashokkumar have provided additional information on the sonochemical preparation of monometallic, bimetallic and metal loaded semiconductor nanoparticles. In continuation with these reviews, Associate Prof. Sonawane and Dr. Kulkarni have described the sonochemical synthesis of nanocalcium carbonate through the acoustic and hydrodynamic cavitations. Associate Prof. Sivakumar has summarized various kinds of simple and mixed oxides and sulphides obtained in the last few years through sonochemical processes. Prof. Pankaj has discussed the effect of ultrasound propagation in aqueous solutions in the atmospheres of inert and reactive gases and the precipitation behavior of hydroxides of several di- and tri-valent metal ions, besides reporting the results of nephelometric and conductometric studies of sonicated solutions of these metal ions. Prof. Pankaj and Dr. Chauhan further reported the redox characteristics of ferrous and ferric ions in aqueous solutions and a comparative account of the oxidizing power of permanganate and dichromate ions, under the influence of ultrasound. In the next two chapters, Mr Verma and Prof. Pankaj have advanced the description of sonophotocatalytic degradation of phenol and several amines and also found a very interesting improvement of such degradation in the presence of rare earth ions, co-added with the photocatalyst, titanium dioxide. Other conventional methods for the degradation of these species in aqueous solutions have been compared with the sonochemical treatment processes. To explain a relatively difficult but equally fascinating consequence of high intensity ultrasound, Prof. Choi has discussed the phenomenon of sonoluminescence from aqueous solutions containing inorganic ions, especially alkali metal atom emission in aqueous solutions in various environments and described the emission mechanism, supporting the gas phase origin of the emission. Finally, Dr. Brotchie, Prof. Grieser and Prof. Ashokkumar have discussed the role of salts in acoustic cavitation and the use of inorganic complexes as cavitation probes to infer invaluable quantitative information regarding the temperature and pressure at the time of cavitation bubble collapse. Few basic experiments of sonochemistry and sonoluminescence have also been described in the last segment of the book.

Besides the contributors of various chapters, we also wish to acknowledge the support and critical evaluation of the chapters by several professionals (cannot be named due to confidentiality) who reviewed the articles in a timely manner.

We sincerely hope that this book is immensely beneficial to graduate students and researchers to learn the fundamental aspects of cavitation and to launch new research activities in the sonochemistry research field. The readers will also realize that sonochemistry is not just limited to "chemistry" but has the potential to incorporate in other areas including physics, engineering, biochemistry and medicine.

Agra, India
Melbourne, Australia
June 2010

Pankaj
Muthupandian Ashokkumar

About the Editors

Professor Pankaj is a graduate and Ph.D. from Lucknow University, India (1982) with specialization in Inorganic Chemistry and a victor of M Raman Nayer Gold Medal. From his initial work on the studies of solvent properties of non-aqueous solvents and later on the measurement of ultrasonic velocity, Prof. Pankaj switched over to sonochemical studies in aqueous solutions involving inorganic systems, after his European Community Post-Doctoral Fellowship (1990 – 91) at the Department of Physics, University of Surrey, UK. He has published ~50 papers in peer reviewed national and international journals and contributed chapters to 5 books. He is a recipient of grants from agencies such as UGC, AICTE, DST & DAE-BRNS. Prof. Pankaj is also the *Executive Editor* of the Journal of Indian Council Chemists and reviewer for several national and international journals like Canadian J Chemical Engineering; CLEAN – Soil, Air, Water; Ind. J. Chem. and Ind. J. Pure Appl. Ultrasonics. He is a Fellow of Ultrasonic Society of India and Indian Council of Chemists.

Professor Muthupandian Ashokkumar (Ashok) is a Physical Chemist who specializes in Sonochemistry, teaches undergraduate and postgraduate Chemistry and is a senior academic staff member of the School of Chemistry, University of Melbourne. Ashok is a renowned sonochemist who has developed a number of novel techniques to characterize acoustic cavitation bubbles and has made major contributions of applied sonochemistry to the Food and Dairy industry. His research team has developed a novel ultrasonic processing technology for improving the functional properties of dairy ingredients. Recent research also involves the ultrasonic synthesis of functional nano- and biomaterials including protein microspheres that can be used in diagnostic and therapeutic medicine. He is an Editorial Board Member of *Ultrasonics Sonochemistry*, an international journal devoted to sonochemistry research. He has edited/co-edited several books and special issues for journals; published ~200 refereed papers in high impact international journals and books; and delivered over 100 invited/keynote/plenary lectures at international conferences and academic institutions. Ashok is the recipient of several prizes, awards and fellowships, including the Grimwade Prize in Industrial Chemistry. He is a Fellow of the Royal Australian Chemical Institute.

Acknowledgement

Prof. Ram Gopal, my Ph.D. supervisor had rightly said about 30 years back that science could not be pursued lifelong unless enjoyed, without me realizing that it is eventually coming to happen in my life too. Towards the end of my Ph.D. work, in early eighties, I read about ultrasound – a word which fascinated me and I drifted towards measuring ultrasonic velocity in non-aqueous solvents of high dielectric constant and their solutions, interpreting their variations in terms of thermodynamic parameters. But the real contact with sonochemistry was through Prof. R.C. Chivers, my supervisor for European Community Post Doctoral fellowship at the University of Surrey, UK, who fixed my appointment with Prof. T.J. Mason in early nineties, when sonochemistry was still in its infancy. However, limitations of finance, space and culture of working in an apprehensive small Indian University later was always only an impediment but, of course, never a barrier to my motivation. The initial sonochemical results of my students, different from conventional chemical reactions of aqueous wet chemistry always stimulated me to stick to this field.

The idea of writing a book on inorganic sonochemistry originated way back in 2000 but turned to reality only recently when discussed with Dr. Sonia Ojo from Springer UK. She thankfully introduced me to Prof. Muthupandian Ashokkumar, University of Melbourne, Australia, a very well known name in the area of sonochemistry and sonoluminescence, to accomplish the task. Furthermore, I would also like to acknowledge Mrs. Claudia Culierat from Springer UK for her instantaneous and supportive attention to all my queries related to editorial assistance.

With all humility, I acknowledge the initial strength derived for this book from Dr. Ashok Kumar, Head, Ultrasonics Division, National Physical Laboratory, New Delhi, besides the unwavering encouragement from Dr. G.C. Saxena, Ex VC, Awadh University, Faizabad and Dr. BR Ambedkar University, Agra and Prof. P. Muruthamuthu, Ex VC Madurai Kamraj University, Dr. G.N. Pandey, BRNS and Dr. A.K. Tripathi, BARC.

I cannot close the eyes to admit a very patient, silent and supportive co-operation of my compassionate wife, Dr. Hemlata Srivastava and two considerate sons, Abhijit Srivastav and Arpit Srivastava, who suffered seclusion and neglect due to my invariable involvement with the book for about 1 year. Last, but not the least my gratitude to all my friends and well wishers, who through their admiration or criticism added directly or indirectly to my strength and inspired vigorously to complete the task in the form as it is today.

Agra, India Prof. Pankaj
June 2010

Contents

Chapter 1
Fundamentals of Acoustic Cavitation and Sonochemistry

Kyuichi Yasui

Abstract Acoustic cavitation is the formation and collapse of bubbles in liquid irradiated by intense ultrasound. The speed of the bubble collapse sometimes reaches the sound velocity in the liquid. Accordingly, the bubble collapse becomes a quasi-adiabatic process. The temperature and pressure inside a bubble increase to thousands of Kelvin and thousands of bars, respectively. As a result, water vapor and oxygen, if present, are dissociated inside a bubble and oxidants such as OH, O, and H_2O_2 are produced, which is called sonochemical reactions. The pulsation of active bubbles is intrinsically nonlinear. In the present review, fundamentals of acoustic cavitation, sonochemistry, and acoustic fields in sonochemical reactors have been discussed.

1.1 Introduction

An acoustic wave (sound) is a propagation of pressure oscillation in medium such as air or liquid water with the sound velocity [1]. Ultrasound is inaudible sound and its frequency of pressure oscillation is above 20 kHz (20,000 oscillations per second) [2]. For convenience, an acoustic wave above 10 kHz in frequency is sometimes called an ultrasonic wave.

When the pressure amplitude of an acoustic wave in liquid or solid exceeds the ambient pressure (atmospheric pressure), the instantaneous pressure becomes *negative* during the rarefaction phase of an acoustic wave. *Negative* pressure is defined as the force acting on the surface of a liquid (or solid) element per surface area to expand the element [3, 4]. For example, consider a closed cylinder filled with liquid

K. Yasui (✉)
National Institute of Advanced Industrial Science and Technology (AIST), 2266-98 Anagahora, Shimoshidami, Moriyama-ku, Nagoya 463-8560, Japan
e-mail: k.yasui@aist.go.jp

Pankaj and M. Ashokkumar (eds.), *Theoretical and Experimental Sonochemistry Involving Inorganic Systems*, DOI 10.1007/978-90-481-3887-6_1,

with a movable piston. When a piston is pulled strongly, the liquid volume slightly increases. At this moment, the pressure in the liquid is *negative*. *Negative* pressure is possible only in liquid or solid.

When the instantaneous local pressure becomes *negative* in liquid irradiated by ultrasound, bubbles are generated because gas such as air dissolved in the liquid can no longer be dissolved in the liquid under *negative* pressure, which is called acoustic cavitation [5, 6]. For a static condition, vapor bubbles are generated when the static pressure is lower than the saturated vapor pressure, which is called boiling. In many cases of acoustic cavitation, the instantaneous local pressure should be negative because the duration of low pressure is short.

The difference between acoustic cavitation and boiling is the collapse of bubbles in acoustic cavitation. Under ultrasound, a generated bubble expands during the rarefaction phase and collapses during the compression phase. The speed of the bubble collapse increases to the sound velocity in liquid. Accordingly, the bubble collapse is a quasi-adiabatic process where "quasi" means that considerable thermal conduction takes place between the interior of a bubble and the surrounding liquid. The temperature and pressure inside a bubble increase to thousands of Kelvin and thousands of bars, respectively at the end of the bubble collapse [7]. Furthermore, a bubble emits a shock wave into the surrounding liquid just after the end of the bubble collapse [8-11]. This bubble collapse is absent in boiling.

As the temperature and pressure dramatically increase inside a bubble at the end of the collapse, water vapor and oxygen, if present, are dissociated inside a bubble and oxidants such as OH, O, and H_2O_2 are created [12, 13]. They dissolve into the liquid and solutes are oxidized by them. This is called sonochemical reaction. For example, potassium iodide (KI) in aqueous solution is oxidized by the irradiation of ultrasound ((1.1)), and the solution is gradually colored by the product (I_3^-) as the irradiation time increases.

$$3I^- + 2OH \rightarrow I_3^- + 2OH^- \tag{1.1}$$

1.2 Acoustic Cavitation

1.2.1 Transient and Stable Cavitation

There are two types in acoustic cavitation. One is transient cavitation and the other is stable cavitation [14, 15]. There are two definitions in transient cavitation. One is that the lifetime of a bubble is relatively short such as one or a few acoustic cycles as a bubble is fragmented into daughter bubbles due to its shape instability. The other is that bubbles are active in light emission (sonoluminescence (SL)) or chemical reactions (sonochemical reactions). Accordingly, there are two definitions in stable cavitation. One is that bubbles are shape stable and have a long lifetime. The other is that bubbles are inactive in SL and chemical reactions. There exist some bubbles which are both shape stable and active in SL or chemical reactions. They are classified into stable cavitation bubbles by the former definition and called "high-energy stable cavitation" bubbles. On the other hand, they are classified into transient cavitation bubbles by the latter definition and called "repetitive transient cavitation" bubbles. Whenever the terms transient and stable cavitation are used, it is necessary to indicate which definition is used, shape stability or activity.

In Fig. 1.1, the parameter space for transient and stable cavitation bubbles is shown in R_0 (ambient bubble radius) – p_a (acoustic amplitude) plane [15]. The ambient bubble radius is defined as the bubble radius when an acoustic wave (ultrasound) is absent. The acoustic amplitude is defined as the pressure amplitude of an acoustic wave (ultrasound). Here, transient and stable cavitation bubbles are defined by their shape stability. This is the result of numerical simulations of bubble pulsations. Above the thickest line, bubbles are those of transient cavitation. Below the thickest line, bubbles are those of stable cavitation. Near the left upper side, there is a region for bubbles of "high-energy stable cavitation" designated by "Stable (strong nf_0)". In the brackets, the type of acoustic cavitation noise is indicated. The acoustic cavitation noise is defined as acoustic emissions from

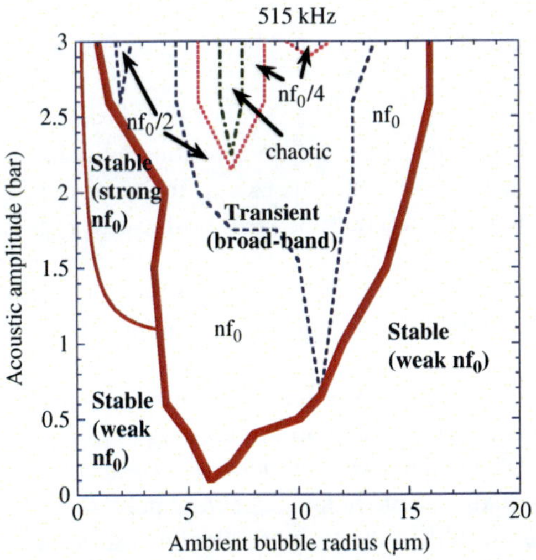

Fig. 1.1 The regions for "transient" cavitation bubbles and "stable" cavitation bubbles when they are defined by the shape stability of bubbles in the parameter space of ambient bubble radius (R_0) and the acoustic amplitude (p_a). The ultrasonic frequency is 515 kHz. The thickest line is the border between the region for "stable" cavitation bubbles and that for "transient" ones. The type of bubble pulsation has been indicated by the frequency spectrum of acoustic cavitation noise such as nf_0 (periodic pulsation with the acoustic period), $nf_0/2$ (doubled acoustic period), $nf_0/4$ (quadrupled acoustic period), and chaotic (non-periodic pulsation). Any "transient" cavitation bubbles result in the broad-band noise due to the temporal fluctuation in the number of bubbles. Reprinted from Ultrasonics Sonochemistry, vol. 17, K.Yasui, T.Tuziuti, J. Lee, T.Kozuka, A.Towata, and Y. Iida, Numerical simulations of acoustic cavitation noise with the temporal fluctuation in the number of bubbles, pp. 460–472, Copyright (2010), with permission from Elsevier

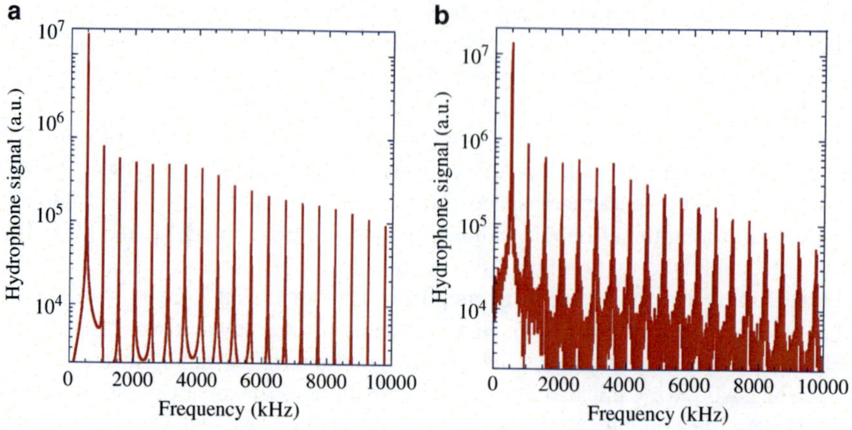

Fig. 1.2 Numerically simulated frequency spectra of the hydrophone signal due to acoustic cavitation noise. The driving ultrasound is 515 kHz in frequency and 2.6 bar in pressure amplitude. (**a**) For stable cavitation bubbles of 1.5 μm in ambient radius. (**b**) For transient cavitation bubbles of 3 μm in ambient radius. Reprinted from Ultrasonics Sonochemistry, vol. 17, K. Yasui, T. Tuziuti, J. Lee, T. Kozuka, A. Towata, and Y. Iida, Numerical simulations of acoustic cavitation noise with the temporal fluctuation in the number of bubbles, pp. 460–472, Copyright (2010), with permission from Elsevier

acoustic cavitation. Every pulsating bubble under ultrasound radiates a secondary acoustic wave, which is the origin of the acoustic cavitation noise. "Strong nf_0" means that strong harmonics components are to be observed in the frequency spectrum of acoustic cavitation noise (Fig. 1.2a) [15]. In Fig. 1.2a, the harmonics components are seen as sharp peaks. From the other space in Fig. 1.1 for stable cavitation bubbles designated by "stable (weak nf_0)", the harmonics components are very weak as the bubble pulsation is much milder. From transient cavitation bubbles, broad-band noise is to be observed as well as the harmonics components (Fig.1.2b). The broad-band noise is the continuum component in the frequency spectrum of the acoustic cavitation noise. According to Ref. [15], temporal fluctuation in the number of bubbles results in the broad-band noise. In transient cavitation, bubbles occasionally fragment into daughter bubbles, coalesce each other, and are nucleated, which results in the temporal fluctuation in the number of bubbles. Some bubbles pulsate with the period of 2T or 4T where T is the acoustic period, which results in the subharmonic and ultraharmonic components in the acoustic cavitation noise. The relationship between the type of cavitation and the acoustic cavitation noise is listed in Table 1.1. Some bubbles pulsate non-periodically as designated by "chaotic", which also results in the broad-band noise without any peaks. However, its contribution to the total broad-band noise is minor at least under the condition studied in Fig. 1.1 [15].

From Fig. 1.1, it is seen that stable cavitation bubbles are tiny bubbles of a few μm in ambient radius or relatively large bubbles of about 10 μm or more in radius at 515 kHz. The range of ambient radius for transient cavitation bubbles becomes

Table 1.1 The relationship between the type of cavitation bubbles and that of the cavitation noise spectrum in the parameter space shown in Fig. 1.1. "Chaotic (initial transient)" means non-periodic pulsation only at the initial transient stage although the pulsation becomes periodic at the steady-state. Reprinted from Ultrasonics Sonochemistry, vol. 17, K.Yasui, T. Tuziuti, J. Lee, T. Kozuka, A. Towata, and Y. Iida, Numerical simulations of acoustic cavitation noise with the temporal fluctuation in the number of bubbles, pp. 460–472, Copyright (2010), with permission from Elsevier

Type	Pulsation	Noise spectrum
Stable (Low energy)	Periodic (period T)	Weak nf_0
(High energy)	Periodic (period T)	Strong nf_0
Transient	Periodic (period T)	nf_0 + broad-band
	Periodic (period 2T)	$nf_0/2$ + broad-band
	Periodic (period 4T)	$nf_0/4$ + broad-band
	Chaotic (steady-state)	Broad-band
	Chaotic (initial transient)	Broad-band

wider as the acoustic amplitude increases and ultrasonic frequency decreases [16]. Roughly speaking, stable cavitation bubbles are more frequently seen when acoustic amplitude is lower and ultrasonic frequency is higher [17]. In other words, transient cavitation bubbles are more frequently seen when acoustic amplitude is higher and ultrasonic frequency is lower. It is possible to control the population of stable cavitation bubbles relative to that of transient ones by the addition of a surfactant to the liquid because coalescence of bubbles is strongly retarded by a surfactant and the ambient radius of bubbles becomes sufficiently small for stable cavitation bubbles as described in 1.3.4 [15, 18, 19].

There is another experimental method to distinguish between transient and stable cavitation bubbles. According to the experiment of Guan and Matula [20], light emission (sonoluminescence (SL)) from a pulsating bubble is completely quenched in the presence of methanol in liquid water after about 8000 acoustic cycles. It is because quenching requires the repetitive injection of alcohol molecules into a pulsating bubble resulting in the accumulation of hydrocarbon products within the bubble. It means that only stable cavitation bubbles which have a long lifetime exhibit complete SL quenching in the presence of alcohol. In other words, transient cavitation bubbles do not exhibit complete SL quenching. Thus, by the experimental observation of SL quenching by the addition of alcohol, transient and stable cavitation bubbles can be distinguished [17]. It should be noted that transient and stable cavitation are defined here by the lifetime of bubbles (shape stability).

1.2.2 Nucleation of Bubbles

How is a bubble created in acoustic cavitation? There are three mechanisms in nucleation of a bubble in acoustic cavitation [14]. One is the nucleation at the surface of solids such as a liquid container, motes or particles in liquid, if present. Nucleation takes place especially at crevices of motes, particles or a liquid container (Fig. 1.3).

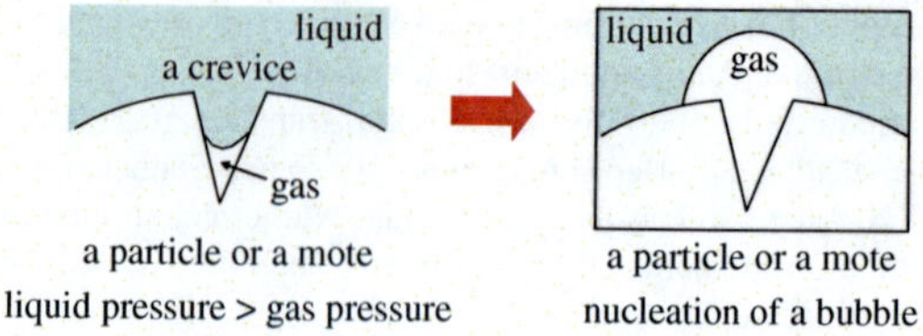

Fig. 1.3 Nucleation of a bubble from a crevice

In a crevice, the surface of a gas pocket is concave and the surface tension of a gas pocket *reduces* the pressure inside a pocket. It means that a gas pocket is stabilized against dissolution into the liquid because the partial pressure of dissolved gas in the liquid is possibly higher than that in a gas pocket. When the liquid is irradiated by ultrasound, a gas pocket in a crevice expands during the rarefaction phase of ultrasound and gas diffuses into the pocket from the surrounding liquid as the pressure inside a pocket further decreases. During the compression phase of ultrasound, a gas pocket shrinks and the gas pressure increases in a pocket. It results in the diffusion of gas out of a pocket into the liquid. Nevertheless, a gas pocket grows as the amount of gas diffusing into the pocket during the expansion is larger than that diffusing out of the pocket during the compression. This is because the surface area of a pocket is larger during expansion than that during compression (the area effect). Furthermore, the boundary layer for gas diffusion in the liquid is thinner during expansion than that during compression because the volume of the boundary layer is nearly constant and the surface area is larger during expansion (the shell effect). It results in the higher rate of diffusion during expansion because the gradient in concentration of the gas in the liquid is larger. Finally, a gas bubble is created from a crevice when the gas pocket sufficiently grows. The presence of particles in liquid reduces the threshold acoustic pressure for cavitation due to the above mechanism [21–23].

The second mechanism for nucleation is the initially present bubble nuclei which are stabilized against dissolution by the coverage of its surface with surfactants slightly present in the liquid as impurities. Without surfactants, bubbles with radius smaller than 1 μm should dissolve within a few seconds in the absence of ultrasound unless the liquid is supersaturated with gas [24]. It is because the gas pressure inside a bubble is larger than the partial pressure of the dissolved gas in the liquid due to surface tension of a bubble (The excess pressure of the gas inside a bubble is $\Delta p = 2\sigma/R$, where σ is the surface tension and R is the bubble radius). Thus the gas inside a bubble gradually dissolves into the surrounding liquid. On the other hand, bubbles larger than 1 μm in radius should float to the liquid surface by a buoyant force. Nevertheless, bubble nuclei (tiny bubbles of a few μm in radius) have been experimentally observed in liquids even in the absence of an acoustic wave (ultrasound) [6]. It suggests that such bubble nuclei are stabilized by surfactants which strongly retard the mass (gas) diffusion across the bubble surface. Under ultrasound, these stabilized nuclei grow by coalescence and gas diffusion, which initiates acoustic cavitation.

The third mechanism for nucleation is the fragmentation of active cavitation bubbles [16]. A shape unstable bubble is fragmented into several daughter bubbles which are new nuclei for cavitation bubbles. Shape instability of a bubble is mostly induced by an asymmetric acoustic environment such as the presence of a neighboring bubble, solid object, liquid surface, or a traveling ultrasound, or an asymmetric liquid container etc. [25–27] Under some condition, a bubble jets many tiny bubbles which are new nuclei [6, 28]. This mechanism is important after acoustic cavitation is fully started.

1.2.3 Growth of a Bubble

There are two mechanisms in growth of a bubble in acoustic cavitation [14]. One is coalescence of bubbles. The other is the gas diffusion into a bubble due to the area and shell effects described before. This is called rectified diffusion.

The coalescence of bubbles is driven by the two mechanisms. One is the attractive radiation force between bubbles called secondary Bjerknes force. The other is the other radiation force called the primary Bjerknes force which drives active bubbles to the pressure antinode of a standing wave field. It should be noted, however, too strong acoustic wave repels bubbles from the pressure antinode as described in the next section [29, 30].

The bubble growth rate due to rectified diffusion strongly depends on acoustic amplitude and frequency. For a very weak driving such as 0.2 bar at 20 kHz, the bubble growth rate is in the order of a few μm per 100s for the initial radius of 35 μm [31, 32]. For a much stronger driving such as 2 bar at 30 kHz, it ranges from 10 to a few hundred μm per second depending on the initial radius [33]. It decreases as ultrasonic frequency increases for the same acoustic pressure amplitude.

Relative importance of coalescence and rectified diffusion in the bubble growth is still under debate. After acoustic cavitation is fully started, coalescence of bubbles may be the main mechanism of the bubble growth [16, 34]. On the other hand, at the initial development of acoustic cavitation, rectified diffusion may be the main mechanism as the rate of coalescence is proportional to the square of the number density of bubbles which should be small at the initial stage of acoustic cavitation. Further studies are required on this subject.

1.2.4 Radiation Forces on a Bubble (Primary and Secondary Bjerknes Forces)

Both the primary and secondary Bjerknes forces are originated from the pressure gradient across a bubble [35].

$$\vec{F}_B = -\langle V(t)\nabla p(\vec{x},t)\rangle \tag{1.2}$$

where $\vec{F}_B$ is the primary or secondary Bjerknes force, $V(t)$ is the instantaneous bubble volume, $\nabla = \left(\frac{\partial}{\partial x}, \frac{\partial}{\partial y}, \frac{\partial}{\partial z}\right)$ in xyz-coordinate, $p(\vec{x},t)$ is the instantaneous local pressure at position $\vec{x}$, and $\langle\,\rangle$ denotes the time average. For the primary Bjerknes force, $p(\vec{x},t)$ is the driving ultrasound. On the other hand, for the secondary Bjerknes force, $p(\vec{x},t)$ is the acoustic wave radiated by a neighboring bubble.

When the driving ultrasound is a standing wave, $p(\vec{x},t)$ is expressed as follows for the primary Bjerknes force.

$$p(\vec{x},t) = p_a \cos(\vec{k}\cdot\vec{x}) \sin \omega t \tag{1.3}$$

where p_a is the acoustic pressure amplitude, $\vec{k}$ is the wave vector, and ω is the angular frequency. Then, the primary Bjerknes force ($\vec{F}_{PB}$) is given by the following equation.

$$\vec{F}_{PB} = p_a\vec{k}\sin(\vec{k}\cdot\vec{x})\langle V(t)\sin\omega t\rangle \tag{1.4}$$

When the bubble pulsation is in phase with the driving ultrasound, a bubble is attracted to the pressure antinode of a standing wave field. For a very low driving such as less than 0.1 bar in acoustic amplitude, bubble pulsation is nearly linear and this condition coincides with smaller ambient radius than the linear resonance radius. For active bubbles, pulsation is strongly nonlinear, and the situation is more complex. For example, at 20 kHz, an active bubble of smaller ambient radius than the linear resonance radius is repelled from the pressure antinode when the acoustic pressure amplitude is larger than about 1.8 bar [29, 30]. This is because a bubble continues expanding even during the compression phase of ultrasound due to the inertia of the surrounding liquid. To active bubbles ("transient cavitation bubbles" by the definition based on activity of bubbles), the linear theory can not be applied.

In a traveling wave of ultrasound, most active bubbles are pushed toward the direction of the wave propagation by the primary Bjerknes force. Furthermore, there is a fluid flow in the direction of the wave propagation called acoustic streaming [36]. Acoustic streaming is caused by the attenuation of a traveling wave resulting in the net radiation force in the direction of the wave propagation. The attenuation is caused by both viscosity of the liquid and the cavitation bubbles. Furthermore, moving bubbles driven by the primary Bjerknes force drag the surrounding fluid. The resulting fluid flow is called quasi acoustic streaming [37].

While the secondary Bjerknes force is always attractive if the ambient radius is the same between bubbles, it can be repulsive if the ambient radius is different [38]. The magnitude as well as the sign of the secondary Bjerknes force is a strong function of the ambient bubble radii of two bubbles, the acoustic pressure amplitude, and the acoustic frequency. It is calculated by (1.5).

$$\vec{F}_{1\to 2} = \frac{\rho}{4\pi d^2}\langle \ddot{V}_1 V_2\rangle \vec{e}_r \tag{1.5}$$

where $\vec{F}_{1\to2}$ is the secondary Bjerknes force acting on bubble 2 from bubble 1, ρ is the liquid density, d is the distance between the bubbles 1 and 2, $\ddot{V}_1$ is the second time derivative of the volume of bubble 1, V_2 is the volume of bubble 2, $\langle\,\rangle$ denotes the time average, and $\vec{e}_r$ is the radial unit vector directed from bubble 1 to bubble 2. For a very low driving (such as less than 0.1 bar in acoustic amplitude) or for very large bubbles, the bubble pulsation is nearly linear and bubbles with the ambient radii both less (more) than the linear resonance radius pulsate in phase resulting in the attractive secondary Bjerknes force [14]. On the other hand, when the ambient radius of a bubble is less than the linear resonance radius and that of the other bubble is more than it, it is repulsive. For active bubbles, however, the bubble pulsation is strongly nonlinear and the situation is much more complex [38]. The theory for linear pulsation of bubbles should not be applied to active bubbles as already noted [39].

1.2.5 Bubble Radial Dynamics

Bubble radial dynamics is well described by the Rayleigh-Plesset equation or its modified version such as Keller and Herring equations [40]. The Rayleigh-Plesset equation is derived as follows [14]. Consider a liquid volume surrounding a pulsating bubble such that the liquid volume is much larger than the bubble volume and that the radius of the liquid volume is much smaller than the acoustic wavelength. The kinetic energy (E_K) of the liquid volume is given by (1.6).

$$E_K = \frac{1}{2}\rho \int_R^{R_L} \dot{r}^2 4\pi r^2 dr = 2\pi\rho R^3 \dot{R}^2 \tag{1.6}$$

where ρ is the liquid density, R_L is the radius of the liquid volume, R is the instantaneous bubble radius, r is the radial distance from the bubble center, the dot denotes the time derivative, and the liquid incompressibility condition ($\dot{r}/\dot{R} = R^2/r^2$) as well as the condition $R \ll R_L$ has been used. The work done by a pulsating bubble (W_{bubble}) to the surrounding liquid is given by (1.7).

$$W_{bubble} = \int_{R_0}^{R} 4\pi r^2 p_B dr \tag{1.7}$$

where R_0 is the ambient bubble radius, and p_B is the liquid pressure at the bubble wall. If the liquid is incompressible, the liquid volume does some work to the surrounding liquid (W_{liquid}) as it moves outward associated with the bubble expansion.

$$W_{liquid} = \int_{R_0}^{R} 4\pi r^2 p_\infty dr \tag{1.8}$$

where p_∞ is the pressure at the surface of the liquid volume including the acoustic pressure. The conservation of energy requires the following relationship.

$$W_{bubble} = E_K + W_{liquid}. \tag{1.9}$$

Differentiation of (1.9) with respect to R results in the following equation.

$$\frac{p_B - p_\infty}{\rho} = \frac{3\dot{R}^2}{2} + R\ddot{R} \tag{1.10}$$

where the following relationship has been used.

$$\frac{\partial\left(\dot{R}^2\right)}{\partial R} = \frac{1}{\dot{R}}\frac{\partial\left(\dot{R}^2\right)}{\partial t} = 2\ddot{R} \tag{1.11}$$

The liquid pressure at the bubble wall (p_B) is related to the gas pressure inside a bubble (p_g) as follows [14].

$$p_B = p_g - \frac{2\sigma}{R} - \frac{4\mu\dot{R}}{R} \tag{1.12}$$

where σ is the surface tension, and μ is the liquid viscosity. The pressure at the surface of the liquid volume is the sum of the acoustic pressure ($p_s(t)$) and the ambient static pressure (p_0). Then the Rayleigh-Plesset equation is derived from (1.10).

$$R\ddot{R} + \frac{3\dot{R}^2}{2} = \frac{1}{\rho}\left(p_g - \frac{2\sigma}{R} - \frac{4\mu\dot{R}}{R} - p_0 - p_s(t)\right) \tag{1.13}$$

In this equation, liquid has been assumed as incompressible. In the following Keller and Herring equations, the liquid compressibility has been taken into account to the first order of $\dot{R}/c_\infty$, where c_∞ is the sound velocity in the liquid far from a bubble [41].

$$\begin{aligned}&\left(1 - (\lambda + 1)\frac{\dot{R}}{c_\infty}\right)R\ddot{R} + \frac{3\dot{R}^2}{2}\left(1 - \frac{1}{3}(3\lambda + 1)\frac{\dot{R}}{c_\infty}\right)\\&= \frac{1}{\rho}\left(1 + (1 - \lambda)\frac{\dot{R}}{c_\infty}\right)\left[p_B - p_s\left(t + \frac{R}{c_\infty}\right) - p_0\right] + \frac{R}{c_\infty\rho}\frac{dp_B}{dt}\end{aligned} \tag{1.14}$$

where $\lambda = 0$ or 1 for Keller or Herring equation, respectively, and $p_s\left(t + \frac{R}{c_\infty}\right)$ is the instantaneous acoustic pressure at time $t + \frac{R}{c_\infty}$. As a similar equation, Gilmore equation has also been widely used [41].

1.2.6 *Inertial Collapse (Rayleigh Collapse)*

Now the bubble collapse is discussed using the Rayleigh-Plesset equation. After the bubble expansion, a bubble collapses. During the bubble collapse, important terms in the Rayleigh-Plesset equation are the two terms in the left hand side of (1.13). Then, the bubble wall acceleration is expressed as follows.

$$\ddot{R} = -\frac{3\dot{R}^2}{2R} \tag{1.15}$$

Thus, it is always negative. It means that the speed of the bubble collapse increases with time (The negative bubble-wall velocity further decreases). As the speed of the bubble collapse increases, the magnitude of the bubble wall acceleration increases according to (1.15). It means that the speed of the bubble collapse automatically increases more and more with time. This is caused by the inertia of the surrounding liquid ingoing into a collapsing bubble as well as the spherically shrinking geometry. Such a bubble collapse is called inertial collapse or Rayleigh collapse [40].

It has been shown theoretically that the speed of the bubble collapse is limited by the sound speed in the liquid at the bubble wall [42]. The sound speed is a function of pressure and density of the liquid as follows.

$$c_{L,B} = \sqrt{7.15(p_B + B)/\rho_{L,i}} \tag{1.16}$$

where $c_{L,B}$ is the sound speed in the liquid at the bubble wall, $B = 3.049 \times 10^8$ Pa, and $\rho_{L,i}$ is the liquid density at the bubble wall. The sound speed ($c_{L,B}$) increases as the bubble collapses up to about 3000 m/s (about two times of the ambient sound speed (1500 m/s) in water), which is the upper limit for the speed of the bubble collapse.

Finally, the bubble collapse stops when the pressure inside a bubble (p_g) in the right hand side of (1.13) dramatically increases as the density inside a bubble nearly reaches that of a condensed phase (A bubble is almost completely occupied by the van der Waals hard-cores of gas and vapor molecules at that moment). At the same time, the temperature and pressure inside a bubble dramatically increase.

In Fig. 1.4a, an example of the radius-time curve for a stably pulsating bubble calculated by the modified Keller equation is shown for one acoustic cycle [43]. After the bubble expansion during the rarefaction phase of ultrasound, a bubble strongly collapses, which is the inertial or Rayleigh collapse. After the collapse, there is a bouncing radial motion of a bubble. In Fig.1.4b, the calculated flux of OH

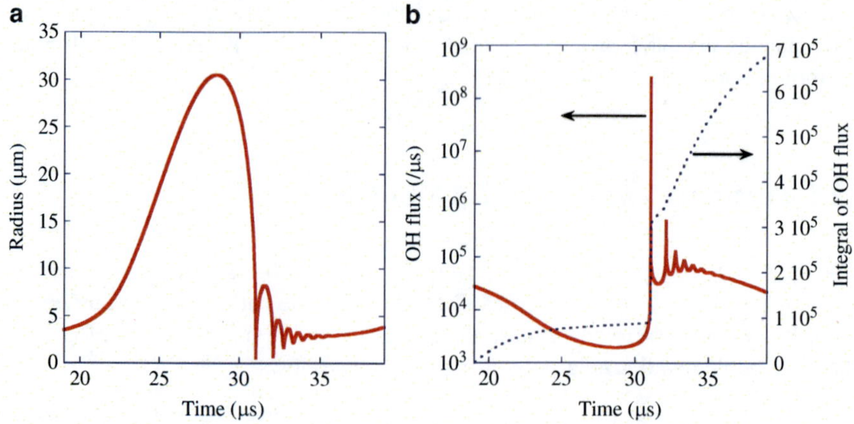

Fig. 1.4 The calculated results for one acoustic cycle when a bubble in water at 3 °C is irradiated by an ultrasonic wave of 52 kHz and 1.52 bar in frequency and pressure amplitude, respectively. The ambient bubble radius is 3.6 μm. (**a**) The bubble radius. (**b**) The dissolution rate of OH radicals into the liquid from the interior of the bubble (solid line) and its time integral (dotted line). Reprinted with permission from Yasui K, Tuziuti T, Sivakumar M, Iida Y (2005) Theoretical study of single-bubble sonochemistry. J Chem Phys 122:224706. Copyright 2005, American Institute of Physics

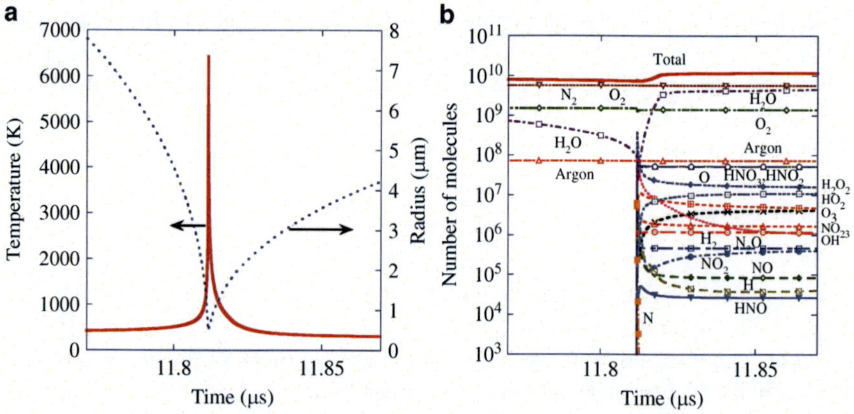

Fig. 1.5 The calculated results for an air bubble at around the end of the inertial collapse only for 0.1 μs. (**a**) The bubble radius and the temperature inside a bubble. (**b**) The number of molecules inside a bubble. Reprinted with permission from Yasui K, Tuziuti T, Sivakumar M, Iida Y (2005) Theoretical study of single-bubble sonochemistry. J Chem Phys 122:224706. Copyright 2005, American Institute of Physics

radicals from a bubble is shown as a function of time for one acoustic cycle [43]. OH radicals are created at each strong collapse, which diffuses out of a bubble into the surrounding liquid especially at the end of the strong collapse. In Fig. 1.5, the details of the inertial or Rayleigh collapse are shown [43]. In Fig. 1.5a, the bubble radius and the temperature inside an air bubble are shown as a function of time only for 0.1 μs at around the end of the collapse. The temperature dramatically increases at the end of the collapse to 6500 K. As a result, water vapor, oxygen and nitrogen are dissociated inside a bubble and many chemical species are created as shown in Fig. 1.5b. The production of oxidants inside a bubble such as OH, O, and H_2O_2 is one of the major origins of sonochemical reactions. In other words, the inertial or Rayleigh collapse is essential for sonochemistry.

1.3 Sonochemistry

1.3.1 Single-Bubble Sonochemistry

In most of the experiments of sonochemical reactions, there are a lot of cavitation bubbles in liquid, which is called a multibubble system. It is a complicated system because bubbles interact with each other and the local acoustic field temporally changes because moving bubbles attenuate and scatter an acoustic wave. Bubbles frequently coalesce each other, fragment into daughter bubbles, grow by rectified diffusion, shrink by gas diffusion, and spatially move due to radiation forces. New bubbles are also created. The bubble pulsation as well as the acoustic wave propagation is strongly nonlinear [44, 45]. The motions of bubbles are also complex [35]. Thus, as a first step, we shall consider a much simpler system; a single-bubble system. The single-bubble system is that a single bubble is trapped at the pressure antinode of a standing wave field [40]. It can be stably driven when the acoustic pressure is lower than the threshold for the shape instability of a bubble. The acoustic pressure should also be lower than the threshold for the bubble repulsion from the pressure antinode (about 1.8 bar at 20 kHz). The light emission from a stable bubble in the single-bubble system has been intensively studied, which is called single-bubble sonoluminescence (SBSL) [40]. The light emission of SBSL in water is due to several light emission processes in weakly ionized gases inside a bubble such as electron-atom bremsstrahlung which is the light emission from free electrons accelerated by collisions with neutral atoms [46–48]. It is not an electric discharge but thermal plasma in which free electrons are created by high temperature and pressure inside a bubble. On the other hand, the mechanism of SBSL in sulfuric acid, which is much brighter than that in pure water, has not yet been solved [49].

Hatanaka et al. [50], Didenko and Suslick [51], and Koda et al. [52] reported the experiment of chemical reactions in a single-bubble system called single-bubble sonochemistry. Didenko and Suslick [51] reported that the amount of OH radicals produced by a single bubble per acoustic cycle was about $10^5 \sim 10^6$ molecules at 52 kHz and $1.3 \sim 1.55$ bar in ultrasonic frequency and pressure amplitude, respectively. The result of a numerical simulation shown in Fig. 1.4 [43] is under the condition of the experiment of Didenko and Suslick [51]. The amount of OH radicals calculated by the numerical simulation sufficiently agrees with the experimental data. It indicates that the theoretical model used in the numerical simulation is sufficiently accurate at least under the condition.

According to Didenko and Suslick [51], the amount of nitrite ions (NO_2^-) produced by a bubble per acoustic cycle was about $10^6 \sim 10^7$ molecules. Koda et al. [52] also reported a similar value.

1.3.2 Optimal Bubble Temperature for Oxidant Production

Using a chemical kinetics model for flames, numerical simulations of chemical reactions inside an air bubble have been performed under various conditions. The theoretical model of a bubble including the chemical kinetics model has been validated through a study of single-bubble sonochemistry [43]. It has been clarified by numerical simulations that there exits an optimal bubble temperature for the production of oxidants inside an air bubble such as OH, O, H_2O_2, and O_3 [53, 54]. The optimal temperature is about 5,500 K because at higher temperature oxidants are strongly consumed inside an air bubble by oxidizing nitrogen (Fig. 1.6a). On the other hand, for an oxygen bubble, the amount of oxidants increases as the bubble temperature increases as oxidants are not consumed inside an oxygen bubble (Fig. 1.6b).

Experimentally, Brotchie et al. [55] have shown that the range of ambient radius of sonoluminescing (SL) bubbles in which the temperature is relatively high

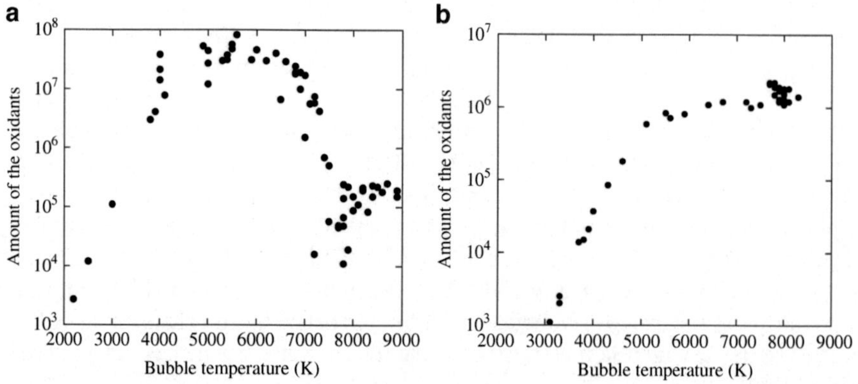

Fig. 1.6 The correlation between the bubble temperature at the collapse and the amount of the oxidants created inside a bubble per collapse in number of molecules. The calculated results for various ambient pressures and acoustic amplitudes are plotted. The temperature of liquid water is 20 °C. (**a**) For an air bubble of 5 μm in ambient radius at 140 kHz in ultrasonic frequency. (**b**) For an oxygen bubble of 0.5 μm in ambient radius at 1 MHz. Reprinted with permission from Yasui K, Tuziuti T, Iida Y, Mitome H (2003) Theoretical study of the ambient-pressure dependence of sonochemical reactions. J Chem Phys 119:346–356. Copyright 2003, American Institute of Physics

completely differs from that of chemically active bubbles which produce oxidants and glow by chemiluminescence in an aqueous luminol solution saturated with air. It suggests that the temperature inside chemically active bubbles is lower than that inside SL bubbles for an air bubble. It agrees with the results of the numerical simulations described above. The experimental method to measure the range of ambient radius of SL bubbles (or sonochemiluminescing (SCL) bubbles in an aqueous luminol solution) is based on the dissolution of bubbles during the pulse-off time of pulsed ultrasound [19]. Bubbles with larger ambient radius need longer pulse-off time for complete dissolution. As the pulse-off time increases, the SL (or SCL) intensity decreases because more bubbles dissolve during the pulse-off time and the number of bubbles decreases. From the dependence of the SL (or SCL) intensity on the pulse-off time, the range of ambient radius of SL (or SCL) bubbles is deduced [19].

1.3.3 Three Sites for Chemical Reactions

There are three sites for chemical reactions for a cavitation bubble as shown in Fig. 1.7 [56]. One is the interior of a bubble. Another is the interface region at around the bubble surface. The other is the liquid region outside the interface region. The liquid region is at the ambient temperature where chemical species with a relatively long lifetime such as H_2O_2 diffusing out of the interface region chemically react with solutes. In the interface region, the temperature dramatically increases due to the thermal conduction from the heated interior of a bubble where radicals with a relatively short lifetime such as OH and O react with solutes or radicals themselves. However, the actual temperature in the interface region is not known while several authors have estimated it [57–59]. Some researchers have suggested that in the interface region there is supercritical water [60]. Surfactants absorbed at the bubble surface can dissociate at the interface region due to both heat and radical attack [61]. In the interior of a bubble, volatile solutes which evaporate into the region are dissociated by high temperature [62].

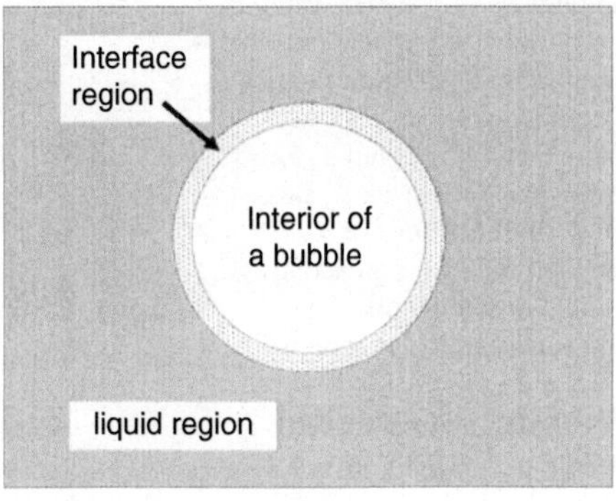

Fig. 1.7 Three sites for chemical reactions for a cavitation bubble

As stated above, more studies are required in future with regard to the interface region. What is the temperature and pressure in the interface region? What is the lifetime of OH radicals and O atoms in the interface region [63]? Is there supercritical water in the region?

1.3.4 Size of Active Bubbles

In some literature, there is a description that a bubble with linear resonance radius is active in sonoluminescence and sonochemical reactions. However, as already noted, bubble pulsation is intrinsically nonlinear for active bubbles. Thus, the concept of the linear resonance is not applicable to active bubbles (That is only applicable to a linearly pulsating bubble under very weak ultrasound such as 0.1 bar in pressure amplitude). Furthermore, a bubble with the linear resonance radius can be *inactive* in sonoluminescence and sonochemical reactions [39]. In Fig. 1.8, the calculated expansion ratio (R_{max} / R_0, where R_{max} is the maximum radius and R_0 is the ambient radius of a bubble) is shown as a function of the ambient radius (R_0) for various acoustic amplitudes at 300 kHz [39]. It is seen that the ambient radius for the peak in the expansion ratio decreases as the acoustic pressure amplitude increases. While the linear resonance radius is 11 μm at 300 kHz, the ambient radius for the peak at 3 bar in pressure amplitude is about 0.4 μm. Even at the pressure amplitude of 0.5 bar, it is about 5 μm, which is much smaller than the linear resonance radius.

In Fig. 1.9, the results of numerical simulations at 300 kHz and 3 bar in ultrasonic frequency and pressure amplitude, respectively are shown as a function of ambient radius [39]. In Fig. 1.9a, the temperature inside a bubble at the end of the bubble collapse is shown with the molar fraction of water vapor inside a bubble.

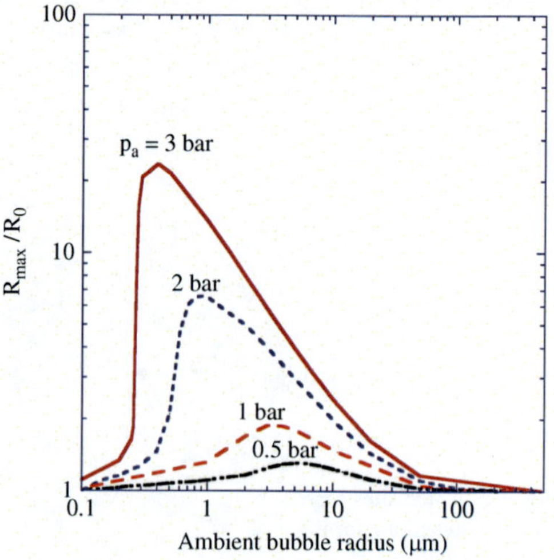

Fig. 1.8 The calculated expansion ratio (R_{max}/R_0) as a function of ambient bubble radius for various acoustic amplitudes at 300 kHz. Both the horizontal and vertical axes are in logarithmic scale. Reprinted with permission from Yasui K, Tuziuti T, Lee J, Kozuka T, Towata A, Iida Y (2008) The range of ambient radius for an active bubble in sonoluminescence and sonochemical reactions. J Chem Phys 128:184705. Copyright 2008, American Institute of Physics

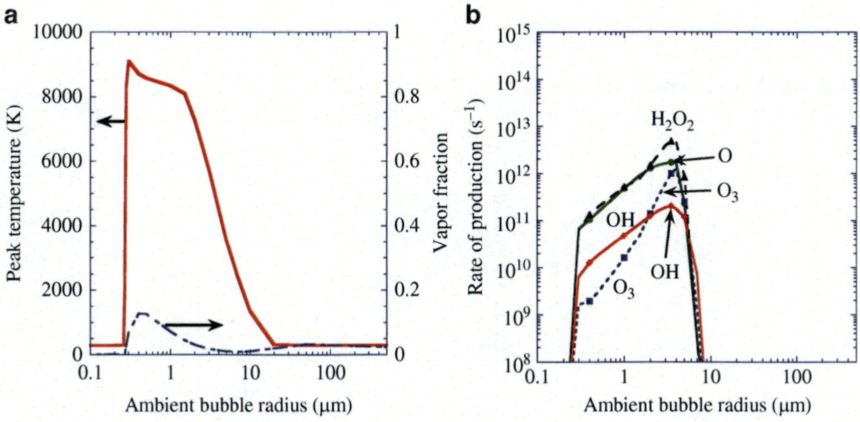

Fig. 1.9 The calculated results as a function of ambient radius at 300 kHz and 3 bar in ultrasonic frequency and pressure amplitude, respectively. The horizontal axis is in logarithmic scale. (**a**) The peak temperature (solid) and the molar fraction of water vapor (dash dotted) inside a bubble at the end of the bubble collapse. (**b**) The rate of production of oxidants with the logarithmic vertical axis. Reprinted with permission from Yasui K, Tuziuti T, Lee J, Kozuka T, Towata A, Iida Y (2008) The range of ambient radius for an active bubble in sonoluminescence and sonochemical reactions. J Chem Phys 128:184705. Copyright 2008, American Institute of Physics

The temperature is higher than 5000 K for bubbles of 0.28 $\sim$ 3.5 μm in ambient radius. For the linear resonance radius of 11 μm, it is only about 1000 K. In Fig. 1.9b, the rate of production of chemical species is shown. It is above 10^8 s^{-1} for bubbles of 0.28 $\sim$ 8 μm in ambient radius. The minimum ambient radius coincides with the Blake threshold radius (R_{Blake}) for transient cavitation (active bubbles). It is calculated by the following formula [6].

$$p_{Blake} = p_0 + \frac{8\sigma}{9}\sqrt{\frac{3\sigma}{2R_{Blake}^3(p_0 + (2\sigma/R_{Blake}))}} \tag{1.17}$$

where p_{Blake} is the pressure amplitude of ultrasound (the Blake threshold acoustic pressure for transient cavitation), p_0 is the ambient static pressure, and σ is the surface tension. (1.17) gives the threshold acoustic pressure or ambient radius for a bubble to expand strongly by overcoming the surface tension.

The maximum ambient radius for an active bubble is smaller than or comparable to the linear resonance radius (R_{res}) given by (1.18) [64].

$$f_{res} = \frac{1}{2\pi}\sqrt{\frac{1}{\rho R_{res}^2}\left(3\kappa p_0 + \frac{2\sigma}{R_{res}}(3\kappa - 1)\right)} \tag{1.18}$$

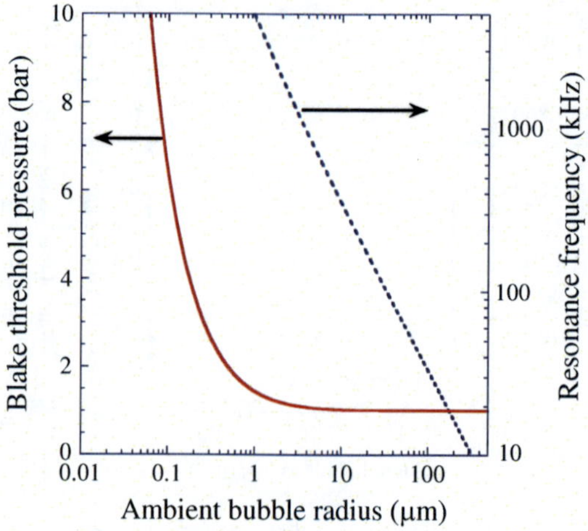

Fig. 1.10 The Blake threshold pressure (1.17) and the linear resonance frequency (1.18) as a function of ambient radius

where f_{res} is ultrasonic frequency (linear resonance frequency), ρ is the liquid density, κ is the polytropic exponent (it is 1.4 for an air bubble), and p_0 is the ambient static pressure. In Fig. 1.10, the numerical solutions of Eq. (1.17) and (1.18) are shown. As the acoustic pressure increases, the Blake threshold radius decreases. As the ultrasonic frequency increases, the linear resonance radius decreases. A distribution of ambient radius for an active bubble in actual experiments is narrower than the above predicted one due to the shape instability of a bubble [19].

As an approximation of (1.18), a simpler equation (1.19) can be used to calculate the linear resonance radius [6].

$$R_{res} = \sqrt{\frac{3\kappa p_0}{4\pi^2 \rho f_{res}^2}} \tag{1.19}$$

1.3.5 Effect of a Surfactant

Bubble-bubble coalescence is strongly retarded by the presence of surfactants, salts, alcohols, and glucose above a certain concentration [65]. There are two mechanisms in the retardation of coalescence. One is a steric effect and the other is the electrostatic repulsion if there is an electric charge on the bubble surface. The steric effect is originated from the gradients in interfacial tension in a thin liquid layer between two attaching bubbles. The gradients in interfacial tension are originated from that in surface concentration of surfactant which is lowered by the thinning of the liquid layer. It stops further thinning of the liquid layer, and the coalescence of bubbles is retarded. The electric repulsion is caused by the charged ionic surfactant on the bubble surface, which also retards coalescence.

Thus, in aqueous solutions of surfactants, the typical ambient radius of a bubble is smaller than that in pure water as bubble-bubble coalescence is retarded [19, 66]. In other words, in aqueous surfactant solutions, the number of large inactive bubbles is smaller than that in pure water. As a result, the spatial distribution of active bubbles is more homogeneous in aqueous surfactant solutions compared to that in pure water [67]. This is probably because large bubbles strongly scatter and attenuate ultrasound and the spatial distribution of acoustic amplitude becomes strongly inhomogeneous although it is not fully understood at present. Then, the total sonoluminescence intensity in an aqueous surfactant solution is larger than that in pure water for a certain range of surfactant concentration [68, 69].

Similar spatial distribution of active bubbles has been observed in partially degassed water and in pure water irradiated with pulsed ultrasound [67]. For both the cases, the number of large inactive bubbles is smaller than that in pure water saturated with air under continuous ultrasound, which is similar to the case of a surfactant solution. As a result, enhancement in sonochemical reaction rate (rate of oxidants production) in partially degassed water and in pure water irradiated with pulsed ultrasound has been experimentally observed [70, 71]. With regard to the enhancement by pulsed ultrasound, a residual acoustic field during the pulse-off time is also important [71].

1.3.6 Nucleation of Particles by Ultrasound

Cavitation bubbles work as nucleation sites of particles. For example, in a supercooled sucrose solution, nucleation of ice crystals induced by cavitation bubbles has been experimentally observed [72]. This phenomenon has been called sonocrystallization [73]. Although there are some papers on the mechanism of sonocrystallization, it has not yet been fully understood [74, 75]. It has been reported that the distribution of crystal size in sonocrystallization is narrower than that without ultrasound [73]. It may be related to the narrower size distribution of sonochemically synthesized particles compared to that without ultrasound [76, 77]. Further studies are required for the mechanism of particle nucleation by ultrasound.

1.3.7 Enhancement of Mass Transfer

Acoustic cavitation increases a rate of mass transfer toward or from a solid surface. When a solute gradually diffuses onto a solid surface as in the case of electrolysis, a diffusion layer is formed near the solid surface in which the concentration of a solute changes from the saturated one at the solid surface to nearly the ambient one at the edge of the layer. A similar diffusion layer is formed when the solid material gradually dissolves into the liquid. Acoustic cavitation makes a diffusion layer

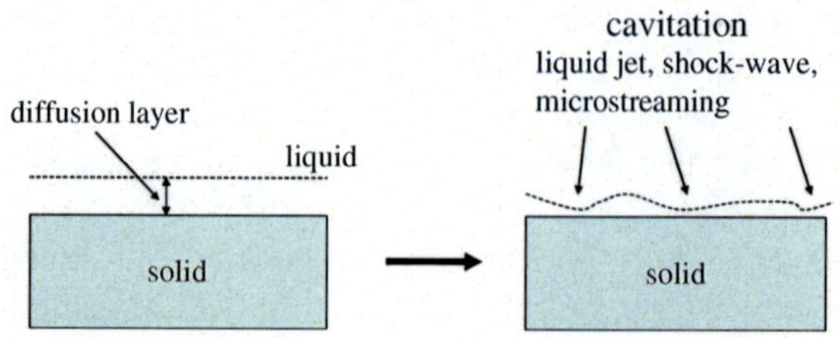

Fig. 1.11 Thinning of a diffusion layer by acoustic cavitation

thinner and the mass transfer rate is increased as it is proportional to the gradient in concentration (Fig. 1.11).

The effect of acoustic cavitation is originated from the following three phenomena. One is liquid jets impacting on the solid surface from collapsing bubbles [14]. Another is shock waves emitted from collapsing bubbles [8–10]. The other is microstreaming induced by pulsating bubbles near the solid surface [14, 78]. When a bubble collapses near a solid surface, the speed of the collapse is higher for the liquid side than that for the solid side. As a result, a liquid jet penetrates into a bubble toward the solid surface and finally hits it. When a bubble symmetrically collapses in liquid, on the other hand, a spherical shock wave is emitted from a bubble into the surrounding liquid. Furthermore, a pulsating bubble induces a liquid flow around the bubble called microstreaming. These effects enhance a mass transfer toward or from the solid surface.

An example of enhancement in mass transfer by acoustic cavitation is the increase in the limiting current density in electrolysis [79]. The electrochemistry with ultrasound is called sonoelectrochemistry. Another example is ultrasonic cleaning [80]. Soluble contaminants on a solid surface dissolve into the liquid faster with acoustic cavitation. Insoluble contaminants are also removed from a solid surface with ultrasound. This is also induced by acoustic cavitation in many cases, but in some other cases it is by acoustic streaming [81–85].

1.4 Conventional Ultrasonic Reactors

There are two types in sonochemical reactors. One is a bath-type reactor, and the other is an ultrasonic horn.

1.4.1 Bath-type Reactor

In Fig. 1.12, a typical experimental set-up of a bath-type reactor is shown. An electric signal with sinusoidal wave of a chosen ultrasonic frequency is generated by a function generator. The signal is amplified by a power amplifier. Then it is

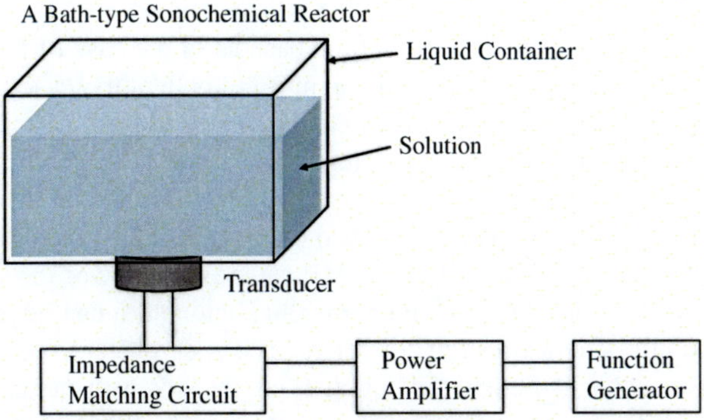

Fig. 1.12 A typical experimental set-up for a bath-type sonochemical reactor

Fig. 1.13 A damped standing wave under a condition of resonance. k and x in the horizontal axis are wave number and distance from the base of a bath-type reactor, respectively. The right end of the horizontal axis is the liquid surface. $\alpha/k = 0.04$ is assumed where α is the absorption coefficient

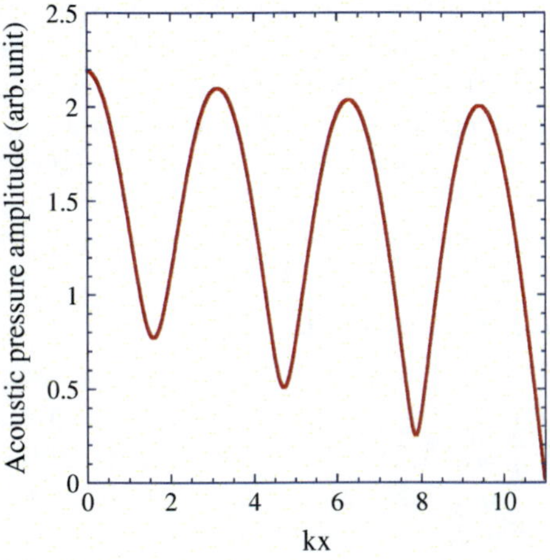

supplied to an ultrasonic transducer through an impedance matching circuit. An impedance matching circuit reduces the reflection of the electric power from the transducer.

In a bath-type sonochemical reactor, a damped standing wave is formed as shown in Fig. 1.13 [1]. Without absorption of ultrasound, a pure standing wave is formed because the intensity of the reflected wave from the liquid surface is equivalent to that of the incident wave at any distance from the transducer. Thus the minimum acoustic-pressure amplitude is completely zero at each pressure node where the incident and reflected waves are exactly cancelled each other. In actual experiments, however, there is absorption of ultrasound especially due to cavitation bubbles. As a result, there appears a traveling wave component because the intensity of the incident wave is higher than that of the reflected wave. Thus, the local minimum value of acoustic pressure amplitude is non-zero as seen in Fig. 1.13. It should be noted that the acoustic-pressure amplitude at the liquid surface (gas-liquid interface) is always zero. In Fig. 1.13, there is the liquid surface at the right end and there is a base of the liquid container at the left end. At resonance, there is a pressure antinode at the base as in the case of Fig. 1.13. The condition for resonance is that the liquid height is equivalent to $\lambda/4 + n\lambda/2$, where λ is the wavelength of ultrasound and n is a natural number or zero. At anti-resonance, on the other hand, there is a pressure node at the base. The condition for anti-resonance is that the liquid height is equivalent to $n\lambda/2$. It should be noted that a (damped) standing wave is always formed in a bath-type reactor even if the ultrasonic frequency is different from the resonance frequency of the system [86].

When the side wall of a liquid container is thick enough, it can be regarded as a rigid wall. When the side wall is too thin, there is considerable vibration of the side wall caused by an acoustic field in the liquid. Then, the side wall can be regarded as a free surface and the acoustic-pressure amplitude near the side wall becomes nearly zero [86].

As ultrasonic frequency increases, the acoustic field is more restricted above an ultrasonic transducer. Roughly speaking, when the wavelength of ultrasound ($\lambda = c/f$, where c is the sound velocity in the liquid and f is the ultrasonic frequency) is much smaller than the radius of the transducer, the acoustic field is restricted above the transducer. It should be noted that the sound velocity in a bubbly liquid is smaller or occasionally larger than that in liquid without bubbles [87, 88].

1.4.2 Ultrasonic Horn

In order to generate a high intensity ultrasound, an ultrasonic horn has been used (Fig. 1.14). The electric system is the same as that in a bath-type reactor.

An ultrasonic horn has a small tip from which high intensity ultrasound is radiated. The acoustic intensity is defined as the energy passing through a unit area normal to the direction of sound propagation per unit time. Its units are watts per square meter (W/m^2). It is related to the acoustic pressure amplitude (P) as follows for a plane traveling wave [1].

$$I = \frac{P^2}{2\rho c} \tag{1.20}$$

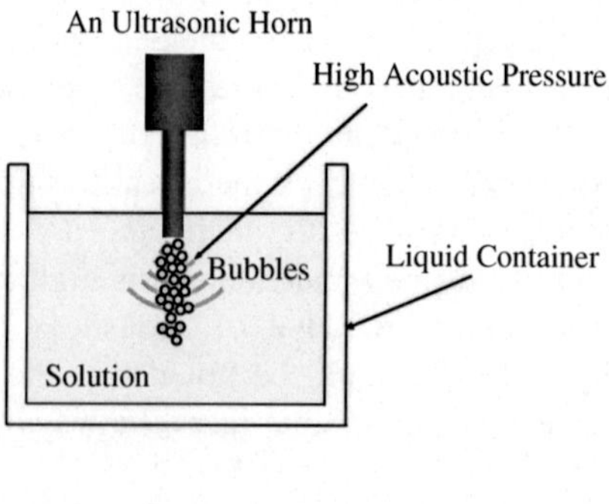

Fig. 1.14 An ultrasonic horn immersed in the liquid

where I is the acoustic intensity, ρ is the liquid density, and c is the sound velocity in the liquid. Thus, as the acoustic intensity increases, the acoustic pressure amplitude increases. The acoustic intensity increases as the surface area decreases if the total acoustic energy is the same. Thus, a horn tip produces an intense ultrasound.

The ultrasound radiated from a horn tip, however, is not a plane wave. The acoustic pressure amplitude is more accurately calculated by Eq. (1.21) along the symmetry axis [1, 89].

$$P(x) = \rho c v_0 \left| 2\sin\left(\frac{\pi}{\lambda}\left(\sqrt{x^2 + a^2} - x\right)\right)\right| \tag{1.21}$$

where $P(x)$ is the acoustic pressure amplitude at distance x from the horn tip along the symmetry axis, v_0 is the velocity amplitude of the circular piston (horn tip), λ is the wavelength of ultrasound, and a is the radius of the circular piston (horn tip). In Fig. 1.15, numerical values of Eq. (1.21) are shown [89]. With cavitation bubbles, however, the acoustic pressure amplitude is much lower than that estimated by Eq. (1.21) as shown in Fig. 1.15 due to the lower value of ρc in a bubbly liquid. Further studies are required on this topic.

For a plane traveling wave, a simpler relationship holds [1].

$$P = \rho c u \tag{1.22}$$

where u is the velocity amplitude of a plane traveling wave.

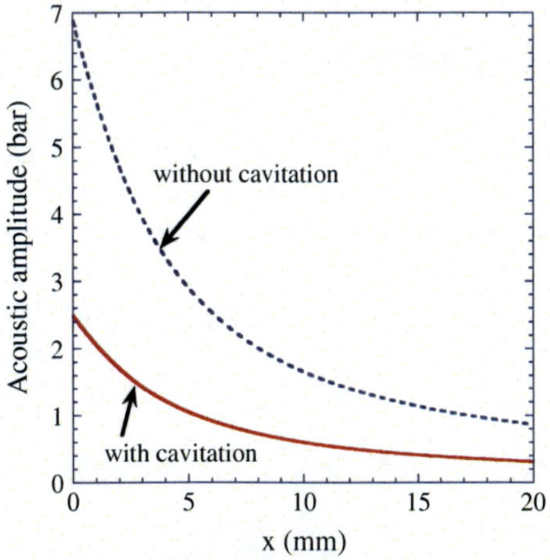

Fig. 1.15 Calculated acoustic amplitude under an ultrasonic horn as a function of the distance from the horn tip on the symmetry axis. The dotted curve is the calculated result by (1.21) when $v_0 = 0.77$ m/s, $\lambda = 51.7$ mm (29 kHz), and $a = 5$ mm. The solid curve is the estimated one in a bubbly liquid. Reprinted figure with permission from Yasui K, Iida Y, Tuziuti T, Kozuka T, Towata A (2008) Strongly interacting bubbles under an ultrasonic horn. Phys Rev E 77:016609 [http://link.aps.org/abstract/PRE/v77/e016609]. Copyright (2008) by the American Physical Society

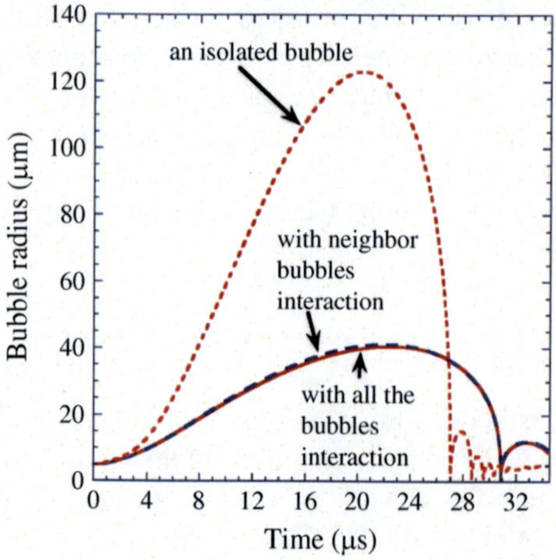

Fig. 1.16 Calculated radius of a bubble in a bubble cloud as a function of time for one acoustic cycle at 29 kHz and 2.36 bar in frequency and pressure amplitude of ultrasound, respectively. The ambient radius is 5 μm. The dotted curve is the calculated result for an isolated bubble. The dashed one is the calculated result with the interaction only with neighboring bubbles. The solid one is the calculated result taking into account all the interactions with surrounding bubbles. Reprinted figure with permission from Yasui K, Iida Y, Tuziuti T, Kozuka T, Towata A (2008) Strongly interacting bubbles under an ultrasonic horn. Phys Rev E 77:016609 [http://link.aps.org/abstract/PRE/v77/e016609]. Copyright (2008) by the American Physical Society

1.5 Bubble-Bubble Interaction

In a multibubble field, every pulsating bubble radiates secondary acoustic wave called acoustic cavitation noise. The pulsation of a bubble is driven by both the primary ultrasound and the acoustic cavitation noise. The influence of the latter on the bubble pulsation is called bubble-bubble interaction [89, 90]. Generally speaking, the bubble-bubble interaction suppresses the bubble expansion as shown in Fig. 1.16 [38, 89–91]. Further studies are required on this topic.

1.6 Conclusion

In acoustic cavitation, some bubbles dramatically expand and violently collapse, which is called the inertial collapse or Rayleigh collapse. It is caused by both the spherically shrinking geometry and the inertia of the surrounding liquid which inwardly flows into the bubble. The bubble collapse is similar to that in hydrodynamic cavitation which is induced by a sudden drop of pressure below the saturated vapor pressure due to a fluid flow through an orifice [92, 93]. At the end of the violent bubble collapse, the temperature and pressure inside a bubble increase to thousands of Kelvin and thousands of bars, respectively due to a quasi-adiabatic collapse [7]. As a result, water vapor and oxygen, if present, are dissociated inside a bubble and oxidants such as OH, O and H_2O_2 are created inside a bubble. The oxidants gradually dissolve into the surrounding liquid and solutes are oxidized by them, which is called sonochemical reactions. The bubble pulsation with the violent collapse is intrinsically nonlinear. As a result, the ambient bubble radius for an active bubble in sonochemical reactions is usually much smaller than the linear resonance radius. There are two types in acoustic cavitation; transient and stable cavitation. There are two definitions for transient and stable cavitation. One is by the shape stability of bubbles and the other is by the activity of bubbles in light emission or chemical reactions. Acoustic cavitation promotes nucleation of particles and enhances mass transfer. There are mainly two types in sonochemical reactors. One is a bath-type reactor and the other is an ultrasonic horn.

Acknowledgments I would like to thank my coworkers T.Tuziuti, J.Lee, T.Kozuka, A.Towata, and Y.Iida for useful discussions. I would also like to thank H.Mitome for his encouragement.

References

1. Kinsler LE, Frey AR, Coppens AB, Sanders JV (1982) Fundamentals of acoustics. Wiley, New York
2. Cheeke JDN (2002) Fundamentals and applications of ultrasonic waves. CRC Press, Boca Raton
3. Maris H, Balibar S (2000) Negative pressures and cavitation in liquid helium. Phys Today 53:29–34
4. Yasui K, Tuziuti T, Sivakumar M, Iida Y (2004) Sonoluminescence. Appl Spectrosc Rev 39:399–436
5. Neppiras EA (1980) Acoustic cavitation. Phys Rep 61:159–251
6. Young FR (1999) Cavitation. Imperial College, London
7. Suslick KS, Flannigan DJ (2008) Inside a collapsing bubble: Sonoluminescence and the conditions during cavitation. Ann Rev Phys Chem 59:659–683
8. Pecha R, Gompf B (2000) Micoimplosions: cavitaion collapse and shock wave emission on a nanosecond time scale. Phys Rev Lett 84:1328–1330
9. Holzfuss J, Rüggeberg M, Billo A (1998) Shock wave emissions of a sonoluminescing bubble. Phys Rev Lett 81:5434–5437
10. Weninger KR, Camara CG, Putterman SJ (2001) Observation of bubble dynamics within luminescent cavitation clouds: sonoluminescence at the nano-scale. Phys Rev E 63:016310
11. Fujikawa S, Akamatsu T (1980) Effects of the nonequilibrium condensation of vapor on the pressure wave produced by the collapse of a bubble in a liquid. J Fluid Mech 97:481–512
12. Henglein A (1993) Contributions to various aspects of cavitation chemistry. In: Mason TJ (ed) Advances in Sonochemistry, vol. 3:17–83, JAI Press, London
13. Riesz P, Kondo T (1992) Free radical formation induced by ultrasound and its biological implications. Free Radic Biol Med 13:247–270
14. Leighton TG (1994) The acoustic bubble. Academic Press, London
15. Yasui K, Tuziuti T, Lee J, Kozuka T, Towata A, Iida Y (2010) Numerical simulations of acoustic cavitation noise with the temporal fluctuation in the number of bubbles. Ultrason Sonochem 17:460–472

16. Yasui K (2002) Influence of ultrasonic frequency on multibubble sonoluminescence. J Acoust Soc Am 112:1405–1413
17. Ashokkumar M, Lee J, Iida Y, Yasui K, Kozuka T, Tuziuti T, Towata A (2009) The detection and control of stable and transient acoustic cavitation bubbles. Phys Chem Chem Phys 11:10118–10121
18. Ashokkumar M, Hodnett M, Zeqiri B, Grieser F, Price G (2007) Acoustic emission spectra from 515 kHz cavitation in aqueous solutions containing surface-active solutes. J Am Chem Soc 129:2250–2258
19. Lee J, Ashokkumar M, Kentish S, Grieser F (2005) Determination of the size distribution of sonoluminescence bubbles in a pulsed acoustic field. J Am Chem Soc 127:16810–16811
20. Guan J, Matula TJ (2003) Time scales for quenching single-bubble sonoluminescence in the presence of alcohols. J Phys Chem 107:8917–8921
21. Madanshetty SI, Apfel RE (1991) Acoustic microcavitation: Enhancement and applications. J Acoust Soc Am 90:1508–1514
22. Tuziuti T, Yasui K, Sivakumar M, Iida Y, Miyoshi N (2005) Correlation between acoustic cavitation noise and yield enhancement of sonochemical reaction by particle addition. J Phys Chem A 109:4869–4872
23. Borkent BM, Arora M, Ohl CD, Jong ND, Versluis M, Lohse D, Morch KA, Klaseboer E, Khoo BC (2008) The acceleration of solid particles subjected to cavitation nucleation. J Fluid Mech 610:157–182
24. Yount DE, Gillary EW, Hoffman DC (1984) A microscopic investigation of bubble formation nuclei. J Acoust Soc Am 76:1511–1521
25. Bremond N, Arora M, Dammer SM, Lohse D (2006) Interaction of cavitation bubbles on a wall. Phys Fluids 18:121505 (10 pages)
26. Calvisi ML, Lindau O, Blake JR, Szeri AJ (2007) Shape stability and violent collapse of microbubbles in acoustic traveling waves. Phys Fluids 19:047101 (15 pages)
27. Wang E, Chen W, Lu M, Wei R (2003) Bubble oscillations driven by aspherical ultrasound in liquid. J Acoust Soc Am 114:1898–1904
28. Lee J, Tuziuti T, Yasui K, Kentish S, Grieser F, Ashokkumar M, Iida Y (2007) Influence of surface-active solutes on the coalescence, clustering, and fragmentation of acoustic bubbles confined in a microspace. J Phys Chem C 111:19015–19023
29. Matula TJ, Cordry SM, Roy RA, Crum LA (1997) Bjerknes force and bubble levitation under single-bubble sonoluminescence conditions. J Acoust Soc Am 102:1522–1527
30. Yasui K (2001) Temperature in multibubble sonoluminescence. J Chem Phys 115:2893–2896
31. Crum LA (1980) Measurements of the growth of air bubbles by rectified diffusion. J Acoust Soc Am 68:203–211
32. Lee J, Kentish S, Ashokkumar M (2005) Effect of surfactants on the rate of growth of an air bubble by rectified diffusion. J Phys Chem B 109:14595–14598
33. Louisnard O, Gomez F (2003) Growth by rectified diffusion of strongly acoustically forced gas bubbles in nearly saturated liquids. Phys Rev E 67:036610
34. Iida Y, Ashokkumar M, Tuziuti T, Kozuka T, Yasui K, Towata A, Lee J (2010) Bubble population phenomena in sonochemical reactor: II Estimation of bubble size distribution and its number density by simple coalescence model calculation. Ultrason Sonochem 17:480–486
35. Mettin R (2005) Bubble structures in acoustic cavitation. In Doinikov AA (ed) Bubble and particle dynamics in acoustic fields: modern trends and applications, pp. 1–36. Research Signpost, Trivandrum
36. Beyer RT (1997) Nonlinear acoustics. Acoustical Society of America, New York
37. Mitome H, Kozuka T, Tuziuti T, Wang L (1997) Quasi acoustic streaming induced by generation of cavitation bubbles. IEEE Ultrason Sympo Proc 1:533–536
38. Mettin R, Akhatov I, Parlitz U, Ohl CD, Lauterborn W (1997) Bjerknes forces between small cavitation bubbles in a strong acoustic field. Phys Rev E 56:2924–2931
39. Yasui K, Tuziuti T, Lee J, Kozuka T, Towata A, Iida Y (2008) The range of ambient radius for an active bubble in sonoluminescence and sonochemical reactions. J Chem Phys 128:184705

40. Brenner MP, Hilgenfeldt S, Lohse D (2002) Single-bubble sonoluminescence. Rev Mod Phys 74:425–484
41. Prosperetti A, Lezzi A (1986) Bubble dynamics in a compressible liquid. Part I. First-order theory. J Fluid Mech 168:457–478
42. Yasui K (2001) Effect of liquid temperature on sonoluminescence. Phys Rev E 64:016310
43. Yasui K, Tuziuti T, Sivakumar M, Iida Y (2005) Theoretical study of single-bubble sonochemistry. J Chem Phys 122:224706
44. Vanhille C, Campos-Pozuelo C (2008) Nonlinear ultrasonic propagation in bubbly liquids: A numerical model. Ultrasound Med Biol 34:792–808
45. Tuziuti T, Yasui K, Lee J, Kozuka T, Towata A, Iida Y (2009) Influence of surface active solute on ultrasonic waveform distortion in liquid containing air bubbles. J Phys Chem A 113:8893–8900
46. Hilgenfeldt S, Grossmann S, Lohse D (1999) A simple explanation of light emission in sonoluminescence. Nature (London) 398:402–405
47. Yasui K (1999) Mechanism of single-bubble sonoluminescence. Phys Rev E 60:1754–1758
48. An Y (2006) Mechanism of single-bubble sonoluminescence. Phys Rev E 74:026304 (14 pages)
49. Flannigan DJ, Suslick KS (2005) Plasma formation and temperature measurement during single-bubble cavitation. Nature (London) 434:52–55
50. Hatanaka S, Mitome H, Yasui K, Hayashi S (2002) Single-bubble sonochemiluminescence in aqueous luminol solutions. J Am Chem Soc 124:10250–10251
51. Didenko YT, Suslick KS (2002) The energy efficiency and formation of photons, radicals and ions during single-bubble cavitation. Nature (London) 418:394–397
52. Koda S, Tanaka K, Sakamoto H, Matsuoka T, Nomura H (2004) Sonochemical efficiency during single-bubble cavitation in water. J Phys Chem A 108:11609–11612
53. Yasui K, Tuziuti T, Iida Y, Mitome H (2003) Theoretical study of the ambient-pressure dependence of sonochemical reactions. J Chem Phys 119:346–356
54. Yasui K, Tuziuti T, Iida Y (2004) Optimum bubble temperature for the sonochemical production of oxidants. Ultrasonics 42:579–584
55. Brotchie A, Grieser F, Ashokkumar M (2009) Effect of power and frequency on bubble-size distributions in acoustic cavitation. Phys Rev Lett 102:084302 (4 pages)
56. Mason TJ (1999) Sonochemistry. Oxford University Press, Oxford
57. Suslick KS, Hammerton DA, Cline RE, J (1986) The sonochemical hot spot. J Am Chem Soc 108:5641–5642
58. Yasui K (1996) Variation of liquid temperature at bubble wall near the sonoluminescence threshold. J Phys Soc Jpn 65:2830–2840
59. Storey BD, Szeri AJ (2000) Water vapour, sonoluminescence and sonochemistry. Proc R Soc Lond A 456:1685–1709
60. Hua I, Hochemer RH, Hoffmann MR (1995) Sonolytic hydrolysis of p-nitrophenyl acetate: The role of supercritical water. J Phys Chem 99:2335–2342
61. Sostaric JZ (1999) Interfacial effects on aqueous sonochemistry and sonoluminescence. PhD thesis, University of Melbourne, Australia
62. Yasui K (2002) Effect of volatile solutes on sonoluminescence. J Chem Phys 116:2945–2954
63. Yasui K, Tuziuti T, Kozuka T, Towata A, Iida Y (2007) Relationship between the bubble temperature and main oxidant created inside an air bubble under ultrasound. J Chem Phys 127:154502
64. Parlitz U, Mettin R, Luther S, Akhatov I, Voss M, Lauterborn W (1999) Spatio-temporal dynamics of acoustic cavitation bubble cloud. Philos Trans R Soc London A 357:313–334
65. Oolman TO, Blanch HW (1986) Bubble coalescence in stagnant liquids. Chem Engnrg Commun 43:237–261
66. Iida Y, Ashokkumar M, Tuziuti T, Kozuka T, Yasui K, Towata A, Lee J (2010) Bubble population phenomena in sonochemical reactor: I Estimation of bubble size distribution and its number density with pulsed sonocation – laser diffraction method. Ultrason Sonochem 17:473–479

67. Lee J, Yasui K, Tuziuti T, Kozuka T, Towata A, Iida Y (2008) Spatial distribution enhancement of sonoluminescence activity by altering sonication and solution conditions. J Phys Chem B 112:15333–15341
68. Ashokkumar M, Hall R, Mulvaney P, Grieser F (1997) Sonoluminescence from aqueous alcohol and surfactant solutions. J Phys Chem B 101:10845–10850
69. Segebarth N, Eulaerts O, Reisse J, Crum LA, Matula TJ (2002) Correlation between acoustic cavitation noise, bubble population, and sonochemistry. J Phys Chem B 106:9181–9190
70. Tuziuti T, Yasui K, Iida Y, Sivakumar M, Koda S (2004) Laser-light scattering from a multibubble system for sonochemistry. J Phys Chem A 108:9011–9013
71. Tuziuti T, Yasui K, Lee J, Kozuka T, Towata A, Iida Y (2008) Mechanism of enhancement of sonochemical-reaction efficiency by pulsed ultrasound. J Phys Chem A 112:4875–4878
72. Chow R, Blindt R, Chivers R, Povey M (2005) A study on the primary and secondary nucleation of ice by power ultrasound. Ultrasonics 43:227–230
73. Luque de Castro MD, Priego-Capote F (2007) Ultrasound-assisted crystallization (sonocrystallization). Ultrason Sonochem 14:717–724
74. Kordylla A, Krawczyk T, Tumakaka F, Schembecker G (2009) Modeling ultrasound-induced nucleation during cooling crystallization. Chem Engnrg Sci 64:1635–1642
75. Saclier M, Peczalski R, Andrieu J (2010) A theoretical model for ice primary nucleation induced by acoustic cavitation. Ultrason Sonochem 17:98–105
76. Xu M, Lu Y, Liu Y, Shi S, Qian T, Lu D (2006) Sonochemical synthesis of monosized spherical $BaTiO_3$ particles. Powder Technol 161:185–189
77. Testinon A, Buscaglia MT, Viviani M, Buscaglia V, Nanni P (2004) Synthesis of $BaTiO_3$ particles with tailored size by precipitation from aqueous solutions. J Am Ceram Soc 87:79–83
78. Elder SA (1959) Cavitation microstreaming. J Acoust Soc Am 31:54–64
79. Walton DJ, Phull SS (1996) Sonoelectrochemistry. In: Mason TJ (ed) Advances in Sonochemistry, vol. 4:205–284, JAI Press, Greenwich
80. Hacias KJ, Cormier GJ, Nourie SM, Kubel EJ (1997) Guide to acid, alkaline, emulsion, and ultrasonic cleaning. ASM International, Materials Park
81. Lamminen MO, Walker HW, Weavers LK (2006) Cleaning of particle-fouled membranes during cross-flow filtration using an embedded ultrasonic transducer system. J Membrane Sci 283:225–232
82. Ohl CD, Arora M, Dijkink R, Janve V, Lohse D (2006) Surface cleaning from laser-induced cavitation bubbles. Appl Phys Lett 89:074102 (3 pages)
83. Kim W, Kim TH, Choi J, Kim HY (2009) Mechanism of particle removal by megasonic waves. Appl Phys Lett 94:081908 (3 pages)
84. Bakhtari K, Guldiken RO, Busnaina AA, Park JG (2006) Experimental and analytical study of semicrometer particle removal from deep trenches. J Electrochem Soc 153:C603–C607
85. Bakhtari K, Guldiken RO, Makaram P, Busnaina AA, Park JG (2006) Experimental and numerical investigation of nanoparticle removal using acoustic streaming and the effect of time. J Electrochem Soc 153:G846–G850
86. Yasui K, Kozuka T, Tuziuti T, Towata A, Iida Y, King J, Macey P (2007) FEM calculation of an acoustic field in a sonochemical reactor. Ultrason Sonochem 14:605–614
87. Dahnke S, Keil F (1998) Modeling of sound fields in liquids with a nonhomogeneous distribution of cavitation bubbles as a basis for the design of sonochemical reactors. Chem Eng Technol 21:873–877
88. Wilson PS, Roy RA, Carey WM (2005) Phase speed and attenuation in bubbly liquids inferred from impedance measurements near the individual bubble resonance frequency. J Acoust Soc Am 117:1895–1910
89. Yasui K, Iida Y, Tuziuti T, Kozuka T, Towata A (2008) Strongly interacting bubbles under an ultrasonic horn. Phys Rev E 77:016609
90. Yasui K, Lee J, Tuziuti T, Towata A, Kozuka T, Iida Y (2009) Influence of the bubble-bubble interaction on destruction of encapsulated microbubbles under ultrasound. J Acoust Soc Am 126:973–982

91. Ida M, Naoe T, Futakawa M (2007) Suppression of cavitation inception by gas bubble injection: A numerical study focusing on bubble-bubble interaction. Phys Rev E 76:046309
92. Gogate PR, Shirgaonkar IZ, Sivakumar M, Senthilkumar P, Vichare NP, Pandit AB (2001) Cavitation reactors: Efficiency assessment using a model reaction. AIChE J 47:2526–2538
93. Wang X, Zhang Y (2009) Degradation of alachlor in aqueous solution by using hydrodynamic cavitation. J Haz Mater 161:202–207

Chapter 2
Theory of Cavitation and Design Aspects of Cavitational Reactors

Parag R. Gogate

Abstract Cavitation is a phenomenon by which large magnitude of energy is dissipated locally due to the violent collapse of the cavities creating effects suitable for intensification of chemical and physical processing applications. The current chapter focuses on the basic aspects related to generation of cavitation in different reactors and explains the mechanistic details by which the cavitational effects bring about desired physicochemical transformations. Design aspects related to understanding of the cavitational activity distribution using experimental and theoretical investigations, optimization of equipment operating parameters such as frequency and intensity of ultrasonic irradiation, geometry of the reactor, liquid physicochemical properties and the operating temperature have been discussed. Some of the novel ways to intensify the cavitational activity using additives and combination with other techniques, with an objective of minimizing the cost of operation and maximizing the yields from processes, have also been highlighted. Overall, it appears that cavitational reactors show considerable promise for industrial applications in the area of chemical processing and combined efforts of scientists and engineers are required to successfully accomplish this intent.

2.1 Introduction

Cavitation can in general be defined as the phenomena of the formation, growth and subsequent collapse of the microbubbles or cavities occurring in extremely small interval of time (milliseconds) releasing large magnitudes of energy [1–3]. Cavitation occurs at millions of locations in the reactor simultaneously and generates

P.R. Gogate (✉)
Chemical Engineering Department, Institute of Chemical Technology, Matunga, Mumbai 400019, India
e-mail: pr.gogate@ictmumbai.edu.in

Pankaj and M. Ashokkumar (eds.), *Theoretical and Experimental Sonochemistry Involving Inorganic Systems*, DOI 10.1007/978-90-481-3887-6_2,

conditions of very high temperatures and pressures (few thousand atmospheres pressure and few thousands K temperature) locally with overall ambient conditions [3]. Cavitation also results in generation of local turbulence and liquid micro-circulation (acoustic streaming) in the reactor. Cavitation can be considered as one of the methods of introducing discrete energy input, where the energy dissipated per unit volume (pockets) for the same amounts of average energy input are few orders of magnitude higher than the conventional processes.

The method of efficiently producing cavities can be taken as the main criterion in distinguishing among different types of cavitation. The four principle types of cavitation and their causes can be summarized as follows [2]:

1. *Acoustic cavitation*: In this case, the pressure variations in the liquid are effected using the sound waves usually ultrasound (16 kHz to 100 MHz). The chemical changes taking place due to the cavitation induced by the passage of sound waves are commonly known as sonochemistry.
2. *Hydrodynamic cavitation*: Cavitation is produced by pressure variations, which is obtained using geometry of the system creating velocity variation. For example based on the geometry of the system, the interchange of pressure and kinetic energy can be achieved resulting in the generation of cavities as in the case of flow through orifice, venturi, etc.
3. *Optic cavitation*: It is produced by photons of high intensity light (laser) rupturing the liquid continuum.
4. *Particle cavitation*: It is produced by the beam of the elementary particles, e.g. a proton, rupturing a liquid, as in the case of a bubble chamber.

It can be said that acoustic and hydrodynamic cavitation are the result of the tensions prevailing in a liquid, while optic and particle cavitation are the consequence of the local deposition of energy. The classification scheme for the phenomena of cavitation has been shown schematically in Fig. 2.1. Once the cavities are generated, subsequent expansion of the minute cavity may be achieved by reducing the ambient pressure using static or dynamic means. The surrounding

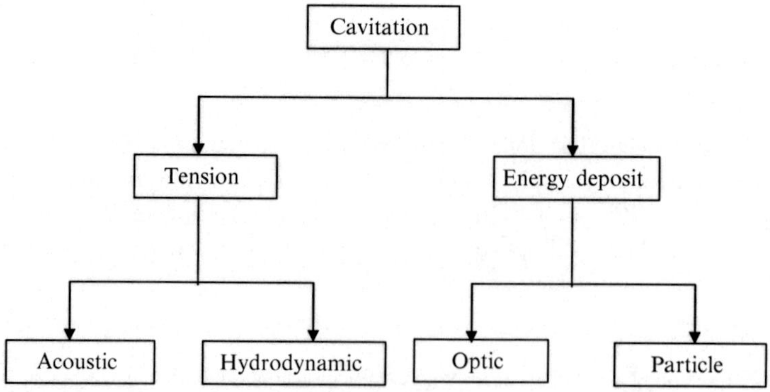

Fig. 2.1 Classification scheme for the different types of Cavitation phenomena

fluctuating pressure field along with the type of the cavity produced (i.e. cavity containing gas or vapour or mixture of the gas and vapour) affects the final intensity of cavitation.

Out of the four types of cavitation, only acoustic and hydrodynamic cavitations generate desired intensity suitable for chemical or physical processing. In the case of cavitational reactors, two aspects of cavity dynamics are of prime importance; the maximum size reached by the cavity before a violent collapse and the life of the cavity. The maximum size reached by the cavity determines the magnitude of the pressure/temperature pulse produced on the collapse and hence the cavitational intensity that can be obtained in the system. The life of the cavity determines the distance travelled by the cavity before the collapse from the point where it is generated and hence it is a measure of the active volume of the reactor in which the actual cavitational effects are observed. The aim of the equipment designer should be to maximize both these quantities by suitably adjusting the different parameters including the methodology used for the generation of cavities (type of the cavity generated is a crucial parameter in deciding the intensity of the cavitation phenomena).

Apart from the classification based on the mode of generation of cavities, cavitation can also be classified as transient cavitation and stable cavitation [3]. The classification is based on the maximum radius reached (resonant size), life time of cavity (which decides the extent of collapse) in the bulk of liquid and the pattern of cavity collapse. Generation of transient or stable cavitation usually depends on the set of operating parameters and constitution of the liquid medium. Depending on the specific application under question, it is very important to select particular set of operating conditions such that maximum effects are obtained with minimum possible energy consumption.

Transient cavitation is generally due to gaseous or vapor filled cavities, which are believed to be produced at ultrasonic intensity greater than 10 W/cm^2. Transient cavitation involves larger variation in the bubble sizes (maximum size reached by the cavity is few hundred times the initial size) over a time scale of few acoustic cycles. The life time of transient bubble is too small for any mass to flow by diffusion of the gas into or out of the bubble; however evaporation and condensation of liquid within the cavity can take place freely. Hence, as there is no gas to act as cushion, the collapse is violent. Bubble dynamics analysis can be easily used to understand whether transient cavitation can occur for a particular set of operating conditions. A typical bubble dynamics profile for the case of transient cavitation has been given in Fig. 2.2. By assuming adiabatic collapse of bubble, the maximum temperature and pressure reached after the collapse can be estimated as follows [2].

$$T_{\max} = T_0 \left(\frac{P_m(\gamma - 1)}{P} \right) \tag{2.1}$$

$$P_{\max} = P \left(\frac{P_m(\gamma - 1)}{P} \right)^{\frac{\gamma}{\gamma - 1}} \tag{2.2}$$

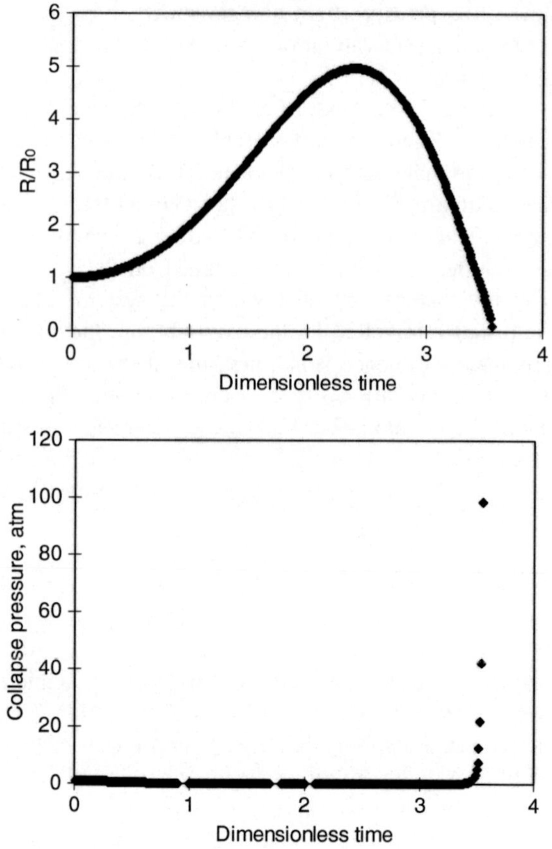

Fig. 2.2 Radius and pressure profiles in the case of Transient Cavitation (Typical profile at frequency of irradiation = 20 kHz, Intensity of irradiation = 0.12 W/m^2 and initial radius of the nuclei = 0.001 mm)

where, T_o is ambient temperature, γ is polytrophic coefficient of gas or vapour, P is pressure in the bubble and P_m is pressure in the liquid at bubble collapse.

The bubble formed in stable cavitation contains gas (and very small amount of vapor) at ultrasonic intensity in the range of 1–3 W/cm^2. Stable cavitation involves formation of smaller bubbles with non linear oscillations over many acoustic cycles. The typical bubble dynamics profile for the case of stable cavitation has been shown in Fig. 2.3. The phenomenon of growth of bubbles in stable cavitation is due to rectified diffusion [4] where, influx of gas during the rarefaction is higher than the flux of gas going out during compression. The temperature and pressure generated in this type of cavitation is lower as compared to transient cavitation and can be estimated as:

$$\frac{T_0}{T_{\max}} = 1 + Q\left(\frac{P_h}{P_m}\right)^{\frac{1}{3\gamma}} - 1\Big)^{3(\gamma-1)} \qquad (2.3)$$

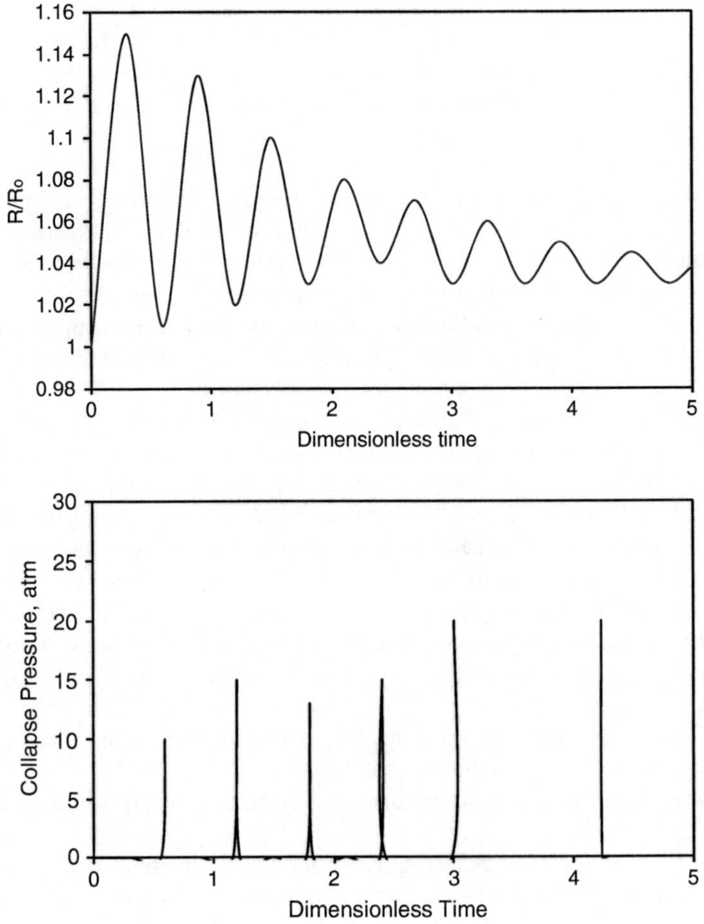

Fig. 2.3 Radius and pressure profiles in the case of Stable Cavitation (Typical profile at frequency of irradiation = 300 kHz, intensity of irradiation = 5 W/m^2 and initial radius of the nuclei = 5 mm)

where Q is the ratio of amount of resonance amplitude to static amplitude and P_m is the pressure of bubble.

2.2 Mechanism of Cavitational Effects for Chemical Processing

In order to understand the way in which cavitational collapse can affect chemical transformations [5–7], one must consider the possible effects of this collapse in different systems. In the case of homogeneous systems, there are two major effects.

First, the cavity that is formed is unlikely to enclose a vacuum (in the form of void) – it will almost certainly contain vapor from the liquid medium or dissolved volatile reagents or gases. During the collapse, these vapors will be subjected to extreme conditions of high temperatures and pressures, causing molecules to fragment and generate highly reactive radical species. These radicals may then react either within the collapsing bubble or after migration into the bulk liquid. Secondly, the sudden collapse of the bubble also results in an inrush of the liquid to fill the void producing shear forces in the surrounding bulk liquid capable of breaking the chemical bonds of any materials, which are dissolved in the fluid or disturb the boundary layer facilitating the transport. When considering the reaction conditions for a cavitational process, the choice of the solvent and bulk operating temperature are significant factors and often interrelated. Any increase in the solvent vapor pressure decreases the maximum bubble collapse temperature and pressure. Thus, for a reaction where cavitational collapse is the primary cause of the activation, a low operating temperature is recommended particularly if a low boiling solvent is used. Conversely, for a reaction requiring elevated temperatures, high boiling solvent is recommended. It is very important to decipher the controlling mechanism in the overall intensification of the chemical processing applications and then appropriate selection of the operating parameters needs to be done. Sometimes, when only mass transfer is the rate limiting step in deciding the overall rate of chemical reaction, use of cavitational reactors may not be needed and similar effects can be achieved using mechanical agitation at higher speeds of rotation or by using proper geometry/type of the stirring device. Cavitation phenomena need to be effectively used for chemical processing applications limited by intrinsic kinetics or where greener chemical synthesis routes need to be established.

The cavitational activation in heterogeneous systems is mainly as a consequence of the mechanical effects of cavitation. In a heterogeneous solid/liquid system, the collapse of the cavitation bubble results in significant structural and mechanical defects. Collapse near the surface produces an asymmetrical inrush of the fluid to fill the void forming a liquid jet targeted at the surface. This effect is equivalent to high-pressure/high-velocity liquid jets and is the reason why ultrasound is used for cleaning solid surfaces. These jets activate the solid catalyst and increase the mass transfer to the surface by the disruption of the interfacial boundary layers as well as dislodging the material occupying the inactive sites. Collapse on the surface, particularly of powders, produces enough energy to cause fragmentation (even for finely divided metals). Thus, in this situation, ultrasound can increase the surface area for a reaction and provide additional activation through efficient mixing and enhanced mass transport. For heterogeneous reactions, a balance must be struck between ensuring enough cavitation to ensure reagent/catalyst activation without overly disturbing the thermodynamics of the reaction. It must also be remembered that in some cases where extended reaction time are required, the solvent itself may not be totally inert and discoloration and charring may occur. In this case, optimization needs to be done on the time of exposure to the cavitating conditions. However, in most synthetic reactions, solvent reaction can be ignored as the reaction times are substantially small.

In heterogeneous liquid/liquid reactions, cavitational collapse at or near the interface will cause disruption and mixing, resulting in the formation of very fine emulsions. When very fine emulsions are formed, the surface area available for the reaction between the two phases is significantly increased, thus increasing the rates of reaction. The emulsions formed using cavitation, are usually smaller in size and more stable, than those obtained using conventional techniques and often require little or no surfactant to maintain the stability [8]. This is very beneficial particularly in the case of phase-transfer catalyzed reactions or biphasic systems.

2.3 Design Aspects of Cavitational Reactors

Design information is usually required from diverse fields such as chemical engineering (gas-liquid hydrodynamics and other reactor operations), material science (for construction of transducers efficiently operating at conditions of high frequency and high power dissipation) and acoustics (for better understanding of the sound and pressure field existing in the reactor). We now discuss different aspects related to design of cavitational reactors concentrating on the ultrasound based reactors more commonly described as sonochemical reactors.

The basic component of any sonochemical reactor is the transducer, a device capable of converting the supplied electrical energy into the mechanical energy in terms of vibrations and finally to the sound energy described as ultrasound, which propagates though the liquid medium. The dissipation of the sound energy into the medium results in the generation of cavitation field which coupled with the fluid-flow pattern existing in the reactor will decide the degree of expected effects. The stepwise transfer of the supplied electrical energy has been schematically depicted in the Fig. 2.4.

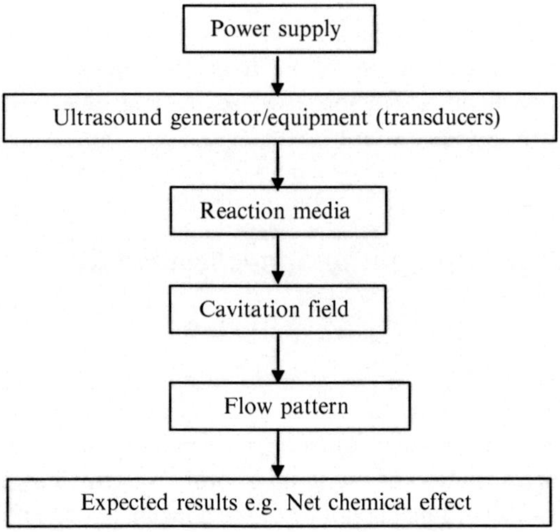

Fig. 2.4 Steps for energy conversion in the case of sonochemical reactors

The transducers operate at a fixed frequency of irradiation emitting radiations through a fixed area of irradiation. Thus, the type of transducer coupled with the total area of irradiation and the operating frequency are the key factors in the efficient design of the sonochemical reactors. The three main types of transducers are gas driven, liquid driven and electromechanical transducers out of which the electromechanical transducers are by far the most versatile and widely used.

The gas-driven transducers are simply whistles with high frequency output. Dog whistles and sirens can be given as the two examples of gas-driven transducers. These transducers can be used to break down foams and agglomerates of dust and for the acceleration of drying processes. However, these types do not have any significant chemical applications, as it is not possible to achieve a sufficiently high-pressure intensity in airborne ultrasound by this method.

Liquid-driven transducers (i.e. a liquid whistle) can be used to produce efficient homogenization. The majority of the chemical effects observed using whistle-type transducers for the sonication of non-homogeneous reactions can be attributed mainly to the generation of very fine emulsions leading to increase in the interfacial phenomena rather than the ultrasonic irradiation itself.

The two main types of electromechanical transducers used in applications are the piezoelectric and the magnetostrictive transducers. Piezoelectric transducers are constructed using a piezoelectric material, such as quartz, which expands and contracts in an oscillating electric field producing sound waves (pressure waves) from the electric signal. Magnetostrictive transducers are constructed from materials, such as nickel alloys, which expand and contract in an alternating magnetic field. The relative advantages and disadvantages of these two types of transducers have been given in Table 2.1.

Traditionally, all the new investigations are carried out on a small scale in the laboratory. Ultrasonic energy furnished by the transducer probes or the standard cleaning baths employing single or multiple transducers is only a logical starting point. It is important that this energy should be transformed into useful effects even at larger scales of operation. Thus the designs used successfully at laboratory scale operations may or may not be equally efficient in giving expected results on an industrial scale operation. It is worthwhile to overview the different designs of sonochemical reactors used so far and investigate the applicability at industrial scales of operation.

2.3.1 Designs of Sonochemical Reactors

2.3.1.1 Probe Systems

Probe systems, also called as the ultrasonic horn are being most frequently used for the sonochemical research at laboratory scale of operation. A typical schematic representation of the setup of probe systems has been given in Fig. 2.5. These are typically immersion type of transducers and the most important advantage of using

Table 2.1 Comparison of piezoelectric and magnetostrictive transducers

Property	Piezoelectric transducers	Magnetostrictive transducers
Material	Quartz and ceramic materials (Barium titanate ($BaTiO_3$), Lead metaniobate ($PbNb_2O_6$) and the mixed crystal Lead-zirconate titanate)	Nickel or an alloy of Nickel. Also, some other high-tech alloys with ferrite materials (MFe_2O_4, M = divalent metal like Ni, Zn and Pb)
Cost (for similar power rating)	Relatively inexpensive	More cost (nickel based)
Size and weight	Small and light	Bulkier and heavier
Operational temperature	Damaged by temperatures above 150°C	Can withstand up to 180°C and with special precautions, temperatures > 250°C can be handled
Mechanical damage	May be damaged by the impact forces or dry operation without liquid medium	Extremely resistant to mechanical damage or dry operation
Gain	High gain transducers operating efficiently over a narrow range of frequencies. It is difficult and less efficient to operate large numbers of piezoelectric transducers in synchronism, a necessary requirement for large power applications	Low gain transducers operating relatively near their peak performance. This is a distinct advantage when applied to large systems
Ageing effect (causes a decrease in ultrasonic power output)	Continuous operation at high temperatures for long periods leads to gradual ageing effect. Loss of performance and even replacement of all transducers with time may be required	Ageing effect is not observed and hence continuous operation is possible. There are many commercial systems operating successfully even after over 20 years of use

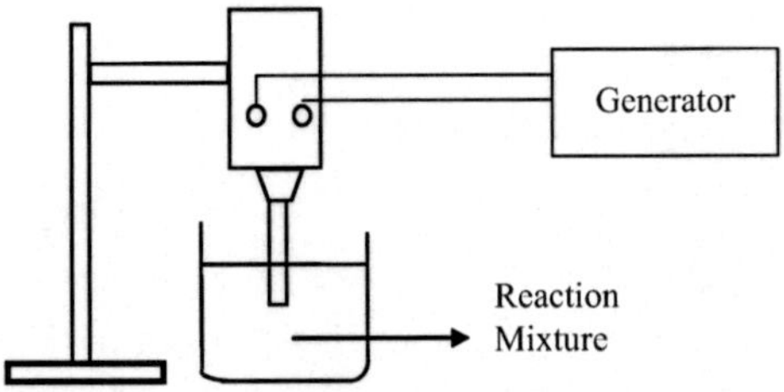

Fig. 2.5 Schematic representation of assembly for probe systems

a horn system is that these systems are capable of delivering large amounts of power directly to the reaction mixture, which can also be regulated by varying the amplitude delivered to the transducer resulting in variable power dissipation into the system. Maximum powers of several hundred W/cm^2 (depending on the size of the reactor) can easily be achieved using the probe systems. Ultrasonic streaming from the tip of the probe may be powerful to provide efficient bulk mixing. Most modern units are provided with a pulse facility that allows the operator to sonicate the reactions repeatedly for fractions of a second. This gives adequate time for cooling between sonic pulses. With such systems, the probe (horn) can be tuned to give optimum performance, which is important in terms of the reproducibility of the results.

Apart from the conventional designs as illustrated in Fig. 2.5, some modifications with an aim of increasing the cavitationally active volume have also been reported. Horst et al. [9] have reported a novel modification in terms of using high intensity ultrasound from a concentrator horn. It has been shown that the concept of a conical funnel fits the demands for nearly perfect radiation effectiveness and a good reaction management. The design used by Dahlem et al. [10, 11] also needs a special mention here. Telsonic horn, which has radial vibrations as against conventional longitudinal vibrations for the immersion system gives dual advantages of higher irradiating surface (lower intensity of irradiation resulting in better yields) coupled with good distribution of the energy in the radial direction. Moreover even if the horn is radially vibrating, local measurements just below horn in axial direction also give high cavitational activity, which will be again more beneficial in enhancing the global sonochemical yields. Very recently Dion [12] have described a new continuous reactor design based on the high power converging acoustic waves in a tube to produce a relatively large volume confined acoustic cavitation zone in flowing liquid reagents under pressure. It has been reported that the new cylindrical sonoreactor design does not contaminate the processed liquids with erosion products since the cavitation zone is maintained away from the wall of the tube. The processing capacity of the largest models may be up to several tons per hour, depending on the required cavitation energy per unit volume to produce the desired process enhancement, using an electric power input of about 50 kW.

The scale up prospects of horn type systems are very poor as it cannot effectively transmit the acoustic energy into large process volume. Usually, ultrasonic horn type systems are generally recommended for laboratory scale investigations. The disadvantages of using the horn systems at large scales of operation are:

1. Continuous operation at high power dissipation leads to the erosion and pitting of the horn tip, which may contaminate the reaction medium. Also continuous replacement of the horn tips may be required affecting the overall economics of the process.
2. Very high intensities (pressures of the order of few thousands atmosphere) are observed very near to the horn. The intensity decreases exponentially as one moves away from horn and vanishes at a distance of as low as 2–5 cm depending on the maximum power input to the equipment and also on the operating frequency [13]. It should be also noted that power dissipation cannot be increased indefinitely with an aim of increasing the cavitationally active volumes as increased power dissipation may lead to decoupling effect resulting in reduced transfer of energy.
3. Stalling of the probe in reaction mixtures of high viscosity

2.3.1.2 Ultrasonic Baths

Ultrasonic baths were originally manufactured and used for cleaning purposes. Typical baths have the transducers attached to the bottom and either single or multiple transducers can be used. A schematic representation of the assembly has been given in Fig. 2.6. In this case, the active zone is restricted to a vertical plane just above the transducers. Thus the area of irradiating surface should be increased (maximum possible) so as to get better distribution/dissipation of energy in the reactor. This has a twin advantage in terms of the decreased ultrasonic intensity (defined as power dissipation per unit area of the irradiating surface), which will increase the magnitude of the pressure pulse generated at the end of the cavitation events [14]. To increase the active zones existing in the reactor, one can easily modify the position of the transducers (if multiple transducers have been used which is likely to be the case at large scale operation due to the fact that it is quite difficult to successfully operate single transducer with very high power and frequency due to limitations over the material of construction for the transducers) so that the wave patterns generated by the individual transducers will overlap, also resulting into uniform and increased cavitational activity.

The reaction vessel is typically immersed in the coupling fluid contained in the bath (indirect sonication) or the bath itself can be used as the reaction vessel but may require additional mechanical agitation for uniform mixing. In addition, in the case of direct sonication, bath walls would be exposed to the reaction mixture and/or cavitational events making them susceptible to corrosion and/or erosion. When indirect sonication is used, the ultrasonic power dissipated into the system is low as compared to other ultrasonic systems, such as a probe. Also, reproducible results may be difficult to obtain because the amount of power reaching the reaction mixture is highly dependent upon the location of the reactant sample in the bath. Also same vessel should be used each time as the thickness of vessel base affects the transfer of power into the reaction system [13]. Another disadvantage in using a bath system is that the coupling fluid surrounding the reaction vessel(s) will eventually be heated due to transfer of ultrasonic energy, making the maintenance of isothermal conditions difficult. Cooling coils can be placed within the bath, but they will have an effect of the attenuation of the sound field and may reduce the amount of power reaching the vessel containing the reactants. Use of water as the coupling fluid limits the upper range for operating temperature to below 100°C but

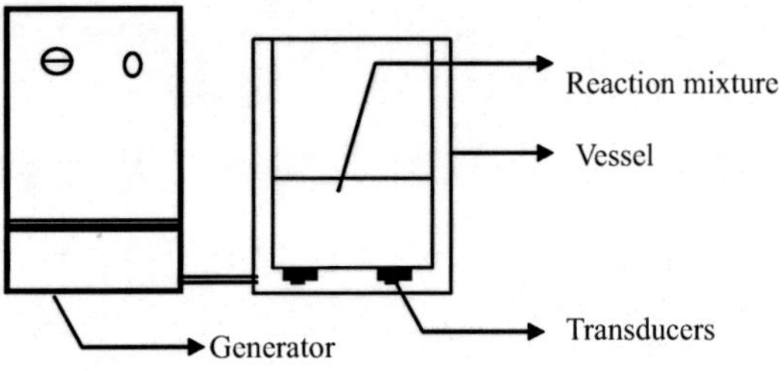

Fig. 2.6 Schematic representation of assembly for ultrasonic bath

change in the medium (high boiling liquids) also decreases the transfer of energy from the transducer to the reaction vessel.

2.3.1.3 Flow Systems

Flow systems can be based on either the use of large number of probes or the use of bath type reactors in sequential operation. The simple flow cell using an inbuilt probe system is an excellent means for processing relatively large volumes and hence generally regarded as the best approach to industrial-scale sonochemistry. One of the simplest methods is to couple the probe transducer into a flow pipe by means of a T section [15]. A number of such transducers can be employed in this manner to give extended treatment times. A schematic representation of the assembly has been given in Fig. 2.7. For larger scale operations where high intensity ultrasound is required, the sonic horn is best utilized as part of a flow loop outside of the main reactor. This permits the continuous processing of large volumes. Temperature control is provided through the circulating reaction mixture. The reactor configuration also allows the residence time of the reacting system in the flow to be varied by changing the flow rate. The reactor, however, has a general disadvantage of the probe system i.e. tip erosion, corrosion of the material and possible production of the radicals leading to side reactions. Pumping is required for the circulation into the flow cell and so the system is not suitable for either very viscous reaction media or heavily particulate systems. An alternative arrangement to using T sections is the use of number of probes inserted through the walls of a pipe, allowing much larger sonochemical treatment zone. Such a system would suffer from problems of individual flow cell except that the system will continue to function even if one or two probe units fail. Another disadvantage of such systems is that sealing of reactor/probe joints particularly for the operation of high-pressure

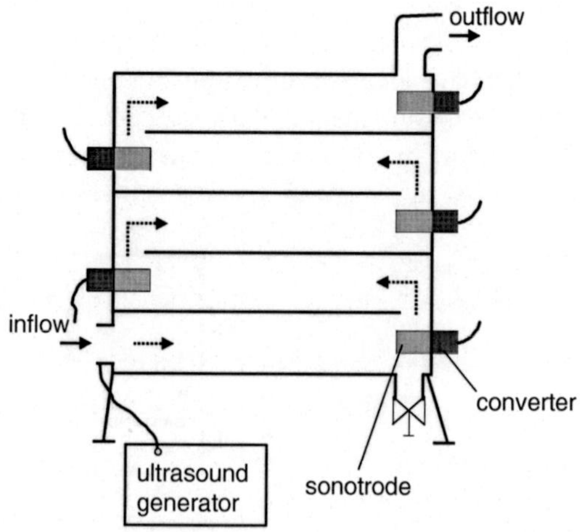

Fig. 2.7 Schematic representation of flow systems based on ultrasonic horn

reactions is a crucial factor, as the small radial vibrations associated with the probe can lead to frequent leakages.

Bath type reactors can also be used as a flow system. In this case the reacting liquids can be continuously fed into an ultrasonic tank with overflow over a weir toward the next process step. Such treatment could be intensified by recycling or by connecting a number of such sonicated tanks in line; the number of tanks or the recycle loops will be strongly dependent on the irradiation time required for the application.

More recent developments have employed direct bonding of the transducer to the surface of the vessel. Improvements in the bonding method, and a move to transducers with lower individual outputs, have enabled the move to systems with large numbers of transducers to give an acoustic pattern that is uniform and noncoherent above the cavitational threshold throughout the working volume. A photographic view of the bonding of multiple transducers to the continuous flow cell has been given in Fig. 2.8 [16]. Arrangements such as tubular reactors with two ends either irradiated with transducers or one end with transducer and

Fig. 2.8 A 20 L Prosonitron multiple 20 kHz bonded transducer flow cell (WO 00/35579). Photograph courtesy of Prosonix Ltd, UK, www.prosonix.co.uk

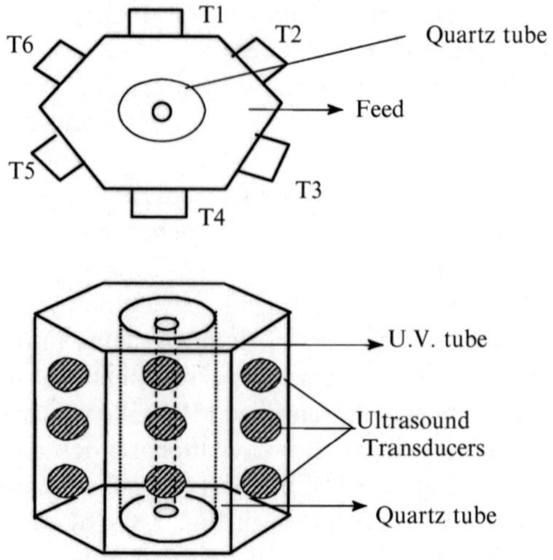

Fig. 2.9 Schematic representation of hexagonal flow cell. Hexagonal Reactor with 10 cm sides. The central quartz tube is kept as provision for simultaneous irradiation with UV light

other with a reflector [17], parallel plate reactors with each plate irradiated with either same or different frequencies [18,19] and transducers on sides of hexagon [20,21] can be constructed. A schematic representation of a hexagonal flow cell has been given in Fig. 2.9. It is of utmost importance to have uniform distribution of the ultrasonic activity in order to get increased cavitational effects. The use of low-output transducers gives the additional advantage of avoiding the phenomenon of cavitational blocking (acoustic decoupling), which arises where power densities close to the delivery point are very high. In addition these multi-transducer units very effectively concentrate ultrasonic intensity towards the central axis of the cylinder and away from the vessel walls, thus reducing problems of erosion and particle shedding.

2.3.2 Understanding Cavitational Activity Distribution

Cavitation is a dynamic phenomenon and there can be a significant variation in the cavitational activity over the entire operating volume. Distribution of the cavitational activity in the sonochemical reactors is a very important issue considering the net overall effects in any processing applications. The cavitational activity is not uniformly distributed in all the conventional designs of the sonochemical reactors and is mostly concentrated nearer the transducers especially in the case of low frequency reactors [13]. There is a strong dependence of the cavitational activity on the operating parameters and geometry of the reactor system including the positions of transducers. It is very important to consider the distribution of cavitational activity in the design of industrial scale cavitational reactors and the optimum design should be such that near uniform activity at all the locations is obtained.

Characterization of the cavitational phenomena and its effects in sonochemical reactors are generally described through mapping. Mapping of sonochemical reactor is a stepwise procedure where cavitational activity can be quantified by means of primary effect (temperature or pressure measurement at the time of bubble collapse) and/or secondary effect (quantification of chemical or physical effects in terms of measurable quantities after the bubble collapse) to identify the active and passive zones.

The available techniques can be classified into two groups viz., experimental techniques for measurement and theoretical prediction of the cavitational activity distribution. Depending on different effects generated by cavitation, experimental techniques can be further classified into techniques based on quantification of primary effects and secondary effects (Fig. 2.10). The effects generated at the same time as the bubble collapse are called as primary effects such as temperature pulse, pressure pulse, generation of free radicals (in the cavity) and micro-circulation in the vicinity of bubble. The effects generated after the bubble collapse are called as secondary effects such as oxidation reactions, intensification of mass transfer coefficients, enhanced electrochemical effects etc. Also it should be noted that, even though generation of free radical is considered as primary effect, they are usually utilized in chemical reaction after bubble collapse. Determination of variation in the cavitational activity by experimental investigation is not always feasible due to the following reasons:

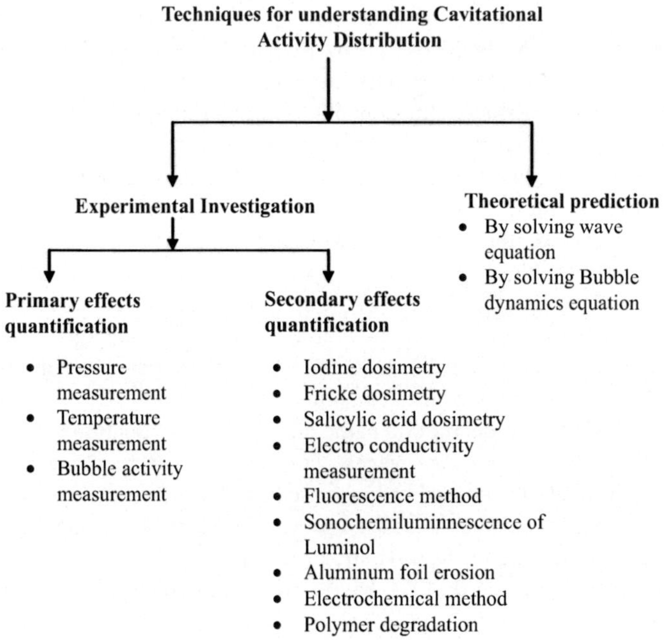

Fig. 2.10 Classification of different types of techniques for understanding cavitational activity distribution

- Experimental techniques are usually quite expensive and time consuming
- Cavitation medium gets disturbed due to the presence of external instrument such as thermocouple, hydrophone, aluminum foil, test tube etc. and hence we may not get a realistic picture of the cavitational activity distribution
- Obtained results may not be reproduced due to the dynamic nature of cavitation phenomena

Theoretical predication of cavitational activity in terms of pressure field gives an efficient alternative to experimental mapping techniques. Based on theoretical analysis, one can obtain the pressure field distribution in any new sonochemical reactor with different geometries and operating conditions, which can aid in optimization for maximum/ uniform cavitational activity. The modeling studies can be extended to quantification of other useful parameters such as distribution of temperature, mass transfer coefficient etc., which can be controlling parameters depending on the type of application. The group of Keil [22-26] have done pioneering work in this area and have investigated the variation in cavitational activity in different sonochemical reactors by solving the Helmholtz and Kirchhoff integral equation with homogenous and inhomogeneous bubble distribution. It has been shown that the cavitational activity is maximum very near to the transducer and decreases away from transducer, which is line with experimental measurements. It has been also reported that the influence of the bubble fraction on the distribution of pressure field can be neglected and damping effects are more dominant for the homogenous distribution of bubbles as compared to the inhomogeneous distribution [24]. It should be also noted that there is some scope for improvement in the models developed by Keil and coworkers. The homogenous density distribution of the bubbles as assumed in the work [24,25] is difficult to be achieved even in a well stirred ultrasound bath. In addition, bulk movement of the liquid in the bath due to stirring and also acoustic streaming can disturb the pressure fields due to the scattering of ultrasound waves. Also the presence of bubbles can scatter the incident sound waves and hence the bubble concentration would have a significant influence on the pressure field. In the simulation with inhomogeneous distribution of the bubbles [24-26], the assumption of homogenous bubble distribution in one plane (radial variation of bubble size and number has been neglected) is also not possible practically since the bubble volume fraction changes continuously with sound wave propagation due to continuous generation and collapse of the cavities and also the number of these cavities vary with time. Also the assumption of small-amplitude sinusoidal motion in the modified wave equation is questionable especially in the case of pressure waves with much higher amplitudes ($>$1 atm), which are again likely to be used at higher scales of operation. Nevertheless, this work surely cannot be underestimated and can be taken as a starting point for obtaining a clearer picture of the cavitation phenomena.

More recently, Saez et al. [27] have carried out numerical simulations to characterize the ultrasonic field propagation and to obtain the spatial distribution of the mechanical effects. The model is based on the assumption of linear wave propagation in a homogeneous media and the results are based on the solution of the

Helmoltz equation using a finite element method considering an optimized mesh size. Comparison of theoretical results with those obtained with different experimental mapping techniques (calorimetry, aluminium foil erosion and use of thermal probes) represents a good agreement with the obtained trends in terms of cavitational activity distribution. Klima et al. [28] have used similar simulation analysis and have highlighted an approach for optimizing the geometry of the sonochemical reactor. It has been reported that appropriate selection of the reactor geometry, in terms of the reactor dimensions, liquid volume and the extent of immersion of ultrasonic horn in the reactor, can result in much better uniformity of the cavitational activity with intensification as compared to the conventional approach, possibly attributed to the multiple reflections of the incident sound waves from the reactor walls and liquid surface leading to resonance effects. The work truly indicates the approach to be used for utilizing the theoretical techniques for optimization of the reactors as discussed earlier. Horst et al. [9] have also reported similar utility of theoretical analysis of the sound wave propagation in the reactor for the optimization of the geometry of the reactor.

Yasui et al. [29] have used solution of wave equation based on finite element method for characterization of the acoustic field distribution. A unique feature of the work is that it also considers contribution of the vibrations occurring due to the reactor wall and have evaluated the effect of different types of the reactor walls or in other words the effect of material of construction of the sonochemical reactor. The work has also contributed to the understanding of the dependence of the attenuation coefficient due to the liquid medium on the contribution of the vibrations from the wall. It has been shown that as the attenuation coefficient increases, the influence of the acoustic emission from the vibrating wall becomes smaller and for very low values of the attenuation coefficient, the acoustic field in the reactor is very complex due to the strong acoustic emission from the wall.

2.3.3 Design Related Information Based on Mapping Investigations

In the past, attempts have been made to understand the uneven distribution of the cavitational activity existing in the reactor, both using experimental techniques as well as using numerical simulations. Based on a critical analysis of the existing literature on mapping of sonochemical reactors, following important considerations for developing an effective scale-up strategy can be made:

1. There exists a wide distribution of the cavitational activity in the sonoreactor with the maximum intensity observed just near the transducer (for the standard arrangements such as ultrasonic horn/bath). The intensity varies both axially as well as in the radial direction with decreasing trends as we go away from the transducer (in axial direction) and away from the axis passing through the center

of the transducer (in radial direction). The intensity has been found to increase by a small amount at distances corresponding to $\lambda/2$ in the axial direction, where λ is the wavelength of the driving sound source [10,11,30,31].

2. Immersion types of transducers are poorest, when scale-up possibilities are considered, though very high intensities (pressures of the order of few thousands atmosphere) are observed very near to the horn. The intensity decreases exponentially as one moves away from horn and vanishes at a distance of 1–3 cm both in axial as well as radial directions depending on the maximum power input to the equipment and also on the operating frequency [11,32].
3. In the case of ultrasonic bath, when the bottom of the reactor is irradiated with a single transducer, the active zone is restricted to a vertical plane just above the transducer with maximum intensity at the centre line of transducer. A recent study of Balasubrahmanyam and Pandit [33] has clearly demonstrated different zones of cavitation at different planes away from the transducers with the help of mapping the erosion patterns of aluminum foil under the influence of ultrasonic irradiations. The area of irradiating surface should be increased so as to get better distribution/dissipation of energy in the reactor. This has a twin advantage in terms of the decreased ultrasonic intensity (defined as power dissipation per unit area of the irradiating surface), which will increase the magnitude of the pressure pulse generated at the end of the cavitation events [14].
4. To increase the active zones existing in the reactor, one can easily modify the position of the transducers (if multiple transducers have been used which is likely to be the case at large scale operation due to the fact that it is quite difficult to successfully operate single transducer with very high power and frequency due to limitations over the material of construction of the transducers) so that the wave patterns generated by the individual transducers will overlap, also resulting into uniform and increased cavitational activity. It is possible to design flow systems with different geometries such as tubular reactors with two ends either irradiated with transducers or one end with transducer and other with a reflector [17], parallel plate reactors with each plate irradiated with either same or different frequencies [18,19,21] and transducers on every side of hexagon [20]. It is of utmost importance to have uniform distribution of the ultrasonic activity in order to get increased cavitational effects. Romdhane et al. [20] have compared different ultrasonic equipments and have shown that hexagonal reactor with single transducer at each face of hexagon gives better homogeneity as compared to conventional ultrasonic horn and parallelepiped geometry. Kumar et al. [34] have also reported better distribution of the cavitational activity in a large scale reactor based on hexagonal configuration. Thus the need for the future is to develop newer designs giving uniform and enhanced ultrasonic activity over the entire region of the cavitational reactor.
5. Prediction of cavitational activity distribution based on theoretical analysis of the bubble dynamics equations can be used to identify the regions with maximum pressure fields in a large scale reactor and then may be small reactors can

be placed strategically at these locations in order to get maximum benefits. It might happen that the threshold required for certain transformation application is obtained at these locations but if considered globally these effects will be marginalized resulting into much lower yields based on the total volume of the cavitation reactors. Thus, the location of the transducers on the irradiating surface and the location of micro-reactors will also depend on the type of application, which decides the required cavitational intensity. It should be noted that one to one correspondence between the simulated pressure fields and yields of applications must be established before reaching to any firm conclusions.

6. Use of multiple frequencies result into relatively better distribution of cavitational activity and provide a better scale-up option. As pointed out by Dahnke and Keil [24], for obtaining better distribution of energy one has to modify the sound source in terms of number and properties and in this context multiple frequencies offer enhanced flexibility. Moholkar et al. [35] have also shown that by adjusting the phase difference and magnitude of two waves, transient cavitation can be obtained at larger number of reactor sites, which results in better overall cavitational activity (based on the total reactor volume). Moreover, due to the use of multiple frequencies the violent collapse of the cavities is restricted near the transducer surface resulting into decreased erosion of the surface [35]. Tatake and Pandit [36] have also indicated, with the help of numerical simulations of bubble dynamics, using Rayleigh-Plesset equation, that the collapse of cavities is more violent for the case of multiple frequency systems as compared to single operating frequency at similar power dissipation.
7. Dahnke and Keil [24] have shown that merely increasing the acoustic power input to the reactor or increasing the transducer frequency cannot enhance/ improve the distribution of the cavitational activity. Romdhane et al. [31] have reported that the local ultrasonic intensity as measured with the help of thermoelectric probes remains unchanged with increasing the power input (in the range 12.5–39 W). Thus increasing the power input to the reactor in order to achieve higher cavitational yields certainly cannot be a good scale-up strategy. Confirmation can also be obtained from the experimental evidences [37–39] which indicate that an optimum power input exists for a particular reaction beyond which the beneficial effects are not observed. The presence of optimum intensity has been attributed to the decoupling effect which restricts the effective propagation of the incident sound waves into the system thereby decreasing the total available energy for the cavitation events.
8. The local ultrasonic activity is also dependent on the agitation, presence of solids (size and volume fraction) or the flow regimes used in circulating type of reactors (liquid is being pumped at a definite rate which passes through a zone of ultrasonic irradiation and back to the storage tank). Romdhane et al. [31] have shown that the local ultrasonic activity decreases with agitation and also with an increase in the Reynolds number (defined as $= d \times v \times \rho/\mu$, where ρ is the density of liquid, μ is the viscosity, v is the re-circulation velocity and d the inner

diameter of the equipment used). Moreover the peaks observed at distances of $\lambda/2$ in the case of stagnant liquid medium (standing wave pattern) diminishes in the case of agitated medium thereby confirming the disturbance caused in the propagation of the sound wave in a specified direction. Similar results were obtained in another set-up where the liquid was re-circulated through the irradiation zone [31]. The effect of flow regime can also be linked to the standing wave pattern existing in the reactor due to reflection of sound waves, which contribute to the enhancement of cavitational activity. At higher operational flow regimes, standing waves will necessarily be disturbed or destructed. However, the exact effect of the flow regime on the propagation of the sound wave and sound field existing in the reactor is far from well understood and quantified, still it can be said that for getting enhanced benefits from the continuous operation of sonochemical reactors, one should operate with as low flow rates as possible. It might be worthwhile to do a computational fluid dynamics study to understand the changes in the pressure field distribution due to the presence of flow. Nevertheless, this is a very important point which needs to be considered particularly for applications involving multiple phases such as chemical reactions and solid liquid extraction etc. On one hand, where the increased agitation or Reynolds number results in less cavitational activity, it also results in better mixing of the two phases, which might enhance the rates. Same trend was observed with the size and the volume fraction of the solid particles present. On one hand where the lower size particles were observed to attenuate the wave propagation to a higher extent, it also gives better interfacial areas for chemical reactions as well as extraction. Also, increase in the volume fraction results in lower local ultrasonic activity but may have a positive effect on the cavitational activity (additional surface and hence enhanced number of nuclei for cavitation phenomena) and also on chemical reactions where solid particles used are acting as catalyst for the reaction. Thus optimization needs to be done considering these opposing effects.

2.4 Optimization of Operating Parameters

The magnitudes of collapse pressures and temperatures as well as the number of free radicals generated at the end of cavitation events and hence the intensification obtained due to the cavitation phenomena for any physical or chemical processing applications are strongly dependent on the operating parameters of the equipment namely, frequency and intensity of irradiation along with the geometrical arrangement of the transducers, geometry of the reactor, the liquid phase physicochemical properties, which affect the initial size of the nuclei and the nucleation process and the operating temperature. We now overview some of the literature illustrations giving the effect of each of the operating parameters and present some guidelines for the optimum set of operating parameters.

2.4.1 Frequency of Ultrasound

Frequency of ultrasonic equipment is usually fixed and can not be varied over wide range for a particular equipment as the maximum transfer efficiency is obtained only when the transducer is driven at the resonating frequency. Generally it has been observed that an optimum frequency specific to the application under question exists at which maximum beneficial effects are observed. Koda et al. [40] have investigated the effect of frequency in seven different types of reactors in the range 19.5 kHz to 1.2 MHz with power rating in the range of 35–200 W. Comparison was made in terms of sonochemical efficiency (mole/J) for *KI* and Fricke solution. It has been reported that sonochemical efficiency increases till 200 kHz and then decreases due to controlling attenuation of ultrasound at high frequency, leading to lower energy dissipation. Similar type of variation in sonochemical efficiency is observed for experiments with Fricke solution with maximum cavitational effects being observed at 130 kHz frequency. Wayment and Casadonte [41] have described the design of a single-transducer variable-frequency (in the range 20–500 kHz) sonication system operating at constant acoustic power. They have observed the effect of frequency of ultrasound on oxidation of potassium iodide by varying concentration in the range 0.10–1 M and have reported that rate of decomposition is maximum at 300 kHz (for 1 M concentration). Beckett and Hua [42] with studies on sonoluminescence and chemical reactions have reported an existence of optimum operating frequency as 358 kHz. Kang et al. [43] have reported lower hydrogen peroxide formation at higher frequencies of irradiation viz. 618 and 1,078 kHz as compared to 358 kHz due to the controlling effect of the higher power dissipation per unit volume.

It can be seen from the above discussion, that usually increasing the frequency of irradiations for getting better results, is not a correct operational strategy especially for chemical processing applications. The power requirement for inception of cavitation events in a high frequency operation is higher. Use of multiple frequency operation can be considered as an efficient alternative to the drawbacks associated with the single frequency operation especially when higher cavitational intensities are required for the application.

Tatake and Pandit [36] have reported the theoretical and experimental investigation of the effect of combination of two frequencies by using a combination of two reactors viz. ultrasonic horn and ultrasonic bath. It has been reported that cavitational activity is a function of the location of reaction vessel in the bath and cavitational activity is less at locations where nodes are formed (at distance of $n\lambda/2$ from the bottom of bath where $n = 2, 4, 6$, etc.). The combination gives better control over the cavitational activity with enhanced reaction rates due to higher resonance effect on bubble growth as compared to the single frequency operation. They have also investigated the effect of frequency on the maximum size of bubble reached during cavitation (with initial size of bubble as 2 μm), life time and collapse pressure. It has been reported that maximum size reached by the bubble for combination of frequencies is higher as compared to the single frequency operation, which results in enhanced cavitational activity and hence the sonochemical reaction yields. Servant et al. [44] have also reported that under the dual frequency operation, cavitation bubble volume fractions are higher as compared to that observed in mono-frequency sonochemical reactors. It has also been reported that the cavitation medium is intensely disturbed due to the combination of frequencies as more breakage of surface continuity in the liquid medium occurs resulting in overall higher cavitational activity due to generation of more cavities and stronger bubble-bubble, bubble-sound field interaction due to primary and secondary Bjerkens forces. Prabhu et al. [45] investigated the dependency of bubble dynamics on the combination of frequency in the case of single, dual and triple frequency operation at a fixed 10 W/cm^2 intensity of irradiation and initial radius of gas filled cavity as 2 μm. It has been reported that the size and life time of cavity are maximum for triple frequency operation which indicates that cavitationally active volume in the reactor is higher for triple frequency operation as compared to the dual and single frequency operations.

Overall, it can be summarized that, use of multiple frequency irradiations based on the use of multiple transducers gives much higher cavitational activity in the reactor and hence enhanced results. It is also recommended that a combination of low frequency irradiation (typically 20 kHz) with other frequencies in the range of 50–200 kHz should be used for obtaining maximum benefits from the cavitational reactors.

2.4.2 *Intensity of Irradiation*

The experimentally observed variation of the sonochemical effects with the intensity of irradiation depends on the manner in which the intensity is varied. Intensity of irradiation is defined as the power dissipation into the system per unit area of irradiation and hence can be changed either by changing the power dissipation or by the changing area of irradiation. If the intensity of irradiation is increased by decreasing the area of irradiating surface, the collapse pressure effect is the controlling factor (number of cavities generated remains the same due to constant power dissipation into the system) resulting in a decrease in the cavitational effects. Entezari and Kruus [46] have shown that the rate of iodine liberation at constant power input for equipments of different area of irradiation decreases with the increase in the intensity of irradiation i.e. decreasing the area of irradiation for constant power input. Thus it is recommended for a given power dissipation level, higher area of transducers should be used.

If the power dissipated into the system is increased, although the collapse pressure, as predicted using bubble dynamics analysis [14], decreases with an increase in the intensity, the number of cavitation events also increases (increase is substantial as compared to the negative effect of decreasing collapse pressure) thereby increasing the overall cavitational activity and hence enhanced effects can be observed. Usually the increase in number of cavities generated seizes after a particular limit of power dissipation beyond which the yield decreases (now the negative collapse pressure effect is dominating and optimum value of intensity is observed). Saez et al. [47] have performed experiments in a sonochemical reactor (of diameter 68 mm, height of 84 mm with frequency of 20 kHz and maximum power rating of 100 W) to investigate the effects of intensity of irradiation over the range of 0.5–8 W/cm^2 on the sonochemical yields. Results show that as intensity of irradiation increases, the yield of reaction increases up to threshold value and then decreases. The optimum intensity of irradiation has been found to be 2.8 and 3.5 W/cm^2 for iodine liberation and Fricke dosimerty respectively. Xie et al. [48] and Sivakumar and Pandit [49] have also reported similar existence of optimum intensity of irradiation.

Prabhu et al. [45] have carried out studies using bubble dynamics analysis to determine the effect of intensity of irradiation (in the range of 1–50 W/cm^2) for triple frequency operation with individual frequency as 30 kHz each, on the maximum bubble size and collapse pressure. It has been observed that as the intensity of irradiation increases, ratio of maximum radius to initial radius increases; thus at higher intensity of irradiation more cavitationally active volume is achieved with higher life time of cavity. Also with an increase in intensity of irradiation, collapse pressure increases slightly, leading to enhanced cavitational effects. Similar results have been reported by Feng et al. [50] though on the basis of experimental investigations. Feng et al. [50] have investigated the effect of intensity of irradiation on the cavitational activity in terms of liberation of iodine, change in thermal conductivity and fluorescence intensity, in a sonochemical reactor operating at a combination of 0.87 MHz and 28 kHz. It has been observed that, as intensity of irradiation increases the cavitational activity also increases. The observed effects are attributed to the resonance effect and enhanced bubble-bubble interaction through primary and secondary Bjerknes forces.

Overall, it can be said that for single frequency operation an optimum intensity of irradiation should be selected, whereas for multiple frequency operation, existence of optimum intensity has not been observed but this cannot be generalized.

2.4.3 Geometrical Design of the Reactor

The reactor design in terms of ratio of the diameter of the immersion transducer to reactor diameter, liquid height, position of the transducers and characteristics of the cell plays a important role in deciding the cavitational activity distribution and hence the efficacy of sonochemical reactors for the specific application. Based on a critical analysis of the existing literature, following important design related information can be recommended:

- With an increase in the diameter of immersion transducer relative to the reactor diameter, the cavitational activity increases till an optimum ratio, usually dependent on the application [51,52]. The ratio mainly affects the level of turbulent

dissipation of energy and the intensity of the acoustic streaming and hence would be more crucial in the applications where physical effects are more important.
- The extent of immersion of the transducer in an ultrasonic horn or the extent of liquid height, which affects the extent of reflection of the incident sound waves from the liquid surface as well as the reactor bottom, also shows an optimum value [53].
- The position of the transducers in reactors based on the multiple frequency arrangement should be done in such a way that maximum and uniform cavitational activity is obtained. Theoretical analysis of the cavitational activity distribution as discussed earlier aids in arriving at an optimum location of the transducers. Similar argument holds true for the geometry of the reactor.

2.4.4 Liquid Phase Physicochemical Properties

The important liquid phase physicochemical properties which affect the cavitation phenomena and hence the extent of cavitational effects for the given application include vapor pressure, viscosity and surface tension.

Usually lower vapor pressure of the liquid is preferred, attributed to the fact that with an increase in the vapor pressure of the liquid, the vapor content of the cavity increases thereby lowering the energy released during the collapse. Thus the net cavitational effects will be lower for liquids with higher vapor pressure. Rae et al. [54] have investigated the effect of vapor pressure on the cavitational intensity and reported that the mean bubble temperature (indicating the cavitational intensity) was found to decrease with increasing concentration of alcohol with the effect being more pronounced for the higher molecular weight (lower vapor pressure) of the alcohol. Vichare et al. [55] have also reported similar results with energy analysis of the collapsing cavities.

For cavitation to occur in a liquid, it has to overcome the natural cohesive forces present in the liquid. Any increase in these forces will tend to increase the threshold pressure and hence the energy required to generate cavitation. In highly viscous liquids, severe attenuation of the sound intensity occurs and the active cavitating zone gets reduced substantially. Moholkar et al. [56] have confirmed this fact with experiments with different liquids and reported that for highly viscous liquids, cavitational effects are not observed.

Liquids, with the highest values of surface tension such as water, generally result in higher cavitational intensity. Plesset [57] have demonstrated that the aqueous liquids and organic liquids such as glycerol show higher cavitating effect as compared to acetone, ethanol and formamide. Presence of surfactants in the system as additives can further enhance the cavitational activity. Ashokkumar and Grieser [58] have reported that the presence of 1 mM SDS enhances the sonoluminescence and sonophotoluminescence signals above that seen in pure water by a factor of about 2.

2.4.5 *Bulk Temperature of Liquid Medium*

Unlike conventional reaction systems, the rate of the sonochemical reactions gets drastically reduced with the rise in the operating liquid temperature beyond a certain value. The results of the simulation of the effect of bulk liquid temperature on collapse intensities of the individual cavities for water and single frequency operation indicate almost linear decrease in the collapse pressure with an increase in the temperature [59]. For multiple frequency operation, Prabhu et al. [45] have analyzed the effect of temperature (in the range of 20–60°C) by cavity dynamics model and observed that an increase in the operating temperature does not have any effect on the growth of the cavity as well as on the total lifetime of the cavity. However, a significant effect on the collapse temperature is observed (collapse temperature decreases with an increase in the operating temperature). This can be attributed to the fact that as the temperature of the liquid is increased, its viscosity and/or surface tension decreases but more importantly vapour pressure increases substantially. The extent of increase in liquid vapour pressure (P_i) is much higher as compared to liquid temperature (T_i). This result in a decrease in final liquid temperature (T_f) as explained by following mathematical relationship:

$$\frac{T_f}{T_i} = \frac{P_f\,(\gamma - 1)}{P_i}$$

where T_f is final temperature, P_i is vapour pressure and γ is poly tropic coefficient of gas. Thus, it is expected that the cavitational activity will be reduced at higher operating temperature in the reactor. In other words, for a reaction where cavitational collapse is the primary cause of the activation, a low operating temperature is recommended. In other cases, where chemical reactions are also occurring, an optimum operating temperature might exist. This is attributed to the fact that there is always a possibility that the higher concentration of chemical species is present in the cavitating bubble due to higher vapor pressure at higher operating temperature and this generates much higher amounts of free radicals in the system leading to higher reaction rates. Many sonochemical reactions such as isomerisation of maleic acid exhibit this behavior i.e. maxima with respect to the temperature [60].

2.5 Intensification of Cavitational Activity in the Sonochemical Reactors

At times the net rates of chemical/physical processing achieved using ultrasonic irradiations are not sufficient so as to prompt towards industrial scale operation of sonochemical reactors. This is even more important due to the possibility of uneven distribution of the cavitational activity in the large scale reactors as discussed earlier. It is thus important to look into supplementary strategies with an aim of intensification of the cavitational intensity. Two types of operating strategies can be recommended depending on the type of applications:

1. Use of process intensifying parameters such as presence of dissolved gases and/or continuous sparging of gases such as air, ozone and argon (to a limited extent), presence of salts such as NaCl, $NaNO_2$ and $NaNO_3$, presence of solid particles such as TiO_2, CuO and MnO_2 which can also act as a catalyst in some cases.
2. Use of combination of cavitation and advanced oxidation processes such as ozonation, chemical oxidation using hydrogen peroxide and photocatalytic oxidation and use of combination of ultrasound and microwave irradiations.

2.5.1 Use of Process Intensifying Parameters

The presence of gases and solid particles mainly provide additional nuclei for the cavitation phenomena and hence the number of cavitation events occurring in the reactor are enhanced resulting in a subsequent enhancement in the cavitational activity and hence the net chemical/physical effects. It must be noted that the presence of both these parameters also have a negative effect on the cavitational activity (aeration gives cushioning effect and incomplete collapse resulting in a decrease in the collapse pressure whereas solid particles result in scattering of the sound waves thereby decreasing the focused energy transferred into the system). The net effect of these two phenomena will be dependent on the system in question and hence optimization is a must before operating parameters are selected for actual operation.

2.5.1.1 Use of Gases

The nature of the dissolved gas i.e. its polytropic constant (γ), solubility in the cavitating medium etc. severely affects the cavitational activity due to its direct effect on the final collapse conditions. The magnitude of temperature reached at the collapse is affected by the amount of gas dissolved in the liquid medium. Final temperature reached by the adiabatically collapsing bubble (mainly depends on polytropic coefficient of the gas (γ)) can be given by following mathematical relationship:

$$T_f = T_i \left(\frac{R_i}{R_f} \right)^{3(\gamma - 1)}$$

where T_i is the initial temperature of liquid, T_f is final collapse temperature, γ is polytropic coefficient, R_i and R_f are the initial and final radius of the cavitating bubble respectively. The temperature of bubble during collapse not only depends on the polytropic coefficient of gas but also on the thermal conductivity and solubility of the gas in the liquid medium.

Gogate et al. [59] have performed simulations for various gases having different polytropic indices (in the range 1.2–1.6) and reported that the pressure developed during the collapse of the cavity increases with an increase in the polytropic constant of the gas. Thus monoatomic gases like Argon, Helium will result in maximum cavitation intensity as compared to the polyatomic gases. Moreover monoatomic gases will also offer added advantage of lower thermal conductivity resulting in lower heat losses and hence larger magnitudes of collapse temperatures. For the gases having the same γ value, the cavitational effect may vary due to its dependence on gas solubility. Usually lower gas solubility results in more violent collapse of cavities.

Segebarth et al. [61] have performed experiments in Janus sonochemical reactor, with an operating frequency of 20 kHz and maximum power rating of 100 W, to investigate the effect of oxygen, nitrogen and noble gases such as Argon (*Ar*) and Helium (*He*) on decomposition of *KI*. It has been reported that the rate of *KI* decomposition is maximum for Ar gas. The trends can be explained from the fact that monoatomic gases give higher collapse temperature than polyatomic gases. This is due to the fact that polyatomic gases have low polytropic coefficient. The variation in rate of iodine liberation as a function of the mole fraction of Argon (*Ar*) in a mixture of helium (*He*) and argon has also been investigated and it has been reported that the rate of reaction increases with an increase in the mole fraction of *Ar*.

Shimizu et al. [62] have investigated the effect of dissolved gases (Xe, Ar, O_2 and N_2) on the generation of *OH* radicals in the presence of TiO_2 catalyst. Experiments were performed in a sonochemical reactor of capacity 5.8 L with operating frequency of 36 kHz and power rating of 200 W. It has been reported that maximum rate is observed for *Xe* followed by *Ar*, O_2 and N_2. It should be also noted that though Ar and Xe have same value of polytropic constant, the product yield in the presence of *Xe* is significant as compared to *Ar* which can be explained as follows:

- Thermal conductivity of *Ar*(0.01772 W/mK) is higher than *Xe* (0.00565 W/mK) and hence the loss of heat from bubble containing *Ar* is more rapid as compared to *Xe*. So the bubble temperature with *Xe* is always higher than bubble with Ar gas, resulting in higher cavitational intensity.
- Also, it should be noted that solubility of *Xe* in water is three times higher than *Ar*, which provides additional nucleation site for cavitation.

2.5.1.2 Use of Solid Particles

Considering the specific application of chemical synthesis, the presence of solid catalyst (particles/salts in a typical concentration range of 1–10% by weight of the reactants; optimization is recommended in majority of the cases using laboratory scale studies) in the sonochemical reactors results in intensification due to the following mechanisms;

1. Formation of increased cavitation nuclei due to more number of discontinuities in liquid continuum as a result of the presence of particles to give larger number of collapse events resulting in increase in the number of free radicals.
2. In a biphasic solid-liquid medium irradiated by power ultrasound, major mechanical effects are the reduction of particles size leading to an increased surface area and the formation of liquid jets at solid surfaces by the asymmetrical inrush of the fluid into the collapsing voids. These liquid jets not only provide surface cleaning but also induce pitting and surface activation effects and increase the rate of phase mixing, mass transfer and catalyst activation.
3. Enhanced generation of free radicals due to some catalysts such as $FeSO_4$ or elemental iron.
4. Better distribution of the organic pollutants increasing the concentration at reaction sites.
5. Alteration of physical properties (vapor pressure, surface tension) facilitating generation of cavities and also resulting in more violent collapse of the cavities.
6. Presence of salts might also result in preferential accumulation of the reactants at the site of cavity collapse thereby resulting in an intensification of the cavitational reactions [63]. However it should be noted that an optimum salt concentration can exist beyond which the cavitational activity decreases. Wall et al. [64] have reported that the optimum concentration is usually in the range 1–2 M and above this optimum, the sonoluminescence intensity was observed to reduce drastically.

2.5.2 Use of Combination of Cavitation and Advanced Oxidation Processes

Intensification can be achieved using this approach of combination of cavitation and advanced oxidation process such as use of hydrogen peroxide, ozone and photocatalytic oxidation, only for chemical synthesis applications where free radical attack is the governing mechanism. For reactions governed by pyrolysis type mechanism, use of process intensifying parameters which result in overall increase in the cavitational intensity such as solid particles, sparging of gases etc. is recommended.

Usually, the use of hydrogen peroxide in conjunction with ultrasound is beneficial only till an optimum loading [65-67]. The optimum value will be dependent on the nature of the chemical reactions and the operating conditions in terms of power density/operating frequency (these decide the rate of generation of the free radicals) and laboratory scale studies are essential to establish this optimum for the specific application in question. Literature reports may not necessarily give correct solutions (for optimum concentration) even if matching is done with respect to the primary components as it is the auxiliary components of the streams (most notably radical scavengers such as bicarbonate, carbonate ions, t-butanol, naturally occurring material, humic acids) that are crucial factors in deciding the effectiveness of the free radical attack.

The synergistic effect of combining ozonation with ultrasonic irradiation is observed only when the free radical attack is the controlling mechanism and the rate of generation of free radicals due to ultrasonic action alone is somewhat lower (at lower frequencies of operation and power dissipation levels). At higher frequencies of operation, optimum for the ozone concentration exists and should be established by laboratory scale studies for the application under question. The effect of the presence of radical scavengers such as bicarbonate, chloride, sulfate ions, naturally occurring material will be again dependent on the specific reaction in question. For volatile materials such as chloroform, MTBE, the reactions takes place in the cavitating bubble and hence the effect will not be felt. But for majority of the other reactants, the reaction takes place usually in the liquid bulk or at gas-liquid interface and hence the decrease in the rates of synthesis processes will be significant. Acid pretreatment (only with a weak acid for the selective removal of ionic radical scavengers; also other type of treatment may be required if some of the non-ionic radical scavengers are present even under acidic conditions) along with sparging with neutral gas can be done for increasing the reactivity. It must also be noted that, large concentration of the weak acid will be required for achieving similar acidic conditions as compared to the stronger acid but if the selection of the weak acid is made in such a way that radical scavenging action of the dissociated acid is not there (Strong acid counter-ions i.e. Cl^-, SO_4^{-2} indeed have a scavenging action of the hydroxyl radicals), large number of free radicals will be present in the system available for the specific application. Thus, the use of proper weak acid is justified for the pretreatment. The extensive work done by the group of Hoffmann [68-70], though in the area of wastewater treatment, is recommended for understanding more details about using combination of ozone and cavitation synergistically.

The synergistic effects of combining photocatalytic oxidation with cavitation can be possibly attributed to:

1. Cavitational effects leading to an increase in the temperatures and pressure at the localized microvoid cavity implosion sites.
2. Cleaning and sweeping of the TiO_2 surface due to acoustic microstreaming allows for an access to more active catalyst sites at any given time.
3. Mass transport of the reactants and products is increased at the catalyst surface and in the solution due to the facilitated transport as a result of shockwave propagation.
4. Surface area is increased by fragmentation or pitting of the catalyst.
5. Cavitation induced radical intermediates participate in the destruction of organic compounds.
6. The organic substrate reacts directly with the photogenerated surface holes and electrons.

7. Cavitation induced turbulence also enhances the rates of the desorption of intermediate products from the catalyst active sites and helps in continuous cleaning of the catalyst surface.

Toma et al. [71] have given a brief overview of different studies pertaining to the effect of ultrasound on photochemical reactions concentrating on the chemistry aspects i.e. mechanisms and pathways of different chemical reactions. It should be noted that in the situations where the adsorption of reactants at the specific sites is the rate controlling step, ultrasound will play a profound role due to substantial increase in the number of active sites and also due to the increased surface area available due to fragmentation of the catalyst agglomerates under the action of turbulence generated by acoustic streaming along with an increase in the diffusional rates of the reactants. It is of utmost importance to operate simultaneously rather than having sequential irradiation of ultrasound followed by photocatalytic oxidation. This is because the catalyst surface is kept clean continuously due to the cleaning action of ultrasound in the operation as a result of which maximum sites are available for the photocatalytic reaction. Also, the number of free radicals generated in simultaneous irradiation (more dissociation of water molecules due to more energy dissipation and also the presence of solid particles enhances the ultrasound effects due to surface cavitation and adsorption of contaminants) will be the maximum as compared to the sequential operation. The details regarding the operating and design strategies for maximizing the synergistic effects have been described by Gogate and Pandit [72].

2.5.3 Combined Use of Microwave Irradiation and Sonochemistry

Combined use of microwave irradiation and sonochemistry is a recent advance that accelerates the chemical reactions and other processing applications significantly, especially in the case of heterogeneous systems. The dramatic acceleration effect may be attributed to a combination of enhanced heat transfer due to microwave irradiation and intensive mass transfer at phase interfaces caused by ultrasonic irradiation. In many heterogeneous reactions, the rate-determining step is the mass transfer at the interface between two (or more) phases. With sonication, the liquid jet caused by cavitation propagates across the bubble towards the phase boundary at a velocity estimated at several hundreds of m/s, and violently hits the surface. The intense agitate leads to the mutual injection of droplets of one liquid into the other, resulting in the formation of fine emulsions. These ultrasonically produced emulsions are smaller in size and more stable than those obtained conventionally and require little or no surfactant to maintain stability. Microwave irradiations compliment these effects by way of enhanced heating as well as by way of absorption by the molecules resulting in significantly enhanced rates of reactions. The main advantage of microwave heating is the instantaneous heating of solids and liquids, which is not the case of conventional heating. The details about the use of microwave irradiations in intensifying the chemical reactions can be referred in the earlier published work [73-75].

Peng and Song [76] reported the use of combination technique for hydrazinolysis of methyl salicylate and compared the yields of reaction under different operating combinations comprising of conventional reflux, sonication alone, microwave irradiation alone and combination technique. The yield of reaction was reported to be 73% in 9 h of reaction time under conventional reflux conditions whereas use of sonication (50 W power dissipation) resulted in a reaction yield of 79% in only 1.5 h. Microwave irradiation at 200 W power dissipation further intensified the process resulting in an yield of 80% in only 18 min of reaction time. Combination technique was the fastest of all investigated processes giving about 84% yield in only 40 s. The combination technique was also applied to different methyl esters and yields reported were around 80–85% in less than a minute depending on the type of the ester investigated.

Cravatto et al. [77] also reported that the combined use of microwave irradiation and ultrasound resulted in intensification of the Suzuki reactions. The system for combined effect consisted of an air-cooled probe equipped with a titanium horn, working at about 20.5 kHz and an US-reaction vessel (made of titanium) in which the horn is inserted. The reaction vessel is cooled by a flow of either tap-water or refrigerated oil. The reaction mixture was pumped using two lengths of thermostatted coaxial tubing and a peristaltic pump for circulation between the US-cell and another vessel (the MW-cell, made of Teflon) placed inside a MW oven. The coupling reaction between 3-Bromoanisole and Phenylboronic acid resulted in a reaction yield of 54% using ultrasonic irradiation and 64% using microwave irradiation when operated individually. The combined operation resulted in a yield of 88% in similar reaction periods. Trotta et al. [78] and Cravotto et al. [79] have also reported the superiority of using a combination of ultrasound and microwave irradiations.

2.6 Qualitative Considerations for Reactor Choice, Scaleup and Optimization

When deciding on the type of the reactor required for a particular chemical or physical transformation, the first question that needs to be addresses is whether the cavitation enhancement is the result of an improved mechanical process (due to enhanced mixing). If this is the case, then cavitation pretreatment of a slurry may be all that is required before the system is subjected to conventional type transformation scheme and the scale up of the pretreatment vessel would be a relatively simpler task.

If, however, the effect is truly based on the cavitation chemistry, then cavitation must be provided during the transformation itself either operated in continuous manner or in suitable pulsed operation. The scientific database from the laboratory study typically provide the parametric window found to be most effective at the laboratory scale and the cause-effect-relationships between the operating parameters and the observed cavitational effects. This could also involve generating design correlations for the prediction of the cavitational yield as a function of different operating parameters as discussed earlier. These design equations will help in selection of the operating parameters to achieve desired level of transformations. The first step is to try to understand the mechanisms of interaction from the observed phenomena so that the desired cavitation field can be created on a larger scale to promote similar interactions. The important scale up consideration is then to establish the optimum conditions for the transformation in terms of the operating/design variables that influence cavitation. In this analysis, the nature of the transformation will also have an impact on the suitability of the given cavitational reactor.

Both chemical and physical properties of the reaction medium will dictate the required level of cavitation power. High viscosity media with low vapor pressure will require higher energy to generate cavitation. The presence of entrained or evolved gases will facilitate cavitation, as will the presence or generation of solid particles.

An efficient coupling of the acoustic energy to the material that will provide a transmission path for the ultrasonic energy is also very important. This is usually a major step and requires a thorough understanding of the nature of wave propagation (either in high frequency or multiple frequency multiple transducer application) and radiation from pipe, plate or channel shaped sonochemical reactors. Though a detailed discussion related to these issues is beyond the scope of the present work, the important factors that need to be analyzed have been illustrated below:

- Mechanisms of cavitation and their interaction with the reactor material
- Wave propagation in structures
- Acoustic coupling and mode of excitation
- Transducers and power generator technology
- Integration of ultrasonics into the process system

Mason and Cordemans de Meulenaer [80] have also given the following 10 recommendations/steps in the optimization of an ultrasonic process.

1. Make cavitation easier by the addition of solids or gas bubbles to act as nuclei.
2. Try entraining different gases or mixture of gases.
3. Try different solvents for different temperature ranges and cavitation energies.
4. Optimize the power required for the reaction.
5. When using a solid-liquid system do not charge all the components in the reactor at once.
6. If possible, try to homogenize two-phase systems as much as possible.
7. Try different shapes (diameters and volumes) for the reactor.
8. It can be better (but not always) to avoid standing wave conditions by performing sonochemical reactions under high power conditions with mechanical stirring.
9. Where possible, try to transform a batch system into a continuous one.
10. Choose conditions, which allow comparisons between different sonochemical reactions.

2.7 Concluding Remarks

Cavitation generates conditions of high temperature and pressure along with the release of active radicals, which results in intensification of many of the physical and chemical transformations. The magnitudes of pressure and temperature and number of free radicals can be easily manipulated by adjusting the operating and geometric parameters depending on the desired intensity of cavitation phenomena suiting a particular transformation/application at the same time trying to minimize the processing costs.

In the sonochemical reactors, selection of suitable operating parameters such as the intensity and the frequency of ultrasound and the vapor pressure of the cavitating media is an essential factor as the bubble behavior and hence the yields of sonochemical transformation are significantly altered due to these parameters. It is necessary that both the frequency and intensity of irradiation should not be increased beyond an optimum value, which is also a function of the type of the application and the equipment under consideration. The liquid phase physico-chemical properties should be adjusted in such a way that generation of cavitation events is eased and also large number of smaller size cavities are formed in the system.

Design of sonochemical reactors is a very important parameter in deciding the net cavitational effects. Use of multiple transducers and multiple frequencies with possibility of variable power dissipation is recommended. Theoretical analysis for predicting the cavitational activity distribution is recommended for optimization of the geometry of the reactor including the transducer locations in the case of multiple transducer reactors. Use of process intensifying parameters at zones with minimum cavitational intensity should help in enhancing the net cavitational effects.

Overall it can be said that, cavitation phenomena offers a novel means for intensification of a variety of physical/chemical transformations including chemical synthesis, biotechnology, environmental engineering, polymer engineering etc. and the rates of transformations can be at times, order of magnitude higher as compared to the conventional approach and also the energy consumption is relatively less. At this stage of development of sonochemistry/cavitation, it seems that there are some technical, economical limitations and very limited processing on an industrial scale is being carried out though some efforts have been made with success in pilot scale application of cavitational reactors by few research groups. More insight into intensification studies using process intensifying parameters and/or combination of different reactor configurations/processes based on the guidelines established in the present chapter should also help in achieving the goal of industrial scale application. Undoubtedly, combined efforts of chemists, physicists, chemical engineers and equipment manufacturers will be required for the Chemical Process Industry (CPI) to harness cavitation as a viable option for process intensification.

References

1. Mason TJ (1992) Practical sonochemistry: Users guide in chemistry and chemical engineering. Ellis Horwood Series in Organic chemistry, Chichester, UK
2. Young FR (1989) Cavitation. McGraw-Hill, London, UK
3. Leighton TG (1994) The acoustic bubble. Academic Press, London, UK
4. Suslick KS (1990) Sonochemistry. Science 247:1439–1445
5. Lindley J, Mason TJ (1987) Use of ultrasound in chemical synthesis. Chem Soc Rev 16:275–311
6. Thompson LH, Doraiswamy LK (1999) Sonochemistry: Science and engineering. Ind Eng Chem Res 38:1215–1249
7. Cravatto G, Cintas P (2006) Power ultrasound in organic synthesis: Moving cavitational chemistry from academia to innovative and large-scale applications. Chem Soc Rev 35:180–196
8. Abismaıl B, Canselier JP, Wilhelm AM, Delmas H, Gourdon C (1999) Emulsification by ultrasound: Drop size distribution and stability. Ultrason Sonochem 6:75–83
9. Horst C, Chen Y-S, Kunz U, Hoffmann U (1996) Design, modeling and performance of a novel sonochemical reactor for heterogeneous reactions. Chem Eng Sci 51:1837–1846
10. Dahlem O, Demaiffe V, Halloin V, Reisse J (1998) Direct sonication system suitable for medium scale sonochemical reactors. AIChE J 44:2724–2730
11. Dahlem O, Reisse J, Halloin V (1999) The radially vibrating horn: A scaling up possibility for sonochemical reactions. Chem Eng Sci 54:2829–2838
12. Dion J-L (2009) Contamination-free high capacity converging waves sonoreactors for the chemical industry. Ultrason Sonochem 16:212–220
13. Gogate PR, Tatake PA, Kanthale PM, Pandit AB (2002) Mapping of sonochemical reactors: Review, analysis and experimental verification. AIChE J 48:1542–1560
14. Gogate PR, Pandit AB (2000) Engineering design methods for cavitation reactors I: Sonochemical reactors. AIChE J 46:372–379
15. Nickel K, Neis U (2007) Ultrasonic disintegration of biosolids for improved biodegradation. Ultrason Sonochem 14:450–455
16. Ruecroft G, Hipkiss D, Ly T, Maxted N, Cains PW (2005) Sonocrystallization: The use of ultrasound for improved industrial crystallization. Org Process Res Dev 9:923–932
17. Gonze E, Gonthier Y, Boldo P, Bernis A (1998) Standing waves in a high frequency sonoreactor: Visualisation and effects. Chem Eng Sci 53:523–532
18. Thoma G, Swofford J, Popov V, Som M (1997) Sonochemical destruction of dichloromethane and o-dichlorobenzene in aqueous solution using a nearfield acoustic processor. Adv Env Res 1:178–193
19. Gogate PR, Shirgaonkar IZ, Sivakumar M, Senthilkumar P, Vichare NP, Pandit AB (2001) Cavitation reactors: Efficiency analysis using a model reaction. AIChE J 47:2526–2538
20. Romdhane M, Gourdon C, Casamatta G (1995) Local investigation of some ultrasonic devices by means of a thermal sensor. Ultrasonics 33:221–227
21. Gogate PR, Mujumdar S, Pandit AB (2003) Large scale sonochemical reactors for process intensification: Design and experimental validation. J Chem Tech Biotech 78:685–693
22. Keil F, Dahnke S (1996) Numerical calculation of pressure field in sonochemical reactor. Chem Ing Tech 68:419–422
23. Keil F, Dahnke S (1997) Numerical calculation of Scale Up effects of pressure field in sonochemical reactors – homogenous phase. Hung J Ind Chem 25:71–80
24. Dahnke S, Keil F (1998) Modeling of three dimension linear pressure field in sonochemical reactores with homogenous and inhomogenous density distribution of cavity bubbles. Ind Eng Chem Res 37:848–864
25. Dahnke S, Keil F (1999) Modeling of linear pressure fields in sonochemical reactor considering an inhomogeneous density distribution of cavitation bubble. Chem Eng Sci 54:2865–2872

26. Dahnke S, Swamy K, Keil F (1999) Modeling of three-dimensional pressure fields in sonochemical reactors with an inhomogeneous density distribution of cavitation bubbles. Comparison of theoretical and experimental results. Ultrason Sonochem 6:31–41
27. Saez V, Frýas-Ferrer A, Iniesta J, Gonzalez-Garcý J, Aldaz A, Riera E (2005) Chacterization of a 20 kHz sonoreactor. Part I: Analysis of mechanical effects by classical and numerical methods. Ultrason Sonochem 12:59–65
28. Klima J, Frias-Ferrer A, Gonzalez-Garcý J, Ludvýk J, Saez V, Iniesta J (2007) Optimisation of 20 kHz sonoreactor geometry on the basis of numerical simulation of local ultrasonic intensity and qualitative comparison with experimental results. Ultrason Sonochem 14: 19–28
29. Yasui K, Kozuka T, Tuziuti T, Towata A, Iida Y, King J, Macey P (2007) FEM calculation of an acoustic field in a sonochemical reactor. Ultrason Sonochem 14:605–614
30. Pugin B (1987) Qualitative characterization of ultrasound reactors for heterogeneous sonochemistry. Ultrasonics 25:49–55
31. Romdhane M, Gadri A, Contamine F, Gourdon C, Casamatta G (1997) Experimental study of the ultrasound attenuation in chemical reactors. Ultrason Sonochem 4:235–243
32. Chivate MM, Pandit AB (1995) Quantification of cavitation intensity in fluid bulk. Ultrason Sonochem 2:S19–S25
33. Balasubrahmanyam A, Pandit AB (2009) Oscillating bubble concentration and its size distribution using acoustic emission spectra. Ultrason Sonochem 16:105–115
34. Kumar A, Gogate PR, Pandit AB (2007) Mapping of acoustic streaming in sonochemical reactors. Ind Eng Chem Res 46:4368–4373
35. Moholkar VS, Rekveld S, Warmoeskerken MMCG (2000) Modeling of the acoustic pressure fields and the distribution of the cavitation phenomena in a dual frequency sonic processor. Ultrasonics 38:666–670
36. Tatake PA, Pandit AB (2002) Modeling and experimental investigation into cavity dynamics and cavitational yield: Influences of dual frequency ultrasound sources. Chem Eng Sci 57:4987–4995
37. Couppis EC, Klinzing GE (1974) Effect of cavitation on reacting systems. AIChE J 20:485–491
38. Ratoarinoro C, Wilhelm AM, Delmas H (1995) Power measurements in sonochemistry. Ultrason Sonochem 2:S43–S47
39. Ondruschka B, Lifka J, Hoffmann J (2000) Aquasonolysis of ether: Effect of frequency and acoustic power of ultrasound. Chem Eng Tech 23:588–592
40. Koda S, Kimura T, Kondo T, Mitome H (2003) A standard method to calibrate sonochemical efficiency of an individual reaction system. Ultrason Sonochem 10:149–156
41. Wayment DG, Casadonte DJ (2002) Design and calibration of a single-transducer variable-frequency sonication system. Ultrason Sonochem 9:189–195
42. Beckett M, Hua I (2001) Impact of Ultrasonic Frequency on Aqueous Sonoluminescence and Sonochemistry. J Phys Chem A 105:3796–3802
43. Kang W, Hung H-M, Lin A, Hoffmann MR (1999) Sonolytic destruction of methyl tert-butyl ether by ultrasonic irradiation: The role of O_3, H_2O_2, frequency, and power density. Env Sci Tech 33:3199–3205
44. Servant G, Laborde JL, Hita A, Caltagirone JP, Gerard A (2003) On the interaction between ultrasound waves and bubble clouds in mono and dual-frequency sonoreactors. Ultrason Sonochem 10:347–355
45. Prabhu AV, Gogate PR, Pandit AB (2004) Optimisation of multi frequency sonochemical reactors. Chem Eng Sci 59:4991–4998
46. Entezari MH, Kruss P (1996) Effect of frequency on sonochemical reactions II. Temperature and intensity effects. Ultrason Sonochem 3:19–24
47. Saez V, Frýas-Ferrer A, Iniesta J, Gonzalez-Garcý J, Aldaz A, Riera E (2005) Characterization of a 20 kHz sonoreactor. Part II: Analysis of chemical effects by classical and electrochemical methods. Ultrason Sonochem 12:67–72

48. Xie B, Wang L, Liu H (2008) Using low intensity ultrasound to improve the efficiency of biological phosphorus removal. Ultrason Sonochem 15:775–781
49. Sivakumar M, Pandit AB (2001) Ultrasound enhanced degradation of Rhodamine B: Optimization with power density. Ultrason Sonochem 8:233–240
50. Feng R, Zhao Y, Zhu C, Mason TJ (2002) Enhancement of ultrasonic cavitation yield by multi-frequency sonication. Ultrason Sonochem 9:231–236
51. Nanzai B, Okitsu K, Takenaka N, Bandow H, Tajima N, Maeda Y (2009) Effect of reaction vessel diameter on sonochemical efficiency and cavitation dynamics. Ultrason Sonochem 16:163–168
52. Kumar A, Kumaresan T, Joshi JB, Pandit AB (2006) Characterization of flow phenomena induced by ultrasonic horn. Chem Eng Sci 61:7410–7420
53. Asakura Y, Nishida T, Matsuoka T, Koda S (2008) Effect of ultasonic frequency and liquid height on sonochemical efficiency of large-scale sonochemical reactors. Ultrason Sonochem 15:244–250
54. Rae J, Ashokkumar M, Eulaerts O, von Sonntag C, Reisse J, Grieser F (2005) Estimation of ultrasound induced cavitation bubble temperatures in aqueous solutions. Ultrason Sonochem 12:325–329
55. Vichare NP, Senthilkumar P, Moholkar VS, Gogate PR, Pandit AB (2000) Energy analysis in acoustic cavitation. Ind Eng Chem Res 39:1480–1486
56. Moholkar VS, Sable SP, Pandit AB (2000) Mapping the cavitation intensity in an ultrasonic bath using the acoustic emission. AIChE J 46:684–694
57. Plesset MS (1970) Cation erosion in non-aqueous liquids. Trans ASME (J Fluid Eng) 92:807–818
58. Ashokkumar M, Grieser F (1999) Sonophotoluminescence from aqueous and non-aqueous solutions. Ultrason Sonochem 6:1–5
59. Gogate PR, Wilhelm AM, Pandit AB (2003) Some aspects of the design of sonochemical reactors. Ultrason Sonochem 10:325–330
60. Majumdar S, Pandit AB (1998) Study of catalytic isomerisation of Maleic acid to Fumaric acid: Effect of Ultrasound. Ind Chem Engr 40:187–192
61. Segebarth N, Eulaerts O, Kegelaers Y, Vandercammen J, Reisse J (2002) About the Janus double horn sonicator and its use in quantitative homogenous Sonochemistry. Ultrason Sonochem 9:113–119
62. Shimizu N, Ogino C, Dadjour M, Ninomiya K, Fujihira A, Sakiyama K (2008) Sonocatalytic facilitation of hydroxyl radical generation in the presence of TiO_2. Ultrason Sonochem 15:988–994
63. Seymore JD, Gupta RB (1997) Oxidation of aqueous pollutants using ultrasound- Salt induced enhancement. Ind Eng Chem Res 36:3453–3457
64. Wall M, Ashokkumar M, Tronson R, Grieser F (1999) Multibubble sonoluminescence in aqueous salt solutions. Ultrason Sonochem 6:7–14
65. Chen JR, Xu X-W, Lee AS, Yen TF (1990) A feasibility study of dechlorination of chloroform in water by ultrasound in the presence of hydrogen peroxide. Environ Technol 11:829–836
66. Chemat F, Teunissen PGM, Chemat S, Bartels PV (2001) Sono-oxidation treatment of humic substances in drinking water. Ultrason Sonochem 8:247–250
67. Teo KC, Xu Y, Yang C (2001) Sonochemical degradation of toxic halogenated organic compounds. Ultrason Sonochem 8:241–246
68. Kang J-W, Hoffmann MR (1998) Kinetics and mechanism of the sonolytic destruction of methyl tert butyl ether by ultrasonic irradiation in the presence of ozone. Environ Sci Tech 32:3194–3199
69. Weavers LK, Ling FH, Hoffmann MR (1998) Aromatic compound degradation in water using a combination of sonolysis and ozonolysis. Environ Sci Tech 32:2727–2733
70. Weavers LK, Malmstadt N, Hoffmann MR (2000) Kinetics and mechanism of pentachlorophenol degradation by sonication, ozonation and sonolytic ozonation. Environ Sci Tech 34:1280–1285

71. Toma S, Gaplovsky A, Luche J-L (2001) The effect of ultrasound on photochemical reactions. Ultrason Sonochem 8:201–207
72. Gogate PR, Pandit AB (2004) Sonophotocatalytic oxidation based reactors for wastewater treatment: A critical review. AIChE J 50:1051–1079
73. Toukoniitty B, Mikkola J-P, Murzin DYu, Salmi T (2005) Utilization of electromagnetic and acoustic irradiation in enhancing heterogeneous catalytic reactions. App Cat A Gen 279:1–22
74. Stankiewicz A (2006) Energy matters: Alternative sources and forms of energy for intensification of chemical and biochemical processes. Chem Eng Res Des 84:511–521
75. Strauss CR, Varma RS (2006) Microwaves in green and sustainable chemistry. Top Curr Chem 266:199–231
76. Peng Y, Song G (2001) Simultaneous microwave and ultrasound irradiation: A rapid synthesis of hydrazides. Green Chem 3:302–304
77. Cravotto G, Beggiato M, Penoni A, Palmisano G, Tollari S, Lévêque J-M, Bonrath W (2005) High-intensity ultrasound and microwave alone or combined, promote Pd/C-catalyzed aryl–aryl couplings. Tetrahedr Lett 46:2267–2271
78. Trotta F, Martina K, Robaldo B, Barge A, Cravotto G (2007) Recent advances in the synthesis of cyclodextrin derivatives under microwaves and power ultrasound. J Incl Phenom Macro Chem 57:3–7
79. Cravotto G, Di Carlo S, Curini M, Tumiatti V, Rogerro C (2007) A new flow reactor for the treatment of polluted water with microwave and ultrasound. J Chem Tech Biotech 82:205–208
80. Mason TJ, Cordemans ED (1998) In: Luche JL (ed) Practical considerations for process optimization in Synthetic Organic Sonochemistry. Plenum, New York, USA

Chapter 3
Cavitation Generation and Usage Without Ultrasound: Hydrodynamic Cavitation

Parag R. Gogate and Aniruddha B. Pandit

Abstract Hydrodynamic Cavitation, which was and is still looked upon as an unavoidable nuisance in the flow systems, can be a serious contender as an alternative to acoustic cavitation for harnessing the spectacular effects of cavitation in physical and chemical processing. The present chapter covers the basics of hydrodynamic cavitation including the considerations for the bubble dynamics analysis, reactor designs and recommendations for optimum operating parameters. An overview of applications in different areas of physical, chemical and biological processing on scales ranging from few grams to several hundred kilograms has also been presented. Since hydrodynamic cavitation was initially proposed as an alternative to acoustic cavitation, it is necessary to compare the efficacy of both these modes of cavitations for a variety of applications and hence comparisons have been discussed either on the basis of energy efficiency or based on the scale of operation. Overall it appears that hydrodynamic cavitation results in conditions similar to those generated using acoustic cavitation but at comparatively much larger scale of operation and with better energy efficiencies.

3.1 Introduction

Hydrodynamic cavitation, which generates similar effects as the well established acoustic cavitation, has long being known for its detrimental effects and tackled accordingly. Literature dating back to late nineteenth century report the destruction caused by cavitation in speed boats. English navy, in an attempt to make high

P.R. Gogate (✉)
Chemical Engineering Department, Institute of Chemical Technology, Matunga, Mumbai 400019, India
e-mail: pr.gogate@ictmumbai.edu.in

Pankaj and M. Ashokkumar (eds.), *Theoretical and Experimental Sonochemistry Involving Inorganic Systems*, DOI 10.1007/978-90-481-3887-6_3,

speed boats, developed mechanism to rotate the marine propellers at high speeds to have an edge over the enemy. But sooner they faced challenge called 'Cavitation' which had put brakes on their speed boats. To sort this matter, Lord Rayleigh, on invitation of English navy, studied the phenomena of cavitation and published his famous report in 1917 [1], which discussed the behavior of spherical bubble near a marine impeller. In this report, Rayleigh concluded that gaseous bubbles present near impeller experiences a fluctuating pressure field due to the velocity alterations. When the pressure lowers, the bubble grow in size and as the pressure is recovered/ increased the bubble tend to shrink. As the bubble shrinks, the bubble wall velocity increases, and since the inertia of liquid is much higher as compared to that of bubble, bubble collapses adiabatically resulting in cavitating conditions, which was established as the main cause for erosion of impellers at higher speeds.

Since then, teams of engineers and scientists have put in gigantic efforts to evade cavitation in hydraulic equipments like impellers, valves, pump, turbines, turbomachinery etc. [2] and possibly attain satisfactory performance of the equipments. It was only recently that few research groups around the world have started to look at cavitation not as an 'issue' to be worried about but as a tool for process intensification [3]. Lohse in an interesting study revealed that even nature exploits cavitation for its benefit in an interesting manner. Lohse and his group [4] carried out an interesting study of snapping shrimp, which used cavitation phenomena for self defence and for attacking its prey. From high speed photographic analysis it was evident that this shrimp snapped its jaws repeatedly and swiftly. Each time the jaws were snapped, water was thrown out at velocity high enough to create cavitational bubbles. These bubble travelled along the water jet towards the prey where it collapsed generating a shock wave which stunned the prey for a while, giving enough time to the shrimp to attack it. Similar to nature utilizing the beneficial effects of cavitation phenomena, concentrated efforts have been targeted at exploiting the beneficial effects by research groups worldwide including the pioneering work undertaken at the Institute of Chemical Technology, Mumbai, India [3, 5]. The effects of hydrodynamic cavitation can be successfully harnessed to improve the energy efficiencies of chemical and physical processing applications. Energy efficiency of transformation is dependent on the availability of energy in the required form and at the required location of transformation. For applications involving a chemical change which is limited by mass transfer resistance, energy should be available in the form of fluid turbulence whereas for applications involving a chemical change which is limited by higher activation energy, the reaction energy should be available in the form of pressure and heat. Similarly, for applications involving physical transformations like nanoparticle synthesis, cell disruption or leaching, energy will be required in the form of turbulence to impart stress on the substrate. Supplying the energy in any other form will not assist these transformations and will only result in the wastage of energy. With manipulation of cavitation by simply controlling the dominant type of cavitation i.e. transient or stable cavitation, it is possible to improve the energy efficiency of transformation.

3.2 Generation of Hydrodynamic Cavitation

Hydrodynamic cavitation can simply be generated by using a constriction such as an orifice plate, venturi or throttling valve in a liquid flow [3]. The pressure–velocity relationship of the flowing fluid as explained by Bernoulli's equation can be exploited to achieve this effect. At the constriction, kinetic energy of the liquid increases at the expense of pressure head as depicted schematically in the Fig. 3.1. If the throttling is sufficient to cause the pressure around the point of vena contracta to fall below the threshold pressure for cavitation (usually vapor pressure of the medium at the operating temperature), cavities are generated. Subsequently, as the liquid jet expands reducing the average velocity, the pressure increases, resulting in the collapse of the cavities. During the passage of the liquid through the constriction, boundary layer separation occurs and a substantial amount of energy is lost in the form of a permanent pressure drop due to local turbulence. Very high intensity fluid turbulence is also generated downstream of the constriction; its intensity depends on the magnitude of the pressure drop and the rate of pressure recovery, which, in turn, depend on the geometry of the constriction and the flow conditions of the liquid, i.e., the scale of turbulence [6, 7]. The intensity of turbulence has a profound effect on cavitation intensity. Thus, by controlling the geometric and operating conditions of the reactor, the required intensity of the cavitation for the desired physical or chemical change can be generated with maximum energy efficiency [3]. A commonly used device based on hydrodynamic cavitation phenomena is the high-pressure homogenizer, which is, in essence, a high-pressure positive displacement pump with a throttling valve (single or multistage).

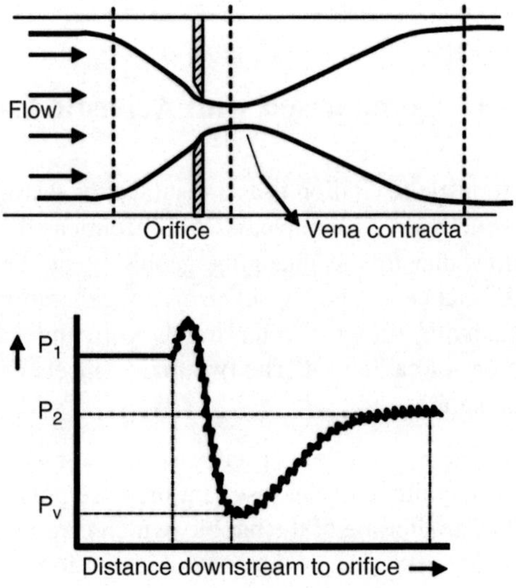

Fig. 3.1 Fluid flow and Pressure variation in hydrodynamic cavitation set-up

A dimensionless number known as the cavitation number (Cv) has generally been used to relate the flow conditions with the cavitation intensity [8]:

$$C_v = \frac{P_2 - P_v}{\frac{1}{2}\rho\, v_o^2}$$

where P_2 is the fully recovered downstream pressure, P_V is the vapor pressure of the liquid and v_o is the velocity of the liquid at the constriction. The cavitation number at which the inception of cavitation occurs is known as the cavitation inception number C_{vi}. Ideally, cavitation inception occurs at $C_{vi} = 1$ and there are significant cavitational effects at C_v less than 1. However, cavitation has been found to occur at a higher cavitation number (in the range 2–4), possibly due to the presence of dissolved gases or some impurities in the liquid medium. Yan and Thorpe [9] have also shown that C_{vi} is a function of the flow geometry and usually increases with an increase in the size of the constriction at comparable fluid velocities. Moreover, comparison of experimental data of Yan and Thorpe [9] for pipe diameter of 3.78 cm with the data of Tullis and Govindrajan [10] for pipe diameters of 7.80 and 15.4 cm indicates that the cavitation inception number is a strong function of the pipe diameter also, and it increases with an increase in the pipe diameter. This can be attributed to the fact that with an increase in the diameter of the pipe, the length scale of fluid turbulence increases, thereby increasing the fluctuating velocity component (level of fluid turbulence downstream of the constriction) at same mean operating pressure drop. Though, cavitation can be achieved even at higher cavitation numbers, for maximum benefit from the reactor, the flow conditions and the geometry should be adjusted in such a way that the cavitation number lies in the range of 0.1–1. Very low operating cavitation numbers are also not recommended as these can lead to supercavitation resulting in vapor locking and no cavitational collapse [9].

3.3 Comparison with Acoustic Cavitation

Acoustic cavitation is as a result of the passage of ultrasound through the medium, while hydrodynamic cavitation occurs as the result of the velocity variation in the flow due to the changing geometry of the path of fluid flow. In spite of this difference in the mechanisms of generation of two types of cavitation, bubble behavior shows similar trends with the variation of parameters in both these types of cavitation. The two main aspects of bubble behavior in cavitation phenomena are:

1. The amplitude of oscillation of cavity/bubble radius, which is reflected in the magnitude of the resultant pressure pulses of the cavity collapse
2. The lifetime of the bubble, which is reflected in the distance travelled and hence the extension of the zone of cavitational influence from point of its inception

Moholkar et al. [11] studied the effect of operating parameters, viz. recovery pressure and time of recovery in the case of hydrodynamic cavitation reactors and the frequency and intensity of irradiation in the case of acoustic cavitation reactors, on the cavity behavior. From their study, it can be seen that the increase in the frequency of irradiation and reduction in the time of the pressure recovery result in an increment in the lifetime of the cavity, whereas amplitude of cavity oscillations increases with an increase in the intensity of ultrasonic irradiation and the recovery pressure and the rate of pressure recovery. Thus, it can be said that the intensity of ultrasound in the case of acoustic cavitation and the recovery pressure in the case of hydrodynamic cavitation are analogous to each other. Similarly, the frequency of the ultrasound and the time or rate of pressure recovery, are analogous to each other. Thus, it is clear that hydrodynamic cavitation can also be used for carrying out so called sonochemical transformations and the desired/sufficient cavitation intensities can be obtained using proper geometric and operating conditions.

In hydrodynamic cavitation reactors, cavitation is produced at the fluid shear layer. The liquid vaporized at vena contracta downstream of the orifice is proportional to the area or volume occupied by this shear layer. This fact enables a designer to control bubble population in the flow. By changing the shape of the orifice (e.g., by making it triangular or hexagonal), the area of shear layer can be varied (obviously it is lowest for a circular orifice), and hence the rate of vaporization and the bubble/cavity population can be controlled. Bubble behavior similar to acoustic cavitation can be obtained in the hydrodynamic cavitation reactor by simple modifications in the orifice design and subsequent pressure recovery. If a rotating valve is installed instead of a permanent orifice, bubbles formed in the shear layer will experience a sinusoidally varying pressure field rather than a constant or linearly increasing field. Also, if two or three orifices are installed one after the other at the downstream of the pump, then the bubbles that are generated experience a highly fluctuating pressure field and collapse is more violent giving rise to higher temperatures and pressure pulses, of magnitudes comparable to those under acoustic cavitation. It is also possible that the cavities generated at the upstream of the orifice or their fragments can act as nuclei for orifices downstream of the first orifice. Moreover, air or steam bubbles of required sizes can be introduced in the flow, and the resultant pressure pulse magnitude can be manipulated.

Several reports appear in literature confirming that hydrodynamic cavitation is more energy efficient than acoustic cavitation [12–14]. Higher energy effectiveness of hydrodynamic cavitation can be mainly attributed to following two factors:

1. Acoustic cavitation based reactors create much higher sound energy/pressure amplitude than that required for cavitation inception. Driving pressure amplitude created in acoustic cavitation is given by

$$P_a = \sqrt{2\, I \rho C}$$

where P_a is pressure amplitude (Pa), I is intensity of ultrasound (W/m^2), ρ is density (kg/m^3) and C is speed of sound in liquid medium (m/s). For ultrasonic irradiations with a moderate intensity of 5–20 W/cm^2, driving pressure amplitude of 3.8–7.6 atm is created. Such a large value of pressure amplitude is required especially for creating nuclei via homogenous nucleation in the liquid. But for most of the practical applications which has a lot of pre-existing nuclei or where heterogeneous nucleation takes place due to impurities present in the system, pressure fluctuations of about 1 atm is also enough to generate cavitating conditions [15] in the medium. Thus, most of the energy delivered by the transducer in acoustic cavitation is wasted in attenuation and viscous heating. This is also further augmented due to poor transfer energy efficiencies for the ultrasonic transducers.

2. In the case of the acoustic cavitation, the active cavitational zone is concentrated very near to the transducer surface and hence there are severe limitations on the scale up of the sonochemical reactors. On the other hand, hydrodynamic cavitation reactors offer much larger active cavitating zones, which can also be adjusted based on the application by modifying the geometry of the cavitation chamber. Generally, higher number of cavitational events can be generated in the reactor based on the adjustments in the geometry of the reactor. Thus, though the collapse pressures due to single cavity is lower in hydrodynamic cavitation as compared to acoustic cavitation, equivalent intensities can be achieved suitable for the desired transformation. The scale up of hydrodynamic cavitation reactors is comparatively easier as vast amount of information about the fluid dynamics downstream of the constriction is readily available and the operating efficiency of the circulating pumps which is the only energy dissipating device in the system is always higher at large scales of operation. Mechanical/hydraulic efficiency of pump, which is the main energy dissipating device in the hydrodynamic cavitation reactor, is much higher (~60%) than that of ultrasonic transducer (<10%) especially at higher scale of operation [16].

3.4 Bubble Dynamics Analysis

Modelling of hydrodynamic cavitation with extensive study of cavity dynamics in the hydraulic systems can aid in a better understanding of the overall cavitation process and also in optimizing the operating parameters as well as the geometry of the cavitation set-up. Typically different bubble dynamics equations valid for explaining the behaviour of cavity in the case of sonochemical reactor such as Rayleigh-Plesset equation, equation considering compressibility of liquid medium such as Tomita and Shima equation, Gilmore equation etc. [17] are equally valid for the case of hydrodynamic cavitation reactors. The only difference is in terms of the fluctuating pressure field which is considered to be driving the various stages of the cavitation phenomena. We will now briefly discuss the various approaches used to quantify the pressure field and use it in the solution of the bubble dynamics equations. The detailed analysis of the bubble dynamics and its implications on the cavitational intensity can be referred in earlier works [6, 7, 18–22].

One of the simplest approaches to quantify the pressure field downstream of the constriction used to generate cavitation is to assume linear pressure recovery profile. Yan et al. [8] have used similar approach also considering a single bubble to be existing independent of the other bubbles. Such an approach may be adequate when the intensity of turbulence is quite low i.e. for a venturi type constriction or any other constriction with a smooth variation in the cross-sectional flow area. The pressure recovery from the point at which cavitation starts to any downstream pipe position can be approximated by a linear expression with respect to the distance downstream of the constriction. In such a case, the local pressure at any downstream position can be estimated as:

$$P_\infty = P_v + \frac{(P_2 - P_v)}{\tau} t$$

where τ is the pressure recovery time and P_∞ is the axial pressure downstream of the orifice, which is to be substituted in the Rayleigh-Plesset equation to obtain a time dependent discrete (at any time 't') solution. P_v is the vapour pressure of the liquid medium under cavitating conditions. The typical radius history i.e. the variation of radius of the cavity/bubble during different stages of cavitation, obtained for such conditions has been depicted in Fig. 3.2. It can be clearly seen from the figure that the bubble just oscillates (does not collapse violently) and produces pressure pulses of very small magnitudes. Such low magnitude pressure pulses are likely to bring about only those transformations based on the physical effects of hydrodynamic cavitation requiring lower intensity. The approach of Yan et al. [8] certainly loses validity when the intensity of turbulence increases and pressure recovery is no longer linear especially for the cases where a sharp edge orifice has been used as a constriction. In this case, significant alteration is likely to occur in the local pressures encountered by the vapor/bubble cavity in the passage along with the fluid flow downstream of the constriction.

A more realistic approach to quantify the pressure field is to consider the effect of turbulence [6]. For a pipe flow, the turbulent pressure fluctuations are due to velocity perturbations as a result of the formation of eddies. The instantaneous turbulent velocity can be calculated by assuming a sinusoidal velocity variation in

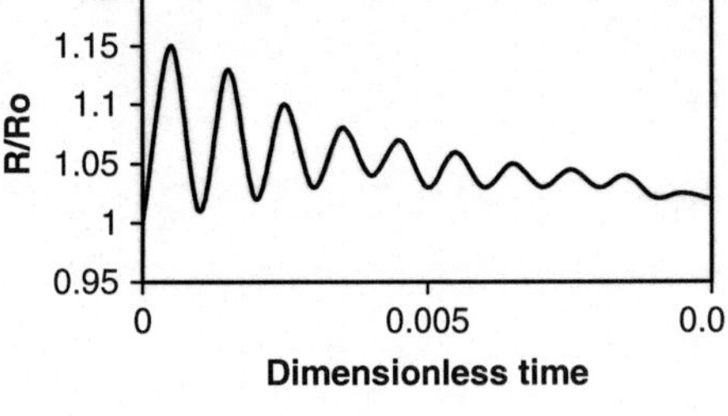

Fig. 3.2 Typical radius profile under conditions of no turbulence in the reactor

the instantaneous local velocity with the frequency of the velocity perturbations and is given by the following equation;

$$V_{tn} = V_t + \bar{v}' \sin(2\pi f_T t)$$

where V_t is the local mean velocity, f_T is the frequency of turbulence and V_{tn} is a function of time with t being the numerical integration step for time. For the estimation of fluctuating velocity ($\bar{v}'$) it is necessary to estimate the length scale of eddy and the power dissipation per unit mass as for the case of isotropic turbulence, length scale of eddy (l) and the fluctuating velocity ($\bar{v}'$) can be related to P_M as follows,

$$P_M = \frac{(\bar{v}')^3}{l}$$

The eddy length scale can be estimated [7] as follows:

$$l = 0.08\left(\frac{d_o + d_p}{2}\right)$$

Power input per unit mass of the system is equal to the rate of energy dissipation per unit mass of the liquid and it is estimated by considering the permanent pressure head loss across the orifice. The rate of energy dissipation due to eddy losses is the product of the head loss and the volumetric flow rate. Frictional pressure drop at downstream of the orifice can be calculated as,

$$\Delta p = \frac{4 f_r L_r v_t^2 \rho_l}{d_p}$$

f_r is friction factor, L_r is the length of pressure recovery zone, d_p is the diameter of pipe, v_t is the velocity of fluid at downstream of the orifice. Friction factor depends on Reynolds number (N_{Re}). For turbulent flow through pipes, f_r is given as follows [23],

$$f_r = \frac{0.079}{(N_{Re})^{0.25}}$$

Dominant frequency of turbulence can be given as:

$$f_T = \frac{\bar{v}'}{l}$$

The instantaneous velocity is then used to estimate the instantaneous local static pressure using Bernoulli's equation of the following form;

$$P_\infty = P_v + 1/2\rho\left(V_o{}^2 - V_{tn}{}^2\right) - \Delta P$$

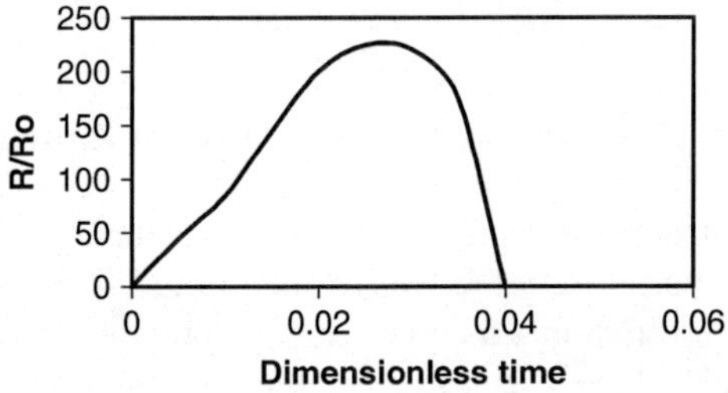

Fig. 3.3 Typical radius profile for simulations considering turbulence in the reactor

where ΔP is the permanent pressure loss across the constriction and P_∞ is the varying pressure field which is used in the solution of Rayleigh-Plesset equation.

Figure 3.3 shows the typical radius history profile obtained as a result of simulations considering the turbulence effects [6, 7]. It can be clearly seen from the figure that there is large growth in the size of the bubble/cavity (around 200 times growth is observed as compared to maximum of 1.2 times in the non turbulent conditions) which also results in significant magnitudes of pressure pulse at the time of the collapse (cavity does not oscillate as in the earlier case but collapses). It should be noted at this stage that the radius history profiles given in Figs. 3.2 or 3.3 are just a schematic representation and is a strong function of the operating and geometric conditions. This important result has shown that the bubble behavior under turbulent conditions is transient and resembles the behavior of a cavity under acoustic cavitation. The change from stable to transient cavitation by the consideration of turbulence have conclusively proved that the hydraulic devices can be used to generate conditions similar to acoustic cavitation in a relatively simple manner if only, one can manipulate the level of fluid turbulence. This conclusion can be given credence if one compares the identical metal erosion rates obtained under acoustic and hydrodynamic conditions [24]. To obtain, identical metal erosion rates, the pressure pulses produced by the collapsing cavities need to be of similar magnitude which can be only obtained with transient cavitation (with turbulence) and not by stable or oscillatory cavitation (without turbulence).

3.5 Hydrodynamic Cavitation Reactor Configurations

It is always important to choose an optimum design configuration of the hydrodynamic cavitation reactor so as to maximize the cavitational effects and achieve cost effective operation. In this section, we will discuss available reactor configurations and give some guidelines, based on theoretical analysis coupled with experimental results, for selection of optimum design and operating parameters for hydrodynamic cavitation reactors.

3.5.1 High Pressure Homogenizer

A commonly used device based on hydrodynamic cavitation phenomena is the high-pressure homogenizer, which is, in essence, a high-pressure positive displacement pump with a throttling device that operates according to the principle of high-pressure relief valve. In this, liquid is pumped at very high pressure through a constriction which converts its pressure energy into kinetic energy. At extremely high kinetic energy levels, the pressure reduces below the liquid vapor pressure generating cavitation. A high pressure homogenizer is designed to operate at pressure ranging from 50 atm to as high as 800 atm. Typically a high-pressure homogenizer reactor consists of a feed tank and two throttling valves, designated as first stage and second stage homogenization, to control the operating pressure in the hydrodynamic cavitation reactor. There is a critical discharge pressure at which cavitation inception occurs, and significant cavitational yields are obtained beyond this discharge pressure [25]. Apart from cavitation, high pressure homogenization also utilizes the intense shear produced due to fluid flow and impingement of high velocity liquid on solid wall.

High pressure homogenizers are especially suitable for the emulsification processes in the food, pharmaceutical and bioprocess industries. A general disadvantage of these type of reactors is that there is no precise control over the cavitationally active volume and the magnitude of the pressure pulses that will be generated at the end of the cavitation events (cavitational intensity), unless the valve seat designs are substantially modified.

3.5.2 High Speed Homogenizer

Cavitation can also be generated in rotating equipments. When the tip speed of the rotating device (impeller) reaches a critical speed, the local pressure near the periphery of the impeller drops and approaches the vapor pressure of the liquid. This results in the generation of vaporous cavities. Subsequently, as the liquid moves away from the impeller to the boundary of the tank, the liquid pressure recovers at the expense of the velocity head. This causes the cavities that have travelled with the bulk liquid to collapse. Again, similar to the high-pressure homogenizer, there exists a critical rotational speed for the inception of cavitation [18, 25]. It should be noted that the energy consumption in these types of reactors is much higher, and flexibility over the design parameters is lesser as compared to reactors based on the use of low pressure cavitation devices like an orifice or a venturi. These devices operate at rotating speeds of 4,000 rpm to as high as 20,000 rpm. A high speed homogenizer usually consists of a stator-rotor assembly preferably made of stainless steel, with flexibility of design of stators and rotor. A plate with holes attached to the stator can be provided for inserting baffles so as to avoid vortex formation and surface aeration in the bulk liquid. The speed of rotation of the homogenizer can be varied by changing the applied voltage on the power supplier.

3.5.3 Low Pressure Hydrodynamic Cavitation Reactor

In this type of reactor, the flow through the main line passes through a constriction where the local velocities suddenly increase due to a reduction in the flow area resulting in lower pressures which may even go below the vapor pressure of the liquid medium generating the cavities. The constriction can be a venturi [18], a single hole orifice [8] or multiple holes on an orifice plate [26]. Multiple hole orifice plates having different combinations of number, diameter and shape of holes can be constructed. Such an arrangement helps to achieve different intensities of cavitation, and also the number of cavitational events generated in the reactor is different. Thus, these reactors present tremendous flexibility in terms of the operating (control of inlet pressure, inlet flow rate, temperature) and geometric conditions (different arrangements of holes on the orifice plates). Depending on the type of application and requirements, geometry and operating conditions can be selected in the hydrodynamic cavitation reactor to maximize the cavitational effects at minimum possible energy consumption. For example, cell disruption requires milder cavitational intensity, whereas microbial disinfection of mixed culture of micro-organisms would require very high cavitation intensities due to the variation in the cell wall strength. Some recommendations for selection of the operating parameters for achieving desired cavitational intensity are discussed later in this chapter.

Sampathkumar and Moholkar [27] recently put forth a conceptual design of a novel hydrodynamic cavitation reactor that uses a converging–diverging nozzle for creating pressure variation in the flow necessary for driving bubble motion instead of the orifice plates as discussed earlier. The cavitation bubbles or nuclei are introduced in the water flow externally, upstream of the nozzle, using a sparger. Different gases can be used for the introduction of the bubbles. Also, the size of the gas distributor (usually a glass frit), flow rate of gas and the pressure of gas in the reservoir (or source) from which gas is withdrawn can be suitably controlled to control the initial size of the cavitational nuclei and the gaseous composition of the cavity, which significantly affects the resultant cavitational intensity. The aim should be to generate the smallest size of nuclei as possible to maximize the intensity and hence the net cavitational effects. Sampathkumar and Moholkar [27] have reported that the design is able to produce a cavitational effect of moderate intensity sufficient for processes such as microbial cell disruption, water disinfections, etc. The results demonstrate that a transient motion of the bubbles can be obtained with the converging–diverging nozzle, giving rise to radical formation in the bubble. The temperature peaks attained in the bubble (on collapse) in a nozzle flow are somewhat lower than those attained in an orifice flow, where temperatures of the order of ~5,000 K are seen. The maximum temperature peak that was seen among all parameter sets of simulations for the converging-diverging nozzle was 2,124 K. Thus, the cavitation intensity generated by bubbles in a nozzle flow is moderate, which is only suitable for conducting milder processes (generally dependent on the physical effects of cavitation). Further, use in applications requiring varied cavitational intensity might be limited, as length and diameter of the nozzle are the only geometric parameters that can be varied, whereas number, size and shape of the holes in the case of the orifice plate can be varied for controlling the intensity of cavitation produced downstream of the orifice plate.

From the above discussion about various hydrodynamic cavitation reactors, it can be easily concluded that the orifice plate set-up offers maximum flexibility and can also be operated at relatively larger scales of operation. It should be also noted that the scale-up of such reactors is relatively easier as the efficiency of the pump increases with an increase in size (flow rate and discharge pressure) which will necessarily result into higher energy efficiencies.

3.6 Guidelines for Selection of Hydrodynamic Cavitation Reactor Configurations

The magnitudes of collapse pressures and temperatures as well as the number of free radicals generated at the end of cavitation events are strongly dependent on the operating parameters and the configuration of hydrodynamic cavitation reactors. Bubble dynamics investigations can aid in obtaining some recommendations regarding the selection of optimum set of parameters. The detailed discussion into the bubble dynamics approaches in cavitational reactors is beyond the scope of present work; however the readers may refer to earlier work [6, 7, 18–20] for better understanding. In the present work only important considerations regarding the selection of operating parameters have been presented (Table 3.1). Liquid phase physicochemical properties also affect the cavitating conditions significantly. Aim should be to use liquids or conditions favouring the process of cavitation inception and also should result in cavities with lower initial size which would grow to a larger extent and give violent collapse and hence greater cavitational activity. Some

Table 3.1 Optimum operating conditions for the hydrodynamic cavitation reactors

No.	Property	Favorable conditions
1	Inlet pressure into the system/Rotor speed depending on the type of equipment	Use increased pressures or rotor speed but avoid super-cavitation by operating below a certain optimum value
2	Diameter of the constriction used for generation of cavities e.g. hole on the orifice plate	Optimisation needs to be carried out depending on the application. Higher diameters are recommended for applications which require intense cavitation whereas lower diameters with large number of holes should be selected for applications with reduced intensity
3	Percentage free area offered for the flow (Ratio of the free area available for the flow i.e. cross-sectional area of holes on the orifice plate to the total cross-sectional area of the pipe)	Lower free areas must be used for producing high intensities of cavitation and hence the desired beneficial effects

Table 3.2 Guidelines for selection of liquid physicochemical properties

No.	Property	Affects	Favourable conditions
1	Liquid vapor pressure (range: 40–100 mm of Hg at 30°C)	Cavitation threshold, Intensity of cavitation, rate of chemical reaction	Liquids with low vapor pressures
2	Viscosity (range: 1–6 cP)	Transient threshold	Low viscosity
3	Surface tension (range: 0.03–0.072 N/m)	Size of the nuclei (cavitation threshold)	Low surface tension
4	Bulk liquid temperature (range: 30–70°C)	Intensity of collapse, rate of the reaction, threshold/ nucleation, almost all physical properties	Optimum value exits, generally lower temperatures are preferable
5	Dissolved gas	Gas content, nucleation,	
	A. Solubility	collapse phase	Low solubility
		Intensity of cavitation events	
	B. Polytropic constant and thermal conductivity		Gases with higher polytropic constant and lower thermal conductivity (monoatomic gases)

recommendations for selection of liquid phase physicochemical properties have been made in Table 3.2.

It is also important to make some recommendations for the selection of a particular type of reactor, desirable for a specific application. Moholkar and Pandit [19] have tried to address this situation and have investigated the comparative effect of several operating parameters on bubble motion in the cavitating flow in two different flow geometries: a venturi tube and an orifice plate. In the case of a venturi tube, a larger fraction of stable oscillatory radial bubble motion is obtained due to a linear pressure recovery gradient, whereas due to an additional oscillating pressure gradient due to turbulent velocity fluctuation, the radial bubble motion in the case of an orifice flow results in a combination of both stable oscillatory and transient cavity behavior. Also, the magnitude of permanent pressure drop across the orifice is much higher as compared to that across the venturi resulting into larger fraction of energy being available for cavitation. Thus, the intensity of cavitation will be higher (due to higher contribution from the transient cavitation) in the case of an orifice system as compared to the classical venturi tube. Bubble dynamics simulations have enabled to establish definite trends in the cavitation intensity produced in the hydrodynamic cavitation reactor with operating/design parameters, which can form a basis for the optimization of hydrodynamic cavitation reactors for a targeted application. The model also enables the quantification of magnitudes of temperature and pressure pulses for a given set of design parameters. The following important strategies for the design of hydrodynamic cavitation reactors have been established:

1. An orifice flow configuration is more suitable for applications requiring intense cavitational conditions, whereas for milder processes (requiring collapse pressure pulses typically between 15 and 20 bar) and for transformations based on physical effects, a venturi configuration is more suitable and energy efficient.

2. In the case of a venturi flow, the most economical technique for increasing cavitation intensity would be to reduce the length of venturi, but for higher volumetric flow rates there could be a limitation due to the possibility of flow instability and super-cavitation. A similar argument can be given for the enhancement in the cavitation intensity by reducing the venturi throat to pipe diameter ratio.
3. In the case of an orifice flow configuration, the most convenient way of controlling the cavitation intensity is to control the orifice to pipe diameter ratio (basically throttling the discharge of a pump through a valve) and to control the cross-sectional flow area through the manipulation of the number and diameter of the holes on the orifice plate, although indiscriminate growth of bubbles downstream of the orifice can lead to splashing and vaporization (super-cavitation) of the flow.
4. Increasing pipe size downstream of the orifice (which offers a faster pressure recovery) is another option to intensify cavitation effects, but using pipes of larger size would require higher volumetric flow rates in order to carry out operation at the same cavitation number and this, results in an increase in the processing cost.

3.7 Overview of Applications of Hydrodynamic Cavitation

3.7.1 Chemical Synthesis

3.7.1.1 Hydrolysis of Fatty Oils

One of the earliest applications of hydrodynamic cavitation reactors for chemical synthesis dates back to 1993 when hydrolysis of vegetable oils was investigated [28]. Conventionally, hydrolysis of fatty oils using steam or water require operating temperature in between 250°C and 350°C and pressures of 30–60 atm. Pandit and Joshi [28] studied the hydrolysis of castor oil and kerdi oil using both acoustic (ultrasonic generator) and hydrodynamic cavitation (in a flow loop at cavitation number less than that required for inception) under ambient operating conditions. The hydrodynamic cavitation setup is a loop reactor with a gate valve, used as a throttling device to cause cavitation. Oil-water mixtures, without any catalyst and of different proportions, totalling to 200 L were subjected to cavitating conditions by circulating the liquid, number of times through the throttling valve with the temperature in the reactor being maintained constant (nearly ambient around 25°C) with the help of cooling coils. The extent of hydrolysis was monitored by determining the acid value of the oil-water emulsion in the tank (AOCS: method 39–63). The obtained trends for the progress of the hydrolysis reaction in terms of decreasing acid value were similar for the two modes of cavitation. Thus, the work was first to depict that the flow loops with some constriction for generation of cavitating conditions produce similar results as that of the acoustic cavitation. It was also found that the energy dissipation required for same degree of hydrolysis using hydrodynamic cavitation (1,080 J/mL of emulsion) was 30% lower than acoustic cavitation (1,384 J/mL of emulsion). The work also clearly illustrated that the rate of hydrolysis increased when the severity of cavitation was increased (decreasing the cavitation number).

3.7.1.2 Depolymerization Reactions

Chivate and Pandit [29] studied the depolymerization reaction using acoustic and hydrodynamic cavitation. The experiments were performed using a throttling valve setup as a means of generating hydrodynamic cavitation and the rate of degradation was followed by measuring the drop in the viscosity of the solution by a rotoviscometer. The viscosity of 0.5% CMC solution (by weight) and 1% polyethylene oxide solution was found to reduce by 6–50 times over 2 h of cavitating conditions. The rate of the decrease of viscosity i.e. rate of depolymerization was observed to be a strong function of the presence of air in the CMC solution. Presence of air in the solution has been reported to decrease the rate of reaction (less than 5% decrease in viscosity has been observed). This has been attributed to the fact that the presence of air as a second phase and not as a dissolved gas decreases the cavitational intensity in the liquid medium and also due to an increase in the average compressibility of gas-liquid mixture some amount of energy released during the collapsing cavity is absorbed by the dispersion. Yan et al. [8] have shown that the measured peak pressure of the collapsing cavity is reduced at least by an order of magnitude when air is introduced in the cavitating flow on the basis of experimental and theoretical studies.

Comparison with acoustic cavitation have also shown that hydrodynamically generated cavitation is far more energy efficient i.e. 76.5 J/mL as against 14,337 J/mL needed for sonically generated cavitation, for the same process, i.e., equivalent viscosity reduction.

3.7.1.3 Oxidation Reactions

Ambulgekar et al. [30] investigated the oxidation of toluene using aqueous $KMnO_4$ as an oxidizing agent in the hydrodynamic cavitation reactor with an objective of optimization of the operating parameters. The reaction scheme can be depicted as follows:

$$2KMnO_4 + H_2O \longrightarrow 2MnO_2 + 2KOH + 3[O]$$

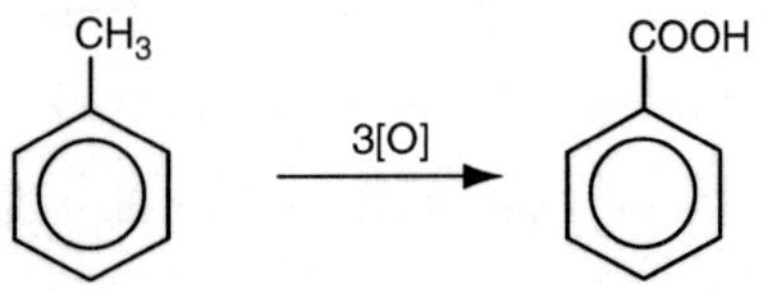

The hydrodynamic cavitation reactor used in the work consists of a reservoir or a collecting tank with 10 L capacity that is connected to the pump (rating of 1.5 kW) which allows re-circulation of the contents through a mainline housing an orifice plate (different configurations of the holes or the orifice plates can be inserted in the main line) and a bypass line (for controlling the inlet pressure and the flow rate into the cavitation chamber). A cooling water jacket is also provided to the reservoir for controlling the temperature of the system as the temperature rises due to the mechanical heat dissipation into the system and higher operating temperature are detrimental for the cavitational reactions. A schematic representation of the experimental setup has been made in Fig. 3.4.

The operating parameters used in the experimental work for the optimization exercise include inlet pressure (1–4 kg/cm^2), concentration of the controlling reactant species (oxidant i.e. $KMnO_4$ in this case with concentration varied in the range 0.2–0.5 mol/L per mol of the toluene), type of the orifice plate (without orifice plate, orifice plates of two different geometries). The optimized conditions as obtained in the study are inlet pressure of 3 kg/cm^2, 0.4 mol/L of the oxidant (beyond these values, the increase in the cavitational yield is only marginal) and orifice plate with more number of holes (cavitational yield is maximum out of all the geometries considered in the work).

In continuation of the work on oxidation reactions, Ambulgekar et al. [31] investigated different chemical reactions (oxidation of toluene, (*o*-/*p*-/*m*)-xylenes,

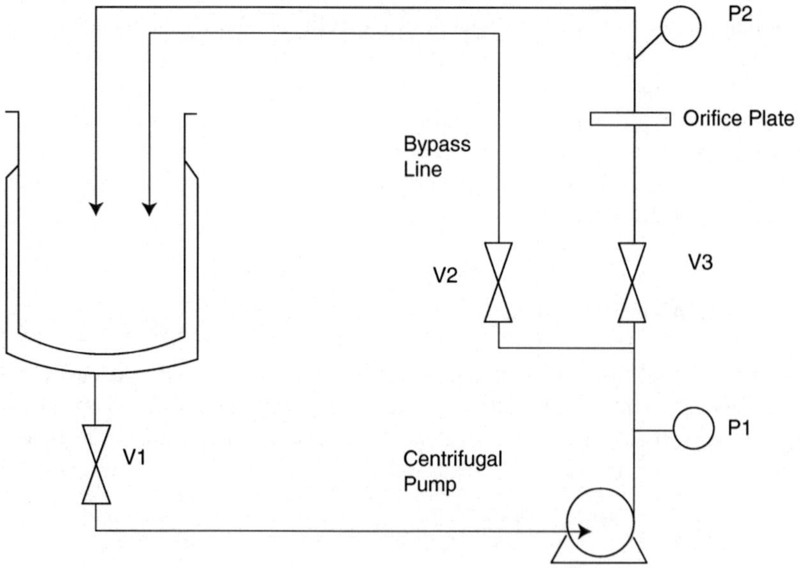

P1, P2 = Pressure Gauges; V1, V2, V3 = Control Valves

Fig. 3.4 Schematic representation for experimental setup for the orifice plate hydrodynamic cavitation reactor mesitylene, (*o*-/*m*)-nitrotoluenes and (*o*-/*p*)- chlorotoluenes and trans-esterification reaction) in the same experimental setup under optimized conditions as described earlier. The same reactions were also investigated in a conventional sonochemical reactor (ultrasonic bath, reaction volume of 55 mL, power dissipation of 120 W and operating frequency of 20 kHz). The objective was to compare the efficacies of the two modes of generation of cavitation and the criteria used for comparison was cavitational yield which is defined as the quantity of product formed per unit of supplied energy. Table 3.3 shows the values of cavitational yields obtained for all the reactions in hydrodynamic and acoustic cavitation reactors (the specific operating conditions are mentioned in the foot note for the table). It can be clearly seen from the table that the cavitational yield values in the hydrodynamic cavitation reactors are order of magnitude higher for all the reactions considered in the work and also the processing volume is about 100 times more as compared to the conventional sonochemical reactor. The results have conclusively proved the better efficacy of the hydrodynamic cavitation reactors as compared to the ultrasonic bath reactor for large scale processing as considered in the work.

Table 3.3 Comparative results for different industrially important reactions in Hydrodynamic[a] and Acoustic[b] cavitation reactors

No	Reactants	Product[c]	Cavitational yield in hydrodynamic cavitation reactor (g/J)	Cavitational yield in acoustic cavitation reactor (g/J)
1	Toluene	Benzoic acid	3.3×10^{-6}	5.6×10^{-7}
2	*p*-Xylene	Terephthalic Acid	2.1×10^{-6}	3×10^{-7}
3	*o*-Xylene	Phthalic acid	1.9×10^{-6}	3×10^{-7}
4	*m*-Xylene[d]	Isophthalic acid	1.9×10^{-6}	–
5	Mesitylene	Trimesic acid	7×10^{-6}	1×10^{-7}
6	*o*-Nitrotoluene	*o*-Nitrobenzoic acid	1.9×10^{-6}	1×10^{-7}
7	*m*-Nitrotoluene	*m*-Nitrobenzoic acid	1.3×10^{-6}	1×10^{-7}
8	*p*-Nitrotoluene[e]	*p*-Nitrobenzoic acid	–	3×10^{-7}
9	*o*-Chlorotoluene	*o*-chlorobenzoic acid	1.1×10^{-6}	1×10^{-7}
10	*p*-Chlorotoluene[d]	*p*-chlorobenzoic acid	2×10^{-6}	–
11	Sunflower Oil	Bio-diesel (methyl ester of sunflower oil)	2.1×10^{-6}[f]	5.1×10^{-7}[f]

[a]Toluene (1 mol), (*o*-/*p*-/*m*)-xylene (0.5 mol), Mesitylene (0.4 mol), (*o*-/*m*)-nitrotoluene (1 mol) and (*o*-/*p*)- chlorotoluene (1 mol), Sunflower oil (1 mol) with excess of methanol, $KMnO_4$ for reactions 1–10 (2 mol), and for all above reactions, Water (5 lit), Pressure 3 kg/cm2, orifice plate No. 1.Time = 5h, except for oxidation of toluene where it is 3 h and trans-esterification where it is 30 min

[b]Toluene (10 mmol), (*o*-/*p*)-xylene (5 mmol), Mesitylene (4 mmol), (*o*-/*m*-/*p*)-nitrotoluene (10 mmol) and *o*- chlorotoluene (10 mmol) Sunflower oil (1 mol) with excess of methanol, $KMnO_4$ for reactions 1–10 (20 mmol), and for all above reactions, Water (50 mL) Time = 5 h except for oxidation of toluene where it is 3 h and trans-esterification where it is 15 min

[c]Identification of compounds was done by TLC and Melting point

[d]Not used in the acoustic cavitation

[e]This compound is not used in hydrodynamic cavitation

[f]In mol/J

3.7.1.4 Synthesis of Biodiesel

Various products derived from vegetable oils have been proposed as an alternative fuel for diesel engines. Today "bio-diesel" is the term applied to the simple alkyl esters of fatty acids used as an alternative to petroleum based diesel fuels. Importance of biodiesel in the recent context increases due to increasing petroleum prices; limited fossil fuel reserves and environmental benefits of biodiesel (decrease in acid rain and emission of CO_2, SO_x and unburnt hydrocarbons during the combustion process). Due to these factors and due to its easy biodegradability, production of biodiesel is considered to be advantageous over that of fossil fuels. The conventional techniques for the synthesis of biodiesel [32–34] refer to a catalysed chemical reaction involving vegetable oil and an alcohol to yield acid alkali esters and glycerol. The conventional techniques typically utilize temperatures in the range of 70–200°C, pressures in the range of 6–10 atm and reaction times of up to 70 h for achieving conversions in the range of 90–95% based on the type of raw material used (usually mixtures of waste fatty acids). The reaction is usually limited by mass transfer rates and mixing of the different phases due to their limited solubility and hence there exists a lot of potential for the application of cavitational reactors. Indeed, cavitation generated using both ultrasound and flow i.e. hydrodynamic cavitation has been reported to significantly intensify the process of synthesis of biodiesel. Gogate [35] have reported the use of cavitational reactors for synthesis of biodiesel with different vegetable oils as starting materials. The results obtained for a variety of oils as a starting material has been shown in Table 3.4. It can be seen from the table that cavitation can be very successfully applied to trans-esterification reactions with more than 90% yield of the product as per stoichiometry in as low as 15 min of the reaction time. The technique hence appears to be very effective as compared to the conventional approach which is also evident from the comparison of different techniques based on quantitative criteria of energy efficiency as shown in Table 3.5. It can be seen from this table, that hydrodynamic cavitation is about 40 times more efficient as compared to acoustic cavitation and 160–400 times more efficient as compared to the conventional agitation/heating/refluxing method.

Usually, the cost of pure vegetable oils is very high and can dramatically affect the overall economics of the synthesis process. Waste vegetable oils or mixed streams of fatty acids can also be used as a starting raw material for synthesis. Kelkar et al. [14] have investigated the esterification of Fatty acid (FA) odour cut (C_8-C_{10}) with methanol in the presence of concentrated H_2SO_4 as a catalyst using hydrodynamic

Table 3.4 Transesterfication of different vegetable oils

No	Vegetable oil	Product	Time		Yield	
			Acoustic	Hydro	Acoustic	Hydro
1	Soyabean oil	Soyabean oil ester	15	15	97	98
2	Castor oil	Castor oil ester	10	10	99	99
3	Peanut oil	Peanut oil ester	10	10	99	90

Acoustic cavitation = Vegetable oil (4 g) Methanol (4 mL), NaOH (0.5%)
Hydrodynamic cavitation = Vegetable oil (4,000 g), Methanol (4,000 mL), NaOH (1%), P = 1 kg/cm^2

Table 3.5 Comparison of energy efficiency for different techniques

No	Technique	Time (min)	Yield (%)	Yield/ kJ of energy
1	Acoustic[a]	10	99	8.6×10^{-5}
2	Hydrodynamic[b]	15	98	3.37×10^{-3}
3	Conventional[c]			
	With stirring	180	98	2.27×10^{-5}
	Under reflux	15	98	7.69×10^{-6}

[a]In acoustic cavitation, 4 mL of methanol is mixed with 4 g of vegetable oil and catalyst concentration (NaOH) used in 0.5% of oil. Ultrasonic bath is the sonochemical reactor with 20 kHz frequency and 85 W as power dissipation

[b]Operation with hydrodynamic cavitation is under optimized conditions: 4:4 ratio (w/v) of oil to alcohol, catalyst concentration (NaOH) is 1% of oil, Orifice plate 1 having 16 holes with 2 mm diameter), Volume of methanol is 4,000 mL with 4,000 g of oil

[c]For conventional approach, 4 mL of methanol is mixed with 4 g of vegetable oil and catalyst concentration (NaOH) used in 0.5% of oil (Case I: a stirrer is used for uniform mixing which consumes energy; Case II: heater is used for maintaining reflux conditions)

cavitation reactor and also in the sonochemical reactor for comparison purpose. Few experiments have also been carried out with other acid/alcohol combination viz. coconut fatty acids with methanol and ethanol and FA odour cut with fatty alcohols with an aim of investigating the efficacy of cavitation for giving the desired yields and also to quantify the degree of process intensification that can be achieved using the same. It has been observed that ambient operating conditions of temperature and pressure and reaction times of less than 3 h, for all the different combinations of acid/alcohol studied in the work, was sufficient for giving more than 90% conversion. This clearly establishes the efficacy of cavitation as an excellent way to achieve process intensification of the bio-diesel synthesis process. To cite a specific illustration as regards to the degree of process intensification achieved using hydrodynamic cavitation reactors, with an operating ratio of FA cut (waste fatty acids) to methanol as 1:5, 0.1% by weight loading of the catalyst and at operating temperature of 30°C, 91% conversion was achieved using hydrodynamic cavitation in only 90 min of reaction time whereas conventional method for the esterification of waste cooking oil using methanol required about 69 h to obtain more than 90% acid oil conversion to methyl esters at 65°C operating temperature and a molar ratio of methanol to oil was 30:1. Comparison of energy efficiencies of hydrodynamic cavitation reactors with sonochemical reactors indicated that hydrodynamic cavitation is more energy efficient as compared to acoustic cavitation. Depending on the type of acid/alcohol combination used in the present work the energy efficiency for hydrodynamic cavitation varied in the range of 1×10^{-4} to 2×10^{-4} g/J whereas for acoustic cavitation it was order of magnitude lower i.e. in the range of 5×10^{-6} to 2×10^{-5} g/J.

3.7.1.5 Synthesis of Rubber Nano-Suspensions

Patil and Pandit [36] investigated the application of hydrodynamic cavitation and acoustic cavitation for synthesis of nano-scale particles of styrene butadiene rubber. The setup used for the generation of hydrodynamic cavitation was essentially similar to that depicted in Fig. 3.4. It has been observed that the mean particle size after circulation for 3 h (72 passes) was 129 nm with less than 2% variation. This was the least size of the rubber particles which could be obtained attributed to the energy dissipation levels in the reactor. A second configuration based on different type of pump and orifice plate was used to treat this solution (processed in the first hydrodynamic cavitation reactor) and after five circulations the particle size was reduced to 20 nm.

Comparison of the reported results with two types of cavitational reactors indicated that the particle size variation is substantially lower in the case of the hydrodynamic cavitation as compared to the acoustic cavitation. The observed effects can be attributed to a very high degree of mixing and the uniformity within the cavitation zone. Also, energy distribution in the cavitation zone appears to be more uniform in the case of the hydrodynamic cavitation than the acoustic cavitation. The reason for this observation may be the better spatial distribution of the collapsing or oscillating cavities in the cavitating medium in the hydrodynamic cavitation reactor. A proper distribution of the orifices on the plate is responsible for improved spatial distribution of the collapsing cavities. One more important point which needs to be considered is that in the case of acoustic cavitation, the fluid velocities due to acoustic streaming are significantly lower (of the order of 0.5 m/s) as compared to hydrodynamic cavitation set up (velocities of the order of 10–30 m/s). This suggests, that rather than impact grinding (cavity collapse pressure, releasing shock wave), shear grinding or turbulent shear as a result of the stable oscillating cavity appears to be the controlling mechanism of size reduction in the hydrodynamic cavitation reactors.

The calculations related to energy requirement per kg of the solids processed clearly indicates that hydrodynamic cavitation is more energy efficient than acoustic cavitation. Quantitatively speaking, for acoustic cavitation setup the energy dissipation per kg of the solids varies from 2.52×10^7 to 1.38×10^8 J/kg depending on the operating parameters, whereas for the hydrodynamic cavitation set up these values vary from 2.026×10^4 to 6.316×10^5 J/kg, showing at least three orders of magnitude reduction.

3.7.1.6 Synthesis of Nanosize Catalyst Particles

Moser et al. [37] applied hydrodynamic cavitation generated by a microfluidiser for the synthesis of large variety of catalysts in the form of nanosized grains. The grains agglomerated into particles of 100 nm to few microns. This process was found to provide high purity catalysts containing several metal ions than classical synthetic methods.

Sunstrom et al. [38] have also used hydrodynamic cavitation generated, with the use of high-pressure fluid system for the generation of the nanocrystalline oxides. The precipitant stream is subjected to a large pressure drop (21,000 psi) across the interaction chamber. Due to the large pressure drops, bubbles are formed and collapse, causing localized heating of the solvent. In addition, the precipitate undergoes a high degree of mechanical shear in the interaction chamber, leading to nanosize particles. This technique of hydrodynamic cavitation offers a good route to the synthesis of homogeneous solid solutions of complex catalysts, where one attempts to understand the relation between ion modification of a host and catalytic performance. Moser et al. [39] have reported the use of larger scale equipment with capacities of 400-2,000 kg/h depending on the desired intensity of the cavitational effect.

3.7.1.7 Synthesis Process for Pulp/Paper Production

Solonitsyn et al. [40] have extensively reviewed the beneficial effect of hydrodynamic cavitation in the manufacture of paper from synthetic fibers, in intensification of pulp bleaching, preparation of highly-disperse sizes, and wastepaper deinking. One of the important applications of hydrodynamic cavitation that has been commercially exploited is the treatment of suspensions of the fibrous materials in the cavitational devices. Refining of wood pulp is an energy intensive process and grossly this account for one-third of the energy consumption in the pulp industry. The quality of the refined pulp has a strong bearing on the efficiency of the downstream processes such as bleaching. Danforth [41] has introduced the concept of impact intensity in the refining theory, showing that the number of cavitational impacts received by the fibers during the refining process critically affects the pulp quality.

The formation of microjets due to the collapse of the cavities near the fibre surface results in conditions analogous to impact forces experienced in conventional refiners. The mechanism of deformation of the wood pulp can be described in terms of increased capillary and swellability of the amorphous regions. On dewatering, these fibers form a dense stock, having strong interfibrillar bonds, which result in enhanced bursting strength of sheets made from such fibres. It has been reported that hypochlorite bleaching rate of sulfite pulp at 20–60°C increased by sevenfold to ninefold after cavitation pre-treatment and with increasing the bleaching temperature to 60°C and decreasing the final pH level to 7.0–8.0 [40]. The cavitation treated hypochlorite bleached pulp had increased tensile strength and brightness rating. Defibration of waste paper under intense hydrodynamic cavitation in continuous setup results in pulp with higher fibre strength, lower amount of crill and a higher retention of water by fibers than common defibration of waste paper in defibrators equipped with knifes [42].

3.7.2 Microbial Cell Disruption

A key factor in the economical production of industrially important microbial components is an efficient large-scale cell disruption process [43, 44]. The need for an efficient microbial cell disruption operation has always hindered the large-scale production of commercial biotechnological products of intracellular derivation [43]. For the large scale disruption of microorganisms, mechanical disintegrators such as high-speed agitator bead mills and high pressure homogenisers are commonly employed but the typical energy efficiencies of the above methods are in the range of 5–10% and the rest of the energy is dissipated in the form of heat which needs to be efficiently removed to retain the integrity of these delicate bio-products [44]. With an aim of improving the efficacy of the cell disruption process, keen interest was developed in the last decade in newer techniques including acoustic and hydrodynamic cavitation. Harrison and Pandit [45] were the first to report the use of cavitational reactors for cell disruption process using a configuration where cavitation was generated using a throttling valve. Later, Shirgaonkar et al. [25] clearly demonstrated the requirement of cavitational effects for the release of significant amount of enzymes/proteins in a high speed and high pressure homogenizer.

Save et al. [46] used hydrodynamic cavitation reactor based on throttling valves for the disruption of baker's yeast and brewer's yeast cells in pressed yeast form and reported that an increase in the time of treatment and number of passes resulted in a corresponding increase in the extent of cell disruption and enzyme release. The concentration of cells in the suspension influenced the disruption process significantly. The growth stage of the yeast cells is another parameter which affects the energy efficiencies. Preliminary experiments with fresh fermentation broth indicated that the cells in an exponential growth phase are far more susceptible to the disruption compared to those which are either stored or frozen. Comparison of the energy efficiencies, for different operations including, hydrodynamic cavitation, mixer-blender and ultrasonication indicated that the energy requirement of hydrodynamic cavitation setup is lower than the other two methods by more than two orders of magnitude for equivalent protein release. In quantitative terms, energy utilization per mL of yeast suspension to observe the same level of protein release was 20.7 J/mL for hydrodynamic cavitation reactor, 1,500 J/mL for ultrasonic irradiation and 900 J/mL for the mixer blender. In a scaled up version of an earlier work, Save et al. [47] and Balasundaram and Pandit [48, 49] investigated the process of cell disruption using hydrodynamic cavitation operating at a capacity of 200 and 50 L respectively and reported similar results for energy efficiency. Also, it was conclusively established that even though cavitation is known to generate conditions of very high temperature and pressure locally along with the generation of free radicals, activity of the released intracellular enzymes from the cells remains unaltered. This can be attributed to the fact that the intense conditions exist for very small intervals of time (typically around few micro-seconds) and hence do not result in any deactivation of the released enzymes. The activity of the glucosidase and invertase enzymes was not affected under normal circumstances. But, a prolonged exposure (60 min treatment at 3 atm pressure) resulted in about 10% decrease in the activity of the enzymes. Thus, it is important to control the intensity of the cavitation phenomena by suitably adjusting the operating and geometric parameters in the system as well as the time of treatment.

It should be also noted here that the mechanism of the cell disruption process is also different depending on the equipment used [48]. Cell disruption process can proceed via complete breakage of the individual cells releasing the intracellular enzymes in certain devices or can be shear driven where only the outer cell wall breaks as a result of which the enzymes present at the wall or periplasm will only be released (leached slowly). Balasundaram and Pandit [48] investigated the release of invertase enzyme by disruption of *S. cerevisiae* cells using sonication, high pressure homogenization and hydrodynamic cavitation. It has been reported that in the case of hydrodynamic cavitation generated using orifice plate, extent of release of the enzyme invertase was found to be higher than total soluble protein. This could be due to the periplasmic location of the enzyme. Based on the release pattern of enzyme and other proteins, a selective release of invertase (periplasmic) is expected in the early stages of disruption by hydrodynamic cavitation before complete mutilation of the cells releasing all the available proteins (cytoplasmic as well). For the case of ultrasonic induced cavitation, the rate of release of enzyme invertase was comparable with other proteins which can be attributed to the severity of the cavitational intensity in the case of acoustic cavitation as compared to hydrodynamic cavitation. Severe cavitation results in complete breakage of the cells whereas mild cavitational intensity in the case of hydrodynamic cavitation reactor results in an impingement/grinding action on the cell due to the shear, giving rise to the breakage of only the cell wall rather than complete cell. Balasundaram and Harrison [50] investigated the application of hydrodynamic cavitation for the partial disruption of *E. coli* cells and reported selective release of periplasmic and cytoplasmic enzymes relative to the total soluble protein as a function of cavitational intensity.

Balasundaram and Pandit [49] have quantified the dependency of the extent of release of enzymes on its location in the cell using a concept of location factor. Location factor can be defined as the ratio of release rate of the enzyme to the release rate of total proteins. Typically, for enzymes located in the periplasm, location factor is greater than 1 and for cytoplasmic enzymes, location factor is less than 1. For the release of invertase and penicillin acylase, the location factor was observed to be greater than 1 for all the cavitation equipments which confirm the periplasmic location of the two enzymes in the yeast and *E. coli* cells respectively. Further, it was observed that the location factor is higher in the case of hydrodynamic cavitation reactor as compared to sonication and high pressure homogenization, confirming that the mechanism of cell disruption in this case is by impingement/grinding action on the cell wall due to the shear as discussed earlier. For the alcohol dehydrogenase (ADH) enzyme, the location factor value was around 0.5 confirming that ADH is present mostly in the cytoplasm of the cell. Balasundaram and Harrison [51] have also reported similar dependency of the location factor of the different enzymes such as, β-glucosidase (periplasmic), invertase (cell wall bound), ADH (cytoplasmic) and glucose-6-phosphate dehydrogenase (G6PDH; cytoplasmic), on the cavitational intensity generated in the system.

It can be established from above studies that the location of the enzyme indeed affects the extent of energy requirement in the cavitational reactors for its release from the cell. Some pretreatment strategies can be used for the modification of the location of the enzyme in the cell before the cell suspension is subjected to the real cell disruption process. Translocation of enzymes due to pretreatment step can be exploited to improve the efficacy of the cell disruption. This is important as most of the target products are usually produced in the cytoplasm of the cells and thus cannot be obtained readily without spending a large amount of energy in completely mutilating the cell wall and the cytoplasm during disruption. More energy is required to recover the cytoplasmic enzymes than the periplasmic enzymes [52]. Thus, translocation of enzymes from the cytoplasmic space to periplasmic space could result in a large saving in energy requirements. Various techniques that are used for translocation as reported in the literature are heat stress, time of culture in the fermentation process, variable pH operation and chemical pre-treatment [53–55].

Overall, it can be said that use of hydrodynamic cavitation for cell disruption has been conclusively proven for large scale applications, that too with much higher energy efficiencies as compared to the acoustic cavitation reactors based on the use of ultrasound. Also, all the cavitational reactors are more energy efficient as compared to the conventional techniques based on the use of mechanical energy. A particular reactor configuration in terms of geometry of the cavitation chamber and operating parameters such as inlet pressure, circulation flow rate can be chosen based on the location of the specific enzymes in the cells and cell concentration in the medium. Pre-treatment strategies such as heat, pH and chemical treatment can aid in enhancing the selectivity of the target enzyme and at the same time significantly decrease the energy requirements by a process of translocation.

3.7.3 Microbial Disinfection

It has been generally observed that the mechanical effects due to cavitational events are more responsible for the microbial disinfection and the chemical and heat effects play only a supporting role [56]. Microstreaming resulting from stable cavitation has been shown to produce stresses, sufficient to disrupt cell membranes [57]. The mechanism proposed is the onset of turbulence which creates vortices with higher shear rates than the shear rates throughout the bulk of the liquid. Doulah [58] has also confirmed that yeast cell disintegration in cavitational reactors occurs by shear stresses developed by viscous dissipative eddies arising from shock waves.

Use of ultrasonic reactors for microbial disinfection has been substantially investigated and excellent reviews on this subject are available [56, 59, 60]. Though, hydrodynamic cavitation has been found to be much more efficient as compared to acoustic cavitation based reactors, its advent in the area of microbial disinfection has been only in the recent past. Jyoti and Pandit [12] investigated the application of cavitation for disinfection of bore well water in different hydrodynamic cavitation reactors (high speed homogenizer, high pressure homogenizer and orifice plate setup) and also compared their efficacy with ultrasonic horn type of reactor (operating at 22 kHz with the power rating of 240 W) generating acoustic cavitation. It has been reported that cavitation is equally effective in the disinfection of bore well water samples and about 90% disinfection can be achieved in less than 30 min of treatment for ultrasonic horn and high speed homogenizer. The extent of disinfection was somewhat lower in the orifice plate type setup which was attributed to higher volumes used in the operation and generation of lower intensity cavitation. Comparison of all the equipments in terms of extent of disinfection per unit energy supplied, however, indicated that orifice plate setup at higher operating pressures was the most efficient among all the cavitational reactors. To put it in quantitative terms, the extent of disinfection in case of orifice plate setup was 310 CFU/J as compared to only 45 CFU/J in the case of ultrasonic horn and 55 CFU/J in high speed homogenizer. In the case of high pressure homogenizer, the rate of disinfection was substantially higher as compared to the other cavitating equipments but the overall energy efficiency was extremely poor (5 CFU/J).

Apart from making contaminated water into potable one for drinking purpose, cavitational reactors can also find utility in a ship to treat ship's ballast water that is being transported from one region to another. Shipping is the backbone of global economy and facilitates transportation of 90% of the commodities. It is estimated that two to three billion tonnes of ballast water is carried around the world each year. Translocation of organisms through ships (bio-invasion) is considered to be one of the important issues that threaten the naturally evolved biodiversity, the consequences of which are being realized increasingly in the recent years. Although, many treatment technologies such as self-cleaning screen filtration systems, ozonation, de-oxygenation, electro-ionization, gas supersaturation, chemical treatments are adopted, they cannot limit the environmentally hazardous effects that may result from such practices. Hydrodynamic cavitation can be effectively applied for ballast water treatment and the design methodology for the incorporation of cavitational reactors in actual ships is available. Experimental investigations indicated that hydrodynamic cavitation (generated using multiple holes sharp edge orifice plate of size 21.5 mm having a circular hole of diameter 2 mm; fraction of open area = 0.75, flow rate = 1.3 lps and pressure = 3.2 kg/cm^2) resulted in 99% destruction of all the bacteria and also resulted in 80% destruction of the zooplanktons. An increase in the recirculation time as well as operating pressure increases the intensity of cavitation and results in a lowering of the treatment time. Use of multiple hole orifice plates arranged sequentially in the system also resulted in an increase in the extent of disinfection. Aim of the designers in this type of application should be to make the process viable in a single pass as it is practically impossible to have multiple passes in the ballast water treatment considering the volume of the liquid to be treated and a typical piping network for the ballast water flow.

The above two studies with actual contaminated water (possibly containing a wide range of bacteria/microorganisms) confirm the suitability of the hydrodynamic cavitation phenomena for microbial disinfection. Cost of the treatment is another important factor, which needs to be ascertained, before cavitation can be recommended as a replacement technique for the conventional methods of disinfection. Jyoti and Pandit [12] estimated the cost of treatment for different types of cavitational reactors and compared with the costs associated with conventional methods of using ozone and chlorine. It has been reported that hydrodynamic cavitation induced using high speed homogenizer or orifice plate setup is the most cost effective treatment strategy as compared to sonochemical reactors or high pressure homogenizer. However, this cost of treatment is still an order of magnitude higher as compared to chlorination or ozonation, estimated based on small scale applications. The detailed cost estimation analysis for different approaches can be referred in the work of Jyoti and Pandit [12]. It can be concluded here that applicability of hydrodynamic cavitation is more suited when bulk treatment is required (e.g. ballast water treatment or large scale municipal corporation water treatment plants) or when end use of treated water does not allow the formation of hazardous by-products (typically THM associated with chlorination) commonly associated with the conventional treatment schemes [61].

Microbial disinfection using hydrodynamic cavitation can be exploited for improvement in the performance of industrial cooling towers. Cooling water systems are an integral part of process operations in many industries. Biofouling has been recognized as an important contributor to impaired heat transfer causing decrease in the thermal efficiency and increased power consumption. When left uncontrolled, biofouling can lead to unscheduled maintenance and turnarounds, production bottlenecks due to corrosion leading to significant capital expenses. Hydrodynamic cavitation reactors can be installed on the site where water can be circulated from the main storage tanks through the cavitation chamber to achieve microbial disinfection. The treatment targets the root cause of biofouling and ensures that the micro-organisms (bacteria, algae and fungi) are eliminated if already present and are prevented from growing so as to avoid the formation of bio-film on the heat transfer surface. Hyca Technologies, India offers one such patented technologies and for more information on the applicability and installation in the circuit, the readers can refer to www.hyca.co.in.

Combining cavitation with conventional techniques for disinfection such as the use of chlorine, ozone, hydrogen peroxide, hypochlorite etc. could be another cost effective approach. Such a combination is expected to give synergistic effects and lead to a reduced requirement of the chemical dosage and at the same time resulting into much faster rates of disinfection. Jyoti and Pandit [62, 63] indeed reported that hybrid techniques are far superior for treating water as compared to any other individual physical treatment technique. The observed intensification has been mainly attributed to de-clumping of flocs of microorganisms. Microbes tend to be present in the form of clumps protecting inner microbes; if these clumps are broken then better disinfection can be achieved, as the exposure of the inner microbes to the disinfectant increases.

Chand et al. [64] have investigated the use of ozone treatment assisted by a liquid whistle reactor (LWR), which generates hydrodynamic cavitation, for water disinfection using a simulated effluent containing *Escherichia coli*. A suspension having an *E. coli* concentration of approximately 10^8 to 10^9 CFU mL^{-1} was introduced into the LWR to examine the effect of hydrodynamic cavitation alone and in combination with ozone. Hydrodynamic cavitation generated using a liquid whistle reactor alone results in only 22% disinfection but ozone-assisted operation with minimum time of ozone treatment produces 75% disinfection. The enhanced mass transfer rates achieved due to the recirculatory flows in the liquid whistle reactor increase the effective utilization of ozone. The combination has been found to be a cost-effective technique for achieving maximum disinfection compared to the individual operation of hydrodynamic cavitation (lower extent of disinfection) and ozonation (higher costs of treatment usually due to higher cost of ozone generation).

Overall, it can be said that use of cavitation in combination with conventional chemical methods is far more suitable as compared to individual operations. It not only results in substantially lower treatment times but also reduces the requirement of the chemicals under optimized conditions.

3.7.4 Wastewater Treatment

Cavitation, due to its spectacular effects in terms of generation of hot spots, highly reactive free radicals and turbulence associated with liquid circulation, offers potential as an effective tool for industrial wastewater treatment though in general the application of hydrodynamic cavitation has been limited possibly due to lower intensities of generated cavitational activity. Suslick et al. [65] confirmed the formation of hydroxyl radicals in hydrodynamic cavitation reactor using KI oxidation as a model reaction. It has been reported that cavitation inception occurs at inlet pressure of 150 bar indicated by significant release of iodine, beyond this operating pressure and the reaction rate further increases linearly with an increase in the upstream liquid pressure over the range of 200 to 1,500 bar as studied in the work using microfluidizer. Iodine liberation also increases with a reduction in the thermal conductivity of the dissolved gases at constant specific heat ratio. The result can be attributed to the principles of hot-spot theory, which suggest that the maximum temperature reached in collapsing bubbles increases with a decrease in the thermal conductivity. Further the rate of iodine liberation was also found to increase with a decrease in the temperature of liquid bulk due to lower vapour pressures and hence operation at lower temperatures is favoured.

Kalumuck and Chahine [66] investigated the decomposition of p-nitrophenol using a hydrodynamic cavitation reactor based on a closed loop circuit with contaminant pumped with a triplex positive displacement pump (17 L/min and 69 bar pressure). Multiple orifice plates in the re-circulation line were used one after another in order to produce intense cavitation. The total capacity of the reactor was 6.5 L operated in re-circulating mode. It has been reported that the cavitating jet set-up gives oxidation efficiency (mg degraded per unit energy supplied), which is about 25 times more as compared to the sonochemical reactor i.e. ultrasonic horn (irradiating frequency of 15.7 kHz and power density of 0.36 W/mL). The optimum operating temperature has been reported to be 42°C. Lower pH and also lower cavitation number (higher inlet pressure at same pressure drop across the nozzle) favors the degradation and increases the oxidation efficiency for the hydrodynamic cavitation setup.

Sivakumar and Pandit [13] reported the use of hydrodynamic cavitation for decolorization of dye effluent stream. It has been observed that for same flow area, plate with larger number of small diameter holes gives higher extent of color removal as compared to the plate with smaller number of larger diameter holes. The observed results have been explained on the basis of frequency of turbulence for the two cases. The hydrodynamic cavitation reactor using multiple hole orifice plates has been found to give cavitational yields, which are two times higher than the best acoustic cavitation device (Dual frequency flow cell with capacity of 1.5 L).

Wang et al. [67] reported the application of swirling jet-induced hydrodynamic cavitation for decomposition of rhodamine B in aqueous solution. It was found that rhodamine B in aqueous solution can be degraded with swirling jet-induced cavitation and the degradation can be described by a pseudo-first-order kinetics. The effects of operating conditions such as pressure, temperature, initial concentration of rhodamine B, pH of solution on the degradation rate of rhodamine B have been discussed. It was found that the degradation rate of rhodamine B increased with increasing pressure and decreased with increasing initial concentration. It was also found that the degradation of rhodamine B was strongly dependent on temperature and pH of aqueous solution. The oxidation efficiency of swirling jet-induced cavitation for rhodamine B degradation has been reported to be higher as compared to sonochemical cavitation.

Brautigam et al. [68] reported the use of hydrodynamic cavitation for degradation of benzene, toluene, ethylbenzene, and xylene in water and investigated the effect of different operation parameters, such as inlet pressure, solution temperature, and concentration of the chosen substrates, and type of restriction. It has been observed that BTEX can be degraded with moderate conversions, where toluene shows the highest and ethylbenzene the lowest conversion. Optimum operating pressure and temperature values have been reported and multiple hole orifice plates gives higher extent of degradation as compared to the single hole orifice.

Wang and Zhang [69] have investigated the use of hydrodynamic cavitation based on the use of swirling jet for the degradation of alachlor aqueous solution at an operating capacity of 25 L. The results showed that alachlor in aqueous solution can be successfully decomposed with swirling jet-induced cavitation. It was found that the degradation can be described by a pseudo-first-order kinetics and the degradation rate constant k is found to be $4.90 \times 10^{-2}\ \text{min}^{-1}$ for 50 mg/L loading of alachlor in aqueous solution at operating temperature of 40°C. The effects of operating parameters such as fluid pressure, solution temperature, initial concentration of alachlor and medium pH on the degradation rates of alachlor have also been established. The results showed that the degradation rates of alachlor increased with increasing pressure and decreased with increasing initial concentration. An optimum temperature of 40°C exists for the maximum degradation rate of alachlor and the degradation rate was also found to slightly depend on the medium pH.

Apart from the research investigations overviewed earlier, there exists some commercial technologies based on hydrodynamic cavitation for wastewater treatment. Hyca Technologies, India offer a system consisting of a reactor where the large chain biorefractory compounds are subjected to controlled magnitude, random and targeted pressure variations so as to generate precise gas, gas-vapor and vapor filled cavities. The cavities are allowed to grow to a desired size and then made to oscillate or collapse under highly controlled conditions. The collapsing cavities generate spherically diverging shock-waves (symmetric collapse) or surface normal liquid jets (asymmetric collapse) of magnitude high enough to break the chains and to get smaller molecular weight compounds.

Another technology offered by Dynaflow Inc, USA employs hydrodynamic cavitation generated by specially designed submerged cavitating liquid jets to trigger widespread cavitation and bubble growth and collapse leading to oxidation in the bulk liquid. Hydrodynamic cavitation based technology have been found to achieve oxidation with energy efficiencies up to two orders of magnitude larger than those obtained with ultrasonic means. Some of the chemicals being successfully oxidized using this technology include, p-Nitrophenol (PNP), acetone, bioreactor effluent, chloro derivates of ethane/ethylene, methyl orange, isopropyl alcohol, simulated hygiene waste, arsenic, 2,4 –D Malathion, chlorhexidine etc.

The similarity between the mechanism of destruction and some of the common optimum operating conditions in the case of different advanced oxidation techniques point towards the synergism between these methods and fact that combination of these advanced oxidation processes should give better results as compared to individual techniques [70]. This indeed is applicable to hydrodynamic cavitation as well and there have been reports where hydrodynamic cavitation has been combined with other advanced oxidation processes with great success.

Chakinala et al. [71] investigated combination of hydrodynamic cavitation and advanced Fenton process (based on the use of iron metal particles) for the treatment of industrial wastewater comprising mainly of phenolic compounds. Optimization, in terms of the extent of dilution, operating pressure and oxidant loading for maximizing the extent of degradation was achieved. A dilution ratio of 50, operating pressure of 1,500 psi (10,340 kPa) and two cycles of addition of hydrogen peroxide (each of 1,900 ppm loading) was found to be optimum and the maximum TOC removal obtained under these conditions was about 336 mg/L (under diluted conditions) within a treatment time of 150 min. Though the work utilizes diluted effluents and substantially higher operating pressures, it is the only work reporting the use of cavitation based techniques for treatment of real industrial effluents. Further optimization may be done in terms of use of multiple orifice plates, lower operating cavitation numbers, hydrogen peroxide loading and manner in which the oxidant is added, loading of iron particles etc. Optimized conditions will also allow use of much lower inlet pressures as well as lower dilution ratio which should make this process industrially more feasible.

Wang et al. [72] investigated the combination of swirling jet-induced hydrodynamic cavitation and hydrogen peroxide for decomposition of rhodamine B in aqueous solution. It has been reported that there is an obvious synergetic effect between hydrodynamic cavitation and hydrogen peroxide. The relative amounts of •OH radicals produced in swirling jet-induced cavitation and hydrodynamic cavitation combined with H_2O_2 were detected by using TA as a fluorescent probe, and the results showed that the production of •OH radicals in hydrodynamic cavitation can be remarkably enhanced by H_2O_2 addition. This result suggests that the synergetic effect between hydrodynamic cavitation and H_2O_2 for the degradation of rhodamine B could be mainly due to the contribution of additional •OH radicals production. It has been also established that increased loading of H_2O_2, lower medium pH, higher fluid pressures and lower dye initial concentration are more favorable for the degradation of rhodamine B. For temperature, increase of temperature from 30°C to 50°C has advantage to degradation of rhodamine B, but less degradation rate is observed at 60°C. The degradation kinetics of rhodamine B was established and reported to follow a pseudo-first-order kinetics.

Pradhan and Gogate [73] have also recently established the utility of combination of hydrodynamic cavitation and Fenton process. In the case of combination of hydrodynamic cavitation with Fe^{2+}/H_2O_2, hydrogen peroxide reacts with ferrous ions to generate active hydroxyl radicals, resulting in the degradation of model pollutant via the usual Fenton chemistry. The resulting Fe^{3+} can react with H_2O_2 to generate an intermediate complex ($Fe - O_2H^{2+}$) which can be effectively dissociated into Fe^{2+} and $HO_2^{\cdot}$ under cavitating conditions. The isolated Fe^{2+} further reacts with H_2O_2 and generates a higher concentration of hydroxyl radicals compared to that in the absence of cavitating conditions, which results in enhanced degradation efficiency of pollutants. The various radical reactions taking part in the combined operation can be given as follows:

$$Fe^{2+} + H_2O_2 \rightarrow Fe^{3+} + OH^- + HO^{\bullet}$$
$$Fe^{3+} + H_2O_2 \rightarrow Fe - O_2H^{2+} + H^+$$
$$Fe - O_2H^{2+}))) \rightarrow Fe^{2+} + HO_2^{\bullet}$$

Amongst the different cavitating devices investigated in the work, venturi results in more intense cavitation as compared to single hole orifice and higher inlet pressures are recommended for maximizing the extent of removal. Step wise addition of oxidant was found to be more beneficial for intensification of the removal of p-nitrophenol for a combination of hydrodynamic cavitation and hydrogen peroxide at higher loadings. For the combined use of hydrodynamic cavitation and Fenton chemistry, an optimum loading of $FeSO_4$ as 1 g/L and H_2O_2 concentration as 5 g/L was observed for an initial p-nitrophenol concentration of 5 g/L and the maximum extent of removal under these conditions was 3.16 g/L. For 10 g/L initial p-nitrophenol solution, the maximum extent of removal was 5.62 g/L under optimized conditions. Efficacy of removal using the combined approach was found to be strongly dependent on the operating pH and pH of 3.75 was found to be optimum.

It should be also noted that there also exists a commercial process (CAV-OX process[74]), which employs ultraviolet radiation, hydrodynamic cavitation and hydrogen peroxide to oxidize organic compounds present in water at ppm level concentrations to non-detectable levels. The system have been used with success for the effective degradation of volatile organic compounds primarily trichloroethane, benzene, toluene, ethyl benzene, and xylene. Hydrogen peroxide is added to the contaminated ground water, which is then pumped through a cavitation nozzle, followed by ultraviolet radiation (Fig. 3.5). Hydrodynamic cavitation produced in a nozzle is used to generate additional hydroxyl radicals, which help in increasing the rates of degradation. Moreover the cavitating conditions along with UV irradiation also result in increased number of free radicals due to the dissociation of hydrogen peroxide. It must be noted at this stage that depending on the type and concentration of the contaminant in question, an optimum amount of hydrogen peroxide exists beyond which the beneficial effects seize due to enhanced rates of recombination of free radicals thereby making them unavailable for the attack on the pollutants. Moreover excess of H_2O_2 also poses itself as additional pollutant in the exit stream. Thus it is important to select an optimum concentration of hydrogen peroxide. The removal efficiency of contaminants is reported to range from 20% to 99% [74].

3.7.5 Flotation

Hydrodynamic cavitation provides a practical method for in situ generation of tiny bubbles in flotation systems. Although cavitation phenomena in flotation have been considered since the beginning of commercial applications, intentionally applying

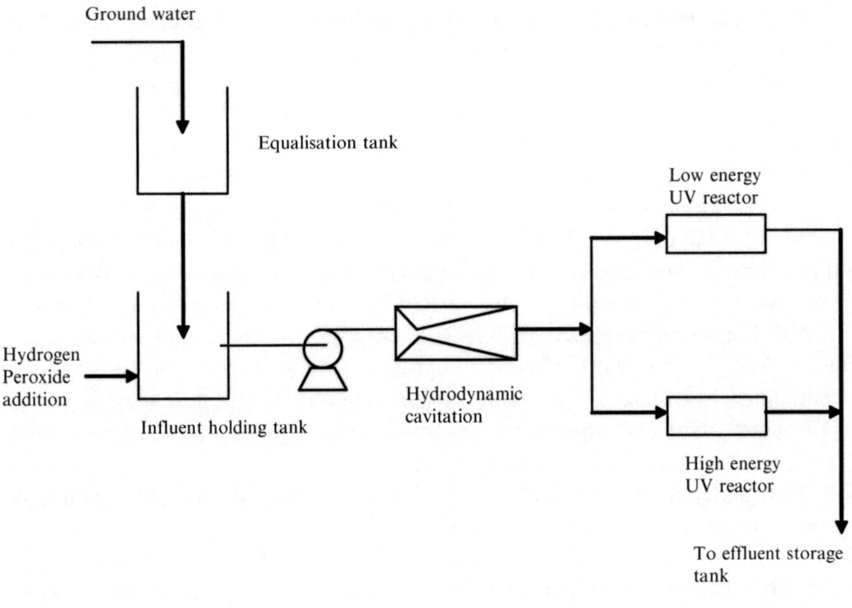

Fig. 3.5 Schematic representation of CAV-OX process

hydrodynamic cavitation for improving flotation performance has only been seen during the last decade or so. Recent developments and understanding of cavitation, gas nucleation phenomena and degassing due to cavitating conditions, have provided a scientific basis for cost-effectively applying hydrodynamic cavitation to flotation operations. Tiny bubbles created by low intensity hydrodynamic cavitation in flotation appear to offer the following:

1. Enhanced particle aggregation by tiny bubble bridging, giving rise to increased collision probability of the enlarged particle aggregates with flotation-sized bubbles
2. Accelerated particle–bubble attachment through coalescence of tiny bubbles frosted on particle surfaces with larger bubbles, instead of the direct particle–bubble contact
3. Increased contact angle and attachment force between particles and bubbles through tiny bubbles frosted on particle surfaces, and increased recovery of fine and coarse particles at reduced reagent consumption
4. Cleaning of particle surfaces through collapse of cavity bubbles (e.g., removal of slime coatings, removal of oxidation films, and removal of collector contaminated layers on gauge particle surfaces)

There are many reports indicating the investigation related to understanding the effect of using hydrodynamic cavitation for improving the performance of flotation devices. Zhou et al. [75] and Rao et al. [76] have highlighted the beneficial effects on the floatation kinetics using experiments on fine silica and zinc sulfide precipitates. Hu et al. [77] using numerical procedures developed for solving the turbulent kinetic energy/dissipation rate model with boundary fitted coordinates, have shown that the cavitation occurs at a pressure drop below ~10 m water. The bubble formation results into particle aggregation through gas nucleus bridging with the consequently larger particles having a higher collection rate. Further, the presence of surfactants and dissolved gas was found to preserve the bubbles generated by cavitation, thus producing more small bubbles. Zhou et al. [78] have also studied the effect of a hydrodynamic cavitation reactor on the flotation performance of a conventional flotation cell in the matte separation plant at Copper Cliff, Inco Ltd., Sudbury, on a larger scale of operation. Samples drawn from the feed to the rougher and first cleaner were tested in a portable continuous mini flotation cell. The use of the reactor improved flotation kinetics on both the samples. The effect on selectivity was mixed: it improved for the cleaner feed, but some decline was observed for the rougher feed. The test work, although limited, illustrates that the reactor offers the potential benefit of increasing circuit capacity while maintaining selectivity.

The concept of cavitation tube technology was extended by CSIRO Energy Technology and Novatech Consulting in Australia for improving fine coal flotation [79]. They generated tiny bubbles by pumping, conditioned flotation slurry through a cavitation tube, which they called picobubbles. Laboratory tests showed that increases in yield up to 10 wt % at the same product ash content were achieved using only 10% the original collector dosage. Such improved performance could be attributed to Harvey nuclei playing a role in accelerating coal particle–bubble attachment. Piloting tests showed an increase in flotation yield at a 50% reduction in diesel fuel (collector) dosage by a combination of mechanical cells and a cavitation unit [79]. Further pilot testing revealed that cavitation enhanced the collection of ultrafine coal particles, while the ash content was almost the same [80]. They attributed this improvement to the mechanism proposed by Zhou et al. [75, 81] that cavitation produces agglomeration of ultrafine particles by bubble bridging, making them as if they are larger particles of higher probability of attachment to the large bubbles in a flotation cell.

The concept of hydrodynamic cavitation was also tested with the Coalberg seam coal in West Virginia, USA [82]. By using cavitation-created tiny bubbles in a 2 in. flotation column fine coal recovery was increased by 10–30% at two-thirds to one half of the normal collector dosage (collector is the fluid such as diesel fuel used for removal of the coal particles).

3.7.6 *Miscellaneous Applications*

There are several other applications already developed on a laboratory scale which has received little attention industrially. The following are few such potential uses of hydrodynamically generated cavitation which can be commercially exploited.

3.7.6.1 Dental Water Irrigator Employing Hydrodynamic Cavitation

Cox [83] has recently described a dental water irrigator employing hydrodynamic cavitation wherein the individual manipulates the handpiece of the irrigator to wash his teeth with cavitated water. A handpiece has an inbuilt cavitation orifice. A pump supplies water from a reservoir through the cavitation orifice to produce cavitation in the water downstream from the orifice generating $OH^{\bullet}$ radicals and ions in the water flow. The water flow is directed onto dental surfaces. The $OH^{\bullet}$ radicals and ions have been proven effective against bacterium. The arrangement used in this application is quite similar to the closed loop orifice plate set-up used for carrying out complex chemical reactions [15, 26] but on a miniaturized level.

3.7.6.2 Preparation of Free Disperse System Using Liquid Hydrocarbons

The work by group of Kozyuk [84–87] has illustrated the use of hydrodynamic cavitation for obtaining free disperse system in liquids, particularly in liquid hydrocarbons. It has been found that, there is substantial improvement in the quality of the obtained free dispersion, even in the absence of any catalyst. Also the geometry of a flow-constricting baffle body [84] effectively increases the degree of cavitation to substantially improve the quality of obtained free disperse system.

This fact is also confirmed by the discussion made earlier regarding the effect of various design parameters on the performance of hydrodynamic cavitation reactors.

3.8 Concluding Remarks

The discussion presented in this chapter has enabled us to understand the different aspects related to generation of cavitation using flow field alterations and its exploitation for beneficial effects. Useful guidelines/recommendations for the design and efficient operation of different hydrodynamic cavitation reactors viz. orifice plate setup, venturi, high speed and high pressure homogenizers have been established. Based on the experimental results given in the chapter, the efficacy of the hydrodynamic cavitation reactors has been conclusively established as compared to the acoustic counterparts. The scale up of hydrodynamic cavitation reactors is comparatively easier as vast amount of information about the fluid dynamics downstream of the constriction is readily available and the operating efficiency of the circulating pumps which is the only energy dissipating device in the system is always higher at large scales of operation. Finally, it can be said that the hydrodynamic cavitation reactors offer immediate and realistic potential for industrial scale applications as compared to the sonochemical reactors for different chemical processing applications as discussed earlier. Technologies based on the hydrodynamic cavitation are already available for some applications such as wastewater treatment and improving the cooling tower performance by avoiding biofouling.

References

1. Rayleigh L (1917) On the pressure developed in a liquid during the collapse of a spherical cavity. Philanthropic Mag 34:94–98
2. Chatterjee D, Arakeri VH (1997) Towards the concept of hydrodynamic cavitation control. J Fluid Mech 332:377–394
3. Gogate PR, Pandit AB (2001) Hydrodynamic cavitation reactors: A state of the art review. Rev Chem Eng 17:1–85
4. Versluis M, Schmitz B, Von der Heydt A, Lohse D (2000) How snapping shrimp snap: Through cavitating bubbles. Science 289:2114–2117
5. Gogate PR, Pandit AB (2005) A review and assessment of hydrodynamic cavitation as a technology for the future. Ultrason Sonochem 12:21–27
6. Moholkar VS, Pandit AB (1997) Bubble behavior in hydrodynamic cavitation: Effect of turbulence. AIChE J 43:1641–1648
7. Gogate PR, Pandit AB (2000) Engineering design methods for cavitation reactors II: Hydrodynamic cavitation reactors. AIChE J 46:1641–1649
8. Yan Y, Thorpe RB, Pandit AB (1988) Cavitation noise and its suppression by air in orifice flow. In: Proceedings of the International Symposium on Flow Induced Vibration and Noise, Chicago, ASME, pp 25–40
9. Yan Y, Thorpe RB (1990) Flow regime transitions due to cavitation in flow through an orifice. Int J Multiphase flow 16:1023–1045

10. Tullis JP, Govindrajan R (1973) Cavitation and size scale effect for orifices. J Hydraul Div HY13:417–430
11. Moholkar VS, Senthilkumar P, Pandit AB (1999) Hydrodynamic cavitation for sonochemical effects. Ultrason Sonochem 6:53–65
12. Jyoti KK, Pandit AB (2001) Water disinfection by acoustic and hydrodynamic cavitation. Biochem Eng J 7:201–212
13. Sivakumar M, Pandit AB (2002) Wastewater treatment: A novel energy efficient hydrodynamic cavitational technique. Ultrason Sonochem 9:123–131
14. Kelkar MA, Gogate PR, Pandit AB (2008) Intensification of esterification of acids for synthesis of biodiesel using acoustic and hydrodynamic cavitation. Ultrason Sonochem 15:188–194
15. Vichare NP, Gogate PR, Pandit AB (2000) Optimization of Hydrodynamic Cavitation Using a Model Reaction. Chem Eng Tech 23:683–690
16. Gogate PR, Shirgaonkar IZ, Sivakumar M, Senthilkumar P, Vichare NP, Pandit AB (2001) Cavitation reactors: Efficiency analysis using a model reaction. AIChE J 47:2326–2338
17. Gogate PR, Pandit AB (2004) Sonochemical reactors: Scale up aspects. Ultrason Sonochem 11:105–117
18. Kumar PS, Pandit AB (1999) Modeling hydrodynamic cavitation. Chem Eng Tech 22:1017–1027
19. Moholkar VS, Pandit AB (2001) Numerical investigations in the behaviour of one-dimensional bubbly flow in hydrodynamic cavitation. Chem Eng Sci 56:1411–1418
20. Moholkar VS, Pandit AB (2001) Modeling of hydrodynamic cavitation reactors: a unified approach. Chem Eng Sci 56:6295–6302
21. Kanthale PM, Gogate PR, Wilhelm AM, Pandit AB (2005) Dynamics of cavitational bubbles and design of a hydrodynamic cavitational reactor: cluster approach. Ultrason Sonochem 12:441–452
22. Sharma A, Gogate PR, Mahulkar A, Pandit AB (2008) Modeling of hydrodynamic cavitation reactors using orifice plates considering hydrodynamics and chemical reactions occurring in bubble. Chem Eng J 143:201–209
23. Davies JT (1972) Turbulence phenomenon. Academic Press, New York
24. Hansson I, Morch KA, Preece CM (1977) A comparison of ultrasonically generated cavitation erosion and natural flow cavitation erosion. In: Proceedings of the Ultrasonics International Conference, Brighton, UK, pp 267–274
25. Shirgaonkar IZ, Lothe RR, Pandit AB (1998) Comments on the mechanism of microbial cell disruption in High Pressure and High speed devices. Biotech Prog 14:657–660
26. Senthilkumar P, Sivakumar M, Pandit AB (2000) Experimental quantification of chemical effects of hydrodynamic cavitation. Chem Eng Sci 55:1633–1639
27. Sampathkumar K, Moholkar VS (2007) Conceptual design of a novel hydrodynamic cavitation reactor. Chem Eng Sci 62:2698–2711
28. Pandit AB, Joshi JB (1993) Hydrolysis of fatty oils: Effect of cavitation. Chem Eng Sci 48:3440–3442
29. Chivate MM, Pandit AB (1993) Effect of hydrodynamic and sonic cavitation on aqueous polymeric solutions. Ind Chem Engr 35:52–57
30. Ambulgekar GV, Samant SD, Pandit AB (2004) Oxidation of alkylarenes to the corresponding acids using aqueous potassium permanganate by hydrodynamic cavitation. Ultrason Sonochem 11:191–196
31. Ambulgekar GV, Samant SD, Pandit AB (2005) Oxidation of alkylarenes using aqueous potassium permanganate under cavitation: Comparison of acoustic and hydrodynamic techniques. Ultrason Sonochem 12:85–90
32. Zhang Y, Dube MA, Mclean DD, Kates M (2003) Biodiesel production from waste cooking oil: 1 Process design and technological assessment. Biores Tech 89:1–16
33. Freedman B, Butterfield RO, Pryde EH (1986) Transesterification kinetics of soyabean oil. J Am Oil Chem Soc 63:1375–1380

34. Freedman B, Pryde EH, Mounts TL (1984) Variables affecting the yields of fatty esters from transesterified vegetable oils. J Am Oil Chem Soc 61:1638–1643
35. Gogate PR (2008) Cavitational reactors for process Intensification of chemical processing applications: A critical review. Chem Eng Proc 47:515–527
36. Patil MN, Pandit AB (2007) Cavitation-A novel technique for nano-suspensions/nanoemulsions. Ultrason Sonochem 14:519–530
37. Moser WR, Marshik-Geurts BJ, Kingsley J, Lemberger M, Willette R, Chan A, Sunstrom JE, Boye AJ (1995) The synthesis and characterization of solid state materials produced by high shear hydrodynamic cavitation. J Mater Res 10:2322–2335
38. Sunstrom JE, Moser WR, Marshik-Guerts B (1996) General route to nanocrystalline oxides by hydrodynamic cavitation. Chem Mater 8:2061–2067
39. Moser WR, Sunstrom JE, Marshik-Guerts B (1996) The synthesis of nanostructured pure-phase catalysts by hydrodynamic cavitation, in: Moser WR (eds.) Proceedings of the Advanced Catalysts and Nanostructured Materials, pp 285-306.
40. Solonitsyn RA, Fumbarev AG, Pilipenko SD (1991) Use of hydrodynamic flow cavitation in pulp and paper technology. Bum Prom-st 8–9:16–19
41. Danforth DN (1986) Effect of refining parameters on paper properties. Proceedings of the International Conference on New Technologies in Refining, 2, PIRA, Birmingham, England, UK
42. Solonitsyn RA, Fumbarov AG, Tomashchuk GL, Grinin TV (1987) Advantages of Cavitation Method for Activation of Waste Paper. Bum Prom-St 1:25–27
43. Geciova J, Bury D, Jelen P (2002) Methods for disruption of microbial cells for potential use in the dairy industry – a review. Int Dairy J 12:541–553
44. Harrison STL (2002) Bacterial cell disruption: A key unit operation in the recovery of intracellular products. Biotech Adv 9:217–240
45. Harrison STL, Pandit AB (1992) The disruption of microbial cells by hydrodynamic cavitation. 9th International Biotechnology Symp. Washington, DC
46. Save SS, Pandit AB, Joshi JB (1994) Microbial cell disruption: Role of cavitation. Chem Eng J 55:B67–B72
47. Save SS, Pandit AB, Joshi JB (1997) Use of hydrodynamic cavitation for large scale cell disruption. Chem Eng Res Des 75:41–49
48. Balasundaram B, Pandit AB (2001) Selective release of invertase by hydrodynamic cavitation. Biochem Eng J 8:251–256
49. Balasundaram B, Pandit AB (2001) Significance of location of enzymes on their release during microbial cell disruption. Biotech Bioeng 75:607–614
50. Balasundaram B, Harrison STL (2006) Study of physical and biological factors involved in the disruption of E. coli by hydrodynamic cavitation. Biotech Prog 22:907–913
51. Balasundaram B, Harrison STL (2006) Disruption of Brewers' yeast by hydrodynamic cavitation: Process variables and their influence on selective release. Biotech Bioeng 94:303–311
52. Chisti Y, Moo-Young M (1986) Disruption of microbial cells for intracellular products. Enz Microb Tech 8:194–204
53. Farkade VD, Harrison STL, Pandit AB (2005) Heat induced translocation of proteins and enzymes within the cells: an effective way to optimize the microbial cell disruption process. Biochem Eng J 23:247–257
54. Farkade VD, Harrison STL, Pandit AB (2006) Improved cavitational cell disruption following pH pretreatment for the extraction of β-galactosidase from Kluveromyces lactis. Biochem Eng J 31:25–30
55. Anand H, Balasundaram B, Pandit AB, Harrison STL (2007) The effect of chemical pretreatment combined with mechanical disruption on the extent of disruption and release of intracellular protein from E. coli. Biochem Eng J 35:166–173
56. Mason TJ, Joyce E, Phull SS, Lorimer JP (2003) Potential uses of ultrasound in the biological decontamination of water. Ultrason Sonochem 10:319–323

57. Scherba G, Weigel RM, O'Brien WD (1991) Quantitative assessment of the germicidal efficacy of ultrasonic energy. App Env Microb 57:2079–2084
58. Doulah MS (1977) Mechanism of disintegration of biological cells in ultrasonic cavitation. Biotech Bioeng 19:649–660
59. Phull SS, Newman AP, Lorimer JP, Pollet B, Mason TJ (1997) The development and evaluation of ultrasound in the biocidal treatment of water. Ultrason Sonochem 4:157–164
60. Piyasena P, Mohareb E, McKellar RC (2003) Inactivation of microbes using ultrasound: A review. Int J Food Microb 87:207–216
61. Cheremissinoff NP, Cheremissinoff PN, Trattner RB (1981) Chemical and nonchemical disinfection. Ann Arbor Science Publishing, Ann Arbor, MI
62. Jyoti KK, Pandit AB (2003) Water disinfection by acoustic and hydrodynamic cavitation. Biochem Eng J 7:201–212
63. Jyoti KK, Pandit AB (2004) Effect of cavitation on chemical disinfection efficiency. Water Res 38:2249–2258
64. Chand R, Bremner DH, Namkung KC, Collier PJ, Gogate PR (2007) Water disinfection using a novel approach of ozone assisted liquid whistle reactor. Biochem Eng J 35:357–364
65. Suslick KS, Mdleleni MM, Reis JT (1997) Chemistry Induced by Hydrodynamic Cavitation. J Am Chem Soc 119:9303–9304
66. Kalumuck KM, Chahine GL (2000) The use of cavitating jets to oxidize organic compounds in water. J Fluids Eng 122:465–470
67. Wang X, Wang J, Guo P, Guo W, Li G (2008) Chemical effect of swirling jet-induced cavitation: Degradation of rhodamine B in aqueous solution. Ultrason Sonochem 15:357–363
68. Braeutigam P, Wu Z-L, Stark A, Ondruschka B (2009) Degradation of BTEX in Aqueous Solution by Hydrodynamic Cavitation. Chem Eng Tech 32:745–753
69. Wang X, Zhang Y (2009) Degradation of alachlor in aqueous solution by using hydrodynamic cavitation. J Haz Mat 161:202–207
70. Gogate PR, Pandit AB (2004) A review of imperative technologies for Waste water treatment II: Hybrid methods. Adv Env Res 8:553–597
71. Chakinala AG, Gogate PR, Burgess AE, Bremner DH (2008) Treatment of industrial wastewater effluents using hydrodynamic cavitation and the advanced Fenton process. Ultrason Sonochem 15:49–54
72. Wang X, Wang J, Guo P, Guo W, Wang C (2009) Degradation of rhodamine B in aqueous solution by using swirling jet-induced cavitation combined with H_2O_2. J Haz Mat 169:486–491
73. Pradhan AA, Gogate PR (2009) Degradation of p-nitrophenol Using Acoustic Cavitation and Fenton Chemistry. J Haz Mat 173:517–522
74. CAV-OX Cavitation Oxidation Process (1994) Application Analysis Report, Magnum Water Technology, Inc., Risk Reduction Engineering Laboratory, Office of Research and Development, U.S.E.P.A., Cincinnati, OH
75. Zhou ZA, Hu H, Xu Z, Finch JA, Rao SR (1997) Role of hydrodynamic cavitation in fine particle flotation. Int J Miner Process 51:139–149
76. Rao SR, Finch JA, Zhou ZA, Xu Z (1998) Relative flotation response of zinc sulfide: mineral and precipitate. Sep Sci Tech 33:819–833
77. Hu H, Zhou ZA, Xu Z, Finch JA (1998) Numerical and experimental study of a cavitation tube. Metallur Mat Trans B 29:911–917
78. Zhou ZA, Langlois R, Xu Z, Finch JA, Agnew R (1997) In-plant testing of a hydrodynamic reactor in flotation. In: Finch JA, Rao SR, Huang LM (eds) Processing of Complex Ores. CIM, Sudbury, Canada, pp 185–193
79. Hart G, Morgan S, Bramall N, Nicol S (2002) Enhanced coal flotation using picobubbles. CSIRO Report-C9048, Australia
80. Hart G, Townsend P, Morgan S, Morgan P, Firth B (2005) Enhanced coal flotation using picobubbles. CSIRO Report-C12049, Australia

81. Zhou ZA, Xu Z, Finch JA (1994) On the role of cavitation in particle collection during flotation – a critical review. Minerals Eng 7:1073–1084
82. Tao Y, Liu J, Yu S, Tao D (2006) Picobubble enhanced fine coal flotation. Sep Sci Tech 41:3597–3607
83. Cox DW (1999) Dental irrigator employing hydrodynamic cavitation. US Patent number US 5860942A.
84. Kozyuk OV (1996) Method and device for obtaining free disperse system in liquid. US Patent application No. 602069
85. Kozyuk OV (1998) Method of obtaining a free disperse system in liquid and device for effecting the same. US Patent US 5810 052
86. Kozyuk OV (1999) Use of hydrodynamic cavitation for emulsifying and homogenizing processes. Am Lab 31:6–8
87. Kozyuk OV (1999) Method and apparatus for producing ultra-thin emulsions and dispersions. US Patent US 5931771A

Chapter 4
Sonoelectrochemical Synthesis of Materials

José González-García

Abstract In the last decade, the sonoelectrochemical synthesis of inorganic materials has experienced an important development motivated by the emerging interest in the nanostructures production. However, other traditional sonoelectrochemical synthesis such as gas production, metal deposits and metallic oxide films have also been improved with the simultaneous application of both electric and ultrasound fields. In this chapter, a summary of the fundamental basis, experimental set-up and different applications found in literature are reported, giving the reader a general approach to this branch of Applied Sonoelectrochemistry.

4.1 Introduction

Sonoelectrochemistry is a relatively new discipline which has achieved considerable international scientific interest and generated many significant theoretical and applied papers spanning a range of topics [1]. The vast majority of this research has been carried out at laboratory scale, with individually designed systems based on ultrasonic transducers dipped into traditional glass electrochemistry vessels [2]. It is remarkable that even with this rudimentary experimental set-up many interesting results have been generated; despite this, sonoelectrochemistry has suffered some drawbacks related to reproducibility, scale-up and design aspects which have slowed its development.

J. González-García (✉)
Grupo de Nuevos Desarrollos Tecnológicos en Electroquímica: Sonoelectroquímica y Bioelectroquímica. Departamento de Química Física e Instituto de Electroquímica, Universidad de Alicante, Ap. Correos 99, 03080 Alicante, Spain
e-mail: jose.gonzalez@ua.es

Pankaj and M. Ashokkumar (eds.), *Theoretical and Experimental Sonochemistry Involving Inorganic Systems*, DOI 10.1007/978-90-481-3887-6_4,

In spite of these drawbacks, the benefits associated to this discipline have been interesting enough to support its continuous development. In a sonoelectrochemical process, two clean energies (electrical and mechanical) are introduced into the reaction system to provide a high-energy microenvironment to perform a range of chemical reactions. In essence, this means the use of electrons as clean reactants thus avoiding the use of dangerous and harmful reagents and solvents. In addition, sonoelectrochemistry offers the ability to control the kinetics of the process by adjustment of additional operational variables over traditional chemistry, including ultrasonic power and frequency and electric current. The sonoelectrochemical reactor will be interfaced with a computer and so, with the appropriate feed-back loops, it will be an efficient device for safe and easy processing over a wide range of applications. Once optimized, a generic reactor can be developed from this to carry out processes on a significantly larger scale than that used in the laboratory.

In academic literature, Sonoelectrochemistry has presented different applications in materials science (e.g. improved electroplating [3], nanoparticle production [4] and organic synthesis [5]), in environmental applications (e.g. electrochemical degradation of organic pollutants [6] or the removal of metals [7]), in the improved performance of photoelectrochemical cells for solar energy and H_2 production [8] and in the field of enhanced electroanalysis [9]. Underlying these applications there are a number of fundamental deliverables related to the deeper study of important aspects of physical chemistry, materials synthesis, fundamental electrochemistry and applied electrochemistry. For example, a better understanding of how primary and secondary radicals generated during acoustic cavitation could be used for nanomaterials synthesis [10], electrodeposition [11] and for the degradation of organic compounds [12].

In the literature we can now find several papers which establish a widely accepted scenario of the benefits and effects of an ultrasound field in an electrochemical process [13–15]. Most of this work has been focused on low frequency and high power ultrasound fields. Its propagation in a fluid such as water is quite complex, where the acoustic streaming and especially the cavitation are the two most important phenomena. In addition, other effects derived from the cavitation such as microjetting and shock waves have been related with other benefits reported for this coupling. For example, shock waves induced in the liquid cause not only an enhanced convective movement of material but also a possible surface damage. Microjets of liquid, with speeds of up to 100 ms^{-1}, result from the asymmetric collapse of cavitation bubbles at the solid surface [16] and contribute to the enhancement of the mass transport of material to the solid surface of the electrode. Therefore, depassivation [17], reaction mechanism modification [18], surface activation [19], adsorption phenomena decrease [20] and the mass transport enhancement [21] are effects derived from the presence of an ultrasound field on electrode processes. We have only listed the main phenomena referring to the reader to the specific reviews [22, 23] and reference therein.

4.2 Experimental Systems

Any sonoelectrochemical experiment requires the necessary devices to produce the ultrasound and electrical fields in the system under study. In the case of an ultrasound field, the world wide extended systems are based on the electromechanical effect derived from the piezoelectric phenomena [24]. In practice, we can produce the mechanical vibration of an ultrasound field by the excitation of piezoelectric ceramics themselves, or by means of a more sophisticated devices composed by several piezoelectric ceramics specifically arranged and coupled with a mechanical amplifier which optimizes the yield of the device [25]. This device is named transducer but in sonochemical community is normally named as ultrasonic horn [26]. In literature, we can find different revisions of the available sonochemical devices [27]. On the other hand, for an electrochemical experiment, galvanostats (or power supplies) and potentiostats allow the experimentalist to carry out electrochemical experiments under current (two electrode system) or electrode potential (three electrode system) control [28] respectively. The following step is to design a practical procedure for coupling both fields. For that, in addition to the standard considerations in the design and development of any experimental system for sonochemical or electrochemical experiments, the development of experimental arrangements for sonoelectrochemical experiments should take into account, in principle, different aspects related to (1) the specific scenarios related to both energy fields, (2) the strategy of application of both energy fields and (3) the inherent characteristics of the ultrasound field related to the electrode nature:

1. As we have mentioned before, acoustic streaming, cavitation and other effects derived from them, microjetting and shock waves take also relevance when the ultrasound field interacts with solid walls. On the other hand, an electrochemical process is a heterogeneous electron transfer which takes place in the interphase electrode-solution, it means, in a very located zone of the electrochemical system. Therefore, a carefully and comprehensive read reveals that all these phenomena can provide opposite effects in an electrochemical process. For example, shock waves can avoid the passivation of the electrode or damage the electrode surface depending on the electrode process and/or strength of the electrode materials [29].
2. Another point to take into account is the different possible strategies of the combined application of the two energy fields. A sonoelectrochemical process can be carried out applying simultaneously both fields in time and space, or using time sequential protocols or different operation units connected in parallel or serial arrangements. Therefore, the number of possible configurations is high.
3. Finally, the intrinsic features of the ultrasonic field, frequency and power, should also be taken into account in the design of the experimental arrangement. It is obvious that the mechanical and chemical effects derived from a low frequency field are quite different than those provided by high frequency fields, and these features should match with features of the electrode materials such dimension, structure and physical and chemical properties [30].

In spite of the wide range of possibilities which could be imaged from the previous paragraph, if we look at literature, only a few of these combinations have been routinely used in Sonoelectrochemistry in general, and in inorganic sonoelectrosynthesis in particular. Due to the mechanical nature of the ultrasound field, the first used, most simple and easy approach is the simultaneous application of both fields, but while the electrical field is established directly in the working solution, the ultrasound field is applied in another medium and is transmitted through the physical separation of a wall. The most easy and fast experimental set-up according with this idea is the immersion of the electrochemical cell in an ultrasonic bath [31], see Fig. 4.1a. In this case, the ultrasound field generated by the piezoelectric ceramics bonded to the tank walls is propagated by the tank liquid, passing the electrochemical cell walls and reaching the electrochemical system. This approach has routinely presented more drawbacks (low reproducibility and the fact that the power transmitted inside de cell is low and depends on the position of the cell inside the ultrasound field), than advantages (its low cost and an easier control of the electrochemical system from an electric point of view). This arrangement lost relevance in the last decade but it presents an emerging use, especially in the synthesis of nanostructures [8, 32]. In spite of its drawbacks, this approach has received continuous attention and reconsideration and some authors, such as Klima et al. [33] have proposed refined approaches with a specific design where some of the previous drawbacks have been overcome, such as the low power transmitted and reproducibility. In this specific design, an elecrochemical cell of cylindrical shape and flat bottom was immersed in a larger glass tube. The horn for sonication was situated at the bottom of this tube. Thermostated water flowed through this tube and served as both cooling agent and coupling medium for the transfer of ultrasound from the horn to the cell, see Fig. 4.1b. This flowing fluid can be recirculated without [34] or with an overpressure of 4–5 atm. [35] in order to lower the transient cavitation outside the electrochemical cell. This design has been later adapted for working with complex chemical systems such as ionic liquids [36]. The physical separation of the electrochemical system from the ultrasound field generation in an optimized device is also being considered as an approach where, not only the overcoming of the lack of reproducibility and of low power transmitted is subject of study, but also the design and development of practical devices [37]. In the earlier design [35], the intermediate fluid, where tip transducer was directly immersed, is under high pressure, trying to avoid the cavitation appearance, and thus, allowing a higher power transmission to the working solution. In the latter, the ultrasound field should be directly applied in the electrochemical working solution, and the propagation of the ultrasound field into the filled adjacent thermostating chambers will also allow to use these chambers for the development of additional sonochemical reactions with the same energy consumption [37].

In fact, this last configuration (a sonochemical cell where the electrodes and the ultrasound transducer tip are directly dipped into the working solution) is the most used experimental configuration [38], see Fig. 4.2a, which we will name electrode-apart-transducer configuration. The relative orientation between the transducer tip and electrode surfaces has been taken into account: when the electrode is placed so

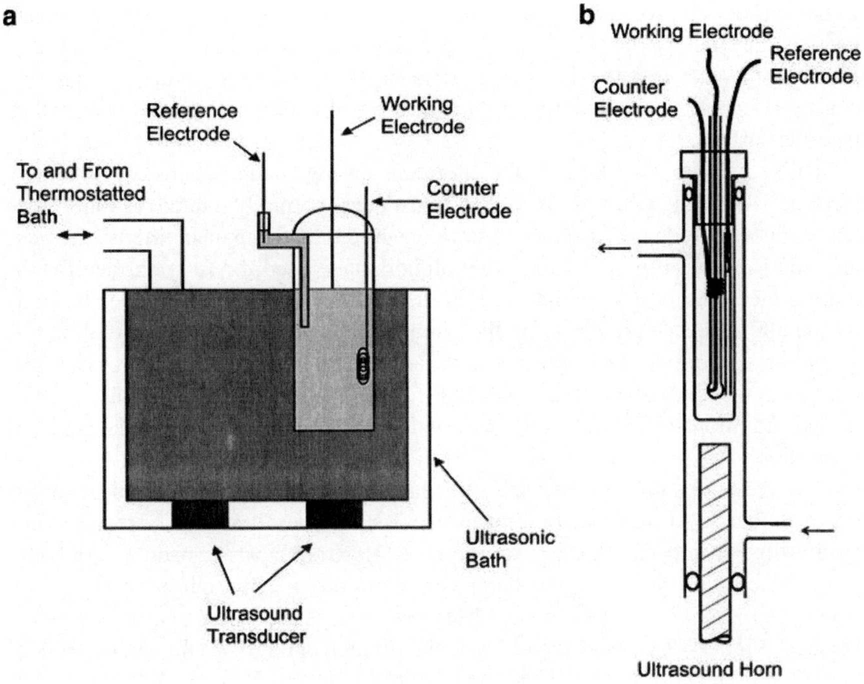

Fig. 4.1 (**a**) Electrochemical cell dipped in ultrasonic bath. (**b**) Klima sonoelectrochemical cell (Reprinted from [35]. Copyright (1999) with the permission from Elsevier)

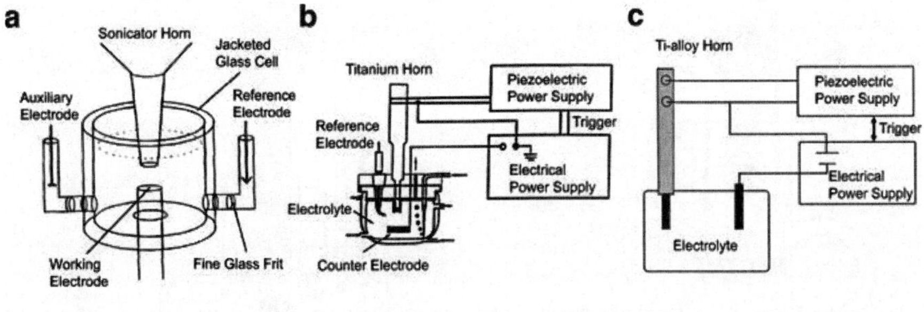

Fig. 4.2 (**a**) Electrode-apart-transducer configuration (Acknowledge [21]. Reproduced by permission of The Electrochemical Society). (**b**) Sonotrode configuration with sleeve (Reprinted from [42]. Copyright (2008) with the permission from Elsevier). (**c**) Sonotrode configuration without sleeve (Reprinted from [43]. Copyright (2006) with the permission from Elsevier)

that its surface is perpendicular to the tip surface, the geometry is named "side-on", and if parallel, the geometry is named "face on" [39]. This arrangement presents again drawbacks (complexity in the electrical arrangement [40], chemical contamination from transducer tip materials, extra cooling system) and also advantages (higher reproducibility and higher radiated ultrasound intensities), being this last one the more determining in their extended use. Another interesting benefit is related to the fact that the ultrasound field at the horn tip in solutions leads to acoustic streaming, which manifest itself as a jet of solution directed towards the electrode surface.

Finally, we can also find in the literature arrangements where the working electrode is also the emitter part of the transducer, normally named as sonotrode [22] or sonoelectrode [41]. Some authors have used only the main emitter surface as electrode [42], see Fig. 4.2b, and other authors have used the fully surface tip as working electrode [43], see Fig. 4.2c. In theory, this arrangement assures that all the specific effects derived from the ultrasound field propagation are directly focused on the surface electrode. Not only the shorted-lived bubbles non-uniformly collapse on the electrode surface but also the electrode surface itself oscillates. This provides additional effects which have been specifically used in the nanoparticles preparation.

Now we show some specific designs which have been routinely used in sono-electrosynthesis. All of them present electrodes and the ultrasound transducer tip directly dipped into the working solution, i.e. electrode-apart-transducer configuration. We can find cell designs in batch configuration, such those introduced by the Compton group [44, 45], which have been preferably used in organic sonoelectro-synthesis. Figure 4.3a shows the low volume divided cell (20–50 mL) where a fast circulation of the liquid is the most important effect obtained. Figure 4.3b shows the

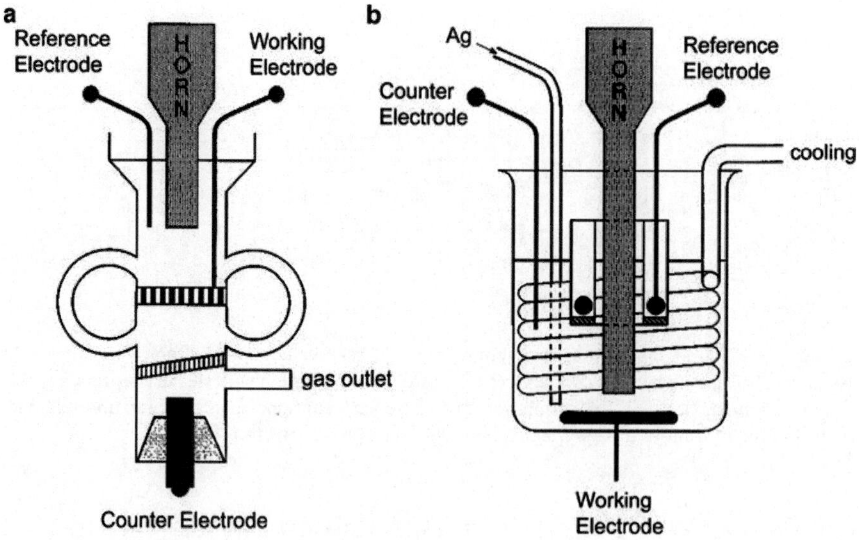

Fig. 4.3 Electrode-apart-transducer configuration (**a**) small volume cell (**b**) preparative cell for bulk electrolysis in the presence of high intensity ultrasound. (Acknowledge [46]. Reproduced by permission of The Royal Society of Chemistry)

"Compton cell" (in undivided configuration) used for larger scale bulk electrolysis (500 mL) and with a lower increase in mass transport. A comparative study of these different laboratory sonoelectrochemical cells [46] has pointed out that this arrangements generate exceptionally high levels of mass transport comparable only to conditions achieved at ultramicroelectrodes. An upper limit for ultrasound induced mass transport at macroelectrodes an order of magnitude beyond that achieved in conventional electrolysis is reported. Therefore, one considerable advantage introduced by ultrasound is the gain achieved for mass transport controlled processes over background processes, enhancing the performance and current efficiencies in electrosynthesis by selectively increasing the rates of the mass transport controlled processes.

Other cell designs, flow cells, can be found in literature [47]. In this design, a flow sonoelectrochemical reactor is the operational unit in a batch recirculation system, see Fig. 4.4. In this, analyzing the performance of the sonoelectrochemical degradation of trichloroacetic acid, influence of the fluid flow, gases evacuation system and, especially, the maintenance of the values of the performance parameters in the scale-up were checked.

In these last researches, a continuous feedback between the process study and the prototype design and development was established. In this way FEM (Finite element method) simulation has provided useful information about geometry, ultrasound intensity distribution and structural material coupling [37, 48, 49] for the design of an optimized sonoelectrochemical reactor.

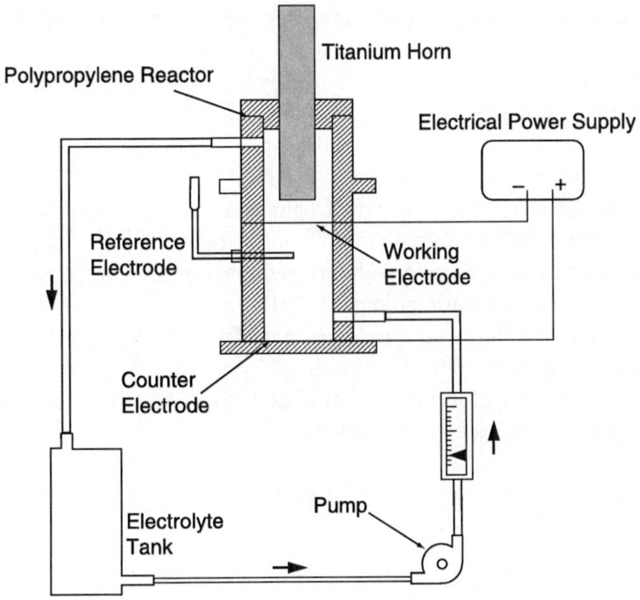

Fig. 4.4 Flow sonoelectrochemical reactor in batch recirculation configuration

4.3 Inorganic Sonoelectrosynthesis

Sonoelectrosynthesis has received poor attention in comparison with other applications, such as sonoelectroanalysis, and few review articles can be found in literature [50], which are mainly focused in organic sonoelectrosynthesis [51, 52]. Therefore, next we show a summary of the investigation focused on sonoelectrosynthesis of inorganic materials.

4.3.1 Gases

To our knowledge, the first reference found in literature related to Sonoelectrochemistry was focused in the influence of an ultrasound field in the electrolysis of water, generating hydrogen and oxygen as gaseous products [53]. Other works followed the research on this field and depolarizing effects of ultrasound were emphasized. It was found that the hydrogen or chloride evolution is affected by intense ultrasonic field operating at 280 or at 1,200 kHz [54]. It was reported that at constant current density, there exists and acoustical intensity above which depolarization suddenly increases and, for example, the evolution potential of hydrogen is decreased by 0.7–0.8 V and reaches the equilibrium potential. A later work analyzed the ultrasonic enhancement in the electrolysis of brine on graphite electrodes [54]. The most important phenomenon reported was a very strong degassing effect of ultrasonic waves owing to improved coalescence of gas bubbles and their mechanical stripping of gas. This effect increases dramatically the yields of chlorine gas produced under sonication. A net depolarization action of ultrasound has been detected, and the possible consequences of the application of ultrasound to industrial chlor-alkali cells have been outlined. Three gas-evolving reactions, hydrogen-evolution, chlorine evolution and oxygen evolution were extensively study by sonovoltammetric studies [55]. This study showed that simultaneous ultrasonic irradiation produces enhanced rates of gas evolution largely due to the removal from the electrode surface of adherent product species in the case of hydrogen and chlorine evolution. In contrast, oxygen gas evolution failed to be enhanced by ultrasound. The sonoelectrochemical enhancement of chlorine evolution without simultaneous increase in oxygen evolution offers the opportunity for chlorakali cells to operate in conditions where parasitic oxygen production would normally interfere, offering greater freedom in selection of cell parameters. The effect could also prolong useful electrode lifetime, since loss of activity for chlorine evolution and increase in oxygen evolution is a common failure of anodes after extensive use.

4.3.2 Hydrogen Peroxide

The electrochemical reduction of oxygen to form hydrogen peroxide is a very interesting process which has not suffered the expected development due to its low kinetics, consequence of the low solubility of oxygen in water and its poor electrocatalysis for most electrode substrates. In this context, sustained high rates of mass transport provided by the combination of acoustic streaming and interfacial cavitational activity have been subject of study. For this hydrogen peroxide sonoelectrosynthesis, not only the suitability of modified electrodes, from a catalytic and stability point of view, has been analyzed [56, 57] but also the scale-up of the process using a flow sonoelectrochemical reactor [58]. The electrodes were a range of carbon-based electrodes modified by various quinones and derivates via solvent evaporation and electrochemical cycling. All of them were reasonably stable under ultrasound and further work was carried out using the flow sonoelectrochemical reactor. With this flow system, the effect of several electrochemical operational variables (pH, volumetric flow, potential) and of the sonoelectrochemical parameters (ultrasound amplitude and horn-to-electrode distance) on the cumulative concentration of hydrogen peroxide and current efficiency were explored. The application of power ultrasound was found to increase both the cumulative concentration of hydrogen peroxide and the current efficiency.

4.3.3 Colloidal Hydrous Metal Oxide Reductions

Sonoelectrochemistry has also been used for the efficient employment of porous electrodes, such as carbon nanofiber-ceramic composites electrodes in the reduction of colloidal hydrous iron oxide [59]. In this kind of systems, the electrode reactions proceed with slow rate or require several collisions between reactant and electrode surface. Mass transport to and into the porous electrode is enhanced and extremely fast at only modest ultrasound intensity. This same approach was checked in the hydrogen peroxide sonoelectrosynthesis using RVC three-dimensional electrodes [58].

4.3.4 Metal Deposits

A wide range of metals are produced by electrochemical routes not only from aqueous solutions of one of their salts but also from fused salts themselves [60]. The electrodeposition technique presents a large number of advantages, being the possibility of producing pure metals and alloys in a wide range of forms (plates, powders) in a cheap way one of the most attractive features. The election of the adequate values of the key operational variables (current density, fluid flow, temperature, additive presence) with an optimized arrangement among cathodes and anodes have provided a well-defined product since the beginning of the twentieth century [61]. However, the continuous evolution of this technique looking for new alternative of production or even new materials has provoked that the combination of ultrasound with electrochemistry has also been employed as an advanced technique since 1950 [62, 63]. Ultrasound is found to have a significant effect on the metal deposition which is under diffusion control [29], while little, if any, effect was observed on systems in which deposition is under interfacial/charge transfer control. It is thought that ultrasound confers various benefits over conventional silent electrodeposition such increased deposit hardness, enlarged film thickness, improved deposition rates and efficiencies, and greater adhesion of the deposit to the electrode [64, 65]. These effects are attributed to factors such as the acoustic streaming increasing transport of active species to the electrode surface, continuous cleaning/activation of the electrode (particularly in passivating media) [66], effects resulting from the appearance and collapse of cavitation bubbles and ultrasonic degassing of solutions.

Much of this work has been carried out from an industrial standpoint, via empirical approaches, particularly on important metals in electroplating such as zinc [67], iron [68], chromium [69], copper [70] mercury [71], lead [7] and nickel [30]. An old review of the different metals deposited under ultrasound can be found in [70]. In this reference, specific details about brightness enhancement by ultrasound are reported. The brightness of a deposit often decreases as the current density is raised above a certain value called the critical current density. Ultrasonic agitation has been observed to considerably increase the critical current density and, for example, good quality deposits have been produced from sulphate bath at current densities whereas spongey deposits are formed under ordinary conditions. It is suggested that the production of a bright deposit is because either the grain size is less than the wavelength of light (fine-grain theory) or because a high degree of preferred orientation exists with the crystal faces being parallel in the deposit (the texture theory). The increase of the brightness of electrodeposits obtained with ultrasonic agitation has been associated with the production of deposits with a fine grain size and higher grain-packing density. This increase is explained by the effects coming from the shock waves followed by cavitation erosion and the removal of lateral growth of the deposit on the cathode surface.

In addition to the sonoelectrodeposition of metals, the surface treatment of metals with an ultrasonic field has also been a subject of development in the inorganic applications of sonoelectrochemistry. Tak et al. [72] studied the etch pits formed during sonoelectrochemical etching of aluminum. The high aluminum surface area for the Al-electrolytic capacitors is prepared by electrochemical oxidation in high temperature and aggressive media to obtain a high density of deep etch pits. During the growth of the tunnel, metal dissolution occurs from its tip surface at high current densities, while the tunnel sidewalls are covered with an oxide film which inhibits corrosion. Ultrasound effect through liquid media induces strong mechanical, physical and chemical changes, which can lead to the activation of chemical and electrochemical reactions. Especially, ultrasound can produce both useful and damaging effects, which result in the accelerated material removal and corrosion [73]. In this way, the authors found that ultrasound contributes to the increase in the pit density by prohibiting anodic oxide film formation and induces uniform tunnel length distribution by enhancing the transport of corrosion products present inside etch tunnels. Lower frequencies of ultrasound field produce higher pit and tunnel density than high frequencies.

Other possibility is the application of sonication during the dissolution of electrodeposited metals such as copper or nickel [74]. For both, the texture coefficient of the dissolved plane is affected, and is dependent on the plastic deformation by the shock wave and jet flow pressures. For both metals, the author sees that the greatest effect of ultrasound is located around 20 and 40 kHz. However, there is a marked difference between the two metals. Copper demonstrated the greatest effects at 45 kHz while nickel was most dramatically affected at the lower frequency of 28 kHz, but the possible reasons for that are not provided by the authors.

Finally, the presence of ultrasound in the electrodeposition of metals can produce both massive metal and metal colloid [75]. The reduction of $AuCl_4^-$ at polycrystalline boron-doped diamond electrodes follows two pathways forming (1) a deposit of gold metal at the surface of the boron-doped diamond and (2) colloidal gold which escapes from the electrode surface into the solution phase. The formation of the colloidal gold appears to be favored at low concentrations of $AuCl_4^-$ and in the presence of power ultrasound. The gold "nanoparticles" collected in the graphite electrode showed a diameter close to 1 μm.

4.3.5 Metal Oxides Deposits and Other Derivatives

Deposits of materials other than metals can be obtained by sonoelectrochemistry and, among them, metal oxides are of particular interest specially in electrochemistry.

One of the oldest sonoelectrodeposition of a metal oxide found in literature is the preparation of lead dioxide anodes [76], trying to avoid the presence of pores in the PbO_2 massive film formed from the oxidation of Pb(II), besides of the improvement of the physical properties. The lead dioxide sonoelectrodeposition on glassy carbon [31] has been extensively analyzed in aspects related to the surface activation [11, 77] and analyzing the influence of the ultrasound power [78] and ultrasound frequency [79]. It has been shown that the functionalization of the surface electrode by the $OH^{\cdot}$ radicals coming from the water sonolysis electrocatalyzes the lead dioxide crystallization. Assuming a progressive three-dimmensional nucleation and growth for the lead dioxide electrodeposition, the influence of the ultrasound intensity and frequency has been quantitatively analyzed and the electrocrystallization process determined by the kinetics parameters of the process. Induction time, i.e., lag time before the nucleation and growth take place, is the main parameter affected by the presence of ultrasound and this effect is more notorious at lower overpotential than at high overpotentials. However, at low frequencies, the ultrasonic intensity seems to affect the nucleation constant but not the constant growth. Similar effects are detected onto titanium [80] but, in this case, it is clear that a direct immersion of the ultrasonic probe in the electrochemcial system is needed to obtain relevant influence of the ultrasound field in the process. On highly boron-doped diamond [81], the electrodeposition follows a stage procedure where the initial nucleation and growth start on active diamond crystal faces, merging of the deposit into a thin layer and full coverage of the diamond surface with PbO_2 takes place. The deposit presents a strong adherence with a good electrical contact with the diamond. The use of power ultrasound in this process enhances the efficiency of the lead dioxide deposition procedure through increased mass transport and uniform deposit distribution. Silver oxide has been also electrodeposited onto boron-doped diamond in presence of ultrasound [82] and similar results of adherence and good electrical conductivity were achieved.

4.3.6 Nanomaterials

The sonoelectrosynthesis of nanostructures has been, without any doubt, the most active research field in Sonoelectrochemistry in the last years. The application of ultrasound with electrochemistry for the production of nanoparticles was first published by Reisse et al [41] and, after them, the nanomaterials, nanostructures and procedures reported have been widely varied. It is also important to point out that nanostructures can be obtained not only as a precipitated powder but also as colloidal solutions of the compound. For this last option, a capping agent that prevents aggregation in the strong electrolyte is necessary. They are extensively used in the shape-selective synthesis of nanostructures, because they manipulate growth by selectively adhering to certain crystallographic planes. The effect of these capping agents is directly related to the presence of the ultrasound field. For larger aggregates, the presence of the ultrasound waves breaks these structures and so, the aggregate size and therefore colloidal stability can also be controlled by the ultrasound intensity. This cooperation determines the type and concentration of capping agents to be used.

Several advantages support the sonoelectrochemical method such as one of the most popular procedures for producing nanomaterials. Along with the attractive features which comes from the electrochemical nature of the method (simple, low cost, easy conditioning of the driving force (electrode potential) which can be tuned continuously and reversibly and can generate novel nanostructures with well-defined shapes [83, 84]), the introduction of ultrasound to electrochemistry provides the cleaning of electrodes and the enhancement of the mass transport to the electrode surface. Ultrasound increases the transport not only for the electroactive species but also for organic capping reagents such poly(vinyl pyrrolidone) (PVP) [85].

Because ultrasound can be an efficient driving force in the synthesis of metal nanoparticles [86], it is important to point out the specific effects of the combination [87] (in contrast to the use of both techniques alone) but the intensity of the effects depends on the experimental set up used:

1. If the direct ultrasonic irradiation by a cathode emitter equipment (sonotrode or sonoelectrode configuration) is used, Fig. 4.2b–c, [88] a strong electrochemical

driving force and prevention of the particle attachment in the electrolyte is obtained. This device exposes only the flat circular area at the end of the sonic tip to the electrodeposition solution. The exposed area acts as both cathode and ultrasound emitter, Fig. 4.2b. A pulse of electric current produces a high density of fine metal nuclei. This is immediately followed by a burst of ultrasonic energy that removes the metal particles from the cathode, cleans the surface of the cathode, and replenishes the concentration near the electrode surface with metal cations by stirring the solution. Current efficiencies are about 75–90%. The metals are obtained as chemically pure, fine crystalline powders of high surface area with an average particle size of 100 nm. The electrochemical and ultrasonic parameters, including the pulse duration can be used to control the particles parameters. Gedanken et al. [89] have reported the use of the pulse sonoelectrochemistry for the preparation of silver nanoparticles with different shapes and uncovered the crucial role of electrolyte composition in the shape formation. According to their explanation, during the reaction, the particles are not completely detached from the sonoelectrode by the ultrasonic oscillations, so that the particles formed in the next pulse can further grow on seeds that are left on the sonoelectrode. The common model of the sonoelectrochemical synthesis involves two successive processes: the silver deposition on the cathode during the electric pulse which is followed by the breaking up of the thin layer into nanoparticles, and their dispersion in solution during the sonic pulse.

Some authors have modified this previous model [90] pointing out two group of facts which cannot be explained by it. First, the particle enlargement is characterized by two stages: growth and agglomeration. The agglomeration occurs when the solution is already depleted from metal ions and only in the presence of the electrical power. The second group of facts involves the abnormal dependence of the solution resistance on the metal ions concentration in the solution. Thus a new model, based on the suspensive electrode concept is suggested and consists on the idea that the particles suspended in solution obtain the electrode potential, thus acting as part of the electrode. The sonoelectrochemical system may be described as follows: initially, when the solution contains only metal ions, the original model holds. A thin metal layer is deposited on the cathode by the electric pulse. The following ultrasound pulse leads to the destruction of this layer, and a suspension of nanosized metal particles is formed. At this point, when the solution contains some suspended metal particles, the suspensive electrode is formed. From that point on, two simultaneous processes occur. First, the sonoelectrode continues in the electrodeposition and break down of the original model. And second, the suspended particles are moved in the solution by acoustic streaming, hit the sonoelectrode, accept its potential, and travel back to the solution. These charged metallic particles acting as part of the cathode can cause electrodeposition of metallic ions and therefore grow. The particle growth stops after all the dissolved metal ions are depleted. At this time, the second stage of particle agglomeration begins. Since the agglomeration occurs only in the presence of the electric pulse, the authors relate the agglomeration mechanism to the charging of the suspended metal particles by the electrodes. That is, the agglomeration allows for stabilization of the small particles carrying the excess charge. Furthermore, it is possible that the agglomeration also involves a reaction between counter-charged particles. Strictly speaking, agglomeration may take place from the very beginning, but its rate is much lower than the rate of electrodeposition on the suspended particles when there is enough dissolved metal.

The suspensive electrode method also explains at least qualitatively the abnormal dependence of the solution resistance on the metal ion concentration in the solution. The system resistance falls at the beginning of the sonoelectrochemical operation, since the metal particles formed increase the effective cathode surface area. In addition, the original charge carriers, the metal ions, are replaced by the charged nanoparticles that travel to long distances. After the solution is fully depleted of the metal ions, the nanoparticles become the sole kind of charge carriers. Thus, after hours of operation, the resistance increases because of agglomeration of the particles that makes this charge transport shuttle less effective. This new mechanism agrees with the following mechanism, established for the use of the other configuration.

2. When ultrasound emitter and electrode are different elements in the system, electrode-apart-transducer configuration, Fig. 4.2a, ultrasound removes the nanoparticles from the cathode to form suspended seeds. The ultrasonic agitation maintains a suspension of these preformed nanoparticles, which move continuously around, hit the electrodes, and these charged particles attach to one another and grow in suspension [90]. In this case, ultrasound keeps the larger structures from agglomeration.

It should also be important to highlight that, in addition to these two more popular experimental set-up, the early electrochemical glass cell dipped in ultrasonic bath configuration is also being used with satisfactory results [8, 32].

In addition to the experimental set-up, the procedure used is also determining for the quality of the obtained product. The proposed protocol consisting on the combination of pulsed electrodeposition and pulsed "out-of-phase" low frequency-high power ultrasound in sonotrode systems has revealed as an effective method, not only for the production of nanoparticles of metals [41], of metallic oxides [91], of metal alloys [92] and polyaniline nanoparticles [93], but also nanostructures [94]. Basically, the ultrasonic power density applied is largely enough to pull up all the metallic atoms during the electrolysis and not to leave any residue on the surface. However, some controversy can be found in the influence of the wide range of operational variables (electrolyte composition, temperature, working current density, duty cycle (pulse on time and ratio between pulse on and off time), power ultrasound and frequency of the ultrasound power) in the nanoparticle size. Some authors suggest that decreasing temperature, shorter pulse duration and higher sonic intensity all lead to a decrease in crystal size [95]. With the exception of ultrasonic intensity in the low intensity region, which has a major effect on crystal size, the effects of these variables on crystal size, while not very large, are appreciable. For example, the plating current density is found to be an important factor in determining crystal size. Lower current density (for the same time) resulted in larger crystal size (ca. 10 nm at 100 mA cm^{-2}) compared with 5 nm at 250 mA cm^{-2}. The same authors, in a later work [96], found that the reaction time almost did not affect the particle size and concluded that decreasing temperature, shorter pulse duration and higher ultrasound intensity, and lower current density all lead to a decrease of the particle size. This procedure protocol is also used in the electrode-apart-transducer configuration for the production of magnesium nanoparticles in organic solvents [97] and nanotubular TiO_2 arrays by sonoelectrochemical anodization [98].

Nanoparticles of different materials can be synthesized by sonoelectrochemistry. Nanoparticles of metals as copper [41, 99], magnesium [97], tungsten [100], iron [101], silver [85, 102, 103], gold [104, 105], Co-Fe alloys [92] gold-silver alloy [106] and CuBr [10] are cathodically produced while metallic oxides such as Cu_2O [91], PbO_2, Ag_2O and MnO_2 [107] are also produced cathodically by reduction of high oxidation state cations followed by precipitation plus dewatering, which yields the corresponding oxide. Other nanoparticles of materials such as metallic sulfides (i.e PbS) [108] have been successfully produced using metals foils as sacrificing anodes. The basis of this sonoelectrochemical technique is again the massive nucleation using high current density electrolysis, followed by the removal of the deposit from the electrode by the ultrasonic pulse. However, now, the electrolysis of lead produces Pb^{2+} in the ultrasonic bath, which immediately reacts with the S^{2-}. The high-density burst of ultrasound removes the PbS particles from the anode, clean the electrode surface and replenishes the double layer with S^{2-} by stirring the solution. If ultrasonic irradiation was not employed, it was observed that PbS was deposited on the lead electrode surface. Other nanoparticles of metallic selenides such as CdSe and PbSe [95, 96] are also produced. MoS_2 fullerene-like nanoparticles synthesis by electrodeposition onto sonotrode system at room temperature has been reported as an efficient method, providing a well-crystallized structure in front of the either electrodeposition or ultrasonic irradiation alone, which provide amorphous Mo-S product [109]. The use of sonoelectrochemistry has also been used in the preparation of silver-coated TiO_2 nanoparticles with extremely high coverage [110].

Other materials synthesis apart from nanoparticles can be obtained by the presence of ultrasound, such as the production of long and thick nanowires. In the case of silver [111], nanowires of 40 nm diameter and a length up to over 6 μm without the use of any template and in presence of EDTA can be obtained with electrolysis under controlled-current, and the nanowires become thicker and longer with controlled-potential electrolysis. Nanorods of copper sulfide have also been efficiently synthesized by using the ultrasound during a template-assisted electrodeposition [112]. The nanoporous structure of the templates, due to its high mass transport resistance, hinders the deposition rate and therefore uniform and high-quality nanostructures. The presence of ultrasound provides higher growth rate of nanorods, due to the lower mass transport resistance obtained. Nanorods of PbTe have also been synthetized by sonoelectrochemistry [113].

TiO_2 nanotubular array templates have also been formed on titanium surface, but in this case by sonoelectrochemical galvanostatic anodization [8]. Several authors [32] have proposed a clear description of the ultrasound presence in the galvanostatic anodization process, especially in the production of nanotubes. Three stages of current transient are observed [8]: (1) formation of a compact oxide layer (2) nucleation of nanopores and formation of nanotubular oxide and (3) growth of the oxide nanotubes which duration determines the length of the nanotubes. In presence of ultrasound the time spent in stage (2) is shorter with higher current densities compared to classical stirring. These higher current densities imply an increase in the diffusion limiting current densities, because of better ionic transport in the electric double layer by bubble implosions/ microjet streaming conditions. The increased current density under ultrasound was associated with the increase in the oxidation and dissolution currents. The increase in the dissolution current increases cation vacancy flux reaching the metal/oxide interface and results in the growth of oxide layer by incorporating more metal cations in the oxide. Another contribution of the ultrasound field presence is the overall distribution of the oxide nanotubes which becomes in a more homogeneous distribution of well ordered oxide nanotubes. We can also find in the literature the synthesis of non-layered compound nanotubes of CdSe [114] and even nanoporous Zn and Ni particles [115].

Finally, it has been reported a wide range of nanostructures of silver [87, 89], platinum [94], copper [116], palladium [117] produced by sonoelectrochemical methods. Time and concentration of electroactive cation have been proposed as key operational variables for obtaining this variety of geometries [89] using the pulsed method and the sonotrode geometry. The explanation is again based on the fact that, during the reaction, the particles are not completely detached from the titanium horn by the ultrasound waves, so that the particles formed in the next pulse can further grow on seed left on the electrode. This growth can promote the formation of different shaped-silver particles. Longer sonoelectrolysis are mechanically breaking the dendrites and promoting aggregation.

4.4 Influence of the Operational Variables

From the previous review, many operational variables reveal influence in the performance of the sonoelectrochemical processes. Therefore, we only point out the most representatives of the technology:

Ultrasound frequency has revealed as the most important operational variable. Low frequency (20–60 kHz) has been most used to obtain mechanical effects such mass transport enhancement, shock waves, microjetting and surface vibration, especially used in the nanostructure preparation. It has been reported [118] that for the cavitation processes, even if they are very important at high frequencies, the observed currents may not be a simple measure of this cavitational activity. Besides, the thermal and degassing effects are as important as reactor geometry and electrode materials, and the effects of the unusual mass transport characteristics in the presence of high frequency is a subject to be explored.

Apart from the expected mass transport enhancement, an increase in the ultrasound intensity has provided specific effects in some processes. In the lead dioxide sonoelectrodeposition, an increase in the ultrasound intensity has led to an increased surface activation (lower induction times) derived from the functionalization by $OH^{\cdot}$ radicals coming from the water sonolysis. However, a lower nuclei density has been detected, showing a more complex influence in the process [78]. In nanoparticles synthesis by pulsed sonoelectrochemistry, due to the fact that the use is based on mechanical effects, the greater the ultrasound intensity, the greater will be the efficiency of removal of the deposit and, therefore, the less chance there will be for crystal growth of existing nuclei. Above a certain intensity where all the deposit is removed, further increase in intensity is not expected to affect growth a lot [95]. In general, the ultrasonic intensity ranges from 100 to 700 $kW\,m^{-2}$ [107].

Current density, which ranges from 2,000 to 300,000 Am^{-2}, has been probed as an important operational variable for the sonoelectrodeposition process of massive metals [70], sonoelectrodeposition of oxide metals [80], sonoelectrosynthesis of gases [54] and also nanomaterials synthesis [96], where current density can affect crystal size in at least two opposing directions. A smaller size would be expected, on the basis of the small amount of material deposited at a lower current. On the other hand, lower current density allows more time for atomic diffusion processes to occur which can lead to larger crystal size. However, the former effect is dominant [85].

4.5 Benefits of the Ultrasound for the Electrochemical Processes

After this revision, it has been clearly established that the enhanced mass transport conditions have been routinely identified as the main benefit coming from the ultrasound field. This benefit has been pointed out in the gas evolution reactions detaching the gas bubbles from the electrode surface [55], hydrogen peroxide sonoelectrosynthesis and especially in metal deposits. The application of ultrasonic agitation to the plating bath increases the surface smoothness and brightness of the electrodeposited metal. This improvement in brightness is considered to be due to the action of ultrasound producing shock waves and cavitation erosion on the growing surface of the deposit. Growth perpendicular to the surface is inhibited and this results in a smoother finish on the deposit. The reduction grain size often observed with ultrasound is not considered to be a satisfactory theory to account for the increased brightness observed in all deposits [70]. The surface activation has been another benefit provided by the application of an ultrasound field. The benefit should be most used, especially in the functionalization of electrode substrates in different applications.

Combination of sonochemistry and electrochemistry has provided a new strategy for the synthesis of nanomaterials. Keeping in mind that the basis of the sonoelectrochemical technique to form nanostructures is massive nucleation using high current density for the electrodeposition pulse followed by removal of the deposit from the sonoelectrode by an ultrasound pulse, it can be concluded that the synthesis of quite sophisticated nanomaterials is possible through careful control of the electrochemistry, sonochemistry and initial composition of the precursor solutions. For example (1) it is possible to achieve 1D nanostructures control without any template, thereby practically overcoming the limitation of generating nanorods with diameters below 10 nm [119]. Size from 2 to 15 nm can be obtained with the increase of the cathodic overpotential even without the addition of any stabilizer [102, 105]. The production of some unexpected nanoparticles such CuBr instead of Cu using CTAB as additive instead of PVP (poly(N-vinyl-2-pyrrolidone)) or PVA poly(vinyl alcohol) as capping agents is explained involving the sonochemistry of water with the production of $OH^{\cdot}$ radicals [10]. (2) The sonoelectrochemical method is a highly efficient technique for quickly synthesizing highly ordered titania nanotubes. The pore diameter and nanotube length can also be tuned by changing the applied potential and anodization time [120]. (3) Another synthesis which has improved its development by sonolectrochemistry is the template-assisted electrochemical deposition at nanoscale [112]. The electrodeposition requires a high salt concentration and fine-tuning of the electrical and structural properties of the template but sonoelectrochemistry being a thermophysical phenomenon can be used regardless of the chemistry, composition of the electrolyte or charge on the electrode to produce high quality single crystalline one-dimensional structures.

Acknowledgments The author would like to thank to his collegues R. Gómez, P. Bonete, T. Lana-Villarreal, O. Louisnard and Ph. D students A. J. Frías-Ferrer, V. Sáez, M. D. Esclapez-Vicente, D. Milán, I. Tudela, M. I. Díez, A. Rico for their help and friendship.

References

1. González-García J, Esclapez MD, Bonete P et al (2010) Current topics on Sonoelectrochemistry. Ultrasonics 50:318–322
2. Compton RG, Eklund JC, Marken F (1997) Dual activation: coupling ultrasound to electrochemistry – an overview. Electrochim Acta 42:2919–2927
3. Winand R (1998) Contribution to the study of copper electrocrystallization in view of industrial applications-submicrocopic and macroscopic considerations. Electrochim Acta 43:2925–2932
4. Saez V, Mason TJ (2009) The synthesis of nanoparticles using Sonoelectrochemistry: a review. Molecules 14:4284–4299
5. Asami R, Atobe M, Fuchigami T (2006) Ultrasonic effects on electroorganic processes. Part 27. Electroreduction of acrylonitrile at suspended lead particle-electrode. Ultrason Sonochem 13:19–23, and the series

6. Gogate PR (2008) Treatment of wastewater streams containing phenolic compounds using hybrid techniques based on cavitation: a review of the current status and the way forward. Ultrason Sonochem 15:1–15
7. Agulló E, González-García J, Expósito E et al (1999) Influence of an ultrasonic field on lead electrodeposition on copper using a fluoroboric bath. New J Chem 23:95–101
8. Mohapatra SK, Misra M, Mahajan VK et al (2007) A novel method for the synthesis of titania nanotubes using sonoelectrochemical method and its application for photoelectrochemical splitting of water. J Catal 246:362–369
9. Compton RG, Hardcastle JL, del Campo J et al (2003) Sonoelectroanalysis: applications. In: Bard AJ, Stratmann M (eds) Encyclopedia of electrochemistry, vol 3. Wiley-VCH, Weinheim
10. Haas I, Gedanken A (2006) Sonoelectrochemistry of Cu^{2+} in the presence of cetyltrimethylammonium bromide: obtaining CuBr instead of copper. Chem Mater 18:1184–1189
11. González-García J, Iniesta J, Aldaz A et al (1998) Effects of ultrasound on the electrodeposition of lead dioxide on glassy carbon electrodes. New J Chem 22:343–347
12. Oturan MA, Sirés I, Oturan N (2008) Sonoelectro-Fenton process: a novel hybrid technique for the destruction of organic pollutants in water. J Electroanal Chem 624:329–332
13. Brett C (2008) Sonoelectrochemistry. In: Antonio Arnau Vives (ed) Piezoelectric transducer and Applications. Springer, Berlin Heidelberg
14. Pollet BG, Phull SS (2001) Sonoelectrochemistry-theory, principles and applications. Recent Res Develop Electrochem 4:55–78
15. Mason TJ, Lorimer JP, Walton DJ (1990) Sonoelectrochemistry. Ultrasonics 28:333–337
16. Suslick KS (1990) Sonochemistry. Science 247:1439–1445
17. Marken F, Compton RG (1998) Sonoelectrochemically modified electrodes: ultrasound assisted electrode cleaning, conditioning, and product trapping in 1-octanol/water emulsion systems. Electrochim Acta 43:2157–2165
18. Rejňák M, Klíma J, Svodoba J et al (2004) Synthesis and electrochemical reduction of methyl 3-halo-1-benzothiophene-2-carboxylates. Collect Czech Chem Commum 69:242–260
19. Zhang H, Coury LA Jr (1993) Effects of high-intensity ultrasound on glassy carbon electrodes. Anal Chem 65:1552–1558
20. Compton RG, Eklund JC, Page SD et al (1994) Voltammetry in the presence of ultrasound. Sonovoltammetry and surface effects. J Phys Chem 98:12410–12414
21. Cooper EL, Coury LA jr (1998) Mass transport in sonovoltammetry with evidence of hydrodynamic modulation from ultrasound. J Electrochem Soc 145:1994–1999
22. Compton RG, Eklund JC, Marken F (1997) Sonoelectrochemical processes. A review. Electroanalysis 9:509–522
23. Compton RG, Hardcastle JL, del Campo J et al (2003) Sonoelectrochemistry: physical aspects. In: Bard AJ, Stratmann M (eds) Encyclopedia of electrochemistry, vol 3. Wiley-VCH, Weinheim
24. Curie J, Curie P (1880) Dévelopment, par pression, de l'électricité polaire dans les cristaux hémièdres à faces inclinées. Compt Rend 91:291–294
25. Gallego-Juárez JA, Rodríguez Corral G, Riera E et al. (2001) Development of industrial models of high power stepped-plate sonic and ultrasonic transducer for use in fluids. IEEE Ultrasonic Sypoos. Proceedings, pp 571–578
26. Mason TJ, Cordemans E (1998) In: Jean-Louis Luche (ed) Practical considerations for process optimization in synthetic organic sonochemistry. Plenum Press, London
27. Mason TJ, Lorimer JP (1988) Ultrasonic equipment and chemical reactor design in: Sonochemistry: theory, applications and uses of ultrasound in Chemistry. Ellis Horwood, Chichester
28. Walsh FC (1993) A first course in electrochemical engineering. The electrochemical consultancy, Romsey
29. Hyde ME, Compton RG (2002) How ultrasound influences the electrodeposition of metals. J Electroanal Chem 531:19–24
30. Touyeras F, Hihn JY, Bourgoin X et al (2005) Effects of ultrasonic irradiation on the properties of coatings obtained by electroless plating and electroplating. Ultrason Sonochem 12:13–19

31. Yegnaraman V, Bharathi S (1992) Sonoelectrochemistry – an emerging area. Bull Electrochem 8:84–85
32. Mohapatra SK, Raja KS, Misra M et al (2007) Synthesis of self-organized mixed oxide nanotubes by sonoelectrochemical anodization of Ti-8Mn alloy. Electrochim Acta 53:590–597
33. Klíma J, Bernard C, Degrand C (1994) Sonoelectrochemistry: effects of ultrasound on voltammetric measurements at a solid electrode. J Electroanal Chem 367:297–300
34. Klíma J, Bernard C, Degrand C (1995) Sonoelectrochemistry: transient cavitation in acetonitrile in the neighbourhood of a polarized electrode. J Electroanal Chem 399:147–155
35. Klíma J, Bernard C (1999) Sonoassisted electrooxidative polymerisation of salicylic acid. Role of acoustic streaming and microjetting. J Electroanal Chem 462:181–186
36. Costa C, Hihn JH, Rebetez M et al (2008) Transport-limited current and microsonoreactor characterization at 3 low frequencies in the presence of water, acetonitrile and imidazolium-based ionic liquids ([BuMIm] [$(CF_3SO_2)_2N$]). Phys Chem Chem Phys 10:2149–2158
37. Louisnard O, González-García J, Tudela I et al (2009) FEM simulation of a son-reactor accounting for vibrations of the boundaries. Ultrason Sonochem 16:250–259
38. Dewald HD, Peterson BA (1990) Ultrasonic hydrodynamic modulation voltammetry. Anal Chem 62:779–782
39. Eklund JC, Marken F, Waller DN et al (1996) Voltammetry in the presence of ultrasound: a novel sono-electrode geometry. Electrochim Acta 41:1541–1547
40. Marken F, Compton RG (1996) Electrochemistry in the presence of ultrasound: the need for bipotentiostatic control in sonovoltammetric experiments. Ultrason Sonochem 3:S131–S134
41. Reisse J, François H, Vandercammen J et al (1994) Sonoelectrochemistry in aqueous electrolyte: a new type of sonoelectroreactor. Electrochim Acta 39:37–39
42. Aqil A, Serwas H, Delplancke JL (2008) Preparation of stable suspensions of gold nanoparticles in water by sonoelectrochemistry. Ultrason Sonochem 15:1055–1061
43. Lei H, Tang YJ, Wei JJ et al (2007) Synthesis of tugsten nanoparticles by sonoelectrochemistry. Ultrason Sonochem 14:81–83
44. Marken F, Kumbhat S, Sanders GHW et al (1996) Voltammetry in the presence of ultrasound: surface and solution processes in the sonovoltammetric reduction at glassy carbon and gold electrodes. J Electroanal Chem 414:95–105
45. Sonovoltametric measurement of the rates of electrode processes with fast coupled homogeneous kinetics: making macroelectrodes behave like microelectrodes: Compton RG, Marken F, Rebbitt TO (1996) Chem Commun 1017–1018
46. Marken F, Compton RG, Davies SG et al (1997) Electrolysis in the presence of ultrasound: cell geometries for the application of rates of mass transfer in electrosynthesis. J Chem Soc (Perkin Trans) 2(10):2055–2059
47. Esclapez MD, Sáez V, Milán-Yáñez D et al (2010) Sonochemical treatment of water poulled with trichloroacetic acid: from sonovoltammetry to pre-pilot plant scale. Ultrason Sonochem 17:1010–1020
48. Sáez V, Frías-Ferrer A, Iniesta J et al (2005) Characterization of a 20 kHz sonoreactor. Part I: analysis of mechanical effects by classical and numerical methods. Ultrason Sonochem 12:59–65
49. Klima J, Frías-Ferrer A, González-García J et al (2007) Optimisation of 20 kHz sonoreactor geometry on the basis of numerical simulation of local ultrasonic intensity and qualitative comparison with experimental results. Ultrason Sonochem 14:19–28
50. Compton RG, Hardcastle JL, del Campo J et al (2003) Ultrasound and electrosynthesis. In: Bard AJ, Stratmann M (eds) Encyclopedia of electrochemistry, vol 3. Wiley-VCH, Weinheim
51. Cognet P, Wilhem AM, Delmas H et al (2000) Ultrasound in organic electrosynthesis. Ultrason Sonochem 7:163–167
52. Walton DJ, Mason TJ (1998) In: Jean-Louis Luche (ed) Organic sonoelectrochemistry in synthetic organic sonochemistry. Plenum Press, London, pp 263–300
53. Morigushi N (1934) The effect of supersonic waves on chemical phenomena (III). The effect of the concentration polarization. J Chem Soc Jpn 55:749–750

54. Cataldo F (1992) Effects of ultrasound on the yield of hydrogen and chlorine during electrolysis of aqueous solutions of NaCl or HCl. J Electroanal Chem 332:325–331
55. Walton DJ, Burke LD, Murphy MM (1996) Sonoelectrochemistry: chlorine, hydrogen and oxygen evolution at platinised platinum. Electrochim Acta 41:2747–2751
56. Šljukić B, Bank CE, Compton RG (2004) The search for stable and efficient sonoelectrocatalysts for oxygen reduction and hydrogen peroxide formation: azobenzene and derivatives. Phys Chem Chem Phys 6:4034–4041
57. Šljukić B, Bank CE, Compton RG (2005) Exploration of stable sonoelectrocatalysis for the electrochemical reduction of oxygen. Electroanalysis 17:1025–1034
58. González-García J, Banks CE, Šljukić B et al (2007) Electrosynthesis of hydrogen peroxide via the reduction of oxygen assisted by power ultrasound. Ultrason Sonochem 14:405–412
59. Murphy MA, Marken F, Mocak J (2003) Sonoelectrochemistry of molecular and colloidal redox systems at carbon nanofiber-ceramic composite electrodes. Electrochim Acta 48:3411–3417
60. Pletcher D, Walsh FC (1993) Industrial electrochemistry. Chapman & Hall, London
61. Kuhn AT (1971) Industrial electrochemical processes. Elsevier, Amsterdam
62. Lindstrom O (1952) Astudy of some electrochemical effects in a field of stationary ultrasonic waves. Acta Chem Scand 6:1313–1323
63. Dereska J, Jaeger E, Hovorka F (1957) Effects of acoustical waves on the electrodeposition of Chromium. J Acoust Soc Am 29:769–769
64. Walker R (1990) Ultrasound and electroplating. Chem Britain 26:251–254
65. Lorimer P, Mason TJ (1999) Sonoelectrochemisry. The application of ultrasound in electroplating. Electrochemistry 67:924–930
66. Hardcastle JL, Compton RG (2001) The electroanalytical detection and determination of copper in heavily passivating media: ultrasonically enhanced solvent extraction by N-benzoyl-N-phenyl-hydroxylamine in ethyl acetate coupled with electrochemical detection by sono-square wave stripping voltammetry analysis. Analyst 126:2025–2031
67. Prasad PBSN, Vasudevan R, Seshadri SK (1994) Effect of ultrasonic agitation on surface finish of electrodeposits Indian. J Eng Mater Sci 1:178–180
68. Walken R, Halagan SA (1985) Electrodeposition of nickel-iron alloys with ultrasound. Plat Surf Finish 72:68–73
69. Namgoong E, Chun JS (1984) The effect of ultrasonic vibration on hard chromium plating in a modified self-regulating high speed bath. Thin Solid Films 120:153–159
70. Walker R, Walker CT (1975) New explanation for the brightness of electrodeposits produced by ultrasound. Ultrasonics 13:79–82
71. Lee C-W, Compton RG, Eklund JC et al (1995) Mercury-electroplated platinum electrodes and microelectrodes for sonoelectrochemistry. Ultrason Sonochem 2:S59–S62
72. Kang J, Shin Y, Tak Y (2005) Growth of etch pits formed during sonoelectrochemical etching of aluminium. Electrochim Acta 51:1012–1016
73. Doche ML, Hihn JY, Mandroyan A et al (2003) Influence of ultrasound power and frequency upon corrosion kinetics of zinc in saline media. Ultrason Sonochem 3:357–362
74. Effects of sonication on the anodic dissolution of copper and nickel electrodeposits: Chiba A (2003) Met Finish 117–122
75. Holt KB, Sabin G, Compton RG et al (2002) Reduction of tetrachloroaureate(III) at boron-doped diamond electrodes: gold deposition versus gold colloid formation. Electroanalysis 14:797–803
76. Narasimham KC, Gomathi PS, Udupa HVK (1976) The influence of ultrasonics on the electrodeposition of lead dioxide. J Appl Electrochem 6:397–401
77. González-García J, Gallud F, Iniesta J et al (2001) Kinetics of electrocrystallization of PbO_2 on glassy carbon electrodes: influence of ultrasound. New J Chem 25:1195–1198
78. González-García J, Sáez V, Iniesta J et al (2002) Electrodeposition of PbO_2 on glassy carbon electrodes: influence of ultrasound power. Electrochem Commun 4:370–373
79. Sáez V, González-García J, Iniesta J et al (2004) Electrodeposition of PbO_2 on glassy carbon electrodes: influence of ultrasound frequency. Electrochem Commun 6:757–761

80. González-García J, Iniesta J, Expósito E et al (1999) Early stages of lead dioxide electrodeposition on rough titanium. This Solid Films 352:49–56
81. Saterlay AJ, Wilkins SJ, Holt KB et al (2001) Lead dioxide deposition and electrocalysis at highly boron-doped diamond electrodes in the presence of ultrasound. J Electrochem Soc 148:E66–E72
82. Saterlay AJ, Wilkins SJ, Goeting CH et al (2000) Sonoelectrochemistry at highly boron-doped diamond electrodes: silver oxide deposition and electrocatalysis in the presence of ultrasound. J Solid State Electrochem 4:383–389
83. Ko WY, Chen WH, Tzeng SD et al (2006) Synthesis of pyramidal copper nanoparticles on gold substrate. Chem Mater 18:6097–6099
84. Fukami K, Nakanishi S, Yamasaki H et al (2007) General Mechanism for the synchronization of electrochemical oscillations and self-organized dendrite electrodeposition of metals with ordered 2D and 3D microstructures. J Phys Chem C 111:1150–1160
85. Jiang LP, Wang AN, Zhao Y et al (2004) A novel route for the preparation of monodisperse silver nanoparticles via a pulsed sonoelectrochemical technique. Inorg Chem Commun 7: 506–509
86. Gedanken A (2004) Using sonochemistry for the fabrication of nanomaterials. Ultrason Sonochem 11:47–55
87. Tang S, Men X, Lu H et al (2009) PVP-assisted sonoelectrochemical growth of silver nanostructures with various shapes. Mater Chem Phys 116:464–468
88. Pileni MP, Lisiecki I (1993) Nanometer metallic copper particle synthesis in reverse micelles. Colloids Surf A 80:63–68
89. Zhu J, Liu S, Palchik O et al (2000) Shape-controlled synthesis of silver nanoparticles by pulse sonoelectrochemical methods. Langmuir 16:6396–6399
90. Socol Y, Abramson O, Gedanken A et al (2002) Suspensive electrode formation in pulsed sonoelectrochemical synthesis of silver nanoparticles. Lagmuir 18:4736–4740
91. Mancier V, Daltin A-L, Leckercq D (2008) Synthesis and characterization of copper oxide (I) nanoparticles produced by pulsed sonoelectrochemistry. Ultrason Sonochem 15: 157–163
92. Dabalà M, Pollet BG, Zin V et al (2008) Sonoelectrochemical (20 kHz) production of $Co_{65}Fe_{35}$ alloy nanoparticles from Aotani solutions. J Appl Electrochem 38:395–402
93. Ganesan R, Shanmugam S, Gedanken A (2008) Pulsed sonoelectrochemical synthesis of polyaniline nanoparticles and their capacitance properties. Synt Met 158:848–853
94. Shen Q, Jiang K, Zhang H et al (2008) Three-dimensional dendritic Pt Nanostructures: sonoelectrochemical synthesis and electrochemical applications. J Phys Chem C 112: 16385–16392
95. Mastai Y, Polsky R, Koltypin Y et al (1999) Pulsed sonoelectrochemical of cadmium selenide nanoparticles. J Am Chem Soc 121:10047–10052
96. Zhu JJ, Aruna ST, Koltypin Y et al (2000) A novel method for the preparation of lead selenide: pulse sonoelectrochemical synthesis of lead selenide nanoparticles. Chem Mater 12:143–147
97. Synthesis of metallic magnesium by sonoelectrochemistry: Hass I, Gedanken A (2008) Chem Commun 1795–1797
98. Mohapatra SK, Misra M, Mahajan VK et al (2008) Synthesis of Y-branched TiO_2 nanotubes. Mat Lett 62:1772–1774
99. Haas I, Shanmugam S, Gedanken A (2006) Pulsed sonoelectrochemical synthesis of size-controlled copper nanoparticles stabilize by poly(N-vinylpyrrolidone). J Phys Chem B 110: 16947–16952
100. Lei H, Tang Y-J, Wei J-J et al (2007) Synthesis of tungsten nanoparticles by sonoelectrochemistry. Ultrason Sonochem 14:81–83
101. Sáez V, González-García J, Kulandainathan MA et al (2007) Electro-deposition and stripping of catalytically iron metal nanoparticles at boron-doped diamond electrodes. Electrochem Commun 9:1127–1133

102. Liu YC, Lin LH (2004) New pathway for the synthesis of ultrafine silver nanoparticles from bulk silver substrates in aqueous solutions by sonoelectrochemical methods. Electrochem Commun 6:1163–1168
103. Liu S, Huang W, Chen S et al (2001) Synthesis of X-ray amorphous silver nanoparticles by the pulse sonoelectrochemical method. J Non-Cryst solids 283:231–236
104. Liu YC, Yu CC, Yang KH (2006) Active catalysts of electrochemically prepared gold nanoparticles for the decomposition of aldehyde in alcohol solutions. Electrochem Commun 8:1163–1167
105. Liu YC, Kin LH, Chiu WH (2004) Size-controlled synthesis of gold nanoparticles from bulk gold substrates by sonoelectrochemical methods. J Phys Chem B 108:19237–19240
106. Liu YC, Yang KH, Yang SJ (2006) Sonoelectrochemical synthesis of spike-like gold-silver alloy nanoparticles from bulk substrates and the application on surface-enhanced Raman scattering. Anal Chim Acta 572:290–294
107. Reisse J, Caulier T, Deckerkheer C et al (1996) Quantitative sonochemistry. Ultrason Sonochem 3:S147–S151
108. Yang YJ (2006) A novel electrochemical preparation of PbS nanoparticles. Mat Sci Eng B 131:200–202
109. Mastai Y, Homyonfer M, Gedanken A et al (1999) Room temperature sonoelectrochemical synthesis of molybdenum sulfide fullurene-like nanoparticles. Adv Mat 11:1010–1013
110. Liu YC, Wang CC, Juang LC (2004) Sonoelectrochemical methods of preparing silver-coated TiO_2 nanoparticles with extremely high coverage. J Electroanal Chem 574:71–75
111. Zhu JJ, Qiu QF, Wang H et al (2002) Synthesis of silver nanowires by a sonoelectrochemical method. Inor Chem Commun 5:242–244
112. Singh KV, Martinez-Morales AA, Senthil Andavan GT et al (2007) A simple way of synthesizing single-crystalline semiconducting copper sulfide nanorods by using ultrasonication using template-assisted electrodeposition. Chem Mater 19:2446–2454
113. Qiu X, Lou Y, Samia ACS et al (2005) PbTe nanorods by sonoelectrochemistry Angew. Chem Int Ed 44:5855–5857
114. Sonoelectrochemical synthesis of CdSe nanotubes: Shen Q, Jiang L, Miao J (2008) Chem Commun 1683–1685
115. Jia F, Hu Y, Tang Y et al (2007) A general nonaqueous sonoelectrochemical approach to nanoporous Zn and Ni particles. Power Tech 176:130–136
116. Haas I, Shanmugam S, Gedanken A (2008) Synthesis of copper dendrite nanostructures by a sonochemical method. Chem Eur J 14:4696–7403
117. Qiu XF, Xu JZ, Zhu JM et al (2003) Controlable synthesis of palladium nanopaticles via a simple sonochemical method. J Mater Res 18:1399–1404
118. Del Campo FJ, Coles BA, Marken F et al (1999) High-frequency sonoelectrochemical process: mass transport, thermal and surface effects induced by cavitation in a 500 kHz reactor. Ultrason Sonochem 6:189–197
119. Qiu XF, Burda C, Fu RL et al (2004) Heterostructured Bi2Se3 nanowires with periodic phase boundaries. J Am Chem Soc 126:16276–16277
120. Raja KS, Misra M, Paramguru K (2005) Formation of self-ordered nano-tubular structure of anodic oxide layer of titanium. Electrochim Acta 51:154–165

Chapter 5
Sonochemical Synthesis of Metal Nanoparticles

Kenji Okitsu

Abstract In this chapter, sonochemical synthesis of nanometer sized metal particles is described and consists of sonochemical reduction of the corresponding metal ions in aqueous solutions. The reduction mechanism is suggested to be due to the reactions with reducing species formed from the sonolysis of organic additives and water. The rate of reduction of metal ions can be changed by changing the types and concentration of organic additives. In addition, various parameters such as ultrasound intensity, ultrasound frequency, dissolved gas, position of reaction vessel, etc. also affect the rate of reduction of metal ions. It is important to control the rate of reduction of metal ions, because the size of the formed metal particles is dramatically affected by the rate of reduction. It is recognized that smaller metal particles are obtained when the rate of reduction is higher. Bimetallic nanoparticles with core/shell structures can be also prepared by the sonochemical reduction of the corresponding metal ions. In addition, the immobilization of metal nanoparticles on metal oxides and the shape control of metal nanoparticles by the sonochemical reduction are described.

5.1 Introduction

Metal nanoparticles have been actively synthesized for applications as catalysts, sensors, adsorbents, analytical probes and optical data storages [1–7]. This is due to the fact that the physicochemical properties of metal nanoparticles can be

---

K. Okitsu (✉)
Graduate School of Engineering, Osaka Prefecture University, 1-1 Gakuen-cho, Naka-ku, Sakai, Osaka 599-8531, Japan
e-mail: okitsu@mtr.osakafu-u.ac.jp

Pankaj and M. Ashokkumar (eds.), *Theoretical and Experimental Sonochemistry Involving Inorganic Systems*, DOI 10.1007/978-90-481-3887-6_5,

controlled by tuning the size, shape and crystal structure of metal nanoparticles. The types and composition of the precursors and preparation methods are also the important factors affecting their properties. Therefore, a number of preparation methods have been developed to synthesize various types of metal nanoparticles: controlled chemical methods, photochemical and radiation chemical methods and metal vaporization methods have been developed. For example, in the controlled chemical method, the size and shape of the metal particles could be controlled by selecting a suitable reductant and stabilizer: various types of stabilizers such as citrate, poly(N-vinyl-2-pyrrolidone), triphenylphosphine, disulfide, dendrimer, sodium polyacrylate and cetyltrimethyl ammonium bromide can be used to govern the particle nucleation and growth processes.

In recent years, it has been recognized that sonochemistry is one of the new techniques for the synthesis of functional nanoparticles and nanostructured materials, because unique reactions can be induced by the irradiation of a liquid even at around room temperature.

Ultrasonic irradiation of a liquid leads to the generation of cavitation phenomenon which comprised of unique reaction fields in addition to physical and mechanical effects: the formation of micro-meter sized bubbles, formation of bubbles with high temperature and high pressure conditions, formation of shock waves, and strong micro-stirring effects are produced. Table 5.1 shows representative ultrasound techniques to synthesize inorganic and metal nanoparticles and nanostructured materials.

Although various techniques have been reported, sonochemical reduction technique for the synthesis of metal nanoparticles in an aqueous solution are reviewed in this chapter.

It is important to understand the reduction mechanism of metal ions, because the reduction processes can be applied to the synthesis of various metal nanoparticles and nanostructured materials. Therefore, the reduction mechanism of metal ions is

Table 5.1 Representative ultrasound techniques to synthesize inorganic and metal nanoparticles and nanostructured materials

	Techniques	Synthesized materials	Reference
1.	Cavitation bubble pyrolysis technique	Fe, Fe/Co, Mo_2C, $CoFe_2O_4$	[8–10]
2.	Thermal treatment synthesis technique of mists formed from ultrasonic atomizer (Ultrasonic spray pyrolysis technique)	Y_2O_3-ZrO_2, NiO, ZnS, $BaTiO_3$-$SrTiO_3$, MoS_2, $BiVO_4$	[11–14]
3.	Micro-stirring and vibration technique (Ultrasonic-assisted technique)	Au, Ag, Bi_2Te_3	[15, 16]
4.	Sonomechanical-assisted metal displacement reduction technique	Au, Pt, Ag, metal oxides	[17, 18]
5.	Sonoelectrochemical technique	Ag, Cu	[19, 20]
6.	Sonochemical reduction technique using reductants formed from hot cavitation bubbles	Au, Pd, Pt, Ag, Ru, Au/Pd	[9]

described in the first section. And then, the details of the sonochemical reduction processes and the formation of metal nanoparticles are shown in the later section.

5.2 Reduction Mechanism of Metal Ions in Aqueous Solution Under Ultrasonic Irradiation

To synthesize metal nanoparticles in an aqueous solution, the reduction reactions of the corresponding metal ions are generally performed. Gutierrez et al. [21] reported the reduction of $AuCl_4^-$ and Ag^+ ions in an aqueous solution by ultrasonic irradiation under H_2-Ar mixed atmosphere. They found that the optimum condition of these reductions was under the 20 vol% H_2 and 80 vol% Ar atmosphere. Following this study, many papers reported the sonochemical reduction of noble metal ions under pure Ar atmosphere to produce the corresponding metal nanoparticles [22–28].

Based on these reports, the following sequence of reactions is suggested:

$$H_2O \rightarrow {}^{\cdot}OH + {}^{\cdot}H \tag{5.1}$$

$$RH + {}^{\cdot}OH\ ({}^{\cdot}H) \rightarrow \text{reducing species} + H_2O\ (H_2) \tag{5.2}$$

$$RH + H_2O \rightarrow \text{pyrolysis radicals and unstable products} \tag{5.3}$$

$$M^{n+} + \text{reductants} \rightarrow M^0 \tag{5.4}$$

$$nM^0 \rightarrow (M^0)_n \tag{5.5}$$

$$M^0 + (M^0)_n \rightarrow (M^0)_{n+1} \tag{5.6}$$

where M^{n+} corresponds to a metal ion and RH corresponds to an organic additive. Reactions (5.1)–(5.3) indicate the sonochemical formation of reductants: (1) ${}^{\cdot}H$ is formed from pyrolysis of water, (2) reducing species and H_2 are formed from the abstraction reaction of RH with ${}^{\cdot}OH$ or ${}^{\cdot}H$, and (3) pyrolysis radicals and unstable products are formed via pyrolysis of RH and water. After the formation of M^0 (Reaction (5.4)), nM^0 gives $(M^0)_n$ (Reaction (5.5)). The sonochemically formed M^0 can be adsorbed on $(M^0)_n$ to give $(M^0)_{n+1}$ (Reaction (5.6)). In this mechanism, it should be noted that general organic stabilizers such as surfactants or water soluble polymers can act as RH under ultrasonic irradiation.

On the other hand, if O_2 existed in the reaction system, the reaction mechanism would be affected by the reactions with O_2: the reaction mechanism is dependent on the types of dissolved gases in the sample solution. The details for the effects of various parameters on the reduction of metal ions and formation of metal nanoparticles are described in the following sections.

5.3 Effects of Various Parameters on the Rates of Reduction of Metal Ions

The effects of various parameters on the rates of reduction of metal ions and formation of metal nanoparticles under ultrasonic irradiation have been investigated with a standing-wave type irradiation system. One example of the irradiation set-up and the characteristic of the reaction vessel are shown in Fig. 5.1 [29]. Here, ultrasonic irradiation was carried out using an ultrasonic generator (Kaijo Co. 4021type, frequency; 200 kHz) and a 65 mm ϕ oscillator (Kaijo Co.), which was operated at ca. 20–200 W in a water bath maintained at a constant temperature. The glass vessel was cylindrical and it had a Teflon valve and a port covered by a silicon rubber septum for gas bubbling. The bottom of the vessel was planar, 1 mm thick and 55 mm in diameter. An aqueous solution of metal ions (0.1–1.0 mM, 60 or 65 mL) was added to the vessel and then it was purged with the appropriate gas. During the course of the irradiation, the vessel was isolated from the air atmosphere. After irradiation, small amounts of the sample solutions were drawn from the silicon rubber septum by a syringe with a stainless-steel needle.

Various irradiation parameters were investigated, such as the types of organic additives, intensity of the ultrasound, dissolved gas and distance between the reaction vessel and the oscillator. In the case of frequency effects, other irradiation systems were used. The details are described in the section of effects of ultrasound frequency on the rate of reduction.

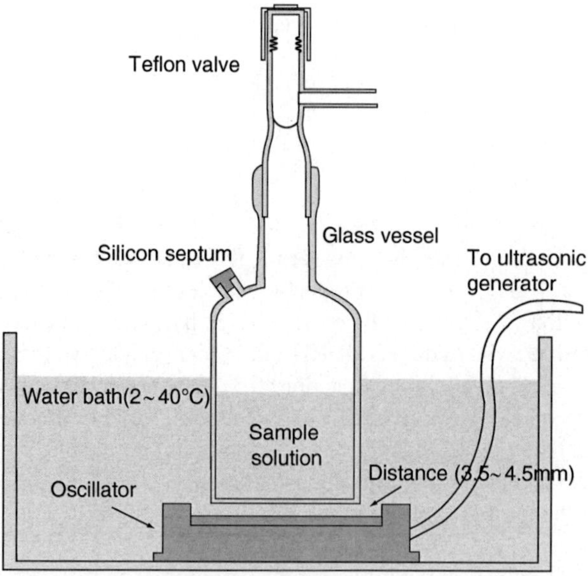

Fig. 5.1 One example of the irradiation set-up and the characteristic of the reaction vessel: a standing-wave type ultrasound irradiation system [29]

5.3.1 *Effect of Organic Additives on the Rate of Reduction*

It has been reported that the sonochemical reduction of Au(III) reduction in an aqueous solution is strongly affected by the types and concentration of organic additives. Nagata et al. reported that organic additives with an appropriate hydrophobic property enhance the rate of Au(III) reduction. For example, alcohols, ketones, surfactants and water-soluble polymers act as accelerators for the reduction of Au(III) under ultrasonic irradiation [24]. Grieser and coworkers [25] also reported the effects of alcohol additives on the reduction of Au(III). They suggested that the rate of the sonochemical reduction of Au(III) is related to the Gibbs surface excess concentration of the alcohol additives.

Okitsu et al. [26, 30, 31] reported the sonochemical reduction of Pd(II) to Pd(0) in the absence and presence of organic and inorganic additives. Figure 5.2 shows the changes in the concentration of Pd(II) during ultrasonic irradiation in the presence of polyvinyl pyrrolidone (PVP, Mw 40,000, 1 g/L), sodium dodecylsulfate (SDS, 8 mM), polyethylene(40)glycol monostearate (PEG-MS, 0.4 mM), and polyoxyethylene(20)sorbitan monolaurate (Tween20, 5 g/L) [26, 32], where the concentration of Pd(II) in the irradiated solutions is determined by an improved colorimetric method [26]. The initial rates of Pd(II) reduction were in the order of no additive: ca.7 μ M/min < PVP: 87 μ M/min < SDS: 130 μ M/min < PEG-MS: 230 μ M/min < Tween 20: 400 μ M/min, at an initial concentration of 1 mM of Pd(II). It can be seen that the rates of Pd(II) reduction are accelerated at least tenfold by the addition of organic stabilizers. As shown in Reactions (5.1)–(5.3), this is due to that organic stabilizers act as not only the stabilizers for the formed metal nanoparticles but also the precursors for the formation of reducing species.

Figure 5.3 shows the changes in the concentration of Pd(II) during ultrasonic irradiation in the presence of 2.02 g/L Al_2O_3 powder [31], where alcohols (20 mM) are included as an organic additive. This condition corresponds to the formation of 5 wt% Pd/Al_2O_3 (weight ratio of Pd to Al_2O_3), when the reduction of Pd(II) to metallic Pd is completed and all of the formed Pd particles are supported on Al_2O_3.

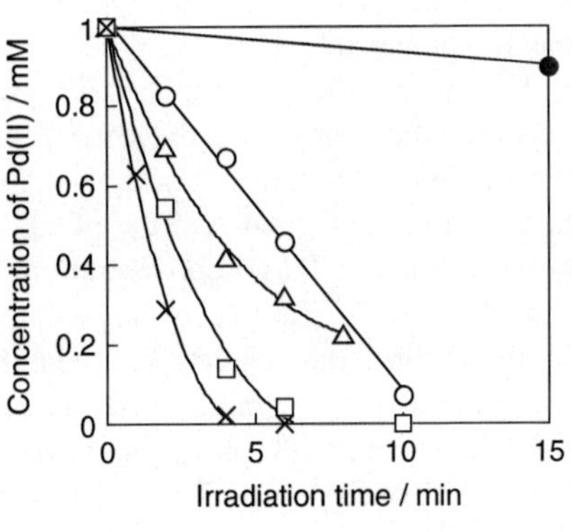

Fig. 5.2 Reduction of Pd(II) under Ar in the absence and presence of various stabilizers. Initial concentration of Pd(II); 1mM, (●); none, (○); PVP, (Δ); SDS, (□); PEG-MS, (X); Tween 20 [32]

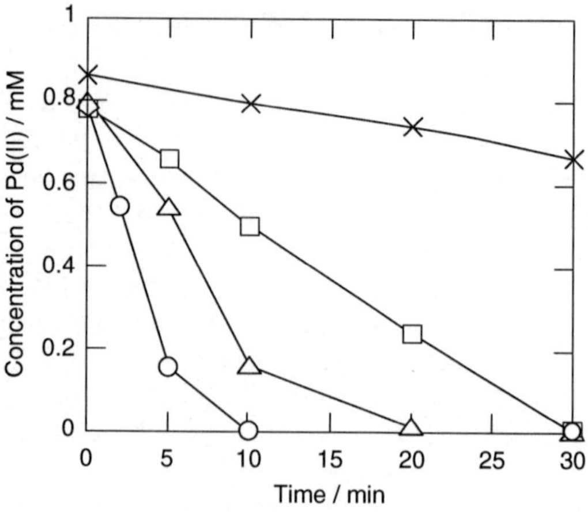

Fig. 5.3 Changes in the concentration of Pd(II) during ultrasonic irradiation under Ar in the presence and absence of alcohol additives. Before irradiation, Ar bubbling is performed. Conditions: 1 mM Pd(II), 2.02 g/L Al_2O_3, 20 mM alcohol. (X) None, (□) methanol, (Δ) ethanol, (○) 1-propanol [31]

Even in the presence of Al_2O_3 powders, the reduction of Pd(II) proceeds as shown in Fig. 5.3. It is clear that the rate of Pd(II) reduction is accelerated by the addition of alcohols. The rate of Pd(II) reduction is strongly dependent on the carbon number of alcohol additives: the rate increases in the order of methanol < ethanol < 1-propanol, although the same concentration of alcohol is present in the solution. This observation is due to the fact that the reductants are more efficiently formed from higher hydrophobic alcohols, because higher hydrophobic molecules more efficiently accumulate at the interface of the cavitation bubbles [31].

It is also observed in Fig. 5.3 that Pd(II) ions are partly adsorbed on Al_2O_3 before ultrasonic irradiation: the concentration of Pd(II) just before irradiation becomes ca. 0.8 mM, although 1 mM Pd(II) was added in the sample solution. From a preliminary adsorption experiment, the rate of Pd(II) adsorption on Al_2O_3 was found to be slow compared with those of Pd(II) reduction in the presence of alcohols. Therefore, it is suggested that the sonochemical reduction of Pd(II) in the presence of alcohols mainly proceeds in the bulk solution. The mechanism of the Pd/Al_2O_3 formation is also described in the section of sonochemical synthesis of supported metal nanoparticles.

Figure 5.4 shows the pH value of the irradiated sample solution, where the experimental condition is the same as in Fig. 5.3. It can be seen that the pH value of the sample solution remarkably decreases with increasing irradiation time. In addition, the behavior of the pH decrease in the presence of alcohols is similar to the change in Pd(II) concentration. Therefore, the concentration of H^+ ions is considered to increase according to the progress of Pd(II) reduction.

To confirm the relation between the Pd(II) reduction and the change in the concentration of H^+ ions, the sonolysis of 1-propanol-water solution without Pd(II) was carried out and the change in the pH value was investigated. The pH value without Pd(II) decreased from 5.7 to 5.2 in the 10 min irradiation, while that with Pd(II) decreased from 3.5 to 2.7 in the 10 min irradiation. This result suggests that the change

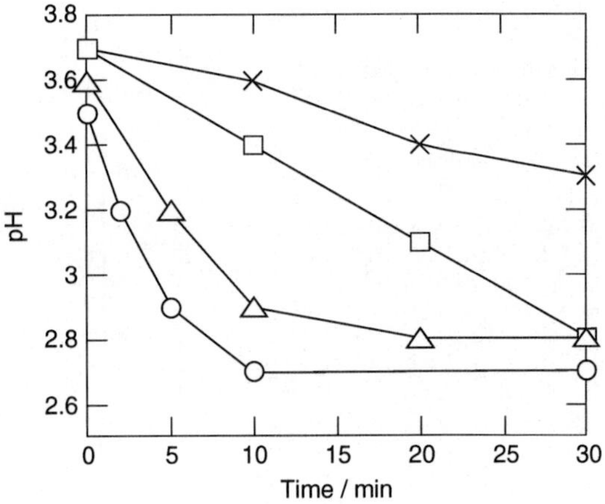

Fig. 5.4 Changes in the pH of the sample solution during ultrasonic irradiation under Ar in the presence and absence of alcohol additives. Conditions: 1 mM Pd(II), 2.02 g/L Al_2O_3, 20 mM alcohol. (X) None, (□) methanol, (Δ) ethanol, (○) 1-propanol [31]

in the concentration of H^+ ions is related to the occurrence of Pd(II) reduction. As seen in Reactions (5.1)–(5.3), various reductants such as ·H, H_2, pyrolysis radicals and unstable products are formed during ultrasonic irradiation. The following reactions should occur.

$$Pd^{2+} + \text{reductants} \rightarrow Pd^+ + H^+ \quad (5.7)$$

$$Pd^+ + \text{reductants} \rightarrow Pd^0 + H^+ \quad (5.8)$$

Based on Reactions (5.7) and (5.8), the progress of the reduction of Pd(II) to Pd(0) could be monitored by measuring the pH change of the irradiated solution.

5.3.2 Effects of Ultrasound Intensity on the Rate of Reduction

The chemical reactions induced by ultrasonic irradiation are generally influenced by the irradiation conditions and procedures. It is suggested that "ultrasound intensity", "dissolved gas", "distance between the reaction vessel and the oscillator" and "ultrasound frequency" are important parameters to control the sonochemical reactions.

Figure 5.5 shows the changes in the concentration of Au(III) at different ultrasound intensities [29], where the intensities are determined by the calorimetric method. It can be seen that the concentration of Au(III) decreases with increasing irradiation time and the reduction behavior is clearly dependent on the ultrasound intensities. At more than 1.20 W cm^{-2}, the reduction of Au(III) was completely finished within the 20 min irradiation. On the other hand, it was also observed that no reduction occurred in a conventional ultrasonic cleaning bath (Honda Electric Co., W-113, 28 kHz, 100 W, bath-volume: ca. 2 L) [29].

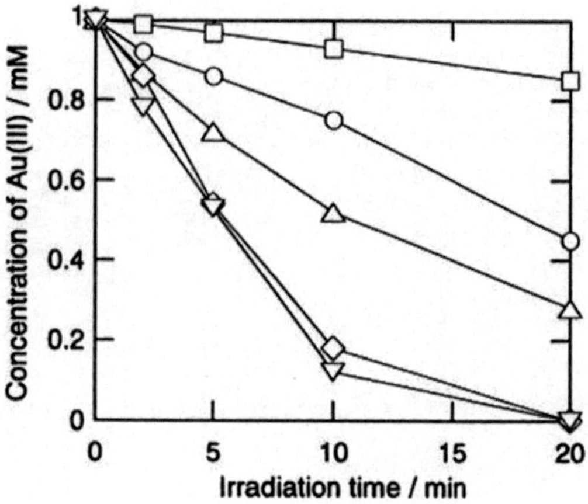

Fig. 5.5 Changes in the concentration of Au(III) during ultrasonic irradiation under various intensities. Condition: 1-propanol; 20 mM, Ar atmosphere, temperature; 20°C, distance; 4.0 mm. (□) 0.15 W cm^{-2}, (○) 0.53 W cm^{-2}, (Δ) 0.92 W cm^{-2}, (◇) 1.20 W cm^{-2}, (▽) 1.43 W cm^{-2} [29]

Based on these results, the reduction of Au(III) requires the formation of hot cavitation bubbles which cause pyrolysis of water and 1-propanol molecules. In addition, it is suggested that the number of hot cavitation bubbles and/or the bubble temperatures increase with increasing ultrasound intensity in the irradiation system.

5.3.3 Effects of Dissolved Gas on the Rate of Reduction

It is widely considered that the physical properties of dissolved gases affect the sonochemical efficiency. The ratio of specific heats, $\gamma = C_p/C_v$, the thermal conductivity, and the solubility in water are the important parameters. The effects of dissolved gas on the reduction of Au(III) under ultrasonic irradiation are shown in Fig. 5.6 [29]. It can be seen that the changes in the concentration of Au(III) are strongly dependent on the types of dissolved gas.

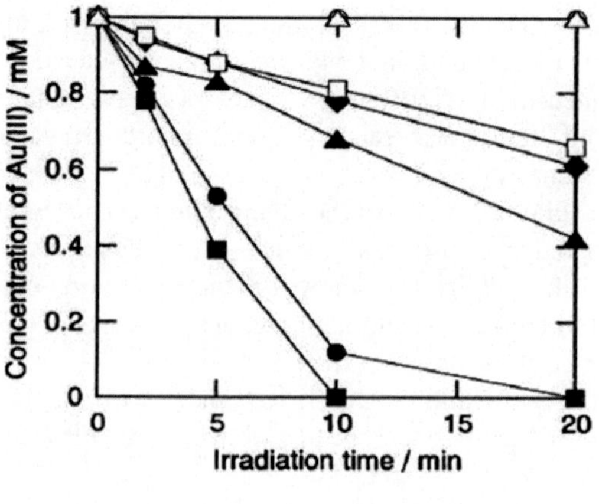

Fig. 5.6 Changes in the concentration of Au(III) during ultrasonic irradiation under various atmospheres. Condition: Intensity; 1.43 W cm^{-2}, 1-propanol; 20 mM, temperature; 20°C, distance; 4.0 mm. (■) Kr, (●) Ar, (▲) Ne, (◆) He, (□) N_2, (○) CO_2, (Δ) CH_4 [29]

The rates of the sonochemical reduction of Au(III) are in the order of $CH_4 = CO_2 << N_2 < He < Ne < Ar < Kr$: the rates of reduction under monoatomic molecule gases are faster than those under polyatomic molecule gases. Under CO_2 and CH_4 gases, no reduction is observed. In general, it is suggested that higher γ value and lower thermal conductivity result in the formation of hot cavitation bubbles with higher temperature. On the other hand, higher solubility in water affects the number of cavitation bubbles during ultrasonic irradiation [29]. Under CO_2 and CH_4 gases, it is suggested that no formation of hot cavitation bubbles so that the reduction of Au(III) does not occur.

5.3.4 Effects of the Distance Between Reaction Vessel and Oscillator on the Rate of Reduction

Figure 5.7 shows the effects of the distance from the oscillator to the bottom of the reaction vessel on the rate of Au(III) reduction, where the distance is changed from 3.5 to 4.5 mm [29]. It is clear that the rates of reduction are affected by the position of the reaction vessel. The rate of reduction became the maximum at a distance of ca. 3.8 mm. This was almost the same as the half-wavelength of the ultrasound (3.71 mm) used in this study. It is suggested that ultrasound is effectively transmitted into the reaction vessel at 3.8 mm distance. It should be noted that the position of the reaction vessel sensitively affects the efficiency of the sonochemical reduction.

5.3.5 Effects of Ultrasound Frequency on the Rate of Reduction

When the ultrasound frequency used changes, the following factors would change: (1) the temperature and pressure inside the collapsing cavitation bubbles, (2) the number and distribution of bubbles, (3) the size and lifetime of bubbles, (4) the

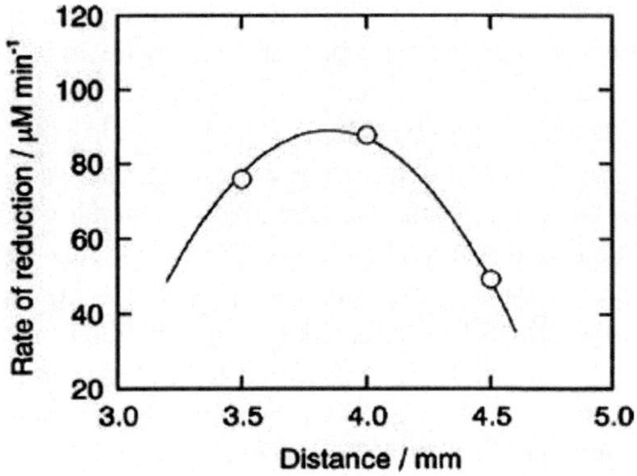

Fig. 5.7 Rate of Au(III) reduction as a function of the distance between the reaction vessel and the oscillater. Conditions: 1 mM Au(III), 20 mM 1-propanol, intensity; 1.43 W cm^{-2}, Ar atmosphere, temperature; 20°C [29]

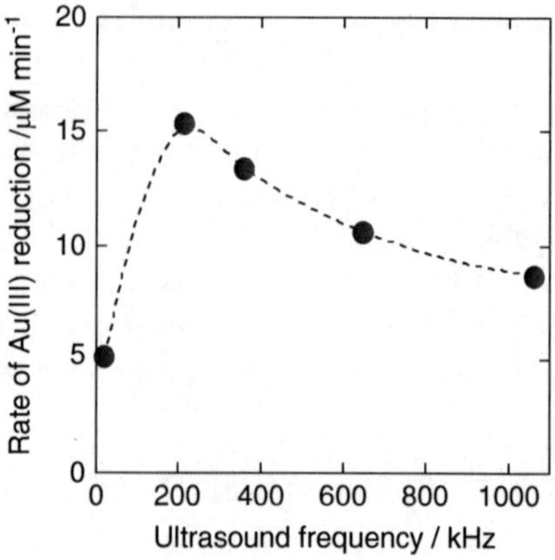

Fig. 5.8 Rate of Au(III) reduction as a function of ultrasound frequency. Conditions: Au(III): 0.2 mM, 1-propanol: 20 mM, atmosphere: Ar, ultrasound intensity: 0.1 W mL^{-1} [33]

dynamics and symmetry (shape) of the bubble collapse, (5) the effect of organic additive on bubble temperature, (6) the formation of primary and secondary radical, intermediate and product, etc.

The sonochemical reduction of Au(III) has been investigated under Ar in the presence of 20 mM 1-propanol at different frequencies, where two types of ultrasound irradiation systems were used: one is a horn type sonicator (Branson 450-D, frequency: 20 kHz, diameter of Ti tip: 19 mm) and the other is a standing wave sonication system with a series of transducers operating at different ultrasound frequencies (L-3 Communication ELAK Nautik GmbH, frequency: 213, 358, 647, and 1,062 kHz, diameter of oscillator: 55mm) [33]. All experiments were performed at a constant ultrasound intensity ((0.1+/−0.01 W mL^{-1}), which was determined by calorimetry.

Figure 5.8 shows the rates of sonochemical reduction of Au(III) at five different frequencies [33]. It can be seen that the rate of reduction increases from 20 to 213 kHz and then decreases with increasing frequency (213 > 358 > 647 > 1,062 kHz).

The rate of Au(III) reduction should have a correlation with the cavitation efficiency at these frequencies. Therefore, the result of Fig. 5.8 suggests that maximum amounts of reductants are sonochemically formed at 213 kHz in the presence of 1-propanol. The existence of an optimum frequency in the sonochemical reduction efficiency would be explained as follows. As the frequency is increased, the number of cavitation bubbles can be expected to increase. This would result in an increase in the amount of primary and secondary radicals generated and an increase in the rate of Au(III) reduction. On the other hand, at higher frequencies there may not be enough time for the accumulation of 1-propanol at the bubble/solution interface and for the evaporation of water and 1-propanol molecules to occur during the expansion cycle of the bubble. This would result in a decrease in the amount of active radicals. Furthermore, the size of the bubbles also decreases with increasing frequency. These multiple effects would result in a very complex frequency effect.

In addition, it should be noted that the experimental setup is different between the 20 kHz and the high frequency systems. Therefore, the type of the cavitation would be different between them. It has been reported that the 20 kHz system consisting of a horn type sonicator generates predominantly transient cavitation whereas the systems used for higher frequencies generate stable cavitation [34].

5.4 Effects of Various Parameters on the Properties of Metal Nanoparticles

The effects of various parameters on the rates of sonochemical reduction of metal ions were described in the previous sections. From this section, the effects of such parameters on the properties of metal nanoparticles are described in relation to the rates of reduction.

5.4.1 Effects of the Rates of Reduction on the Size of the Formed Nanoparticles

A representative TEM image of the sonochemically formed Au particles in the absence of stabilizer is shown in Fig. 5.9 [33]. In this experiment, only 20 mM of 1-propanol is included as an organic additive. It can be seen that the size of Au particles is in the nanometer regime. The sizes of the sonochemically formed metal nanoparticles are dependent on the initial concentration of metal ions and the types of stabilizers [26] as same as the conventional methods: smaller metal particles are formed in the presence of lower initial concentration of metal ions. In addition, if a suitable stabilizer were used, the growth of particles would be suppressed effectively by the stabilizer adsorption on the particle surface, resulting in the formation of smaller metal particles.

To investigate how cavitation bubbles affect the sizes of the formed metal nanopartilces, sonochemical reduction of Au(III) was carried out under various irradiation conditions in an aqueous solution containing only a small amount of

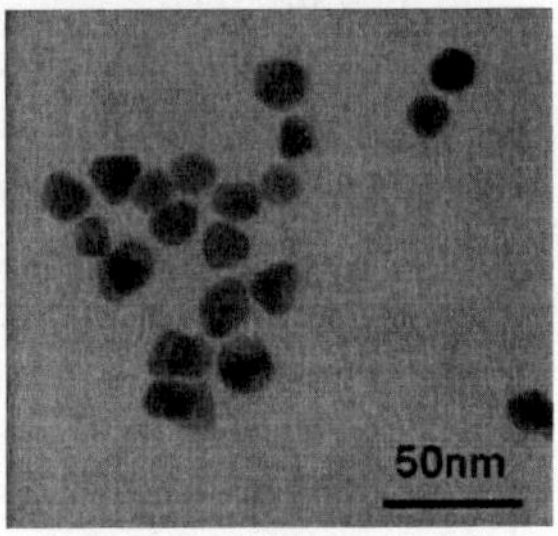

Fig. 5.9 A TEM image of sonochemically synthesized Au nanoparticles under Ar atmosphere [33]

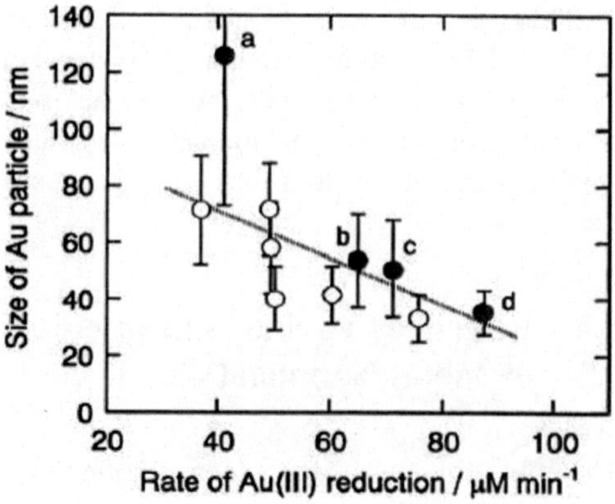

Fig. 5.10 Relation between the rate of Au(III) reduction and the average size of the formed gold particles. Each error bar corresponds to the standard deviation of the size of the gold particles. Closed circles correspond to the dependence of the ultrasound intensities, in which the rate of reduction increases with increasing the intensity. Conditions: 200 kHz. (**a**) 0.53 W cm^{-2}, (**b**) 0.92 W cm^{-2}, (**c**) 1.20 W cm^{-2}, (**d**) 1.43 W cm^{-2} [29]

1-propanol [29]. In addition, identical starting materials and the same initial concentrations were used in each preparation. It was found that the sizes of the sonochemically formed Au particles are sensitively affected by the irradiation parameters such as ultrasound intensity, distance between reaction vessel and oscillator, etc. The relationship between the rate of Au(III) reduction and the size of the formed Au nanoparticles at 20 min irradiation is shown in Fig. 5.10, where the reduction ratio of Au(III) is in the range of 55–100% at 20 min irradiation [29]. Although the reduction ratio is not the same among these samples, it is clear that the size of the Au nanoparticles decreases with an increase in the rate of Au(III) reduction.

In general, it is suggested that shock waves are generated during ultrasonic cavitation. If strong shock waves affected the growth of the Au particles, the size of the particles might become larger when the intensity of shock waves became larger. From the result of Fig. 5.10, however, the size of the Au nanoparticles became smaller at the higher ultrasound intensity as seen in (a) to (d). Therefore, the strength of shock waves would be not so strong to affect the size of the Au nanopartilces. Consequently, the initial rate of Au(III) reduction mainly affects the nucleation process of the Au particles: the size of the Au particles would depend on the number of the formed Au nuclei which correspond to the rate of Au(III) reduction.

The rate of Au(III) reduction is also affected by ultrasound frequency as seen in Fig. 5.8. From TEM analyses, the size of the formed Au nanoparticles was found to be affected by ultrasound frequency. Figure 5.11 shows the relationship between the average size of the formed Au nanoparticles and the rate of Au(III) reduction. It is found that, even in the case of different frequency, the rate of Au(III) reduction affects the size of the formed Au nanoparticles. Based on the obtained results, the nucleation process is important in determining the size of Au particles, because the nucleation process should be closely related to the rate of reduction [29, 33].

If suitable stabilizers were added in this solution system, smaller metal nanoparticles could be obtained as seen in many papers [22, 23, 27, 35]. In addition, it is

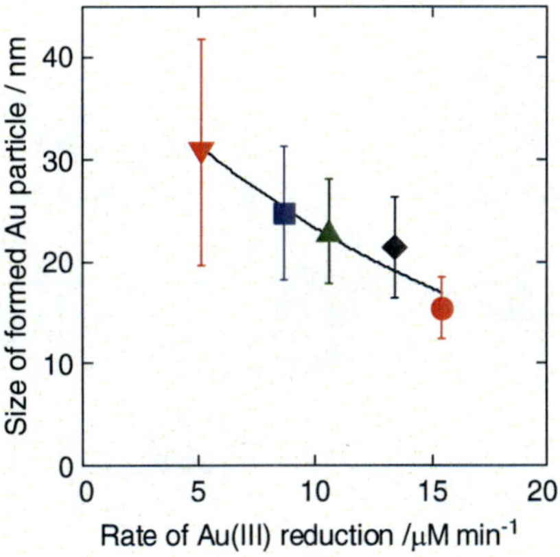

Fig. 5.11 Average size of the formed Au nanoparticles as a function of the rate of sonochemical reduction of Au (III). (▼); 20 kHz, (●); 213 kHz, (◆); 358 kHz, (▲); 647 kHz, (■); 1062 kHz. Each error bar corresponds to the standard deviation of the size of Au nanoparticles. Conditions: Au(III): 0.2 mM, 1-propanol: 20 mM, atmosphere: Ar, ultrasound intensity: 0.1 W mL^{-1}, irradiation time: 120 min [33]

reported that various types of metal nanoparticles such as Ag, Au, Pd, Pt, Ru, Au/Pd, Pt/Ru, etc. can be prepared by applying the sonochemical reduction of the corresponding metal ions in an aqueous solution [22, 23, 25, 27, 35–37].

5.4.2 Sonochemical Synthesis of Supported Metal Nanoparticles

Catalysts are often used in the form of transition metal nanoparticles immobilized on inorganic or organic substrates such as Al_2O_3, SiO_2, TiO_2, Fe_2O_3, zeolite, activated carbon, carbon nanotube, chitosan, etc., because of their convenient utilities such as the easy recovery of the catalysts and the high stability over the long reaction period. In addition, such metal nanoparticles immobilized on various substrates should be one of the key materials in nanoscience and nanotechnology, because an excellent synergy and bifunctional effect would also be expected.

It has been reported that highly dispersed Pd nanoparticles immobilized on Al_2O_3 can be prepared by ultrasonic irradiation of an aqueous solution containing Pd(II), Al_2O_3 and small amount of alcohols. As shown in Fig. 5.3, the reduction of Pd(II) proceeds in the presence of Al_2O_3 powders. Figure 5.12 shows TEM images of the sonochemically prepared Pd/Al_2O_3 [30, 31]. It can be seen that the formed Pd particles are of nanometer size and highly dispersed on the surface of Al_2O_3. In Fig. 5.12b, it is noted that the formed Pd nanoparticles are densely immobilized on Al_2O_3 as a shell of the Pd nanoparticles.

Table 5.2 shows the average size of the formed Pd nanoparticles. As described in the previous section, the added alcohol acts as a precursor for the reduction of Pd(II) under ultrasonic irradiation and affects the rate of the reduction. Therefore, it is considered that smaller Pd particles are formed when the rate of Pd nucleation (corresponding to the rate of Pd(II) reduction) is faster. In addition, it can be seen from Table 5.2 that, in the presence of 1-propanol, the size of the formed Pd particles and its standard deviation become smaller with increasing amount of

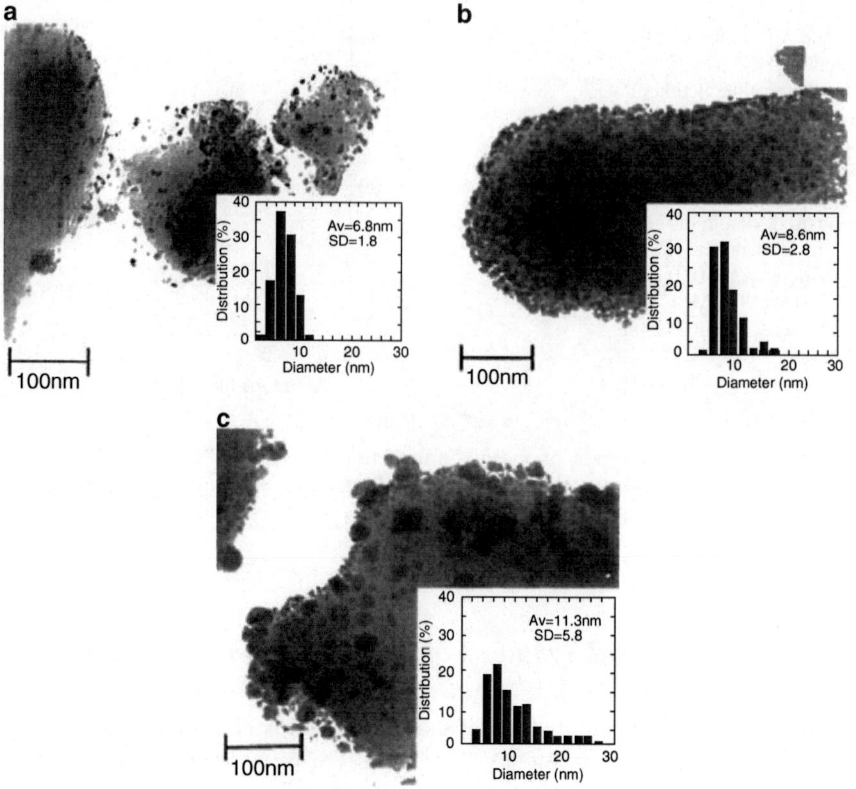

Fig. 5.12 TEM images of sonochemically prepared Pd/Al_2O_3 and size distribution of Pd particles formed by 30 min irradiation of 1mM Pd(II) and 20 mM alcohol. Conditions: 200kHz, Ar, 20°C. (**a**) 1 wt% Pd/Al_2O_3 formed in the presence of 1-propanol, (**b**) 5 wt% Pd/Al_2O_3 formed in the presence of 1-propanol, (**c**) 5 wt% Pd/Al_2O_3 formed in the presence of methanol [30]

Table 5.2 Average size and standard deviation of Pd particles formed on the Al_2O_3 surface and rate of Pd(II) reduction by ultrasonic irradiation [30, 31][a]

Conditions[b] Pd/Al_2O_3, Alcohol	Average size/nm	Standard deviation[c]	Rate[d] ($\mu M\ min^{-1}$)
1 wt%, 1-propanol	6.8	1.8	120
1 wt%, ethanol	7.4	2.5	52
1 wt%, methanol	7.9	3.4	15
3 wt%, 1-propanol	7.1	2.2	130
5 wt%, 1-propanol	8.6	2.8	130
5 wt%, ethanol	10.0	2.9	53
5 wt%, methanol	11.3	5.8	25

[a]Irradiation time: 30 min of 200 kHz ultrasound, Pd(II): 1 mM, alcohol: 20 mM. [b]Weight ratio of Pd/Al_2O_3, [c]Standard deviation of size of Pd particles, [d]Average rate for initial 5 min irradiation.

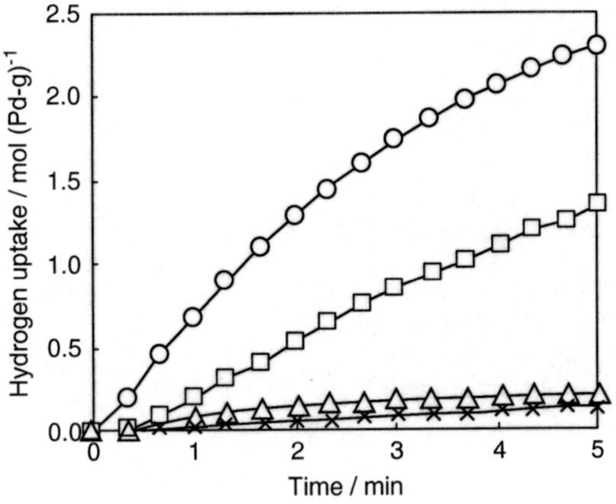

Fig. 5.13 H_2 consumption in the hydrogenation of 1-hexene over various Pd catalysts. Conditions: 1 atm. H_2, 23 ± 0.5°C. (X) Pd black, (Δ) conventional Pd/Al_2O_3, (□) Pd/Al_2O_3 sonochemically prepared in 20 mM methanol, (○) Pd/Al_2O_3 sonochemically prepared in 20 mM1-propanol [31]

Al_2O_3 in the order of 5 wt% Pd (Al_2O_3: 2.02 g/L) < 3 wt% Pd (Al_2O_3: 3.44 g/L) < 1 wt% Pd (Al_2O_3: 10.5 g/L). Since the rate of reduction was almost the same among them, the growth and the agglomeration of the Pd particles in solution would be suppressed by its adsorption onto Al_2O_3 so that the size of the formed Pd particles was smaller as the amount of Al_2O_3 increased. Taking into account this formation mechanism, it is noted that the formed Pd particles are immobilized only on the surface of the Al_2O_3 powders.

It is also reported that the size of the formed Pd particles on Al_2O_3 strongly influences the catalytic activities of hydrogenation of olefins (1-hexene and 3-hexene) in 1-propanol solution [31]. Figure 5.13 shows the amount of hydrogen uptake during the catalytic hydrogenation of 1-hexene. As seen in Fig. 5.13, the catalytic activities of the sonochemically prepared Pd/Al_2O_3 toward 1-hexene were three to seven times higher than that of a conventionally prepared Pd/Al_2O_3 catalyst and 10–23 times higher than that of a commercially available Pd black catalyst.

By using sonochemical reduction processes, supported metal nanoparticles on metal oxides such as Au/SiO_2, Au/Fe_2O_3, Pd/Fe_2O_3, Pt/TiO_2, etc. can be synthesized [38–41].

5.4.3 Sonochemical Synthesis of Bimetallic Nanoparticles

It has been reported that bimetallic nanoparticles with core/shell structure can be prepared by ultrasonic irradiation. Mizukoshi et al. reported the formation of bimetallic nanoparticles of Au core/Pd shell structure [42,43] from the sonochemical reduction of Au(III) and Pd(II), where the stepwise reduction of metal ions was observed to proceed during ultrasonic irradiation. That is, the reduction of Pd(II) started after the reduction of Au(III) finished. Vinodgopal et al. reported

Fig. 5.14 Catalytic activities (●) for the hydrogenation of cyclohexene over catalysts with various Au/Pd composition and average sizes (○) of loading noble metal. The reaction was carried out in a conventional closed system at 23 ± 0.5°C. Conditions: initial H_2 pressure of 1 atm., cyclohexene of 1.18 mmol, 1-propanol solution of 30 mL [44]

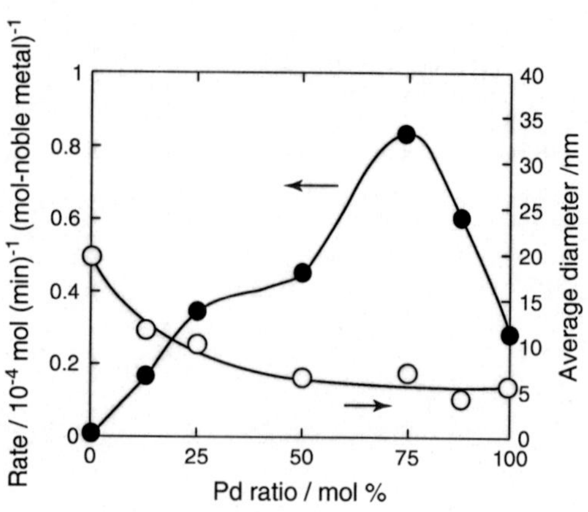

the sonochemical synthesis of Pt-Ru bimetallic nanoparticles to apply it as a methanol oxidation catalyst in direct methanol fuel cells [37].

Catalytic activities of Au core/Pd shell bimetallic nanoparticles have also been investigated [43,44]. Okitsu et al. reported that catalytic activities are closely related to the nanostructure of bimetallic nanoparticles [44], where Au core/Pd shell bimetallic nanoparticles on and inside SiO_2 are prepared by consecutive sonochemical and sol-gel processes.

Figure 5.14 shows the effect of Au/Pd composition on the rate of hydrogenation of cyclohexene in a 1-propanol solution, where the average sizes of supported Au/Pd nanoparticles are also shown. It can be seen that the rates of the hydrogenation over the Au/Pd supported on SiO_2 were distinctly higher than that over the monometallic Pd on SiO_2, although the size of their particles was almost the same among them. Pure Au supported on SiO_2 exhibited no catalytic activity. The catalytic activity was strongly affected by the composition of Au/Pd and the highest activity was at the composition of 75 mol% of Pd and 25 mol% of Au. In this case, the layer of Pd shell and the core of Au could be estimated to be ca. 1.1 nm in thickness and ca. 4.2 nm in diameter, respectively, assuming that monodispersed Au/Pd particles of 6.2 nm could be formed. This thickness of Pd shell corresponds to several atomic layers. Based on the quantum sized structure, the ligand effect might be affected by any quantum size effects [44–46] with respect to the optimum size effect of the Pd layer or the Au cluster of the core part.

5.4.4 Use of Templates for Controlling the Size of Sonochemically Formed Metal Particles

In order to prepare functional particles, it is important to control the size of metal particles accurately. To achieve such size controlled synthesis of metal particles, inorganic or organic templates are useful to suppress the growth of metal particles.

Sonochemical reduction processes can be also used to the synthesis of precise size controlled particles in the presence of various types of inorganic or organic templates. Okitsu et al. reported the preparation of Pd clusters from the sonochemical reduction of $[Pd(NH_3)_4]^{2+}$ complexes inside Y-zeolite pore matrix in an aqueous solution containing 2-propanol at 20°C [47]. The zeolite matrix has regularly ordered pore structures in the angstrom range and plays a role as a template to prevent metal growth and metal aggregation.

To understand a heterogeneous synthesis process, the following preliminary experiments were performed in a homogeneous solution: (1) In the absence of zeolite, the sonochemical reduction of $[Pd(NH_3)_4]^{2+}$ to Pd^0 was found to occur, but the rate of $[Pd(NH_3)_4]^{2+}$reduction was much slower than that of $[PdCl_4]^{2-}$reduction. (2) It was confirmed that 2-propanol acted as a precursor to form reductants under ultrasonic irradiation.

Based on the above results, ultrasonic irradiation to ion-exchanged $[Pd(NH_3)_4]^{2+}$-zeolite powders was performed in an aqueous solution containing 2-propanol. The reduction of $[Pd(NH_3)_4]^{2+}$-zeolite to Pd^0-zeolite was confirmed by XPS analyses. However, from XPS depth analyses of the prepared samples, it was suggested that the $[Pd(NH_3)_4]^{2+}$ complexes in the zeolite pore were not sufficiently reduced even in the presence of 2-propanol. Presumably, the reductants formed from 2-propanol sonolysis could not easily diffuse into the zeolite nano-pore (size: 1.2 nm) and/or reductants undergo recombination reactions and quenching reactions with the walls. In addition, the results of XPS spectral analyses of the sonochemically prepared Pd-zeolite powders indicated that the average size of the Pd clusters on the zeolite surface is roughly estimated to be less than 1 nm and composed of several dozen Pd atoms.

By using other templates, the size of metal nanoparticles can be also controlled. Chen et al. reported the sonochemical reduction of Au(III), Ag(I) and Pd(II) and synthesis of Au, Ag and Pd nanoparticles loaded within mesoporous silica [48,49]. Zhu et al. also reported the sonochemical reduction of MnO_4^- to MnO_2 and synthesis of MnO_2 nanoparticles inside the pore channels of ordered mesoporous cabon [50]. Taking into account these reports, the rigid pore of inorganic materials can be used as a template for the size controlled metal nanoparticle synthesis even in the presence of ultrasound.

Compared to inorganic materials, organic materials such as polymers, surfactant molecules and micelles also act as a capping material or soft template. Figure 5.15 shows TEM images of gold nanorods and nanoparticles synthesized by sonochemical reduction of Au(I) in the presence of cetyltrimethylammonium bromide, Ag(I) and ascorbic acid [51].

It can be seen that the shape and aspect ratio of gold nanoparticles formed are strongly affected by the pH value of the solution. The average aspect ratio of gold nanorods decreases with increasing pH value. It is also confirmed by extinction spectrum analyses that optical properties of gold nanorods are closely related to the aspect ratio of gold nanorods. By choosing appropriate organic materials and ultrasonic parameters, the synthesis of various types of shape controlled metal nanoparticles will be achieved.

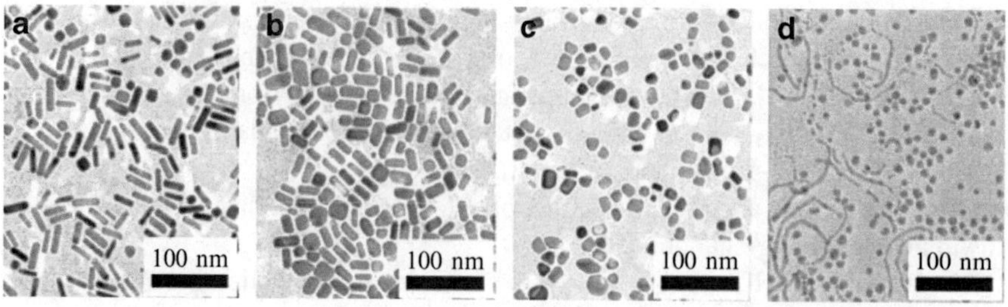

Fig. 5.15 TEM images of gold nanorods and nanoparticles formed in different pH solutions of (**a**) pH 3.5, (**b**) pH 6.5, (**c**) pH 7.7 and (**d**) pH 9.8 after 180 min irradiation under argon [51]

As described in this chapter, the sonochemical reduction technique appears to be a promising method for the preparation of various types of metal nanoparticles in an aqueous solution. By choosing more efficient organic additives, easily-reducible metal precursors, supports and templates with an appropriate role, more advanced functional nanoparticles could be synthesized successfully using the sonochemical reduction technique. In future, it is also possible to develop effective synthetic methods by combining the sonochemical method with other chemical methods.

References

1. Daniel M-C, Astruc D (2004) Gold nanoparticles: assembly, supramolecularchemistry, quantum-size-related properties, and applications toward biology, catalysis, and nanotechnology. Chem Rev 104:293–346
2. Burda C, Chen X, Narayanan R, El-Sayed MA (2005) Chemistry and properties of nanocrystals of different shapes. Chem Rev 105:1025–1102
3. Dahl JA, Maddux BLS, Hutchison JE (2007) Toward greener nanosynthesis. Chem Rev 107:2228–2269
4. Murphy CJ, Gole AM, Hunyadi SE, Stone JW, Sisco PN, Alkilany A, Kinard BE, Hankins P (2008) Chemical sensing and imaging with metallic nanorods. Chem Commun 544–557
5. Wang Z, Ma L (2009) Gold nanoparticle probes. Coordination Chem Rev 253:1607–1618
6. Campelo JM, Luna D, Luque R, Marinas JM, Romero AA (2009) Sustainable preparation of supported metal nanoparticles and their applications in catalysis. ChemSusChem 2:18–45
7. Zijlstra P, Chon JWM, Gu M (2009) Five dimensional optical recording mediated by surface plasmons in gold nanorods. Nature 459:410–413
8. Suslick KS, Fang M, Hyeon T (1996) Sonochemical synthesis of iron colloids. J Am Chem Soc 118:11960–11961
9. Suslick KS, Hyeon T, Fang M (1996) Nanostructured materials generated by high-intensity ultrasound: sonochemical synthesis and catalytic studies. Chem Mater 8:2172–2179
10. Shafi KVPM, Gedanken A (1998) Sonochemical preparation and size-dependent properties of nanostructured $CoFe_2O_4$ particles. Chem Mater 10:3445–3450
11. Xia B, Lenggoro IW, Okuyama K (2001) Novel route to nanoparticle synthesis by salt-assisted aerosol decomposition. Adv Mater 13:1579–1582
12. Xia B, Lenggoro IW, Okuyama K (2002) Nanoparticle separation in salted droplet microreactors. Chem Mater 14:2623–2627

13. Skrabalak SE, Suslick KS (2005) Porous MoS_2 synthesized by ultrasonic spray pyrolysis. J Am Chem Soc 127:9990–9991
14. Dunkel SS, Helmich RJ, Suslick KS (2009) $BiVO_4$ as a visible-light photocatalyst prepared by ultrasonic spray pyrolysis. J Phys Chem C 113:11980–11983
15. Jiang L-P, Xu S, Zhu JM, Zhang JR, Zhu JJ, Chen HY (2004) Ultrasonic-assisted synthesis of monodisperse single-crystalline silver nanoplates and gold nanorings. Inorganic Chem 43:5877–5883
16. Zhou B, Liu B, Jiang LP, Zhu JJ (2007) Ultrasonic-assisted size-controllable synthesis of Bi_2Te_3 nanoflakes with electrogenerated chemiluminescence. Ultrason Sonochem 14:229–234
17. Wu C, Mosher BP, Zeng T (2006) Rapid synthesis of gold and platinum nanoparticles using metal displacement reduction with sonomechanical assistance. Chem Mater 18:2925–2928
18. Wu C, Mosher BP, Zeng T (2008) Chemically-mechanically assisted synthesis of metallic and oxide nanoparticles in ambient conditions. J Nanosci Nanotechnol 8:386–389
19. Jiang L-P, Wang A-N, Zhao Y, Zhang J-R, Zhu J-J (2004) A novel route for the preparation of monodisperse silver nanoparticles via a pulsed sonoelectrochemical technique. Inorg Chem Commun 7:506–509
20. Haas I, Shanmugam S, Gedanken A (2006) Pulsed sonoelectrochemical synthesis of size-controlled copper nanoparticles stabilized by poly(N-vinylpyrrolidone). J Phys Chem B 110:16947–16952
21. Gutierrez MS, Henglein A, Dohrmann JK (1987) H atom reactions in the sonolysis of aqueous solutions. J Phys Chem 91:6687–6690
22. Nagata Y, Watanabe Y, Fujita S, Dohmaru T, Taniguchi S (1992) Formation of colloidal silver in water by ultrasonic irradiation. J Chem Soc Chem Commun 1620–1622
23. Yeung S A, Hobson R, Biggs S, Grieser F (1993) Formation of gold sols using ultrasound. J Chem Soc Chem Commun 378–379
24. Nagata Y, Mizukoshi Y, Okitsu K, Maeda Y (1996) Sonochemical formation of gold particles in aqueous solution. Radiat Res 146:333–338
25. Grieser F, Hobson R, Sostaric J, Mulvaney P (1996) Sonochemical reduction processes in aqueous colloidal systems. Ultrasonics 34:547–550
26. Okitsu K, Bandow H, Maeda Y, Nagata Y (1996) Sonochemical preparation of ultrafine palladium particles. Chem Mater 8:315–317
27. Okitsu K, Mizukoshi Y, Bandow H, Maeda Y, Yamamoto T, Nagata Y (1996) Formation of noble metal particles by ultrasonic irradiation. Ultrason Sonochem 3:S249–S251
28. Caruso RA, Ashokkumar M, Grieser F (2002) Sonochemical formation of gold sols. Langmuir 18:7831–7836
29. Okitsu K, Yue A, Tanabe S, Matsumoto H, Yobiko Y, Yoo Y (2002) Sonolytic control of rate of gold(III) reduction and size of formed gold nanoparticles: relation between reduction rates and sizes of formed nanoparticles. Bull Chem Soc Jpn 75:2289–2296
30. Okitsu K, Nagaoka S, Tanabe S, Matsumoto H, Mizukoshi Y, Nagata Y (1999) Sonochemical preparation of size-controlled palladium nanoparticles on alumina surface. Chem Lett 28:271–272
31. Okitsu K, Yue A, Tanabe S, Matsumoto H (2000) Sonochemical preparation and catalytic behavior of highly dispersed palladium nanoparticles on alumina. Chem Mater 12:3006–3011
32. Maeda Y, Okitsu K, Inoue H, Nishimura R, Mizukoshi Y, Nakui H (2004) Preparation of nanoparticles by reducing intermediate radicals formed in sonolytical pyrolysis of surfactants. Res Chem Intermed 30:775–783
33. Okitsu K, Ashokkumar M, Grieser F (2005) Sonochemical synthesis of gold nanoparticles: effects of ultrasound frequency. J Phys Chem B 109:20673–20675
34. Tronson R, Ashokkumar M, Grieser F (2002) Comparison of the effects of water-soluble solutes on multibubble sonoluminescence generated in aqueous solutions by 20- and 515-kHz pulsed ultrasound. J Phys Chem B 106:11064–11068

35. Mizukoshi Y, Takagi E, Okuno H, Oshima R, Maeda Y, Nagata Y (2001) Preparation of platinum nanoparticles by sonochemical reduction of the Pt(IV) ions: role of surfactants. Ultrason Sonochem 8:1–6
36. He Y, Vinodgopal K, Ashokkumar M, Grieser F (2006) Sonochemical synthesis of ruthenium nanoparticles. Res Chem Intermed 32:709–715
37. Vinodgopal K, He Y, Ashokkumar M, Grieser F (2006) Sonochemically prepared platinum-ruthenium bimetallic nanoparticles. J Phys Chem B 110:3849–3852
38. Mizukoshi Y, Seino S, Okitsu K, Kinoshita T, Otome Y, Nakagawa T, Yamamoto TA (2005) Sonochemical preparation of composite nanoparticles of Au/$\gamma-Fe_2O_3$ and magnetic separation of glutathione. Ultrason Sonochem 12:191–195
39. Mizukoshi Y, Sato K, Konno TJ, Masahashi N, Tanabe S (2008) Magnetically retrievable palladium/maghemite nanocomposite catalysts prepared by sonochemical reduction method. Chem Lett 37:922–923
40. Pol VG, Gedanken A, Calderon-Moreno J (2003) Deposition of gold nanoparticles on silica spheres: a sonochemical approach. Chem Mater 15:1111–1118
41. Sivakumar M, Towata A, Yasui K, Tuziuti T, Kozuka T, Tsujimoto M, Zhong Z, Iida Y (2010) Fabrication of nanosized Pt on rutile TiO_2 using a standing wave sonochemical reactor (SWSR) – observation of an enhanced catalytic oxidation of CO. Ultrason Sonochem 17:213–218
42. Mizukoshi Y, Okitsu K, Yamamoto T, Oshima J, Nagata Y, Maeda Y (1997) Sonochemical preparation of bimetallic nanoparticles of gold/palladium in aqueous solution. J Phys Chem B 101:7033–7037
43. Mizukoshi Y, Fujimoto T, Nagata Y, Oshima R, Maeda Y (2000) Characterization and catalytic activity of core-shell structured gold/palladium bimetallic nanoparticles synthesized by the sonochemical method. J Phys Chem B 104:6028–6032
44. Okitsu K, Murakami M, Tanabe S, Matsumoto H (2000) Catalytic behavior of Au core/Pd shell bimetallic nanoparticles on silica prepared by sonochemical and sol-gel processes. Chem Lett 29:1336–1337
45. Xu C, Lai X, Zajac GW, Goodman DW (1997) Scanning tunneling microscopy studies of the TiO(110) surface: structure and the nucleation growth of Pd. Phys Rev B 56:13464–13482
46. Valden M, Lai X, Goodman DW (1998) Onset of catalytic activity of gold clusters on titania with the appearance of nonmetallic properties. Science 281:1647–1650
47. Okitsu K, Yue A, Tanabe S, Matsumoto H (2002) Formation of palladium nanoclusters on Y-zeolite via a sonochemical process and conventional methods. Bull Chem Soc Jpn 75:449–455
48. Chen W, Cai W, Lei Y, Zhang L (2001) A sonochemical approach to the confined synthesis of palladium nanoparticles in mesoporous silica. Mater Lett 50:53–56
49. Chen W, Zhang J, Cai W (2003) Sonochemical preparation of Au, Ag, Pd/SiO_2 mesoporous nanocomposites. Scripta Materialia 48:1061–1066
50. Zhu S, Zhou H, Hibino M, Honma I, Ichihara M (2005) Synthesis of MnO_2 nanoparticles confined in ordered mesoporous carbon using a sonochemical method. Adv Funct Mater 15:381–386
51. Okitsu K, Sharyo K, Nishimura R (2009) One-pot synthesis of gold nanorods by ultrasonic irradiation: The effect of pH on the shape of the gold nanorods and nanoparticles. Langmuir 25:7786–7790

Chapter 6
Sonochemical Preparation of Monometallic, Bimetallic and Metal-Loaded Semiconductor Nanoparticles

Sambandam Anandan and Muthupandian Ashokkumar

Abstract A convenient method to synthesize metal nanoparticles with unique properties is highly desirable for many applications. The sonochemical reduction of metal ions has been found to be useful for synthesizing nanoparticles of desired size range. In addition, bimetallic alloys or particles with core-shell morphology can also be synthesized depending upon the experimental conditions used during the sonochemical preparation process. The photocatalytic efficiency of semiconductor particles can be improved by simultaneous reduction and loading of metal nanoparticles on the surface of semiconductor particles. The current review focuses on the recent developments in the sonochemical synthesis of monometallic and bimetallic metal nanoparticles and metal-loaded semiconductor nanoparticles.

6.1 Introduction

Materials science goes back to prehistoric times, where people started to utilize rocks, bones, leather, and other materials they found in nature to fabricate tools and clothing. Later, the knowledge evolved and metals, alloys, ceramics, and fabrics replaced the older materials with inferior properties. In recent times, the knowledge of materials and processing improved further and more advanced materials for more sophisticated (or fashionable) applications have become available. The synthesis of inorganic nanomaterials of specific composition and size is a burgeoning area of materials science research.

In recent years, "nano" has become a hot topic in materials science. The common theme here is always "size matters" and over the last decades, researchers have developed exciting new materials with unique and tunable properties. In this

M. Ashokkumar (✉)
School of Chemistry, University of Melbourne, Vic 3010, Australia
e-mail: masho@unimelb.edu.au

Pankaj and M. Ashokkumar (eds.), *Theoretical and Experimental Sonochemistry Involving Inorganic Systems*, DOI 10.1007/978-90-481-3887-6_6,

regard, the interest in using nanocrystalline inorganic materials as building blocks for new nanoscale devices has spawned an enormous research effort aimed at developing selective syntheses of inorganic nanocrystals [1]. Further advances in synthetic techniques have made it possible to produce high-quality samples of metal and semiconductor nanoparticles in a variety of different shapes such as, rods [2–4], triangles [5, 6], cubes and boxes [7] and branched structures [8, 9]. The properties of these particles in applications as diverse as catalysts, sensors and medicine depend critically on the size and composition of the nanoparticles [10–14]. In this regard, significant progress has been made in controlling the chemistry of nanoscale solid-state materials and in understanding their unique size-dependent properties. To attain such properties, inorganic nanomaterials have been synthesized by a variety of methods, including physical evaporation, laser ablation, chemical vapor deposition and solvothermal reactions [15–17]. The mechanism of formation mostly involves either a vapor solid (VS) or vapor-liquid-solid (VLS) growth process. In the vapor solid process, chemical reduction or reaction generates a vapor species, which is transported and condensed on to a substrate whereas in the vapor-liquid-solid process, the growth is promoted by a liquid-solid interface. Recently, high power ultrasound has attracted some interest due to its application for the preparation of inorganic nanostructured materials by acoustic cavitation [18–20].

The chemical effects of ultrasound are caused by the phenomenon of acoustic cavitation which involves the nucleation, growth and collapse of bubbles in a liquid. Cavitation occurs whenever a new surface, or cavity, is created within a liquid [21]. A cavity is any bounded volume, whether empty or containing gas or vapor, with at least part of the boundary being liquid. The collapse of the bubbles induces localized extreme conditions: high temperature, high pressure, etc. It has been reported that the gaseous contents of a collapsing cavity reach temperatures of 5,000°C, and the liquid immediately surrounding the cavity reaches 2,100°C [22, 23]. The pressure is estimated to be about 500 atm., resulting in the formation of transient supercritical water. Thus, cavitation serves as a means of concentrating the diffuse energy of sound into microreactors. Even though the local temperature and pressure conditions created by the cavity implosion are extreme, chemists have managed to use these conditions for achieving useful chemical reactions.

The intensity of cavity implosion, and hence the nature of the reaction, are controlled by such factors as acoustic frequency, acoustic intensity, bulk temperature, static pressure, and the choice of liquid or dissolved gas [22]. The consequences of these extreme conditions are the cleavage of dissolved water molecules into $^{\bullet}H$ atoms and $^{\bullet}OH$ radicals. From the reactions of these entities ($^{\bullet}H$, $^{\bullet}OH$) with each other and with H_2O and O_2 during the quick cooling phase, $HO_2^{\bullet}$ radicals and H_2O_2 are formed. In this molecular environment, organic compounds are decomposed and inorganic compounds are oxidized or reduced. Ultrasound has been widely known to induce radical reactions [24, 25]. This useful property has found its applications in the sonolytic degradation of aqueous organic pollutants [26], and sonochemical synthesis of inorganic nanomaterials [27]. There are a number of review articles available on the sonochemical synthesis of metal nanoparticles [28, 29]. Hence, we have taken due care not to reproduce a similar review article. Our focus in this review is on the recent development on the sonochemical synthesis of different inorganic nanomaterials. Particular attention is given to the size-control of the metal nanoparticles, synthesis of bimetallic nanoparticles and metal-loaded semiconductor nanoparticles. The photocatalytic efficiency of the sonochemically synthesized metal particles-loaded semiconductors has also been discussed.

6.2 Monometallic Nanoparticles

Nanoparticles have been used since time immemorial, predominantly as artifacts until mid nineteenth century, when Faraday studied colloidal gold in an attempt to explain their optical properties [30]. Then, Brust et al. [31, 32] have in the early 1990s triggered an immense interest in gold nanoparticles, which is reflected in the dramatic increase in publications and patents. Although gold nanoparticles can be prepared by various methods, the sonochemical route offers several advantages over other synthetic processes. Yeung et al. [33] produced gold sols using ultrasound by employing propan-2-ol to capture $^{\bullet}H$ and $^{\bullet}OH$ radicals, based on the common procedures used in conventional radiation chemistry. The reactions that lead to the formation of colloidal gold are shown in Reactions (6.1)–(6.4).

$$H_2O + \text{sonication})))))\rightarrow {}^{\bullet}H + {}^{\bullet}OH \tag{6.1}$$

$$^{\bullet}H\ ({}^{\bullet}OH) + (CH_3)_2CHOH \rightarrow (CH_3)_2COH + H_2(H_2O) \tag{6.2}$$

$$3(CH_3)_2COH + AuCl_4^- \rightarrow Au + 3(CH_3)_2CO + 4Cl^- + 3H^+ \tag{6.3}$$

$$nAu \rightarrow (Au)_n \tag{6.4}$$

The formation of Au nanoparticles can be easily monitored by following the appearance of a surface plasma resonance band around 520–540 nm (Fig. 6.1). Yeung et al. [33] observed that the efficiency of gold particle formation was different in different alcohols (n-pentanol > propan-2-ol > methanol). This is due to the air/water surface activity of the alcohols and the ability of the solute to scavenge the primary OH radicals at the bubble/liquid interface.

In addition to the above reactions, some of the pyrolysis intermediates and products can also act as reductants (Reaction 6.5) during sonication. Such reducing species subsequently react with Au^{3+} to form gold atoms (Reaction 6.6), which can agglomerate to produce gold seed particles.

$$(CH_3)_2CHOH + \text{sonication})))))\rightarrow \text{Pyrolysis intermediates and products} \tag{6.5}$$

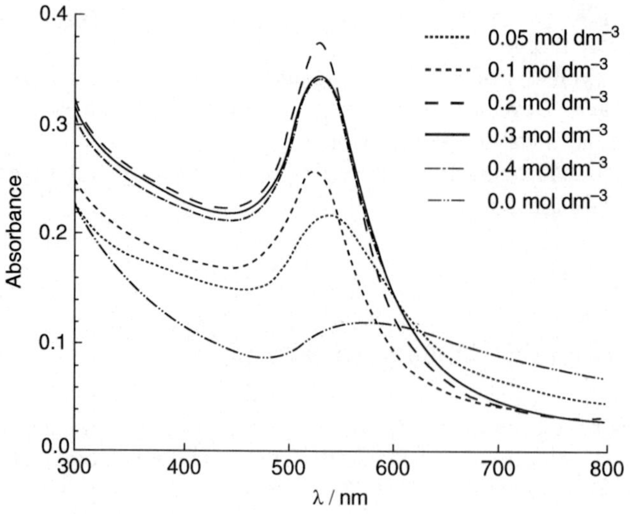

Fig. 6.1 Absorption spectra of gold sols generated from solutions containing 1×10^{-4} mol dm^{-3} $AuCl_4^-$, 0.1 wt% PVP, and various concentrations of propan-2-ol [31]

$$Au^{3+} + {}^{\bullet}H\ /{}^{\bullet}ROH/\text{other reducing species} \rightarrow Au^0 + 3H^+ + \text{ products} \quad (6.6)$$

To obtain stable gold seed particles in a colloidal state, stabilizing agents such as citric acid was added to the above reaction mixture in addition to propan-2-ol [34]. The size and distribution of the Au nanoparticles produced can also be correlated with the rate of Au(III) reduction, which in turn is influenced by the applied frequency. Okitsu et al. [35] performed systematic studies by varying the ultrasound frequency and suggested that maximum amount of reducing radicals (maximum rate of Au (III) reduction) were generated at around 213 kHz in the presence of 1-propanol. They observed that the ultrasound frequency significantly affected the size and size distribution of Au particles, indicating that the frequency is an important parameter that controls the size of the formed particles. In addition, the particle size can also be found to be inversely related to the rate of Au(III) reduction: the size of the particles decreased with an increase in the rate of reduction.

In addition to Au, nanoparticles of other noble metals have also been prepared by the sonochemical technique. Nagata et al. [36] prepared stable colloidal dispersions of silver by the ultrasonic irradiation of aqueous $AgClO_4$ or $AgNO_3$ solutions in the presence of surfactants. In their proposed mechanism the surfactant is active as a reducing agent. Anandan et al. [37] prepared monometallic pure silver nanoparticles from an aqueous solution of $AgNO_3$ (0.2 mM) containing polyethylene glycol (0.1 wt%) and ethylene glycol (0.1 M) at room temperature under argon atmosphere. The solution containing silver nanoparticles was yellow in color and the absorption spectrum showed a band ($\lambda_{max} = 437$ nm) in the visible region

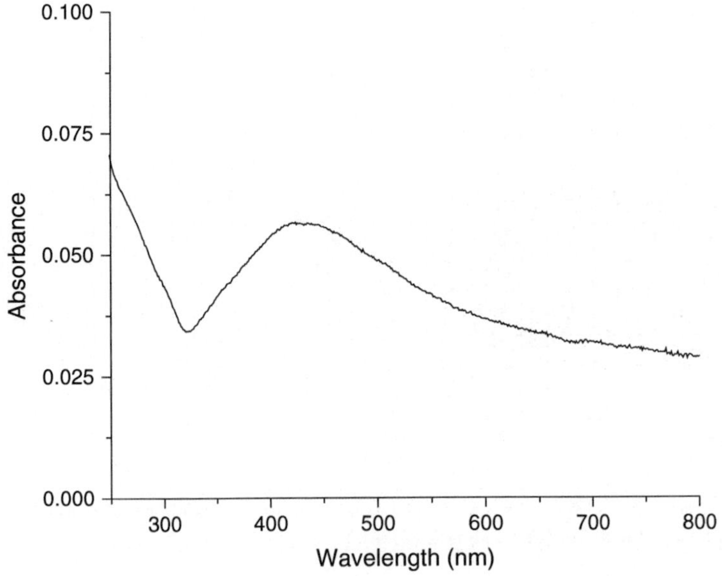

Fig. 6.2 Absorption spectrum of silver nanoparticles prepared from an aqueous solution (70 ml) of $AgNO_3$ (0.2 mM) containing polyethylene glycol (0.1 wt%) and ethylene glycol (0.1 M)

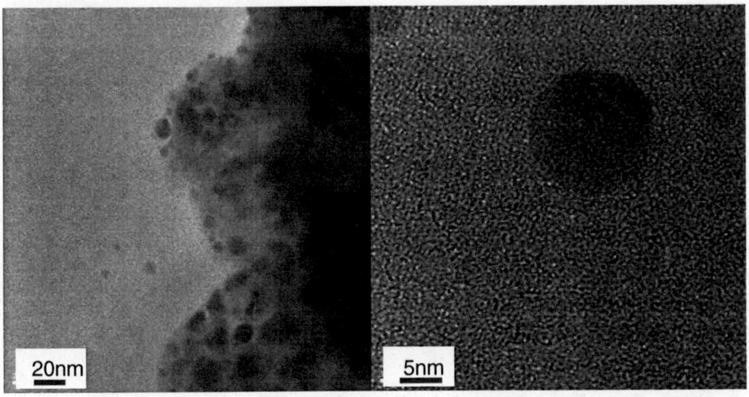

Fig. 6.3 TEM images of monometallic silver nanoparticles prepared from an aqueous solution (70 ml) of $AgNO_3$ (0.2 mM) containing polyethylene glycol (0.1 wt%) and ethylene glycol (0.1 M) [37]

(Fig. 6.2). TEM observations for the as-prepared, silver nanoparticles showed that the nanoparticles of silver were ca.15 nm in size (Fig. 6.3).

Sonochemical reduction processes of Pt(IV) ions in the presence of anionic, cationic or non-ionic surfactants was investigated by Mizukoshi et al. [38]. During the processes, the color of the aqueous solution containing H_2PtCl_6 and surfactants

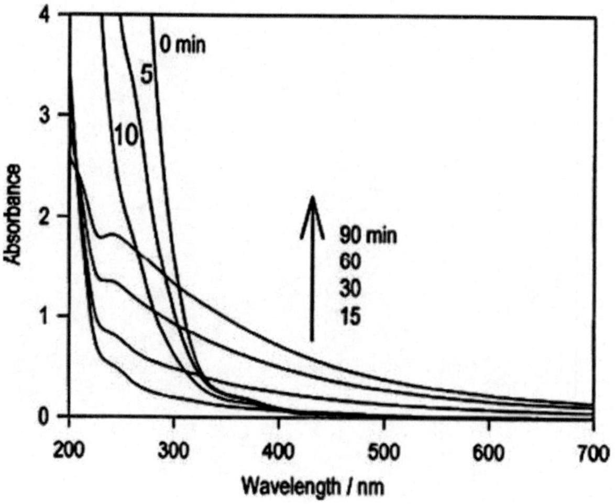

Fig. 6.4 Change in the absorption spectra of aqueous solutions containing 1 mM of H_2PtCl_6 and 8 mM of SDS as a function of sonication time [38]

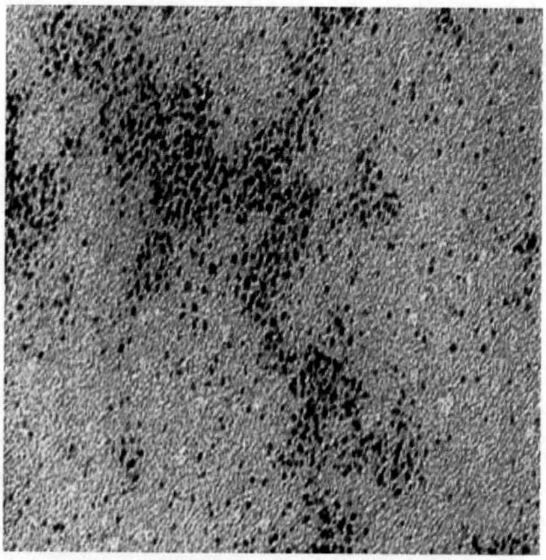

Fig. 6.5 TEM image of Pt nanoparticles synthesized following 3 h of simultaneous sonication of 1mM Pt(II) at 213 kHz. The accelerating TEM voltage was 100 kV [39]

turned from pale yellow, originated from $PtCl_6^{2-}$ complex, into dark brown originated from Pt^0. Figure 6.4 shows the change in the absorption spectra containing 1 mM of H_2PtCl_6 and 8 mM of sodium dodecylsulfate (SDS), indicating the formation of platinum nanoparticles. The absorption in the UV region gradually decreased followed by an increase in the absorbance in the longer wavelength region. Similarly, Vinodgopal et al. [39] applied ultrasound to a tetrachloroplatinate solution containing 8 mM SDS, 200 mM propanol, and 0.1M $HClO_4$ to produce colloidal platinum. TEM observations for the as-prepared, platinum nanoparticles were ca.15 nm in size (Fig. 6.5).

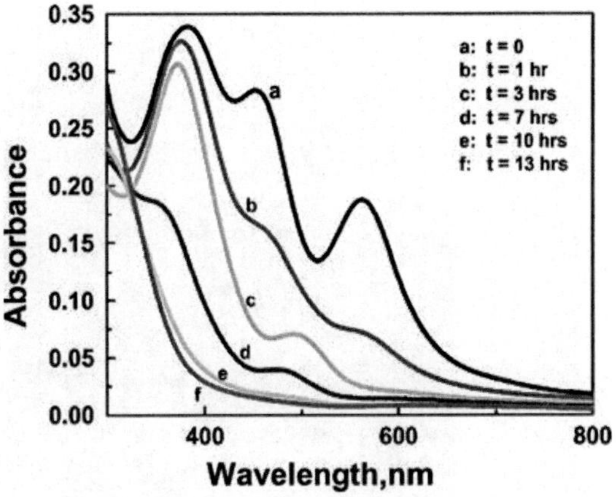

Fig. 6.6 Changes in the absorption spectra of aqueous solutions of 1 mM $RuCl_3$ containing 0.1 M $HClO_4$, 8 mM SDS and 0.2 M propanol as a function of sonication time: (**a**) $t = 0$, (**b**) $t = 1$ h, (**c**) $t = 3$ h, (**d**) $t = 7$ h, (**e**) $t = 10$ h and (**f**) $t = 13$ h. The ultrasound frequency used was 213 kHz [44]

Stable palladium nanoparticles were sonochemically synthesized by Nemamcha et al. [40] starting from palladium (II) nitrate in aqueous solution with ethylene glycol and poly(vinylpyrrolidone) as reducing and stabilizing agents, respectively. The irradiation of palladium (II) nitrate solution using a 50 kHz ultrasonic generator for 180 min changed the color of the solution from an initial pale yellow to a dark brown indicating the formation of palladium nanoparticles.

The synthesis of ruthenium nanoparticles using sonochemistry is however not as facile as other noble metals. The formation of ruthenium nanoparticles took relatively a longer time (13 h). Due to the slow reduction process, the size of Ru particles generated was larger when compared to that of other metal nanoparticles. The reason for this is that the reduction of Ru(III) is a very complex process and a complete reduction of Ru(III) is often difficult to achieve. The sonochemical reduction of Ru^{3+} ions was monitored using UV-visible absorption spectroscopy. The yellow Ru^{3+} solution before sonication showed three absorption bands centered at 560, 450 and 400 nm (Fig. 6.6). Upon ultrasound irradiation, the Ru^{3+} species underwent sequential one-electron reduction to produce Ru^0 [39, 41–43], which was identified by the disappearance of the above mentioned absorption bands. This reduction was slow when PVP was used as a stabilizer [44] whereas it required only 7 h when polyethylene glycol (PEG) was used as a stabilizer [45]. This is because of the dual role of PEG: as a reducing agent and as a stabilizer. The TEM image of the prepared Ru^0 nanoparticles using PVP as the stabilizer is shown in Fig. 6.7.

6.3 Bimetallic Nanoparticles

Recently, metal cluster catalysts composed of two different metallic elements are of interest from both scientific and technological points of view because of their interesting physiochemical properties [46, 47]. Bimetallic clusters are known to exhibit specific reactivity. Their catalytic efficiency is also controlled by their size.

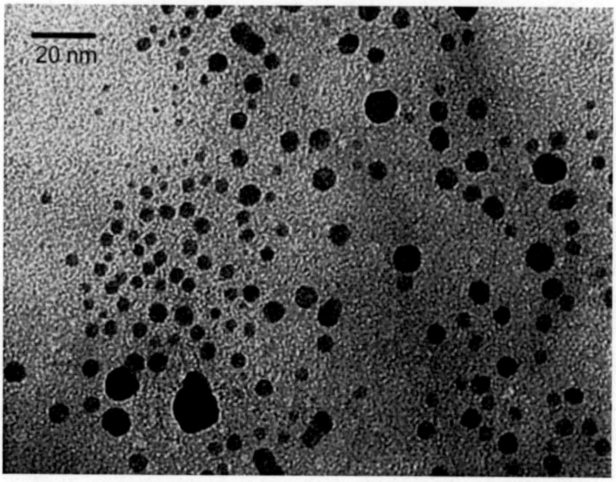

Fig. 6.7 TEM image of Ru nanoparticles synthesized following 13 h of sonication at 213 kHz [44]

One of the most important factors that control the reactivity of metal clusters is the nucleation and growth mechanism of the clusters [48–51]. It seems that most of the features presented by the final clusters depend on the synthetic conditions under which clusters are formed during the early steps of the metal ion reduction. Therefore, it is of interest to develop new and effective preparation methods. Various methods of preparation have been reported; chemical reduction [52], photolytic reduction [53], radiolytic reduction [54] and metal evaporation [55]. As mentioned earlier, the sonochemical synthesis of nanometric materials [29, 30, 56] has been found to be unique due to the unusual experimental conditions generated during acoustic cavitation [57]. Suslick et al. [58] were among the first to use ultrasound to synthesize bimetallic particles in an organic solvent. They used 20 kHz ultrasound to synthesize Fe-Co alloys from non-aqueous solutions containing $Fe(CO)_5$ and $Co(CO)_3(NO)$ complexes. Mizukoshi et al. [59, 60] have synthesized bimetallic particles of gold and palladium with a narrow size distribution by sonochemical reduction of an aqueous solution containing both Au(III) and Pd(III) ions. They reported core-shell morphology for the resulting nanoparticles with a Pd shell over the Au core. Kan et al. [61] synthesized bimetallic particles with Au core and Pd shell by successive sonochemical reduction of the precursors $Pd(NO_3)_2$ and $HAuCl_4$ in ethylene glycol.

Harada et al. [62] achieved Pd core-Au shell nanoparticles by a co-reduction method. The difference in the structure was argued to be due to the difference in the reduction potentials of Pd and Au ions. When Au ions were added in the presence of Pd nanoparticles, some Pd^0 atoms of the nanoparticles were oxidized and reduced Au^{III} ions, the oxidized Pd ions were reduced again by the reductants, such as, alcohols. This process led to the formation of particles with core-shell structure in the co-reduction method.

Anandan et al. [37] reported the sonochemical synthesis of gold-silver bimetallic nanoparticles with core-shell geometry by the sonochemical co-reduction of Au and

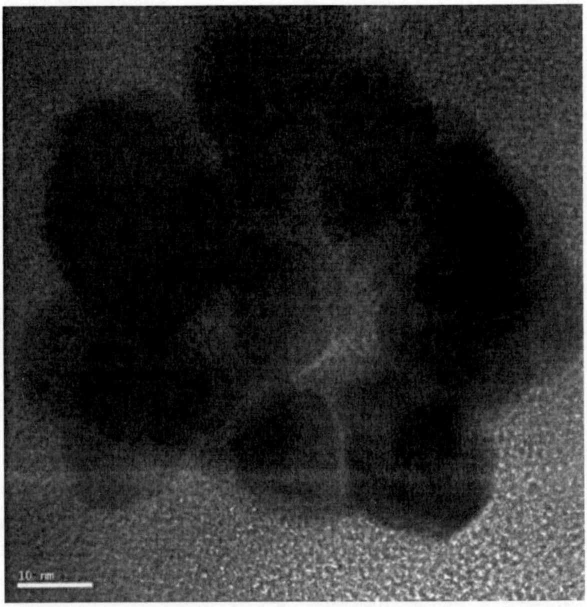

Fig. 6.8 TEM image of sonochemically co-reduced Au-Ag core-shell bimetallic nanoparticles [37]

Ag ions using 20 kHz ultrasound in aqueous solutions. An aqueous solution (70 ml) of $HAuCL_4.3H_2O$ (0.2 mM) and $AgNO_3$ (0.2 mM) containing polyethylene glycol (PEG, 0.1 wt%) and ethylene glycol (0.1 M) was sonicated at room temperature under argon atmosphere for 30 min. TEM image (Fig. 6.8) provided support for the formation of core-shell nanoparticles, gold core with silver shell upon simultaneous sonochemical irradiation of gold and silver ions. In order to understand the core-shell structure formation process, Anandan et al. [37] monitored the growth rates of individual metal nanoparticles under similar sonochemical irradiation conditions. The growth rate of Au was followed by monitoring the absorption changes at 530 nm (Au plasmon absorption band) and at 265 nm ($AuCl_4^-$ absorption band). The rate of reduction of Ag^+ was monitored at 303 nm. A comparison of the rate of formation of individual metal clusters indicated (Fig. 6.9) that the rate of formation of gold nanoparticles was higher compared to that of silver nanoparticles. What these results suggest is that the formation of core-shell morphology might be due to the difference in the reduction rates of the individual metal ions. Since gold ions were easily reduced under the sonochemical conditions, it was suggested that gold particles were first generated followed by the reduction of Ag ions on the surface of the gold particles.

Sathish Kumar et al. [45] prepared bimetallic Au-Ru nanoparticles by the simultaneous reduction of both Au^{3+} and Ru^{3+} ions by ultrasound irradiation at three different molar ratios (Au^{3+}:Ru^{3+}; 1:1, 1:3 and 1:5) in 4 h in the presence of PEG. A significant change in the absorption as a function of sonication time was observed for Au-Ru bimetallic particles (Fig. 6.10), which indicated the

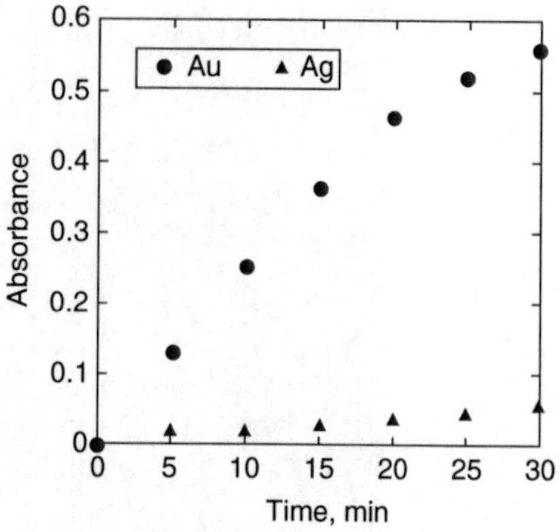

Fig. 6.9 The absorbance versus time plot indicates that the rate of Au nanoparticle formation is significantly greater than the rate of Ag nanoparticle formation [37]

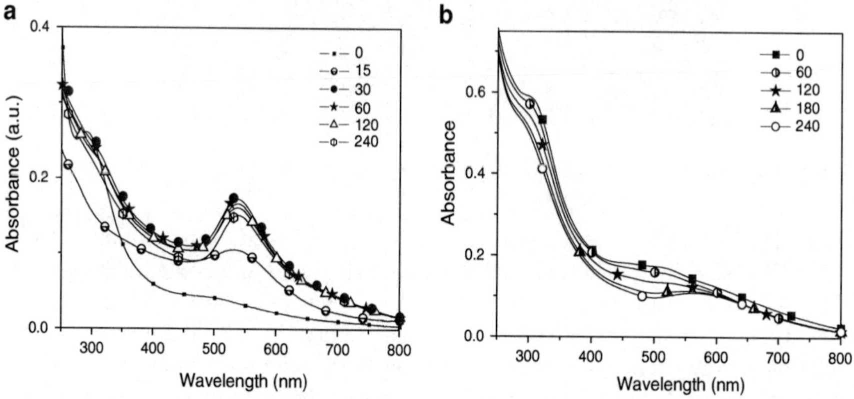

Fig. 6.10 UV-vis absorption spectra of gold – ruthenium bimetallic nanoparticles prepared by the sonochemical co-reduction method using (**a**) 1:1 and (**b**) 1:5 gold – ruthenium compositions, respectively [45]

simultaneous reduction of both Au^{3+} and Ru^{3+} ions. For 1:1 molar ratio sample, the absorption maximum at 536 nm corresponds to the plasmon absorption of gold particles. It can be noticed in Fig. 6.10a that the absorption at 536 nm increases for the first 30 min and then slightly decreases. The increase in absorption was due to the initial formation of gold nanoparticles and the consecutive decrease in absorption was ascribed to the formation of a ruthenium layer on the surface of gold nanoparticles. The reason for the formation of Au particles prior to Ru particles was due to the ease of reduction of gold ions compared to that of Ru ions. The Au particles may also catalyze the reduction of Ru ions leading to the formation of a Ru shell on the surface of the gold particles. Similar absorption characteristics were also observed with 1:3 molar ratio sample. However, the absorption characteristics were significantly different for the 1:5 molar ratio sample (Fig. 6.10b). This was due to the higher ruthenium ion concentration.

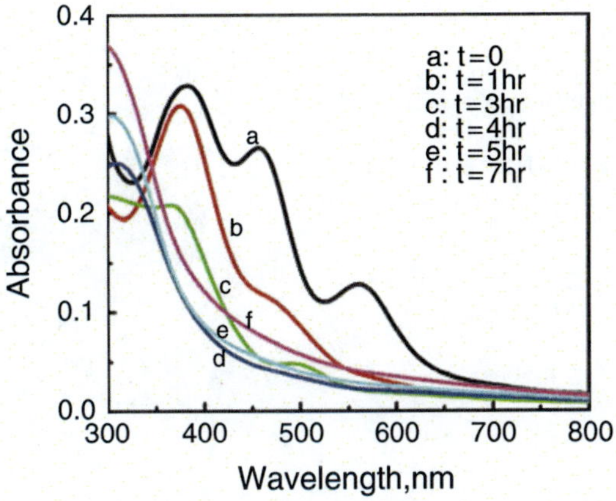

Fig. 6.11 Absorption spectra showing the simultaneous sonochemical reduction of an aqueous solution of 1 mM $RuCl_3$ and 1 mM K_2PtCl_4 containing 0.1 M $HClO_4$, 8 mM SDS, and 0.2 M propanol under argon atmosphere (**a**) $t = 0$. (**b**) $t = 1$ h. (**c**) $t = 3$ h (**d**) $t = 4$ h. (**e**) $t = 5$ h. (**f**) $t = 7$ h. The ultrasound frequency used was 213 kHz. [39]

Among the bimetallic system, bimetallic materials of Pt and Ru have a long history of applications in electrochemistry, primarily for the promotion of CO and CH_3OH oxidation in support of low temperature fuel cell development [63, 64]. In this regard, Vinodgopal et al. [39] synthesized Pt-Ru bimetallic nanoparticles by ultrasonic irradiation of an aqueous solution of Pt(II) and Ru(III) at a frequency of 213 kHz. Figure 6.11 shows the absorption spectra of an aqueous solution containing 1 mM Pt^{2+} and Ru^{3+} along with 8 mM SDS (stabilizer), 200 mM propanol and 0.1 M $HClO_4$ as a function of sonication time. The initial spectrum at time t = 0 shows prominent absorption bands centered at 560, 450 and 400 nm in the visible region of the spectrum arising from Ru(III). However, when sonicated simultaneously with Pt(II), the reduction rate for the ruthenium ions was quite fast. Within 4 h, nearly all of the above mentioned absorption bands disappeared, and a clear brown solution was obtained. The accelerated reduction rate of Ru(III) was most likely due to the excess electrons on the platinum surface donated by the reducing radicals produced by sonolysis. The change in the absorption spectrum also indicated that the reduction of the Ru(III) occured simultaneously with the reduction of the Pt (II). TEM image (Fig. 6.12) showed particles with different sizes (the large ones presumably ruthenium particles and the ultrasmall ones platinum). However, sequential sonolysis (platinum was first reduced followed by ruthenium reduction) yielded core-shell structure, i.e., a platinum core with a ruthenium shell around it (Fig. 6.13). Rukma Basnayake et al. [65] prepared nanometer scale Pt-Ru bimetallic electrocatalysts rich in Ru and possessing alloy properties in order to achieve high CH_3OH oxidation rates through the sonochemical method.

6.4 Metal-Loaded Semiconductor Nanoparticles

The development of combinations of two or more functional phases is one of the most promising approaches to develop novel materials with specific functional properties. It is known that the immobilization of nanosized noble metal particles

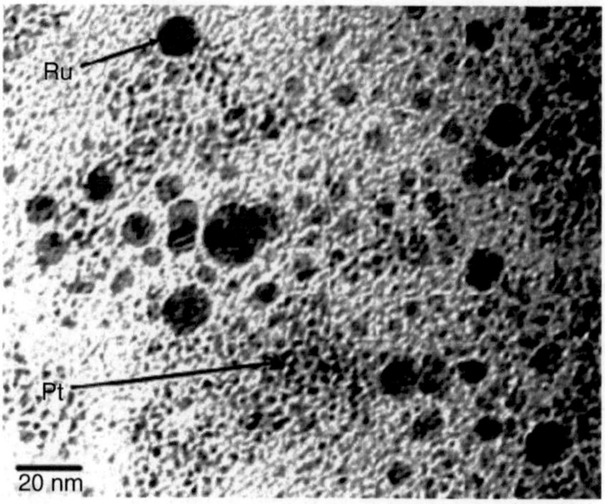

Fig. 6.12 TEM image of Pt-Ru nanoparticles synthesized following 7 h of simultaneous sonication of 1 mM Pt(II) and 1 mM Ru(III) at 213 kHz. The accelerating TEM voltage was 100 kV [39]

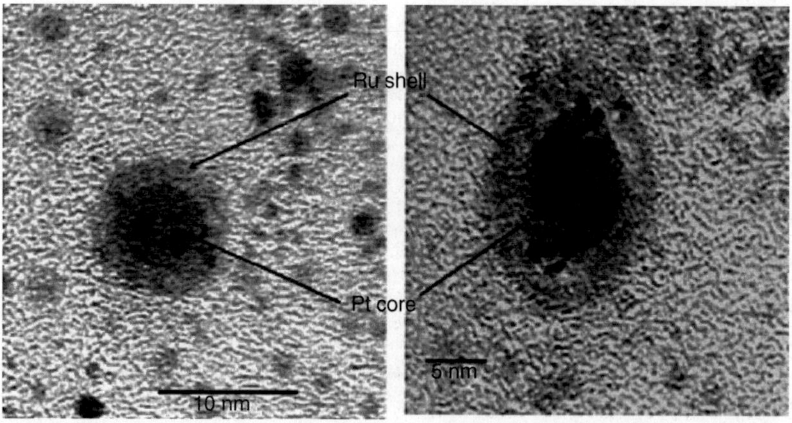

Fig. 6.13 TEM images obtained at an accelerating voltage of 200 kV of Pt-Ru nanoparticles synthesized following sequential sonication of 1 mM Pt(II) and 1 mM Ru(III) at 213 kHz. Two representative particles are shown at different magnifications [39]

on semiconductor enhances their photocatalytic activities [66–68]. Thus, the study of metal-semiconductor nanocomposites has undergone enormous development over the last few decades. This is partly due to the pioneering work of Honda and Fujishima [69] where they demonstrated the splitting of a water molecule into hydrogen and oxygen using platinum and TiO_2 electrode. Further, semiconductor–metal composites have also been used for new processes such as alcohol oxidation to CO [70], urea synthesis [71], gas phase catalytic oxidation (trichloroethylene over TiO_2) [72], and dye degradation [73–78]. These interesting observations have prompted interest in the study of nanocomposites materials involving semiconductor and metals.

Of the various semiconductors tested to date, TiO_2 is the most promising photocatalyst because of its appropriate electronic band structure, photostability, chemical inertness and commercial availability. But currently, a variety of nanostructured TiO_2 with different morphologies including nanorods, nanowires, nanostructured films or coatings, nanotubes, and mesoporous/nanoporous structures have attracted much attention.

Recently, sonochemical processing has been proven to be a useful technique in the synthesis of above composite materials. Utilizing the sonochemical synthetic method, Wang et al. [79] achieved the selective synthesis of anatase and rutile structures of TiO_2 (wormhole-like framework structures) by using titanium isopropoxide and titanium tetrachloride as the precursors. Yu et al. [80] developed a novel method for the preparation of highly photoactive nanometer-sized TiO_2 photocatalyst with anatase and brookite phases by hydrolysis of titanium isopropoxide in pure water or a 1:1 $EtOH/H_2O$ solution. Further, they observed higher photocatalytic activity with this sample (prior to calcination) while compared to that commercial photocatalyst (Degussa P25).

The synthesis of metal-loaded semiconductor oxide materials by conventional physical blending or chemical precipitation followed by surface adsorption usually yields insoluble materials for which control over the size, morphology and dispersion of the metal component still remain inherently difficult. Another approach is based on the mixing of surfactant-capped metal and oxide nanocrystals synthesized separately. A drawback of this method is that organic ligands on the particle surface can block a number of active sites and/or can be susceptible to severe degradation under photolytic conditions. On the other hand, the sonochemical method possesses many advantages: (1) in situ formation of reducing species via acoustic cavitation, (2) effective stirring due to acoustic streaming which enables the preparation of nanoparticles as well as immobilization of the metal nanoparticles on supports and (3) finally it does not also requires any post treatment.

Anandan and Ashokkumar [81] synthesized Au-TiO_2 nanophotocatalysts by three different procedures with and without ultrasound and studied their photocatalytic and sonophotocatalytic degradation efficiencies for a representative organic pollutant, nonylphenol ethoxylate (NPE) surfactant in aqueous solutions. In the first procedure, Au–TiO_2 nanoparticles were prepared by depositing sonochemically synthesized gold nanoparticles on Degussa P25 TiO_2 by stirring in the absence of an ultrasonic field. In the second procedure, Au nanoparticles were sonochemically synthesized and simultaneously deposited on Degussa P25 TiO_2 particles (i.e., sonochemically synthesized Au particles were deposited on commercial TiO_2 with the assistance of low-frequency ultrasound irradiation). In the third procedure, Au-TiO_2 nanoparticles were sonochemically synthesized by the simultaneous irradiation of an aqueous solution containing $AuCl_4^-$ and titanium tetraisopropoxide. Transmission electron micrographs of the Au-TiO_2 nanoparticles prepared by the above three methods show a variation in Au particle size in the range of 3–20 nm (Fig. 6.14). In the conventional physical blending method of Au and TiO_2 nanoparticles (first procedure), Au particle size had been slightly increased due to the agglomeration of Au nanoparticles on TiO_2 surface during the drying/sintering

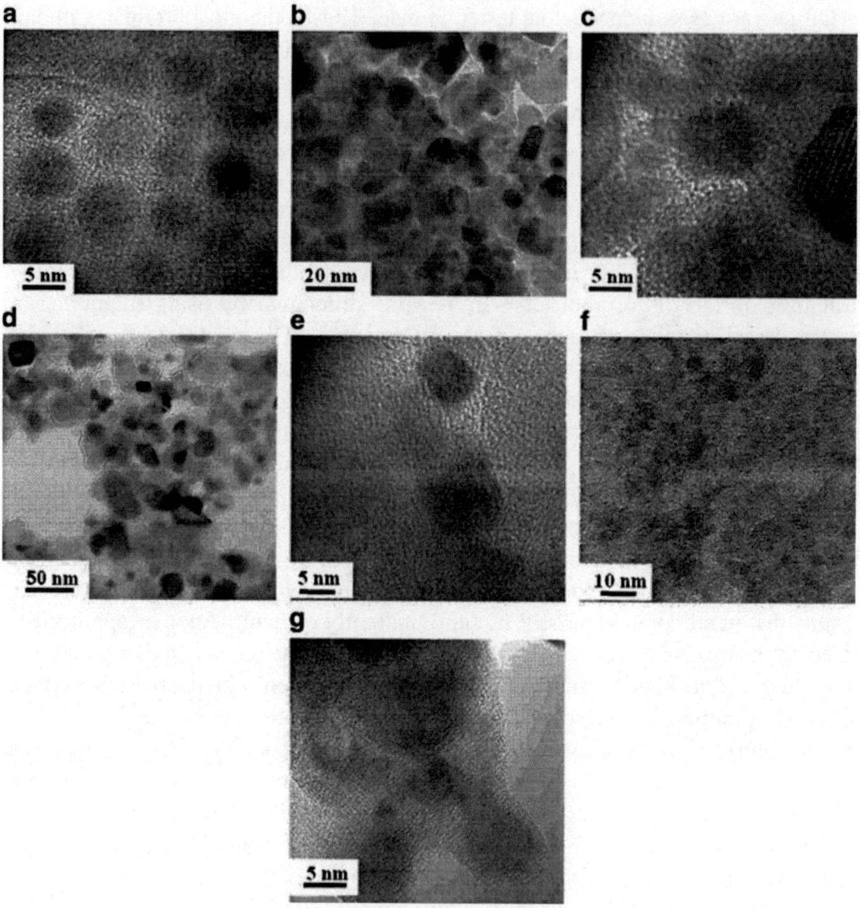

Fig. 6.14 TEM image of (**a**) sonochemically synthesized gold nanoparticles, (**b**) conventional Au–TiO_2 nanoparticles (using first procedure), (**c**) HRTEM image of conventional Au–TiO_2 nanoparticles, (**d**) Au–TiO_2 nanoparticles by ultrasound (using second procedure), (**e**) HRTEM image of Au–TiO_2 nanoparticles by ultrasound, (**f**) Au–TiO_2 nanocolloids (using third procedure) and (**g**) HRTEM image of Au–TiO_2 nanocolloids [81]

process. For the simultaneous reduction/deposition of Au nanoparticles on commercial TiO_2 powder (second procedure) resulted in a slightly larger (>20 nm) Au nanoparticles on the surface of TiO_2 particles. A possible reason for the substantial increase in the Au particle size might be due to the coalescence of gold nanoparticles on the surface of semiconductor particles. The Au-TiO_2 nanoparticles prepared by the simultaneous ultrasonication of titanium tetraisopropoxide and gold chloride ions (that is according to third procedure) clearly showed that 2–3 nm gold nanoparticles were immobilized on the surface of TiO_2 particles without any aggregation. The formation/deposition of significantly smaller Au nanoparticles could be attributed to the in situ reduction of $AuCl_4^-$ on the surface of TiO_2

particles during the sonication process. Similarly, Mizukohsi et al. [82] attained higher photocatalytic activities by immobilizing Pt and Pd nanoparticles onto TiO_2 by sonochemical processes for the generation of H_2 production from aqueous ethanol solutions. Thus, sonochemical method enables to prepare metal nanoparticle-supported TiO_2 photocatalysts possessing high photocatalytic activities.

6.5 Summary

In this review, the potential uses of sonochemistry for the preparation of mono-metallic and bimetallic metal nanoparticles and metal-loaded semiconductor nano-particles have been highlighted. While specific examples available in the literature were discussed, the sonochemical technique seems to offer a platform technique that could be used for synthesizing a variety of functional materials. Most of the studies to date deal with laboratory scale "exploration", it would be ideal if the concepts are tested under large scale experimental conditions involving specific applications. The authors sincerely hope that the information provided in this review would prompt such experimental investigation in a new dimension.

Acknowledgment The research described herein was supported by the Department of Innovation, Industry, Science and Research (DIISR), Australia and Department of Science and Technology (DST), India.

References

1. Murray CB, Kagan CR, Bawendi MG (2000) Synthesis and charcterization of monodisperse nanocrystals and close-packed nanocrystal assemblies. Annu Rev Mater Sci 30:545–610
2. See articles in the (2006) special issue on Anisotropic Nanomaterials. J Mater Chem 16
3. Murphy CJ, San TK, Gole AM, Orendorff GJX, Gou L, Hunyadi SE, Li T (2005) Anisotropic metal nanoparticles: synthesis, assembly, and optical applications. J Phys Chem B 109: 13857–13870
4. Perez-Juste J, Pastoriza-Santos I, Liz-Marzan LM, Mulvaney P (2005) Gold nanorods: synthesis, characterization and applications. Coord Chem Rev 249:1870–1901
5. Jin RC, Cao YW, Mirkin CA, Kelly KL, Schatz GC, Zheng JG (2001) Photoinduced conversion of silver nanospheres to nanoprisms. Science 294:1901–1903
6. Callegari A, Tonti D, Chergui M (2003) Photochemically grown silver nanoparticles with wavelength-controlled size and shape. Nano Lett 3:1565–1568
7. Sun YG, Xia YN (2002) Shape-controlled synthesis of gold and silver nanoparticles. Science 298:2176–2179
8. Manna L, Milliron DJ, Meisel A, Scher EC, Alivisatos AP (2003) Controlled growth of tetrapod-branched inorganic nanocrystals. Nat Mater 2:382–385
9. Kuno M, Ahmed O, Protasenko V, Bacinello D, Kosel TH (2006) Solution-based straight and branched CdTe nanowires. Chem Mater 18:5722–5732
10. Olson LG, Lo YS, Beebe TP Jr, Harris JM (2001) Characterization of silane-modified immobilized gold colloids as a substrate for surface-enhanced Raman spectroscopy. Anal Chem 73:4268–4276

11. Kasumi A, Sako Y, Yamamoto M (1993) Confined lateral diffusion of membrane receptors as studied by single particle tracking (nanovid microscopy. Effects of calcium-induced differentiation in cultured epithelial cells. Biophys J 65:2021–2040
12. Haruta M, Date M (2001) Advances in the catalysis of Au nanoparticles. Appl Catal Gen 222:427–437
13. Zhong X, Yuan R, Chai Y, Liu Y, Dai J, Tang D (2005) Glucose biosensor based on self-assembled gold nanoparticles and double-layer 2d-network (3-mercaptopropyl)-trimethoxy-silane polymer onto gold substrate. Sensor Actuator B 104:191–198
14. Tanahashi I, Manabe Y, Tohda T, Sasaki S, Nakamura A (1996) Optical nonlinearities of Au/SiO_2 composite thin films prepared by a sputtering method. J Appl Phys 79:1244–1249
15. Hodak JH, Henglein A, Giersig M, Hartland GV (2000) Laser-induced inter-diffusion in AuAg Core – shell nanoparticles. J Phys Chem B 104:11708–11718
16. Chen YH, Yeh CS (2001) A new approach for the formation of alloy nanoparticles: laser synthesis of gold–silver alloy from gold–silver colloidal mixtures. Chem Commun 371–372
17. Papavassiliou GC (1976) Surface plasmons in small Au-Ag alloy particles. J Phys F Met Phys 6:L103–L105
18. Suslick KS, Crum LA (1977) Sonochemistry and sonoluminescence, Encyclopedia of Acoustics. Wiley, New York, pp 271–282
19. Putterman SJ, Weninger KR (2000) Sonoluminescence: how bubbles turn sound into light. Annu Rev Fluid Mech 32:445–476
20. Mason TJ, Lorimer JP (1988) Sonochemistry: theory applications and uses of ultrasound in Chemistry. Harword, Chichester, UK
21. Neppiras EA (1984) Acoustic cavitation series: part one: Acoustic cavitation: an introduction. Ultrasonics 22:25–28
22. Suslick KS (1989) The chemical effects of ultrasound. Sci Am 260:80–86
23. Henglein A (1987) Sonochemistry: historical developments and modern aspects Ultrasonics 25:6–16
24. Suslick KS (1990) Sonochemistry. Science 247:1439–1445
25. Kotronarou A, Mills G, Hoffmann MR (1991) Ultrasonic irradiation of p-nitrophenol in aqueous solution. J Phys Chem 95:3630–3638
26. Petrier C, Lamy MF, Francony A, Benahceene A, David B, Renaudin V, Gondreson N (1994) Sonochemical degradation of phenol in dilute aqueous solutions: comparison of the reaction rates at 20 and 487 kHz. J Phys Chem 98:10514–10520
27. Sweet JD, Casadonte DJ (2001) Sonochemical synthesis of iron phosphide. Ultrason Sonochem 8:97–101
28. Daniel MC, Astruc D (2004) Gold nanoparticles: assembly, supramolecular chemistry, quantum-size-related properties, and applications toward biology, catalysis, and nanotechnology. Chem Rev 104:293–346
29. Sonochemical synthesis of inorganic nanoparticles: Ashokkumar M (2008) In: Cozzoli PD (ed) Chapter 4: Advanced we-chemical synthetic approaches to inorganic nanostructures, Transworld Research Network, pp 107–131
30. Grieser F, Ashokkumar M (2004) In: Caruso F (ed) Sonochemical synthesis of inorganic and organic colloids, colloids and colloid assemblies. Wiley-VCH GmbH & Co. KgaA, Weinheim, pp 120–149
31. Brust M, Fink J, Bethell D, Shiffrin DJ, Kiely C (1995) Synthesis and reactions of functionalised gold nanoparticles. J Chem Soc Chem Commun 1655–1656
32. Brust M, Walker M, Bethell D, Shiffrin DJ, Whyman R (1994) Synthesis of thiol-derivatised gold nanoparticles in a two-phase liquid-liquid system. J Chem Soc Chem Commun 801–802
33. Yeung SA, Hobson R, Biggs S, Grieser F (1993) Formation of gold sols using ultrasound. J Chem Soc Chem Commun 378–379
34. Okitsu K, Teo BM, Ashokkumar M, Grieser F (2005) Controlled growth of sonochemically synthesized gold seed particles in aqueous solutions containing surfactants. Aust J Chem 58(2005):667

35. Okitsu K, Ashokkumar M, Grieser F (2005) Sonochemical synthesis of gold nanoparticles: effects of ultrasound frequency. J Phys Chem B 109:20673–20675
36. Nagata Y, Watanabe Y, Fujita S, Dohmaru T, Taniguchi S (1992) Formation of colloidal silver in water by ultrasonic irradiation. J Chem Soc Chem Commun 1620–1622
37. Anandan S, Ashokkumar M, Grieser F (2008) Sonochemical synthesis of Au-Ag core-shell bimetallic nanoparticles. J Phys Chem C 112:15102–15105
38. Mizukoshi Y, Takagi E, Okuno H, Oshima R, Maeda Y, Nagata Y (2001) Preparation of platinum nanoparticles by sonochemical reduction of the Pt(IV) ions: role of surfactants. Ultrason Sonochem 8:1–6
39. Vinodgopal K, He Y, Ashokkumar M, Grieser F (2006) Sonochemically prepared platinum-ruthenium bimetallic nanoparticles. J Phys Chem B 110:3849–3852
40. Nemamcha A, Rehspringer JL, Kharmi D (2006) Synthesis of palladium nanoparticles by sonochemical reduction of palladium(II) nitrate in aqueous solution. J Phys Chem B 110: 383–387
41. Kotranarou A, Mills G, Hoffman MR (1992) Oxidation of hydrogen sulfide in aqueous solution by ultrasonic irradiation. Environ Sci Technol 26:2420–2428
42. Yang J, Lee JY, Too HP (2006) Phase-transfer identification of core-shell structures in bimetallic nanoparticles. Plasmonics 1:67–78
43. Yang J, Lee JY, Too HP (2005) A phase transfer identification of core-shell structures in Au-Ru nanoparticles. Anal Chim Acta 537:279–284
44. He Y, Vinodgopal K, Ashokkumar M, Grieser F (2006) Sonochemical synthesis of ruthenium nanoparticles. Res Chem Intermed 32:709–715
45. Sathish Kumar P, Manivel A, Anandan S, Zhou M, Grieser F, Ashokkumar M (2010) Sonochemical synthesis and characterization of gold-ruthenium bimetallic nanoparticles. Colloids Surf A 356:140–144
46. Mazzone G, Rivalta I, Russo N, Sicilia E (2008) Interaction of CO with PdAu(111) and PdAu(100) bimetallic surfaces: a theoretical cluster model study. J Phys Chem C 112: 6073–6081
47. Chimentao RJ, Cota I, Dafinov A, Medina F, Sueiras JE (2006) Synthesis of silver-gold alloy nanoparticles by a phase-transfer system. Mater Res 21:105–111
48. Xie Y, Ding K, Liu Z, Tao R, Sun Z, Zhang H, An G (2009) In situ controllable loading of ultrafine noble metal particles on titania. J Am Chem Soc 131:6648–6649
49. Gold-platinum bimetallic clusters for aerobic oxidation of alcohols under ambient conditions: Miyamura H, Matsubara R, Kobayashi S (2008) Chem Commun 2031–2033.
50. Hajek J, Maki-Arvela P, Toukoniitty E, Kumar N, Salmi T, Murzin DY (2004) The effect of chemical reducing agents in the synthesis of sol-gel Ru-Sn catalysts: selective hydrogenation of cinnamaldehyde. J Sol-Gel Sci Technol 30:187–195
51. Toshima N, Yonezawa T, Harada M, Asakuara K, Iwasawa Y (1990) The polymer-protected Pd-Pt bimetallic clusters having catalytic activity for selective hydrogenation of diene. Preparation and EXAFS investigation on the structure. Chem Lett 19:815–818
52. Henglein A, Ershov BG, Malow M (1995) Absorption spectrum and some chemical reactions of colloidal platinum in aqueous solution. J Phys Chem 99:14129–14136
53. Yonezawa Y, Sato T, Kuroda S, Kuge KJ (1991) Photochemical formation of colloidal silver: peptizing action of acetone ketyl radical. J Chem Soc Faraday Trans 87:1905–1910
54. Mulvaney P, Giersig M, Henglein A (1992) Surface chemistry of colloidal gold: deposition of lead and accompanying optical effects. J Phys Chem 96:10419–10424
55. Trivino GC, Klabunde KJ, Dale EB (1987) Living colloidal palladium in nonaqueous solvents. Formation, stability, and film-forming properties. Clustering of metal atoms in organic media. 14. Langmuir 3:986–992
56. Ashokkumar M, Grieser F (2006) Sonochemical preparation of colloids, Encyclopedia of surface and colloid science, 2nd edn. Taylor & Francis, New York, pp 5685–5699
57. Sonochemistry: Ashokkumar M, Mason T (2007) In: Kirk-Othmer (ed) Encylcopedia of chemical technology. Wiley, New York

58. Suslick KS, Fang MM, Hyeon T, Mdleleni MM (1999) In: Crum LA, Mason TJ, Riesse JL, Suslick KS (eds) Sonochemistry and sonoluminescence, vol 524. Kluwer, London, p 291
59. Mizukoshi Y, Fujimoto T, Nagata Y, Oshima R, Maeda Y (2000) Characterization and catalytic activity of core-shell structured gold/palladium bimetallic nanoparticles synthesized by the sonochemical method. J Phys Chem B 104:6028–6032
60. Mizukoshi Y, Okitsu K, Maeda Y, Yamamoto YA, Oshima R, Nagata Y (1997) Sonochemical preparation of bimetallic nanoparticles of gold/palladium in aqueous solution. J Phys Chem B 101:7033–7037
61. Kan C, Cai W, Li C, Zhang L, Hofmeister H (2003) Ultrasonic synthesis and optical properties of Au/Pd bimetallic nanoparticles in ethylene glycol. J Phys D Appl Phys 36:1609–1614
62. Harada M, Asakura K, Toshima N (1993) Catalytic activity and structural analysis of polymer-protected gold/palladium bimetallic clusters prepared by the successive reduction of hydrogen tetrachloroaurate(III) and palladium dichloride. J Phys Chem 97:5103–5114
63. Watanabe M, Motoo S (1975) Electrocatalysis by ad-atoms: Part III. Enhancement of the oxidation of carbon monoxide on platinum by ruthenium ad-atoms. J Electroanal Chem 60:275–283
64. Watanabe M, Motoo S (1975) Electrocatalysis by ad-atoms part II. Enhancement of the oxidation of methanol on platinum by ruthenium ad-atoms. J Electroanal Chem 60: 267–273
65. Basnayake R, Li Z, Katar S, Zhou W, Rivera H, Smotkin ES, Casadonte DJ, Korzeniewski C Jr (2006) PtRu nanoparticle electrocatalyst with bulk alloy properties prepared through a sonochemical method. Langmuir 22:10446–10450
66. Takagi E, Mizukoshi Y, Oshima R, Nagata Y, Bandow H, Maeda Y (2001) Sonochemical preparation of noble metal nanoparticles in the presence of various surfactants. Stud Surf Sci Catal 132:335–338
67. Shchukin DG, Ustinovich E, Sviridov DV, Lvov YM, Sukhorukov GB (2003) Photocatalytic microreactors based on TiO2-modified polyelectrolyte multilayer capsules. Photochem Photobiol Sci 2:975–977
68. Sakamoto M, Fujistuka M, Majima T (2009) Light as a construction tool of metal nanoparticles: synthesis and mechanism. J Photochem Photobiol C Rev 10:33–56
69. Fujishima A, Honda K (1972) Electrochemical photolysis of water at a semiconductor electrode. Nature 238:37–38
70. Chen J, Ollis DF, Rulkens WM (1999) Kinetic processes of photocatalytic mineralization of alcohols on metallized titanium dioxide. Water Res 33:1173–1180
71. Kuwabata S, Yamauchi H, Yoneyama H (1998) Urea Photosynthesis from inorganic carbon and nitrogen compounds using TiO2 as photocatalyst. Langmuir 14:1899–1904
72. Driessen MD, Goodman AL, Miller TM, Zaharias GA, Grassian VH (1998) Gas-phase photooxidation of trichloroethylene on TiO_2 and ZnO: Influence of trichloroethylene pressure, oxygen pressure, and the photocatalyst surface on the product distribution. J Phys Chem B 102:549–556
73. Zheng Y, Zheng L, Zhan Y, Lin X, Zheng Q, Wei K (2007) Ag/ZnO heterostructure nanocrystals: synthesis, characterization, and photocatalysis. Inorgnic Chem 46:6980–6986
74. Anandan S, Sathish Kumar P, Pugazhenthiran N, Madhavan J, Maruthamuthu P (2008) Effect of loaded silver nanoparticles on TiO_2 for photocatalytic degradation of Acid Red 88. Solar Energy Mater Solar Cells 92:929–937
75. Georgekutty R, Seery MK, Pillai SC (2008) A highly efficient Ag-ZnO photocatalyst: synthesis, properties, and mechanism. J Phys Chem C 112:13563–13570
76. Sathish Kumar P, Sivakumar R, Anandan S, Madhavan J, Maruthamuthu P, Ashokkumar M (2008) Photocatalytic degradation of Acid Red 88 using Au–TiO_2 nanoparticles in aqueous solutions. Water Res 42:4878–4884
77. Lu W, Gao W, Wang S (2008) One-pot synthesis of Ag/ZnO self-assembled 3d hollow microspheres with enhanced photocatalytic performance. J Phys Chem C 112:16792–16800

78. Pan JH, Zhang X, Du Alan J, Sun DD, Leckie JO (2009) Self-etching reconstruction of hierarchically mesoporous F-TiO_2 hollow microspherical photocatalyst for concurrent membrane water purifications. J Am Chem Soc 130:11256–11257
79. Wang Y, Tang X, Lin L, Huang W, Haochen Y, Gedanken A (2000) Sonochemical synthesis of mesoporous titanium oxide with wormhole-like framework structures. Adv Mater 12:1183–1186
80. Yu JC, Zhang L, Yu J (2002) Direct sonochemical preparation and characterization of highly active mesoporous TiO_2 with a bicrystalline framework. Chem Mater 14:4647–4653
81. Anandan S, Ashokkumar M (2009) Sonochemical synthesis of Au-TiO_2 nanoparticles for the sonophotocatalytic degradation of organic pollutants in aqueous environment. Ultrason Sonochem 16:316–320
82. Mizukozhi Y, Makise Y, Shuto T, Hu J, Tominaga A, Shrionita S, Tanabe S (2007) Immobilization of noble metal nanoparticles on the surface of TiO_2 by the sonochemical method: photocatalytic production of hydrogen from an aqueous solution of ethanol. Ultrason Sonochem 14:387–392

Chapter 7
Acoustic and Hydrodynamic Cavitations for Nano $CaCO_3$ Synthesis

Shirish H. Sonawane and Ravindra D. Kulkarni

7.1 Introduction

Calcium carbonate is a common inorganic compound known as limestone. Calcium carbonate has many applications in industries such as medicine, agriculture, paint plastic and surface coatings etc. The vast majority of calcium carbonate used in industry is extracted by mining process. Pure calcium carbonate (e.g. for food or pharmaceutical use), is synthesized by passing carbon dioxide into a solution of calcium hydroxide slurry. In this process calcium carbonate precipitates out, and this grade of product is referred to as precipitate calcium carbonate (abbreviated as PCC). The common reaction is as follows:

$$Ca(OH)_2\ (l) + CO_2\ (g) \rightarrow CaCO_3\ (s) + H_2O\ (l) \qquad (7.1)$$

Ground calcium carbonate (GCC) is synthesized by wet and dry grinding methods and is used industrially for large production and utilization in the manufacturing of polymer, rubber composites, paper, glass-bottle etc. Polyvinyl chloride (PVC) and fiberglass producers use large quantities of calcium carbonate as a source of calcium, as an essential ingredient in their manufacturing processes. GCC technique has certain disadvantages such as, quality of $CaCO_3$, particle size distribution and brightness, hence industries prefer to use the precipitated calcium carbonate process. The generalized process reaction involves three major steps (1) calcination (2) slaking or hydrolysis (3) precipitation or carbonation. Initially, limestone is converted into calcium oxide by means of calcination at temperatures in excess of 800°C. After the calcination, lime is slaked with water, the resulting milk of lime is purified and carbonation process is carried out with the carbon dioxide. A cake comprising more than 30% solid matter (depending on particle

S.H. Sonawane (✉)
Vishwakarma Institute of Technology, 666 Upper Indira Nagar, Bibwewadi, 411037 Pune, India
e-mail: shirishsonawane@rediffmail.com

Pankaj and M. Ashokkumar (eds.), *Theoretical and Experimental Sonochemistry Involving Inorganic Systems*, DOI 10.1007/978-90-481-3887-6_7,

Schemes 7.1 Industrial precipitation process for manufacturing of calcium carbonate

$$CaCO_3 \xrightarrow[\text{calcination}]{> 800\,^{\circ}C} CaO + CO_2$$

$$CaO \xrightarrow[\text{hydrolysis}]{H_2O} Ca(OH)_2$$

$$Ca(OH)_2 \xrightarrow[\text{carbonation}]{CO_2} CaCO_3$$

Schemes 7.2 Elementary steps involved in kinetics of carbonation process [1, 2]

(i) $Ca(OH)_2(s) \rightarrow Ca^{2+}(aq) + 2OH^-(aq)$

(ii) $CO_2(g) \rightarrow CO_2(aq)$

(iii) $CO_2(aq) + OH^-(aq) \rightarrow HCO_3^-(aq)$

(iv) $HCO_3^-(aq) + OH(aq) \rightarrow H_2O + CO_3^{2-}(aq)$

(v) $Ca^{2+}(aq) + CO_3^{2-}(aq) \rightarrow CaCO_3(s)$

diameter) is then obtained by filtration. This filter cake is then dried and subsequently de-agglomerated in grinders. The reaction scheme is reported in Scheme 7.1.

Juwekar and Sharma [1] described the kinetics of the above reactions. The formation of calcium carbonate is non elementary reaction which involves the number of elementary steps as shown in Scheme 7.2, steps (iv) and (v) assumed to be instantaneous. Absorption of CO_2 gas and dissolution of $Ca(OH)_2$ affects the nucleation step, both are considered as rate controlling steps.

Many of the inorganic nanoparticles such as inorganic salts, fine chemicals are usually prepared by gas-solid-liquid reactive crystallization route. Calcite synthesis using reactive crystallization is a typical example. Morris and Woodburn in 1967 [3] reported that the carbon dioxide dissolution is the major controlling step, while Juvekar and Sharma [1] reported that the reaction between carbondioxide and hydroxyl ion is major the controlling step. Lin et al. [2] concluded that, bulk diffusion and surface integration of growth units are important steps for the formation of morphology and size of $CaCO_3$ nanoparticles. There are generally two methods for synthesis of $CaCO_3$ nanoparticles (1) Spray Carbonation Process (2) Semi-batch Carbonation Process. The Spray process is expensive in comparison to batch process and hence is one of the preferred industrial methods. There are other laboratory processes for the synthesis of nanosize calcite such as miniemulsion method [4], segmented flow tubular reactor [5] and supercritical CO_2 insertion of gas [6]. Each method reported has its own advantages and disadvantages for example the semibatch process includes higher production rate and control, over size and shape during carbonation process. The problems associated with these methods are to obtain smaller particle size of $CaCO_3$ particles, narrow particle size distribution and desired phases. There are three possible phases of $CaCO_3$ such as calcite, vaterlite, and aragonite, and in general the aragonite phase occurs due to secondary nucleation process. Primary nucleation generates phases of calcite and vaterlite. Primary nucleation morphologies (calcite and vaterlite) depend on synthesis parameters such as temperature of reaction mass, super saturation, pH, flow rate of CO_2, initial temperature of CO_2 gas, concentration of $Ca(OH)_2$ slurry and method of synthesis. Ultrasound assisted crystallization is one of the important methods for synthesis of nano $CaCO_3$.

Over past two decades, sonochemistry has evolved as a branch of chemistry. Various homogenous and heterogeneous reactions involving, gas-liquid and gas-liquid-solid phases are characterized by crystal growth [7–9]. Effect of ultrasound on crystallization of different materials has captured interest of researchers in the past decade [10]. During ultrasound assisted synthesis, bubbles are generated during the rarefaction cycle of the waves. Due to the collapse of these cavities pressure of several thousand atmospheres is generated which is used to control the nucleation [11]. However sometimes ultrasonic irradiation, cannot induce primary nucleation at all. This is because the ease with which primary nucleation is induced, differs with substances, solvents, vapor pressure, ultrasonic irradiation conditions and method of preparing super saturated solution. Ultrasound assisted crystallization process is called as "sonocrystallization" in which the nuclei are produced due to cavitation process and ultrasound controls the growth of crystals. Sonocrystallization shows faster and uniform primary nucleation in comparison to conventional crystallization process. During sonocrystallization ultrasound reduces the particle agglomeration resulting in more stable particles. Ultrasound also increases the supersaturation affecting the growth rate of nuclei in primary nucleation, which in turn affects the size, shape and particle size distribution [12]. In comparison with other methods, sonocrystallization process has many advantages such as controlled particle size at nano-meter regime, mono disperse characteristics of particles, simple equipment and mild operating conditions. It is a well known fact that, ultrasound controls primary nucleation of the particles and reduces the width of metastable zone. In addition, sonocrystallization exhibits faster primary nucleation, inhibits secondary nucleation and produces smaller and uniform particle size [13, 14]. Thus ultrasound enables significant control of crystallization process. Several parameters such as reaction time, reaction yield, induction time, nucleation, metastable zone, crystal pattern and orientation have been influenced by acoustic cavitation [15]. Ultrasonic irradiation can also produce •OH and •H radicals through cavitation which further serve as oxidizing and reducing species. Therefore such an environment under ultrasonic irradiation is suitable for synthesis of inorganic materials [16]. Amongst the organic synthesis processes, polymerization using ultrasound is currently gaining a lot of importance [17]. This is because of the positive effect of ultrasound on different parameters of polymerization reaction. The effect of ultrasound on primary nucleation during synthesis of acetylsalicylic acid crystals was studied and a contradictory phenomenon of nucleation inhibition was observed [13]. Similarly a phenomenon of precipitation retardation was observed when growth of calcite crystals was concerned [14]. Sonochemical synthesis of gold nano-whiskers was reported with control of crystal morphology for application in electronics [18]. Similarly synthesis of copper aluminate oxides and salts, particularly those of the transition metals, display the widest and most fascinating range of properties of any single class of materials. Ultrasound assisted synthesis has covered a vast area of organic and inorganic materials.

7.2 Theoretical Aspects: Crystallization and Sonocrystallization to Form Inorganic Nanoparticles

There are obviously two steps involved in the preparation of crystal matter from a solution, the crystals must first form and then grow. The formation of a new solid phase either on an inert particle in the solution or in the solution itself is called nucleation. The increase in size of this nucleus with a layer-by-layer addition of solute is called crystal growth. Both nucleation and crystal growth have supersaturation as a common driving force. Unless a solution is supersaturated, crystals can neither form nor grow. The particle-size distribution of this weight, however, will depend on the relationship between the two processes of nucleation and growth.

The nucleation rate is the sum of contributions by homogeneous nucleation and nucleation, due to the contact between the crystals or between the crystals and the walls of the container. The contact nucleation rate is a function of the input of energy of contact, contact area and supersaturation. At laboratory or bench scale, contact energy level is relatively low and homogeneous nucleation can contribute significantly to the total rate of nucleation. However, in commercial equipment, the contact energy input is intense and contact nucleation is the predominant mechanism. Crystal growth is a layer-by-layer process, and since growth can occur only at the face of the crystal, material must be transported to that face from the bulk of the solution. Nucleation process can be divided into two forms, homogenous and heterogeneous. Homogenous primary nucleation occurs only in the absence of any solid matter in the supersaturated solutions [19–21]. The rate of crystal growth in case of attrition mechanism depends on two factors namely the frequency and intensity of collision and growth rate of attrition fragments [19]. With the use of ultrasound both these parameters are enhanced in the early stages of the reaction, that is, during the primary nucleation. The intensity of ultrasound irradiation is the most important parameter which mainly alters inorganic crystal growth along with formation of number of crystals. In general, when high intensity ultrasound is applied on a system carrying inorganic nanoparticles, approximately 10–10,000 crystals can be formed within a short period of time [22]. This phenomenon can be observed due to the activation of primary nucleation in the presence of ultrasound. On the other hand, at low intensity ultrasound irradiation, the average number of crystals is reduced over the span of time which can also be observed in the case of high speed mechanical agitation. This phenomenon is attributed to disruption of sub-nuclei and molecular clusters [13, 15, 23]. Hence it can be suggested that the intensity of ultrasound not only activates the primary nucleation but also inhibits the same under specific experimental conditions. A study on calcium carbonate sonocrystallization also shows that the crystal formation does not take place until minimum ultrasonic energy is imparted to the reaction system which is necessary to form the stable nuclei [24]. The number of primary nuclei significantly affects the crystal size with due assumption that variation in number of crystals is neglected. Therefore, the control on crystal size of inorganic particles can be gained by controlling primary nucleation.

Supersaturation affects the following parameters during a crystallization process [25].

1. Crystal growth
2. Agglomeration and aggregation
3. Primary and secondary nucleation

Growth rate of inorganic particles during sonocrystallization strongly depends on supersaturation and activation energy; and both provide vital information about the rate determining step. Variation in the activation energy, either due to adsorption on crystal surface or integration into crystal lattice gives idea about the rate determining step in presence of ultrasound. The possibility of surface nucleation mechanism can not be neglected. Homogenous primary nucleation occurs only in the absence of any solid matter. Appropriate application of ultrasound in inorganic particle synthesis helps in obtaining thermodynamically stable polymorph at a certain level of supersaturation. In the sonocrystallization process, the high level of supersaturation rapidly reduces the induction time of nucleation and further reducing spreading of induction regime [26–28]. Supersaturation ratio plays a vital role to breed different morphologies of inorganic particles during sonocrystallization. Various morphologies can originate from the competition of crystal growth and nucleation at different supersaturation ratios. In the absence of ultrasound, nucleation rate is so small that only few nuclei can be generated at the initial time of growth. Therefore nuclei grow in spatial orientations fixed by solute crystallized structures or crystalline orientations of seed in case of seeded growth. When ultrasound is applied, the nuclei break out which leading to the formation of large number of structures limiting the supplementary growth of particles. Different crystal morphologies and distinct phases can be observed during the sonocrystallization of inorganic particles. In the presence of ultrasound, most of the synthesized particles show spherical shape because of limited scope of directional growth of crystal in nucleation region. As the sonication time increases the crystalline nature of the particle becomes dominant. Therefore morphology and desired phases of inorganic particles can be controlled by adjustment of supersaturation ratio which is ultimately controlled by ultrasonication [29].

The effect of ultrasound on crystal growth of inorganic particles arises mainly due to enhanced bulk phase mass transfer. Micromixing generated by ultrasonic cavitation alters the fluid dynamics and increases bulk phase mass transfer of solute to the surface of growing crystal. At low supersaturation, the quantity of growth units near crystal surface is small. Therefore bulk phase mass transfer becomes the rate limiting step in supplying growth units to crystal surface. In these types of circumstances, use of ultrasound dramatically enhances the growth rate.

The reactive crystallization has some peculiar characteristics like insoluble product, initiation of reaction by change in pH and conductivity. In this case the solution becomes saturated and eventually supersaturated with respect to reactant nucleation [30]. The ultrasound assisted decomposition precursors includes dissolving metal organic precursors in organic solvents/water with the assistance of surfactants leads to monodisperse and reduced metal/metal oxide nanoparticles.

Some of the reports are as follows. Mizukoshi et al. [31] reported ultrasound assisted reduction processes of Pt(IV) ions in the presence of anionic, cationic and non-ionic surfactant. They found that radicals formed from the reaction of the surfactants with primary radicals sonolysis of water and direct thermal decomposition of surfactants during collapsing of cavities contribute to reduction of metal ions. Fujimoto et al. [32] reported metal and alloy nanoparticles of Au, Pd and Pt, and MnO_2 prepared by reduction method in presence of surfactant and sonication environment. They found that surfactant shows stabilization of metal particles and has impact on narrow particle size distribution during sonication process. Abbas et al. [33] carried out the effects of different operational parameters in sodium chloride sonocrystallisation, namely temperature, ultrasonic power and concentration sodium. They found that the sonocrystallization is effective method for preparation of small NaCl crystals for pharmaceutical aerosol preparation. The crystal growth then occurs in supersaturated solution. Mersmann et al. (2001) [21] and Guo et al. [34] reported that the relative supersaturation in reactive crystallization is decisive for the crystal size and depends on the following factors.

1. Rate of chemical reaction
2. Number and the size of the particles present in the form of reactants (heterogeneous crystallization)
3. The effectiveness of macro and micromixing

If ultrasound is introduced in this process a major difference can be observed for each of the above constraints which is elaborated in discussion ahead. Nucleation rate which defines the product quality depends on the effective mixing and addition of reactant. The sonication acts as an effective source of micro mixing of reactants. The ultrasonic vibration and cavitation in the liquid play unique role in mixing the solution uniformly and rapidly. Mixing initiated by agitation promotes motion of the macroscopic layers whereas ultrasound initiated mixing occurs at the interface of these layers as mentioned earlier. This interfacial mixing depends on the slow diffusion of the reacting molecules. The shock wave generated from high local pressure of cavitation collapse can accelerate the motion of the liquid molecules and increase molecular impact of the particle present in the bubble [11, 35–38]. This increases the overall chemical potential of the reacting system resulting in the increase in overall nucleation rate of the system. All these facts make ultrasound initiated mixing superior than the conventional ones.

7.3 Cavitation Assisted Synthesis of Nano $CaCO_3$

7.3.1 Effect of Ultrasound on $CaCO_3$ Synthesis

To understand the effect of various parameters during sonocrystallization process following process was adapted, the experimental setup is shown in Fig. 7.1. The carbonation reaction was carried out in a semibatch reactor consisting of

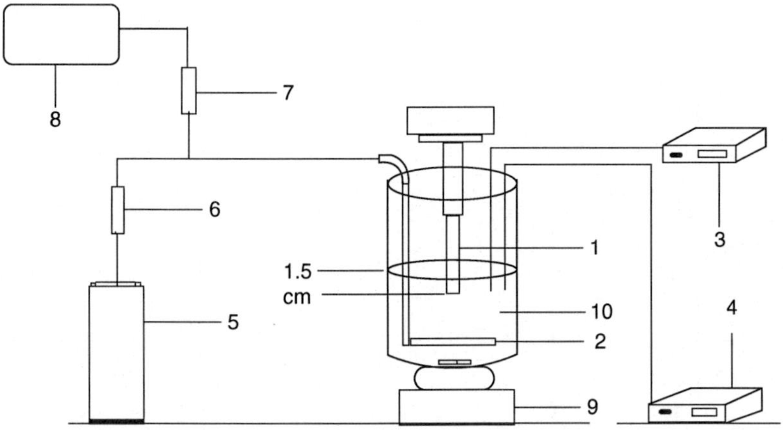

Fig. 7.1 Experimental set up for sonochemical carbonization experiments [24] (*1*) Ultrasound Probe, (*2*) CO_2 gas sparger, (*3*) Conductivity meter, (*4*) pH meter, (*5*) CO_2 gas cylinder, (*6*) CO_2 gas flow meter, (*7*) Air flow meter, (*8*) Air compressor (*9*) Magnetic stirrer, (*10*) $Ca(OH)_2$ slurry

replaceable probes of 10, 14 and 20 mm in size. To understand the effect of conventional carbonation process, some experiments were carried out using the gas distribution plate. In other arrangement CO_2 gas was passed through the probe (drilled with a hole of 4 mm size) in order to get smaller size gas bubbles which can easily take part in reaction and can help to increase the rate of reaction. The reaction was continuously monitored using conductivity and pH meters and the concentration of $Ca(OH)_2$. Optimum concentration, i.e., 4% by wt $Ca(OH)_2$ was used as the slurry concentration. The flow rate of CO_2 was maintained at 50 LPH throughout all experiments. Power dissipation into the reactor was calculated using calorimetric method as reported [24]. Conductivity as a function of time is initially plotted, to know the effect of probe onto induction time. As shown in Fig. 7.2, initially the conductivity value is very high, then sudden drop in conductivity is observed, further it remains stable for a certain period and again conductivity drops to a very low value. Figure 7.2 shows three distinct regions of curve, (1) initiation (2) nucleation (3) precipitation; the first downward peak gives information about the formation of massive nuclei in the reaction mixture. As shown in Table 7.1, lower the induction time lesser is the crystal size. It is also observed that when gas is passed without probe, large induction time is required to complete initiation, which is found out to be 110 min. A 10 min induction time is observed when the gas is passed through the sparger and sonication with probe (20 mm). This shows that passing the gas though the probe is favorable for nucleation process, and hence reduction in time is observed. As shown in Table 7.1, it is also observed that the particle size reduces with increase in probe size. Increasing the probe size, the power dissipation was found to be increasing. About 50 % increase in the power dissipation was observed for 20 mm probe as compared to 10 mm probe size. Power dissipation was found inversely proportion to the reduction in particle size. The XRD data in Fig. 7.3 shows the effect of ultrasound probe on the crystallite size of

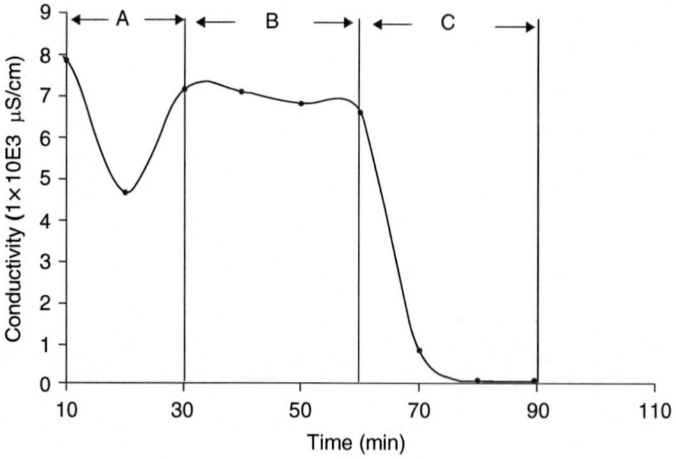

Fig. 7.2 Regions observed in the conductivity plots of 20 mm probe, passing the CO_2 gas through hole [24]

Table 7.1 Effect of ultrasound on crystallite size, induction time and energy during carbonization process [24]

Probe diameter	Description	Total power dissipated into the reactor (W/m^3)	Crystallite size (nm)	Particle size distribution (nm)	Induction time (min)
20	With hole	18.5×10^3	35	11–16	20
20	Without hole	12.5×10^3	51	24–29	30
14	With hole	11.6×10^3	38	26–37	32
14	Without hole	8.9×10^3	53	50–77	35
10	With hole	9.3×10^3	48	41–70	40
10	Without hole	5.0×10^3	60	69–106	60
No probe used	Reaction without ultrasound	–	104	62–117	110

$CaCO_3$ powder. It is found that without ultrasound probe (gas passed though sparger) the calcite phase shows orientation along 101 plane, while probe of 14 and 20 mm diameter with hole shows the orientation of planes to (1 1 4) and (1 1 9), indicating the formation of vaterlite phase. The grain size without ultrasound is 110 nm, while 20 mm diameter probe with a hole gives a crystallite size of 35 nm. As shown in Fig. 7.4, SEM confirms the formation of vaterlite phase for 20 mm probe with hole. The particles are spherical in nature, while from Fig. 7.3, it is found that all other probes show the calcite phase formation with (1 0 1 0) and (1 1 9) planes. Mingzahao et al. [39] reported that with increase in the CO_2 flow rate there is a significant decrease in the grain size. Similar observations are reported by Sonawane et al. [24] by passing the CO_2 though probe and without changing the CO_2 flow rates. This clearly indicates the fact that the rate of reaction increases with increase in the gas exposure area. Further the preferred orientation of $CaCO_3$ powder was found to be changed from (1 0 1 0) plane to (1 1 9) plane for samples synthesized by ultrasonic probe (14 mm and 20 mm diameter) without holes.

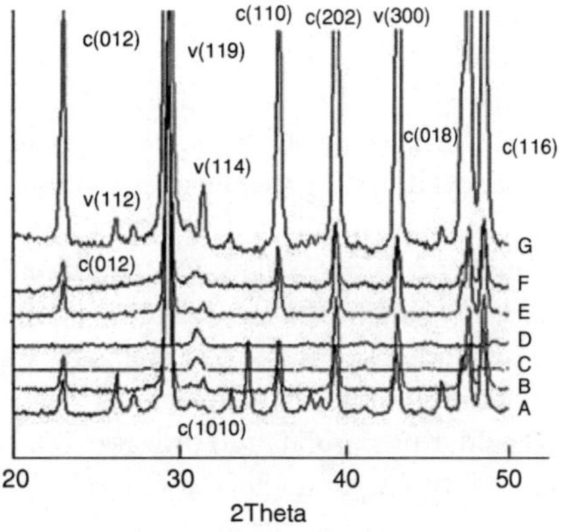

Fig. 7.3 XRD data of nano $CaCO_3$ crystals at different conditions (**a**) without ultrasound (**b**) 10 mm probe without hole (**c**) 10 mm probe with hole (**d**) 14 mm probe without hole (**e**) 14 mm probe with hole (**f**) 20 mm probe without hole (**g**) 20 mm probe with hole [24]

Fig. 7.4 SEM image of nano $CaCO_3$ particles of 20 mm probe with hole [24]

7.3.2 In Situ Functionalization of Nano $CaCO_3$ During Ultrasound Assisted Carbonation Process

The use of surfactants in the carbonation processes provides following advantages (1) inhibition of crystal growth (2) segregation of particles (3) surfactant acts as coupling agent for making compatible alloy during polymer compounding process which leads to improvement in mechanical and rheological properties. Industrially $CaCO_3$ is coupled with stearic acid, other polyelectrolytes or water soluble polymers, which act as inhibiting agents for crystal growth. Xiang et al. [40] reported use of terpinol to absorb CO_2 gas which leads to the production of nano size particles. Wei et al. [41] used 2 wt% calcium lignophosphate, an anionic surfactant and found particle size between 1 and 2 μm. Sheng et al. [42] used Octadecyl dihydrogen phosphate additives for controlling particle size. Lin et al. [2] used calcium tripolyphosphate as inhibition agent and found that increasing the sodium tripolyphosphate concentration; the rate-controlling step could be shifted from the bulk diffusion reaction to the surface-reaction. Generally, dry mixing technique is used for functionalization of $CaCO_3$. Sonawane et al. [43] attempted the addition of myristic acid, stearic acid, polyacrylic acid and sodium tripolyphosphate during carbonation process. The detail methodology of addition and optimization process parameters is reported in literature [43]. The proposed methodology is beneficial for industrial applications and it is possible to eliminate one step from industrial process. The objectives of study were to understand the effect of additives during sonochemical carbonation process and their effect on morphology and distribution of $CaCO_3$ particles.

The proposed mechanism of effect of surfactant and ultrasound is reported in Fig. 7.5. The long chain surfactant molecules attach to surface of nanoparticles due to physical adsorption. Only thin layer is adsorbed onto the $CaCO_3$ nanoparticles. Due to presence of ultrasound and use of surfactant will control the nucleation. Surfactant keeps the particles away from each other by preventing flocculation due to change in surface tension of reaction mass. The concentration of additives was changed from 0.2 to 1.0 g/L. Addition of 0.2 g/L tripolyphosphate shows the increase in the rate of precipitation which is determined from the $Ca(OH)_2$ consumption. Polyacrylic acid shows the least rate of precipitation (0.115 mol/l), which

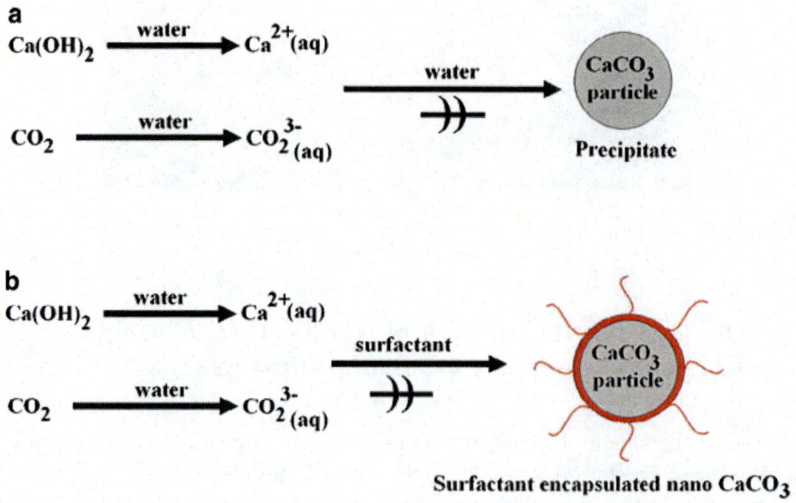

Fig. 7.5 Plausible mechanism of nano $CaCO_3$ particle synthesis during sonocrystallization in presence of surfactants [43]

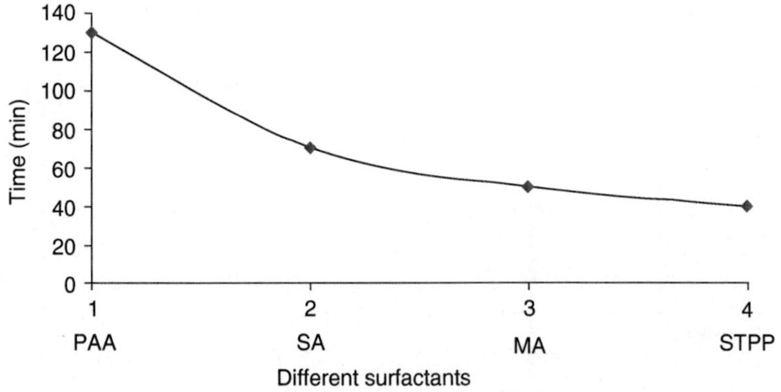

Fig. 7.6 Variation in overall reaction time with different surfactants [43]

indicates that polyacrylic acid hampers the precipitation process. The comparative time data for completion of reaction is reported in Fig. 7.6. Polyacrylic acid takes longer time of reaction in comparison to sodium tripolyphosphate. The polyacrylic acid has longer chains and the interaction between the polymer chains and Ca^{2+} leads to a decrease in super saturation and hence the precipitation process takes longer time. The presence of ultrasound and long chain surfactant slows down the structure formation process of nanoparticles during the precipitation process. Sodium tripolyphosphate keeps high super saturation value and hence lesser induction time, forming smaller size nanoparticles.

Thus, ultrasound and surface active agents together help in reducing the aggregation of particles because of the fact that the bonds between them are extended due to cavitation. Additives inhibit the agglomeration during nucleation process by reducing the surface tension. Ultrasound and additives both reduce population of local nuclei hence reduction in particle size [43].

As shown in Fig. 7.7 and Table 7.2, the crystallite size was found to be in the following order myristic acid > polyacryalic acid > stearic acid > sodium tripolyphosphate. The maximum reduction in crystallite size was obtained for sodium tripolyphosphate. The larger size was obtained for myristic acid, because of its hydrophobic nature and hence myristic acid has negative effect onto the reduction in crystallite size [43]. XRD data reported in Fig. 7.7 shows that the entire surfactant does not affect the $CaCO_3$ phases. All the surfactants lead to form calcite phase, which clearly indicates that orientation of $CaCO_3$ does not change due to the presence of ultrasound and surfactant type. All surfactants show (1 0 1 0) orientation of $CaCO_3$, leading to calcite phase. The calcite phase is confirmed by TEM images. As shown in Fig. 7.8, nano $CaCO_3$ particle synthesized using 0.2 g polyacrylic acid as a surfactant in the presence of ultrasound shows cubic structure of nano $CaCO_3$ size. The figure also confirms the cubic size of particles to be less than 50 nm and it is crystalline in nature. As reported in Table 7.2, the surfactant and use of ultrasound energy has a positive effect resulting in decrease in particle size and narrows down the particle size distribution.

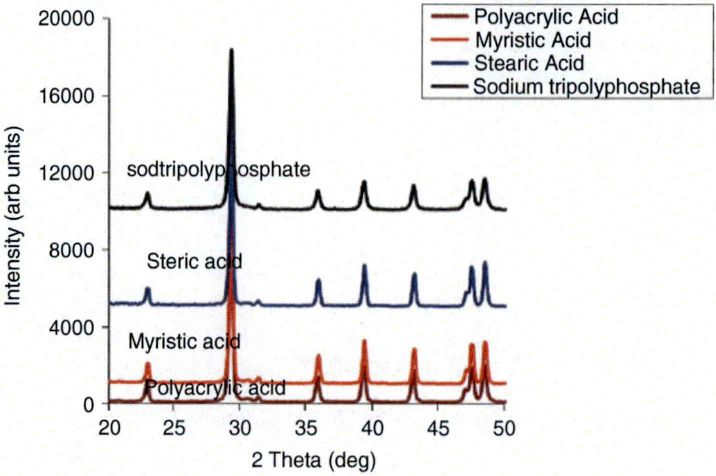

Fig. 7.7 X-ray diffraction patterns of $CaCO_3$ nanoparticles with 0.2 gm of different surfactants [43]

Table 7.2 Effect of surfactant and ultrasound on structure of nano $CaCO_3$ particles (crystallite size, particle size distribution) [43]

Surfactant	Crystallite size (nm) From XRD	Particle size (nm) From particle size analysis
PAA : Polyacrylic acid	44	28–38 Mean particle size 33
SA: Steric acid	39	45–65 Mean particle size 55
MA: Myristic acid	52	45–61 Mean particle size 53
STPP : Sodium tripolyphosphate	28	38–52 Mean particle size 45

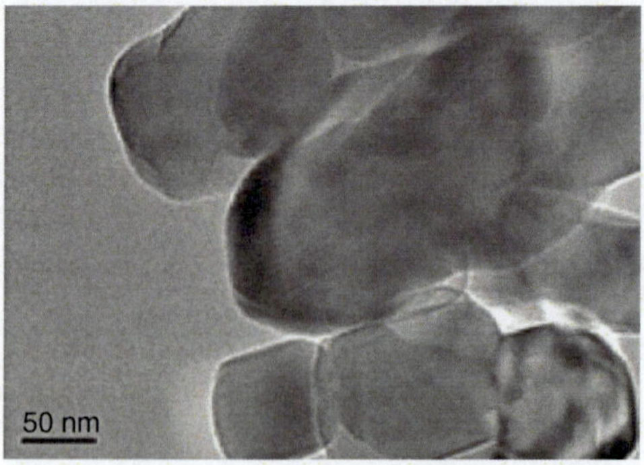

Fig. 7.8 Transmission electron micrograph of nano $CaCO_3$ particles synthesized using polyacrylic acid as surfactant using sonocrystallization method [43]

7.3.3 Hydrodynamic Cavitation Approach for Synthesis of Nano $CaCO_3$ Particles

Hydrodynamic cavitation occurs due to the sudden changes in the pressure of liquid flow in a pipe fitted with orifice or venturi. A liquid experiences a sudden drop in pressure at downstream resulting in the collapse of formed cavities. The collapse of the cavities generates highly reactive radicals, which are responsible for specific chemical reactions. In gas-solid reactions, the dissolution of solids is enhanced due to the turbulent mixing generated by hydrodynamic cavitation. The vigorous mixing enhances the transport of gas solutes to the solid surface that results in an increase in the mass transfer and hence the overall reaction rate [44–46]. Synthesis of $CaCO_3$ nanoparticles using hydrodynamic cavitation under various experimental conditions was carried out to study the effect of experimental parameters on the crystal size and size distribution. The experimental assembly (Fig. 7.9) consists of a closed loop reactor consisting of pump, pipe size and orifice ranging from 2 to 4 mm diameter and different geometry of 1 mm size of five holes was used for the study. The cavitation condition was determined using the equation reported elsewhere [47, 48]

The cavitation number is a dimensionless quantity, defined as

$$C_v = \frac{(P_2 - P_v)}{(0.5 \times \rho \times V_0^2)}$$

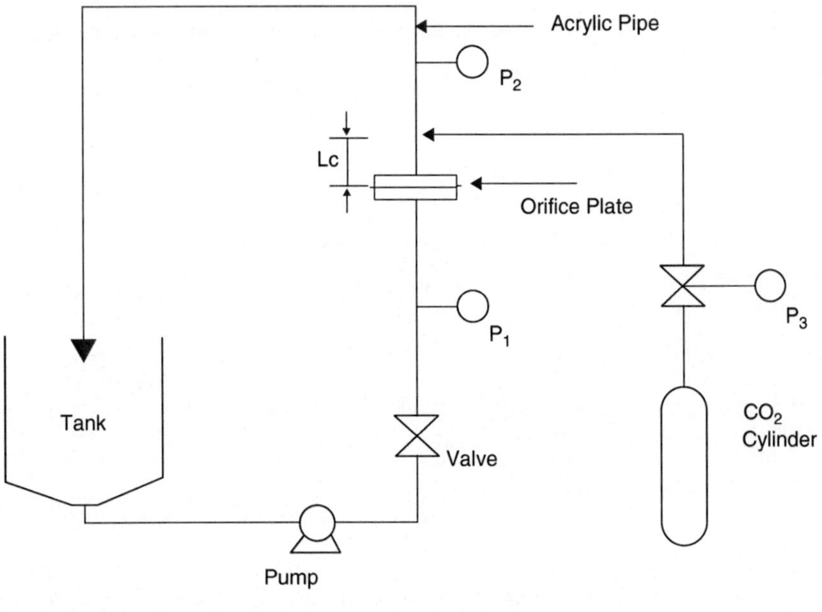

Fig. 7.9 Hydrodynamic cavitation set up for nano $CaCO_3$ production [46]

Where C_v -cavitation number; P_2-downstream pressure; P_v- vapor pressure of water; ρ- density of water at 25°C, V_0-average velocity near orifice. The diameter of the orifice was calculated using C_v values which was calculated from P_2 (downstream pressure), ρ (density of water), V_0 (average velocity near orifice) and Pv (vapor pressure of water) [48]. CO_2 gas was passed near to Lc zone as shown in Fig. 7.9, where cavities collapse in the cavitation zone.

Prior to incorporating orifice into the experimental assembly, initial experiments were carried out using, 4% $Ca(OH)_2$ slurry and 5 l/min CO_2 flow rate to optimize the experimental conditions . pH and conductivity, $Ca(OH)_2$ consumption shows three distinct regions corresponding to an induction period, nucleation and precipitation. It is observed that the completion of reaction takes 30 min. Figure 7.10 shows the consumption profiles of $Ca(OH)_2$ slurry with different orifice diameters. It is interesting to know that the rate of reaction is enhanced by the incorporation of orifice. The reaction rate is enhanced by the hydrodynamic cavitation leading to completion of reaction within 15 min. From Fig. 7.10, it is also concluded that the hydrodynamic cavitation enhances nucleation step and hence precipitation. As shown in Table 7.3, without orifice the crystallite size is 101 nm while it is found that change in geometry of the orifice, i.e. 1 mm × 5 holes orifice generated the smallest crystal size of 37 nm and particle size distribution was found in the range of 29–38 nm. Wide particle size distribution was observed for the sample without orifice ranging from 90 to 168 nm. The XRD data in Fig. 7.11 shows that the flow rates of CO_2 and Calcium hydroxide concentration do not show any effect on the

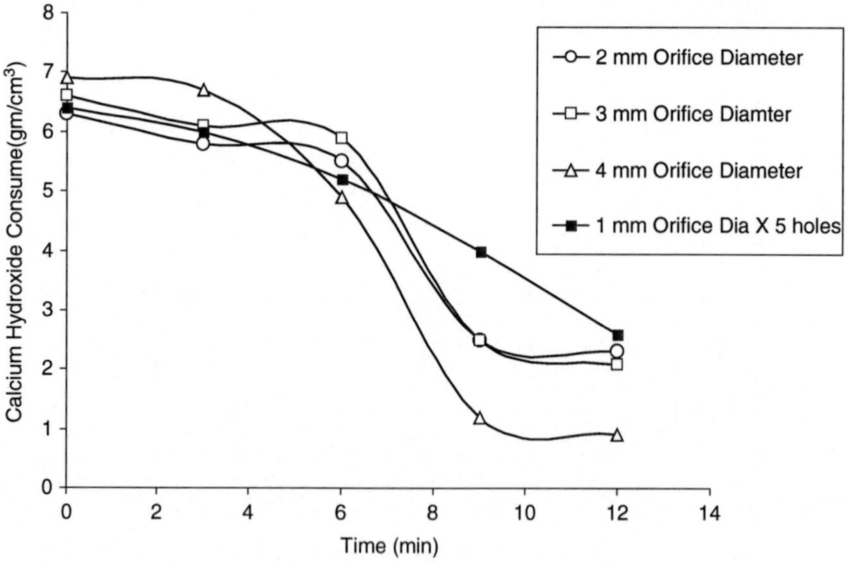

Fig. 7.10 Effect of different orifice diameter on the consumption of $Ca(OH)_2$ slurry at different time (showing constant rate period and falling rate period), (4% $Ca(OH)_2$ and 5 L/min CO2 flow rate) [46]

Table 7.3 Effect of slurry concentration and CO_2 flow rates, Orifice diameter on conversion and cavitation yield [46]

$Ca(OH)_2$ slurry (%)	CO_2 (l/min)	Orifice diameter (mm)	Crystallite size (nm)	Particle size distribution (nm)
Effect of change in CO_2 flow rate				
4	3	4	74	65–92
4	5	4	54	62–53
4	7	4	47	35–55
Effect of change in $Ca(OH)_2$ concentration				
2	5	4	50	69–52
4	5	4	54	62–53
6	5	4	61	50–72
Effect of change in orifice diameter				
4	5	1 × 5 holes	37	29–38
4	5	2	39	3–41
4	5	3	49	43–56
Without orifice				
4	5	4	54	62–53
4	5	–	101	90–168

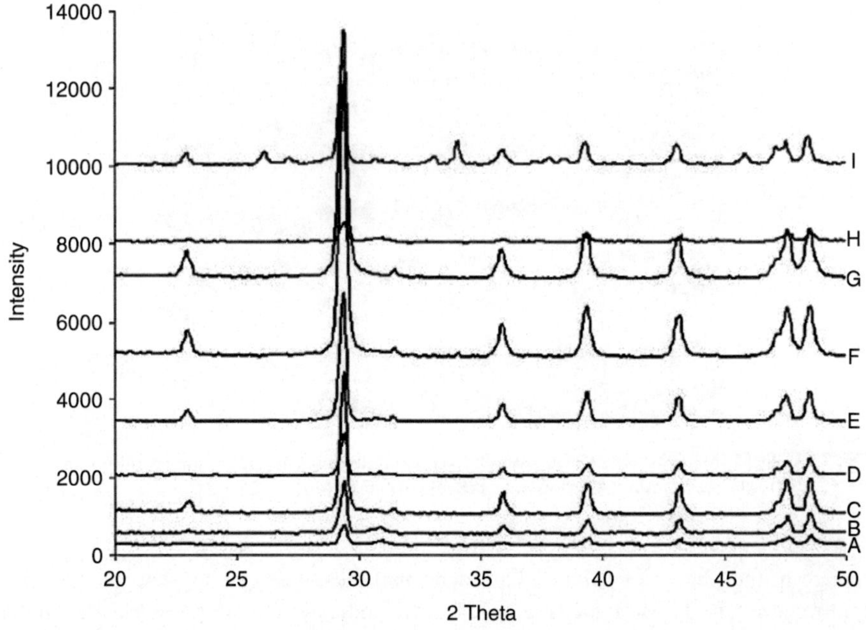

Fig. 7.11 Effect of different experimental conditions on X-Ray diffraction spectra [46]. (**a**) Without orifice, 4% $Ca(OH)_2$ slurry, 5 l/min CO_2 flow rate. (**b**) 1 mm orifice, 4% $Ca(OH)_2$, slurry 5 l/min CO_2 flow rate. (**c**) 2 mm orifice 4% $Ca(OH)_2$, slurry 5 l/min CO_2flow rate. (**d**) 3 mm orifice 4% $Ca(OH)_2$ slurry 5 l/min CO_2 flow rate. (**e**) 4 mm orifice 4% $Ca(OH)_2$, slurry 7 l/min CO_2 flow rate. (**f**) 4 mm orifice 2% $Ca(OH)_2$, slurry 5 l/min CO_2 flow rate. (**g**) 4 mm orifice 6% $Ca(OH)_2$ slurry, 5 l/min CO_2 flow rate. (**h**) 4 mm orifice 4% $Ca(OH)_2$ slurry, 3 CO_2 l/min flow rate. (**i**) 4 mm orifice, 4% $Ca(OH)_2$ slurry, 5 CO_2 l/min flow rate

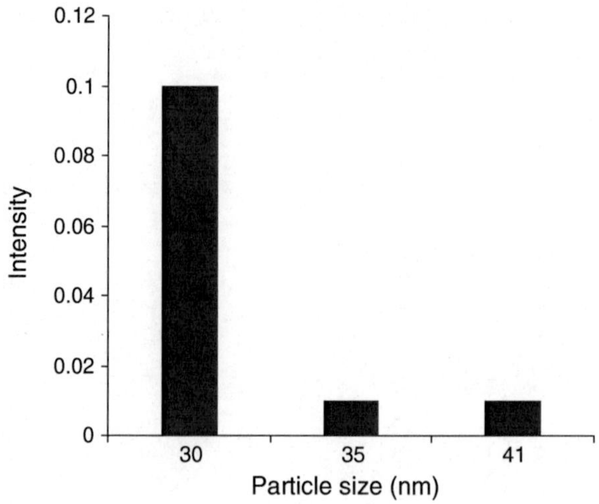

Fig. 7.12 Particle size distribution of $CaCO_3$ particles synthesized using hydrodynamic cavitation (2 mm orifice, 4% $Ca(OH)_2$ slurry, 5 l/min CO_2 flow rate) [46]

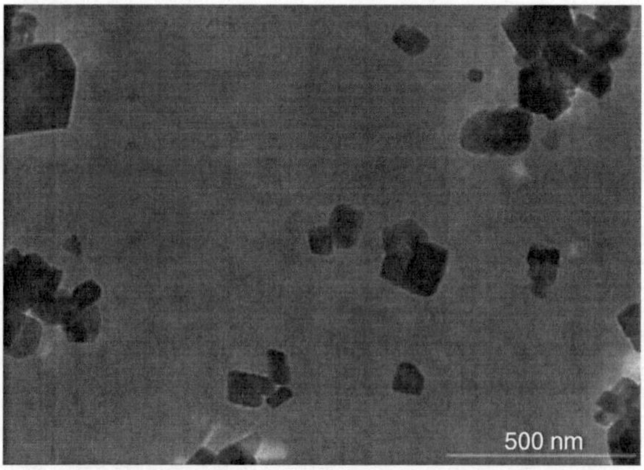

Fig. 7.13 TEM Image of nano $CaCO_3$ synthesized using hydrodynamic cavitation (4 mm orifice, 4% $Ca(OH)_2$ slurry, 5 l/min, CO_2 flow rate) (Scale 500 nm) [46]

change in the phases of calcium carbonate particles. The calcite phase is further confirmed by the TEM image (Fig. 7.12) which indicates that all particles are cubic in nature and size is less than 30 nm when 4 mm orifice, 4% $Ca(OH)_2$ slurry, 5 l/min, CO_2 flow rate was used for synthesis. The particle size distribution of $CaCO_3$ synthesized using hydrodynamic cavitation for 2 mm orifice, 4 % $Ca(OH)_2$ slurry, 5 l/min CO_2 flow rate shows the particle size distribution range of 30–41 nm as shown in Fig. 7.13.

7.4 Summary

Passing CO_2 gas through the probe hole gives significant improvement in the reduction of particle size. Micro mixing of the CO_2 gas through the probe leads to change in the preferred orientation of nano calcite crystals. There is a significant effect of ultrasound power, dissipated in the reactor. It was found that surfactant has significant effect on induction time, pH and conductivity of reaction mixture. The observations made in the present study reveal that nano crystallites of $CaCO_3$ can be easily synthesized by sonocrystallization method using different surfactants. Ultrasound and surfactant both enhance the global process (crystallization and precipitation) of nano calcite synthesis. It is seen that there is a wide distribution of particle size in the range 90–168 nm for the set up without an orifice. The change in the geometry of orifice has resulted in a significant effect on the crystallite size.

References

1. Juvekar VA, Sharma M M (1973) Absorption in of CO_2 in a suspension of lime. Chem Eng Sci 28(3):825–837
2. Lin R, Zhang J, Bai Y (2006) Mass transfer of reactive crystallization in synthesizing calcite nanocrystal. Chem Eng Sci 61(21):7019–7028
3. Morris M, Woodburn ET (1967) S. A. Chemical Process. Mcgraw-Hill, New York
4. Dagaonkar MV, Mehra A, Jain R, Heeres HJ (2004) Synthesis of $CaCO_3$ nanoparticles by carbonation Of lime solutions in reverse micellar systems. Chem Eng Res Des 82 (11):1438–1443
5. Pach L, Duncan S, Roy R, Komarneni S (1996) Morphological control of precipitated calcium carbonates and phosphates by colloidal additives. J Mater Sci 31:6565–6569
6. Wakayama H, Hall SR, Fukushima Y, Mann S (2006) $CaCO_3$/biopolymer composite films prepared using supercritical CO_2. Ind Eng Chem Res 45(10):3332–3334
7. Mason T, Lorimer J (2002) Applied sonochemistry. Wiley-VCH, London
8. Lindley J (1992) Sonochemical effects on syntheses involving solid and supported catalysts. Ultrasonics 30(3):163–169
9. Kim D, Oh S, Lee J (1999) Preparation of ultrafine monodispersed Indium-Tin Oxide particles in AOT-Based reverse microemulsions as nanoreactors. Langmuir 15(5):1599–1603
10. Li H, Wang J, Bao Y, Guo Z, Zhang M (2003) Rapid crystallization in the salting out process. J Cryst Growth 247(1–2):192–198
11. Gogate P, Tatake P, Kanthale P, Pandit A (2002) Mapping of sonochemical reactors: review, analysis, and experimental verification. AIChE J 48(7):1542–1560
12. Castro M, Capote F (2007) Ultrasound-assisted crystallization (sonocrystallization). Ultrason Sonochem 14(6):717–724
13. Miyasaka E, Kato Y, Hagisawa M, Hirasawa I (2006) Effect of ultrasonic irradiation on the number of acetylsalicylic acid crystals produced under the supersaturated condition and the ability of controlling the final crystal size via primary nucleation. J Cryst Growth 289(1):324–330
14. Dalas E (2001) The effect of ultrasonic field on calcium carbonate scale formation. J Cryst Growth 222(1):287–292
15. Nishida I (2004) Precipitation of calcium carbonate by ultrasonic irradiation. Ultrason Sonochem 11(6):423–428

16. Harada H (2001) Isolation of hydrogen from water and/or artificial seawater by sonophotocatalysis using alternating irradiation method. Int J Hydrogen Energy 26(4):303–307
17. Bradley M, Grieser F (2002) Emulsion polymerization synthesis of cationic polymer latex in an ultrasonic field. J Colloids Interface Sci 251(1):78–84
18. Feng Q, Jun Z, Hong C (2003) Controllable synthesis of nanocrystalline gold assembled whiskery structures via sonochemical route. J Cryst Growth 257(3):378–383
19. Mersmann A (2001) Crystallization technology handbook. Marcel Dekker, New York
20. Mullins J (2001) Crystallisation, 4th edn. Buterworth-Heinemann, UK
21. Mersmann A, Bartosch K (1998) How to predict metasable zone width. J Cryst Growth 183(1):240–250
22. Ruecroft G, Hipkiss D, Ly T, Maxted N, Cains PW (2005) Sonocrystallization: the use of ultrasound for improved industrial crystallization. Org Process Res Dev 9(6):923–932
23. Sohnel O, Mullin J (1982) Precipitation of calcium carbonate. J Cryst Growth 60(2):239–250
24. Sonawane S, Shirsath S, Khanna P, Pawar S, Mahajan C, Paithankar V, Shinde V, Kapadnis C (2008) An innovative method for effective micro mixing of CO_2 gas during synthesis of nano calcite crystal using sonochemical carbonization. Chem Eng J 143(1–3):308–313
25. Loffelmann M, Mersmann A (2002) How to measure supersaturation. Chem Eng Sci 57(20): 4301–4310
26. Parsiegla K, Katz J (1999) Calcite growth inhibition by copper (II) Effect of supersaturation. J Cryst Growth 200(1–2):213–226
27. Virone C, Kramer H, Rosmalen G, Stoop A, Bakker T (2006) Primary nucleation induced by ultrasonic cavitation. J Cryst Growth 294(1):9–15
28. Li S, Xu J, Luo G (2007) Control of crystal morphology through supersaturation ratio and mixing conditions. J Cryst Growth 304(1):219–224
29. Keck A, Gilbert E, Koster R (2002) Influence of particles on sonochemical reactions in aqueous solutions. Ultrasonics 40(1–8):661–665
30. Lv W, Luo Z, Yang H, Liu B, Weng W, Liu J (2010) Effect of processing conditions on sonochemical synthesis of nanosized copper aluminate powders. Ultrason Sonochem 17(2):344–351
31. Mizukoshi Y, Takagi E, Okuno H (2001) Preparation of platinum nanoparticles by sonochemical reduction of the Pt(IV) ions: role of surfactants. Ultrason Sonochem 8(1):1–6
32. Fujimoto T, Mizukoshi Y, Nagata Y (2001) Sonolytical preparation of various types of metal nanoparticles in aqueous solution. Scripta Mater 44(8–9):2183–2186
33. Abbas A, Srour M, Tang P, Chiou H, Kim C, Romagnolid J (2007) Sonocrystallisation of sodium chloride particles for inhalation. Chem Eng Sci 62(9):2445–2453
34. Guo Z, Jones A G, Li N (2006) The effect of ultrasound on the homogeneous nucleation of $BaSO_4$ during reactive crystallization. Chem Eng Sci 61(5):1617–1626
35. Amara N, Ratsimba B, Wilhelm A, Delmas H (2001) Crystallization of potash alum: effect of power ultrasound. Ultrason Sonochem 8(3):265–270
36. Amara N, Ratsimba B, Wilhelm A, Delmas H (2004) Growth rate of potash alum crystals: comparison of silent and ultrasonic conditions. Utrason Sonochem 11(1):17–21
37. Ohayon E, Gedanken A (2010) The application of ultrasound radiation to the synthesis of nanocrystalline metal oxide in a non-aqueous solvent. Ultrason Sonochem 17(1):173–178
38. Cravotto G, Omiccioli G, Stevanato L (2005) An improved sonochemical reactor. Ultrason Sonochem 12(3):213–217
39. Mingzhao H, Foressberg E, Wang Y, Han Y (2005) Ultrasonic assisted synthesis of calcium carbonate nanoparticles. Chem Eng Comm 192(11):1468–1481
40. Xiang L, Xiang Y, Wen Y, Wei F (2004) Formation of $CaCO_3$ nanoparticles in the presence of terpineol. Mat Lett 58(6):959–965
41. Wei S-H, Mahuli S K, Agnihotri R, Fan L-S (1997) High surface area calcium carbonate: pore structural properties and sulfation characteristic. Ind Eng Chem Res 36(6):2141–2148
42. Sheng Y, Zhou B, Zhao J, Tao N, Yu K, Tian Y, Wang Z (2004) Influence of octadecyl dihydrogen phosphate on the formation of active super-fine calcium carbonate. J Colloid Interface Sci 272((2):326–329

43. Sonawane S, Khanna P, Meshram S, Mahajan C, Deosarkar M, Gumfekar S (2009) Effect of surfactant on synthesis of calcium carbonate nanoparticles using sonochemical carbonization. Int J Chem Reactor Eng (Be press) 7 A(47):1–15, www.bepress.com/ijcre/vol7/47
44. Senthilkumar P, Sivakumar M, Pandit AB (2000) Experimental quantification of chemical effects of hydrodynamic cavitation. Chem Eng Sci 55(9):1633–1639
45. Moholkar VS, Senthilkumar P, Pandit AB (1999) Hydrodynamic cavitation for sono-chemical effect. Ultrason Sonochem 6(1–2):53–65
46. Nano Calcium Carbonate Crystallization using New Hydrodynamic Cavitation Reactor: Sonawane S, Mahajan C, Gumfekar S, Kate K, Meshram S, Kunte K, Ramajee L, Ashokkumar M (2010) Int J Chem Eng. Article ID 242963,1–8 doi:10.1155/2010/242963
47. Chivate MM, Pandit AB (1993) Effect of sonic and hydrodynamic cavitation on aqueous polymeric solutions. Indian Chem Eng 35:52–57
48. Mishra C, Peles Y (2006) An experimental investigation of hydrodynamic cavitation in micro-Venturis. Phys Fluids 18:103–109

Chapter 8
Sonochemical Synthesis of Oxides and Sulfides

Sivakumar Manickam

Abstract Sonochemical synthesis, an energy efficient processing technique to induce a variety of physical and chemical transformations is on the rise. A variety of simple and mixed metal oxides and sulfides have been obtained using this technique. The present chapter reviews the types of oxides and sulfides obtained in the last few years.

8.1 Introduction

Synthesis of inorganic oxides and sulfides is one of the potential applications and a widely studied one in the area of sonochemistry. The physical and the chemical effects generated after the collapse of the bubbles responsible for the generation of these materials and in many of the cases result in the generation of unusual materials with novel properties. This chapter specifically focuses only on these materials due to the growing interest on them. Suslick and Gedanken groups have made striking contributions in this area. Normally preparation of these materials are carried out in liquid phase using 20 or 40 kHz ultrasound irradiation either with a horn type or with bath type with/without the presence of Ar or some other inert gases. In contrast to the traditional chemical processes, smaller size particles, in specific, could be formed under ultrasonic conditions. In addition to size, shape is also well controlled by this method. Also, it assists in the generation of these materials without using any toxic precursors and without using common additives (in many cases), as is the case with conventional methods and thus could be

S. Manickam
Department of Chemical and Environmental Engineering, University of Nottingham (Malaysia Campus), Jalan Broga, 43500 Semenyih, Malaysia
e-mail: Sivakumar.Manickam@nottingham.edu.my

Pankaj and M. Ashokkumar (eds.), *Theoretical and Experimental Sonochemistry Involving Inorganic Systems*, DOI 10.1007/978-90-481-3887-6_8,

considered as a green chemical processing technique. It is also well known that the processing technique as well as the parameters affect/decide the size and morphology of the materials generated. Conventionally, a multitude of preparative methods are used for these materials which include, mechanochemical, solvothermal, microwave-assisted, sol-gel, microemulsion, photochemical, solid-state, co-precipitation, template-confined synthesis route, hydrothermal, electrochemical, combustion, pyrolysis, gas reaction, etc. Most of the above methods use high temperature, air and temperature sensitive organometallic precursors or environmental pollution agents.

Thus, developing novel and facile methods under mild conditions using green agents is still a challenge both for industry and academia. Sonochemical or ultrasound assisted synthesis is an alternative means to the above-mentioned techniques. The underlying principle for the ultrasound promoted reaction in a liquid medium is due to the dynamics of the bubbles, i.e. formation, growth and implosive collapse. Such an implosive collapse in the medium generates very high intense conditions of 5,000 K and 800 atm. instantaneously but locally and for a short life-time [1]. Such conditions are more than sufficient to induce either a chemical or physical transformations under apparently mild conditions. In case of a solid-liquid system, the bubble collapse occurs near the solid surface and causes interparticle collisions with very high velocity which results in fragmentation of the particles and hence an increase in the available surface area. With an increase in surface area, rate of the reaction is increased in these systems [2]. The chemical effects are dealt with greater emphasis as it leads to the generation of unusual materials with novel properties.

8.2 Formation Mechanism of Crystallinity Versus Amorphicity of Materials

There are three possible regions with the generation of bubbles in the liquid: Interior of the bubble where extreme conditions exist, transient bubble-liquid interface where moderate conditions exist and the bulk liquid medium where ambient conditions dominate. Depending on the vapour pressure, the precursor will have the ability to move to the interface or into the interior of the bubble. Alternatively, if it has very low vapour pressure, it will stay at the bulk liquid. If it reaches the interior of the bubble, it will be exposed to the intense conditions generated with the collapse of the bubble. Although intense conditions are getting generated, but the cooling rates are also very high ($>10^{10}$ K/s) [3]. Due to such cooling rates, the formed products will not have enough time to crystallize and that results in the formation of amorphous products. Alternatively, formation of crystalline products is favored if it reaches the interface. Uniform sized nanoparticles are generally formed during sonochemical processes, which maybe mainly due to the uniform size distribution of bubble which acts as nano-reactor.

8.3 Important Reaction Parameters for Sonochemical Reactions

Following are the important parameters that have to be considered carefully when carrying out sonochemical reactions:

Solvent (reaction medium): In most of the cases, solvent is selected with high boiling point (or) with low vapor pressure in order to generate effective cavitation.
Gas atmosphere: Inert gases like Ar is used in general as cavitation is favored in this atm.
Temperature: Room temperature
Power output: Maximum power goes up to 750 W.
Reaction time: From few minutes to hours

In the present chapter, only the recent work has been reviewed and for some of the interesting previous work, authors are advised to look into the following reviews: [2, 4, 5].

8.4 Characteristics/Advantages with Ultrasonic System

With the utilization of ultrasound, in most of the investigations, the advantages have been clearly demonstrated that sonochemical processes are novel, simple, convenient, elegant, rapid, cost effective, efficient, green and least agglomerative. Sonochemical processes are generally environmentally benign and applicable to a series of compounds, selective in obtaining amorphous/crystalline products with uniform size distribution of particles. Reactions under ultrasound usually give better yield, occur at mild reaction conditions, avoid calcinations conditions, utilize simple and safe precursors besides occurring at lower temperature and require much less time for completion. Sonochemical reactions, due to mixing at atomic level, are the reactions of non-equilibrium nature following different reaction mechanism, often capable of initiating new type of reactions and products.

8.5 Synthesis of Oxides by Ultrasound

In recent years, the growing numbers of publications are concerned with ultra-fine metal oxide structures because of their useful applications as bactericides, adsorbents, energy storage media, magnetic data storage, and ferrofluids and specifically as catalysts [6, 7].

8.5.1 ZnO

Zinc oxide is a versatile smart electronic and optical material that has unique applications in catalysts, sensors, piezoelectric transducers and actuators, photovoltaic, and surface acoustic wave devices [8]. A simple ultrasonic cavitational activation method has been proposed by Sivakumar et al. [9] for the direct conversion of zinc acetate to zinc oxide. By this method, highly monodispersed submicron structures of ZnO have been obtained without using any additives. Normally, 2D ZnO nanosheets or nanodiscs are prepared by vapour phase methods which need a high temperature over 1,500°C and also limited by their low yield. Thus, developing such nanosheets in soft conditions by a simple and template-free method is still a challenge. Using ultrasound and without using any template, ZnO nanosheets have been synthesized [10]. A high pH (12.5) and zinc salt counter ion played a critical role for the formation of ZnO nanosheets.

Bhattacharya and Gedanken [11] have reported a template-free sonochemical route to synthesize hexagonal-shaped ZnO nanocrystals (6.3 $\pm$ 1.2 nm) with a combined micro and mesoporous structure (Fig. 8.1) under Ar gas atmosphere. The higher porosity with Ar gas has been attributed to the higher average specific heat ratio of the Ar which leads to higher bubble collapse temperatures. With an intense bubble collapse temperature, more disorder is created in the product due to the incompleteness of the surface structure that led to greater porosity. Importance of gas atmosphere has been noted; when the same process was carried out in the presence of air which results in the formation of ZnO without any porosity.

Amongst the variety of nanostructures of ZnO, Mishra et al. [12] sonochemically prepared flower-like ZnO, the SEM image of which has been shown in Fig. 8.2,

Fig. 8.1 HRSEM image of the as-prepared ZnO; A single ZnO hexagonal nano-disk (inset) [11]

Fig. 8.2 SEM image of the flower-like ZnO [12]

with the assistance of starch, and it has been found that the gelation of starch plays an important role in controlling the morphology of nanoparticles.

By changing the ultrasound power, changes in the mesoporosity of ZnO nanoparticles (average pore sizes from 2.5 to 14.3 nm) have been observed. In addition to the changes in mesoporosity, changes in the morphology have also been noted [13]. Recently, Jia et al. [14] have used sonochemistry and prepared hollow ZnO microspheres with diameter 500 nm assembled by nanoparticles using carbon spheres as template. Such specific structure of hollow spheres has applications in nanoelectronics, nanophotonics and nanomedicine.

8.5.1.1 Ultrasound and Ionic Liquid

Interestingly, Hou et al. [15] have fabricated well-defined dendritic (branch-shaped) ZnO nanostructures in the presence of an ionic liquid, 1-(2-hydroxyethyl)-3-methylimidazolium tetrafluoroborate, $[C_2OHmim]^+BF_4^-$. Using such ionic liquids are considered to be potential green solvents, in stead of using traditional volatile organic solvents. Such dendritic structures composed of ZnO nanorods of 10–40 nm in diameter and lengths up to several hundred nanometers (Fig. 8.3). Also, it has been observed that either the absence of ultrasound or the ionic liquid did not yield ZnO. Thus, ionic liquid and ultrasound have critical role in the formation of dendritic ZnO nanosheets. The possible mechanism that has been proposed by the authors is based on the coupling of cation of the ionic liquid with anion of the

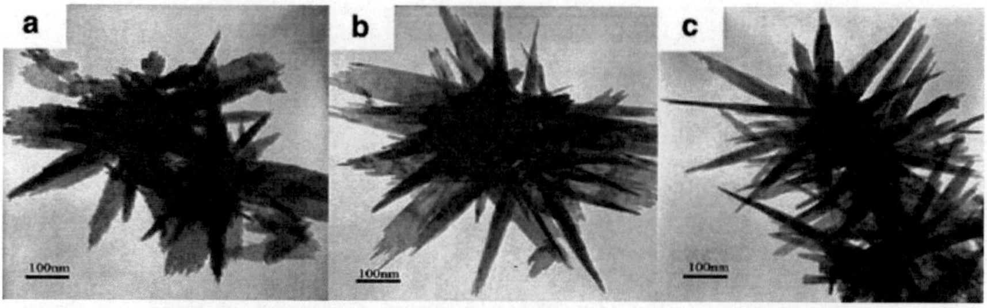

Fig. 8.3 TEM images of the as-prepared ZnO samples at different irradiation times of 0.5 h (**a**), 1 h (**b**) and 2 h (**c**) under the assistance of ultrasound [15]

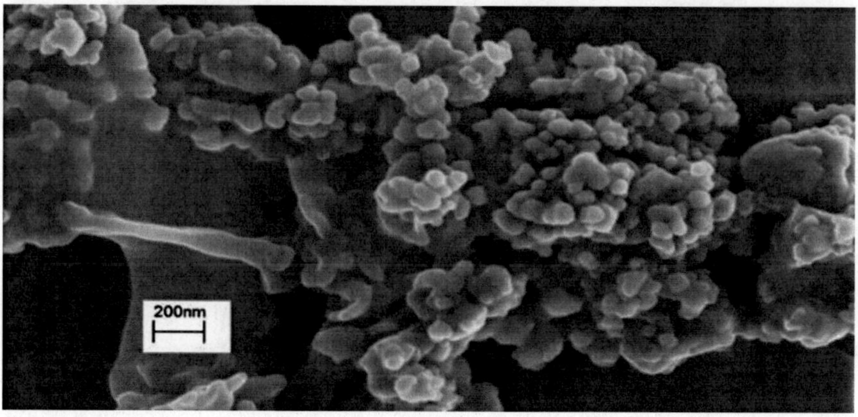

Fig. 8.4 SEM image of the ZnO nanoparticles prepared by ultrasound in the presence of an ionic liquid [16]

precursor, due to which dehydration of the precursor occurs and that leads to the formation of ZnO nuclei and finally results in ZnO nanostructures.

Whereas, Goharshadi et al. [16] have synthesized the ZnO nanoparticles of ~60 nm (Fig. 8.4.) using a room temperature ionic liquid, 1-hexyl-3-methylimidazolium bis (trifluoromethylsulfonyl)imide, Formation of ZnO was not observed when the ionic liquid was replaced by water. Also, in the absence of ultrasound, formation of ZnO was not observed which is very similar to the one as proposed in the previous case of ZnO dendritic nanostructures.

Alammar and Mudring [17] have also synthesized ZnO in the form of nanorods with lengths from 50–100 nm and diameters of about 20 nm (Fig. 8.5.) using the ionic liquid 1-butyl-3-methylimidazolium bis(trifluoromethanesulfonyl)amide, $[C_4mim][Tf_2N]$.

In addition to the processing technique, the properties of the oxides also changed by preparing them in a composite way. Arefian et al. [18] have synthesized SnO/ZnO nanocomposite using the sonochemical method and studied the effects of temperature and power on the morphologies generated. Recently Mg doped ZnO

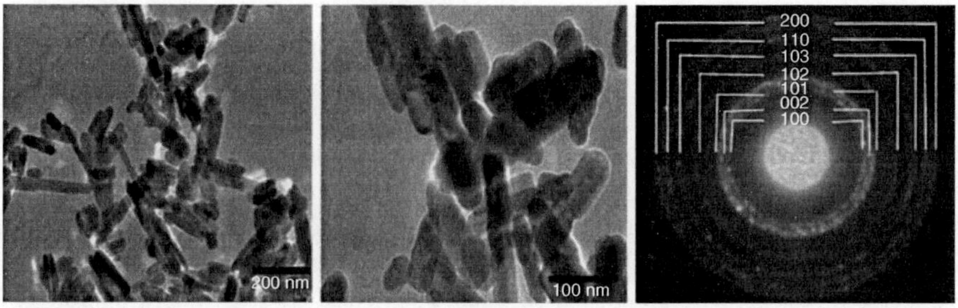

Fig. 8.5 TEM images of ZnO powder (*left* and *middle*), and SAED diffraction pattern of particle (*right*) [17]

Fig. 8.6 SEM of the sonochemically prepared Fe_2O_3 powder [20]

nanoparticles have been obtained by the sonochemical method. Such particles show bright, stable photoluminescence both in the solid state and in the colloidal dispersions [19].

8.5.2 Fe_2O_3

Nanoparticles of iron oxides have applications in diverse areas due to their larger surface area. Major areas of applications include; magnetic liquids, photocatalysis, diagnostic imaging and drug delivery. Amorphous nanoscopic iron (III) oxide (20 nm) with interesting magnetic properties has been prepared by the sonolysis of $Fe(acac)_3$ as precursor and by using tetraglyme as solvent in the presence of Ar gas [20]. The SEM image of the prepared particles has been shown in Fig. 8.6. Depending on the amount of water in the reaction, the surface area of the particles increased from 48 and up to 260 m^2/g.

Nanocrystalline gamma iron oxide (γ-Fe_2O_3) recently been studied as a gas sensing material, has been synthesised at 70°C using sonication-assisted precipitation technique [21]. The synthesised material was then used for fabricating the sensor element and was tested for its electrical property. The response and recovery time of the sensor to1,000 ppm n-butane were less than 12s and 120s, respectively. The enhancement of the gas sensing performance for sonochemically prepared γ-Fe_2O_3 has been mainly attributed to the formation of nanosized form which results in a larger specific surface area. H_2O_2 generated from the evaporation and pyrolysis of water in the gas phase of the collapsing bubble has been attributed to the main mechanism for the formation of γ-Fe_2O_3 nanoparticles [3]. The formed H_2O_2 oxidises Fe^{2+}(aq) to Fe^{3+} (aq). Subsequently, Fe^{3+} ions hydrolyse to $Fe(OH)_3$, the sonication of which causes dehydration and the formation of γ-Fe_2O_3 nanoparticles.

8.5.3 MgO

MgO nanoparticles due to their high specific surface area are useful as destructive adsorbents for toxic chemical agents. Such MgO nanoparticles have been prepared by the sonochemical hydrolysis followed by supercritical drying using $Mg(OCH_3)_2$ and $Mg(OC_2H_5)_2$ as precursors [22]. The fundamental effect of ultrasound on the specific surface of MgO precipitates has been observed. A significantly lower specific surface area and larger particle size have been noted for the samples prepared without passing ultrasound.

8.5.4 PbO

Improving the characteristics with more discharge capacity and more cycle life is important for metal oxides, especially for lead oxide. In addition, obtaining more porous and spongy nanostructured materials is also an area of active research. Uniform and homogeneous lead oxide nanoparticles (20–40 nm) with more spongy morphology have been obtained by Karami et al. [23] using PVP (polyvinyl pyrrolidone) as structure directing agent. It has been found out that the synthesized oxide, as anode and cathode of lead-acid batteries, showed very excellent discharge capacity of 230 $mAhg^{-1}$ and cycle life.

8.5.5 PbO_2

Ghasemi et al. [24] have obtained nanostructured PbO_2 (50–100 nm) using β-PbO precursor and in the presence of ammonium peroxydisulfate as an oxidant. Here, the ultrasonication dispersed and then cracked the β-PbO particles, thereby increasing the contribution of their surface area. Such an ultrasonic treatment resulted in an enhancement in the oxidation of PbO to PbO_2 has been observed. Ultrasonic waves also have been found to inhibit the formation of PbO_2 particles larger than 150 nm.

8.5.6 *SnO and SnO_2*

SnO has received much attention as a potential anode material for the lithium-ion-secondary-battery. The conventional techniques require temperatures above 150°C to form phase pure SnO. Whereas, sonication assisted precipitation technique has been used to prepare phase-pure SnO nanoparticles at room temperature by Majumdar et al. [25]. In this study, ultrasonic power has been found to play a key role in the formation of phase pure SnO as with a reduction in the ultrasonic power authors have observed a mixed phase. For the case of high ultrasonic power, authors have proposed that, intense cavitation and hence intense collapse pressure must have prevented the conversion of SnO to SnO_2.

SnO_2 has been widely used in devices including gas sensors due to the advantages of high sensitivity, simple design and low weight and cost [26]. Gas sensing properties of a material is strongly dependent on its size. Thus, sonication-assisted preparation has been used to fabricate SnO_2 quantum dots (QD) with 3–4 nm to be used as a low temperature sensor with a dual function property [27]. The BET surface area of sonochemically as-prepared product is 257 m^2/g, while the specific surface area of SnO_2 prepared by conventional sol-gel method is about 80 m^2/g. Also, the sonochemically prepared sensor has shown a high response to CO in the whole temperature range of 25–300°C, which is three times higher than that of conventionally fabricated sensor synthesized by sol-gel method. A dramatic increase in response especially at low temperatures has been attributed to the dimension effects. It has also been found to be a highly selective sensor to CO in the presence of methane at temperatures lower than 300°C. Whereas, at temperatures above 300°C the sensor becomes more selective to methane, which clearly establishes a different selectivity at different operating temperature.

8.5.7 *Eu_2O_3*

Europium oxide (Eu_2O_3) nanorods have been prepared by the sonication of an aqueous solution of europium nitrate in the presence of ammonia. In this reaction, ammonium ions adsorbed on the $Eu(OH)_3$ particles (formed due to the collapse of the bubbles) results in the formation of a monolayer which then fuse together by hydrogen bonding leading to the formation of nanorods [28].

8.5.8 *HgO*

Unusual nanostructures with different shapes of mercury oxide have been synthesised by the direct ultrasonic method [29]. Influence of different factors on the size, morphology and crystallinity of HgO nanocrystallites has been reported. The effect of ultrasound on the size and morphology of the nanoparticles has been confirmed by conducting the reaction only in the presence of mechanical stirring. The role of polyvinyl alcohol (PVA) or alkali salts in generating different shapes of the product have also been noted in this study.

8.5.9 Silica

Hollow spheres of nanometer to micrometer dimensions define an important class of shape-fabricated materials which are of interest in the areas of fillers, protective containers, confined reaction vessels and carriers [30]. Hollow spheres are normally obtained by templating method using agents like latex, or gold colloidal to vesicles. But, the disadvantage of using vesicles as templates is that they often spend long time to achieve equilibrium or need pH adjustment [31]. Rana et al. [32] have proved the effectiveness of ultrasound in generating a vesicular hierarchical structure and a rapid synthesis of mesoporous silica vesicles. Authors have proposed that the high intense conditions generated due to ultrasound near the liquid-air interface accelerate the polymerization of inorganic moieties attached to the micelles resulting in a shorter preparation time. Fan and Gao [33] have also proposed an ultrasound method for a simple and effective way of synthesizing silica hollow spheres. In this study, ultrasound induced the formation of vesicles that were formed from oppositely charged sodium dodecyl sulfate (SDS) and tetrapropylammonium bromide (TPAB). Thus formed vesicles then act as templates for the growth of uniform and well-defined silica spheres with the diameter of 200 nm to 5 μm.

8.5.10 V_2O_5

Self-assembled nanorods of vanadium oxide bundles were synthesized by treating bulk V_2O_5 with high intensity ultrasound [34]. By prolonging the duration of ultrasound irradiation, uniform, well defined shapes and surface structures and smaller size of nanorod vanadium oxide bundles were obtained. Three steps which occur in sequence have been proposed for the self-assembly of nanorods into bundles: (1) Formation of V_2O_5 nuclei due to the ultrasound induced dissolution and a further oriented attachment causes the formation of nanorods (2) Side-by-side attachment of individual nanorods to assemble into nanorods (3) Instability of the self-assembled V_2O_5 nanorod bundles lead to the formation of V_2O_5 primary nanoparticles. It is also believed that such nanorods are more active for n-butane oxidation.

8.5.11 TiO_2

Nanostructured anatase with the particle size of 6.2 nm and a specific surface area of 300 m^2/g has been produced with the assistance of sonochemical method [35].

The sonochemically produced anatase subjected to heat treatments under ambient atmospheric conditions and at temperatures from 773 to 1,073 K and times between 1–72 h transformed only to rutile.

Preparation of chiral mesoporous materials has become a great interest for material scientists. Normally chiral property is introduced into chiral mesoporous material via an organic chiral templating component. But, by using a sonochemical method, Gabashvili et al. [36] have prepared mesoporous chiral titania using a chiral inorganic precursor and a non-chiral dodecylamine as a template. Size of the pores was 5.5 nm.

Recently Ohayon and Gedanken [37] have proposed a non-aqueous route for the synthesis of a variety of metal oxides (TiO_2, WO_3 and V_2O_5) in just a few minutes and at a relatively low temperature using ultrasound irradiation. The idea for the non-aqueous route is mainly to control the crystallite size, shape, and the overall dimensionality. In case of TiO_2, a quasi zero-dimensional and a spherical morphology with the size of 3–7 nm has been observed. Whereas for V_2O_5, quasi one-dimensional ellipsoidal morphology has been observed with lengths in the range of 150–200 nm and widths in the range of 40–60 nm. For WO_3, quasi two-dimensional platelets with square shapes having facets ranging from 30 to 50 nm and with the thickness in between 2–7 nm have been obtained.

8.5.12 ZrO_2

Zirconia nanopowders have attracted much attention recently due to their specific optical and electrical properties [38] and as catalysts [39]. Liang et al. [40] have synthesized pure ZrO_2 nanopowders via sonochemical method. In this study, the use of ultrasound has dramatically reduced the temperature of reaction and made the reaction conditions very easy to maintain.

8.5.13 Other Mixed Metal Oxides

Series of scheelite-structured materials with the formula $MMoO_4$ ($M = Ca, Sr, Ba$) have been obtained sonochemically in the nanoregime (8–30 nm) by Thongtem et al. [41]. Monosized spherical particles of $BaTiO_3$ have also been successfully synthesized by the sonochemical method in a strong alkaline environment using $BaCl_2.2H_2O$ as the barium source and $TiCl_4$ as the titanium source. By changing the reactant concentration, particles were obtained in the size range from submicron (600–800 nm) to nanometer (60–70 nm).

Novel single, double and triple doped $ZnAl_2O_4$:M and $ZnGa_2O_4$:M (where $M = Dy^{3+}$, Tb^{3+}, Eu^{3+} and Mn^{2+}) nanophosphors were also synthesized through a simple sonochemical process [42].

$LiCoO_2$, one of the most widely used cathode materials in lithium rechargeable batteries because of its high specific capacity, has been prepared in the form of nanoparticles (20 nm) with very interesting characteristics under ultrasound irradiation at 80°C. More importantly, the above particles were obtained without subjecting them to any further heat treatment at high temperatures [43]. It has been observed that even a slight change in the reaction conditions has a strong influence in the structure and morphology of the resultant particles.

The interest in the synthesis and properties of delafossite structured compounds that have the general formula of ABO_2 have grown due to their p-type conductivity and optical transparency. The application of ultrasound for the synthesis of ternary oxide $AgMO_2$ (M = Fe, Ga) has been investigated by Nagarajan and Tomar [44]. Above materials were obtained in crystalline form within 40–60 min of sonication.

$LiMn_2O_4$ has been attracted as an important cathode material for rechargeable Li^+ ion batteries since it has several advantages such as high potentials, cheap cost, and low toxicity [45]. For this, Mn_3O_4 was used as a precursor, the nanoparticles of which were prepared using a simple sonochemical method at room temperature and by using a reaction time of just 20 min [46]. The size of the particles obtained was in the range of 55–65 nm and the yield was 97% in the above reaction. LiOH was coated onto the resulting nanoparticles, again using the sonochemical method, the heating of which at a relatively low temperature resulted in the formation of phase-pure $LiMn_2O_4$ nanoparticles (50–70 nm).

Homogeneous $LaMnO_3$ nanopowder with the size of 19–55 nm and with the specific surface area of 17–22 m^2/g has been synthesized using a surfactant, sodium dodecyl sulphate (SDS) to prevent agglomeration [47]. The sonochemically prepared $LaMnO_3$ showed a lower phase transformation temperature of 700°C, as compared to the $LaMnO_3$ prepared by other conventional methods which has been attributed to the homogenization caused by sonication. Also, a sintered density of 97% of the powders was achieved for the sonochemically prepared powders at low temperature than that of conventionally prepared powders.

Magnesium aluminate spinel has received much attention as a technological material for its interesting properties such as melting point, high mechanical strength at elevated temperatures, high chemical inertness, and good thermal shock resistance [48]. High surface area $MgAl_2O_4$ spinel has been synthesized by the sonochemical method using two kinds of precursors, alkoxides and nitrates/acetates and by using a surfactant cetyl trimethyl ammonium bromide [49]. The surface area of the material obtained was 267 m^2/g after heat treatment at 500°C and 138 m^2/g at 800°C.

8.5.14 Ultrasound Assisted Techniques

8.5.14.1 Ultrasound and Microwave

Industrially, transformation of syngas is normally carried out using Fisher-Tropsch (FT) method which utilizes either Fe or Co based catalysts to obtain useful fuels and chemicals. For wider applications, the usage of a support for these catalysts improves the mechanical resistance. In this regard, supported iron based FT catalysts with high loading of active metal have been prepared using ultrasound and microwave [50] separately and their catalytic activities have been compared. It has been observed that the catalysts prepared by means of ultrasound found to be the most efficient in terms of both CO conversion and of suitable products yield, particularly when sonication was performed in the presence of Ar atmosphere.

8.5.14.2 Ultrasound and Photochemistry

TiO_2 nanotube array is of a promising and important prospect in solar cells, environmental purification and in bio-application due to its highly ordered array structure, good mechanical and thermal stability [51]. But, the efficiency of photocatalytic degradation is limited due to its high rate recombination of photogenerated electron-hole pairs. One of the ways to suppress that effect is to dope noble metals in the above array. In this regard, ultrasound aided photochemical route has been explored to prepare TiO_2 nanotube array photocatalyst loaded with highly dispersed Ag nanoparticles [51]. The photocurrent and photocatalytic degradation rate of thus prepared Ag-TiO_2 nanotube array were about 1.2 and 3.7 times as that of pure TiO_2 nanotube array respectively.

8.5.14.3 Ultrasound and Electrochemistry (Sonoelectrochemistry)

Reisee et al. [52] first described a pulsed electrodeposition and pulsed out-of-phase ultrasound to prepare copper nanopowders. Such an electrochemical method has since then employed to synthesize a variety of nanoparticles. Mancier et al. [53] have prepared Cu_2O nanopowders (8 nm) with very high specific surface area of 2,000 m^2/g by pulsed ultrasound assisted-electrochemistry.

8.6 Sulfides

Sulfides, in specific due to their nanoparticular form, are important semiconducting group II–IV materials as they have a typical wide band gap energy, for example, cadmium sulphide and zinc sulphide. These materials have excellent optical, photo-, and electroluminescence properties and thus find wide applications in modern technology such as light-emitting diodes, solar cells, optical devices based on the non-linear optical properties, sensors and displays, bio-imaging and catalysis [54]. Following are the sulfides that have been obtained using the sonochemical method.

8.6.1 ZnS

Zinc sulfide was generated *in situ* using an aqueous solution of zinc acetate and thioacetamide in the presence of ultrasonic irradiation and the generated zinc sulfide was coated uniformly further onto the silica microspheres simultaneously [55]. Such a coating of semiconductor nanoparticles was carried out on a solid support to obtain unique optical, electronic and catalytic properties. This is the starting point of utilizing ultrasound irradiation for the surface synthesis of a wide variety of core/shell type materials. Rana et al. [56] have generated ZnS nanoparticles *in situ* using sonication and dodecylamine as the structure-directing agent. Authors have observed a stable mesoporous network with an average pore diameter of 28 Å and with the high surface area of 210 m^2/g, during this method. Also, they have carried out a systematic analysis in order to find out the role of ultrasound on the supramolecular assembling process that leads to the generation of supramolecular structure. Zhou et al. [57] have coupled sonochemistry and bacteria to obtain ZnS hollow nanostructures *in situ* and in a single step. They have used lactobacillus bacteria as a sacrificial template, as they represent a large variety of well-defined stunning morphologies. The authors have predicted that similar structures for different materials could be formed if the precursors have strong interaction with the cell surfaces under ultrasound irradiation.

8.6.2 CdS

CdS nanocrystals have been obtained by precipitation using cadmium carboxylate in dimethyl sulfoxide solution with or without elemental sulfur. Depending on the reaction conditions, 2–7 nm size of the crystals was obtained [58]. Li et al. [59] have synthesized hexagonal CdS nanoparticles (40 nm) using cadmium acetate and S under H_2/Ar atmosphere. Through control experiments, it has been demonstrated that extreme conditions induced by the collapse of the bubbles due to ultrasound could have accelerated the reduction of elemental S by hydrogen. CdS nanocrystals with lamellar morphology with the thickness of few nanometers and with the lengths of micrometer have also been obtained using the complex templates of polyelectrolyte/surfactant by Tao et al. [60]. Jian and Gao [61] have used ultrasound activated liquid-liquid two phase approach for the synthesis of CdS nanocrystals at room temperature and during a reaction time of just 15 min. Figure 8.7 shows the HRTEM images as well as the SAED pattern of the obtained CdS nanocrystals. More importantly this reaction has been scaled-up by 50 times with 90% yield. Formation of CdS nanocrystals was realized through many cycles of diffusion of the large amount of nuclei that have been formed in the first 10 s at the liquid-liquid interface. Aggregation of such nuclei is prevented by the organic amine surfactants and such nuclei then grow at the interface. Using the same approach authors have prepared Au/CdS nanocomposites. Yadav et al. [62] have used an amino acid, histidine as chelating agent, and synthesized CdS nanoparticles under sonochemical

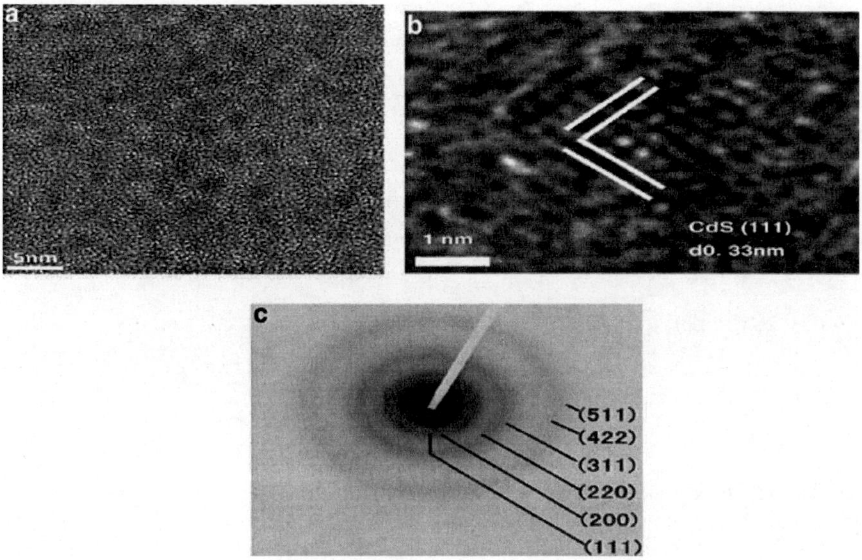

Fig. 8.7 (**a**) and (**b**) HRTEM images and (**c**) SAED pattern of CdS nanocrystals [61]

conditions. To a surprise, band gap was changed along with particle size just by changing the ultrasound irradiation time. But, the yield in this reaction was 72% only, which maybe due to the insufficient amounts of hot-spots formed. In this reaction, the imidazole ring of histidine captures the Cd ions and thereby prevents the agglomeration and later on slowly releasing these ions into the solution in the presence of ultrasound waves.

CdS particles are difficult to be dispersed in water, although they can be dispersed in some organic solvents such as DMF because of the strong interaction between the nanoparticles. A stable colloidal solution of nano-CdS (3–5 nm) has been successfully prepared using simple precursors cadmium chloride and sodium sulfide with PVP as the dispersant in water by a sonochemical method without any sedimentation at least for a month. Figure 8.8. shows the freshly prepared colloidal solution of CdS nanoparticles by the sonochemical method as well as the one kept in air for 1 month. Here, PVP has been used as a capping agent for the nanoparticles to prevent from agglomeration [63].

8.6.3 CuS

High quality one-dimensional copper sulfide (CuS) nanorods (50–200 nm) have been demonstrated using template assisted electrochemical deposition, a sonoelectrochemical method. Thus generated nanorods were also electrically characterized as p-type semiconductors [64]. In this process, ultrasound assists the electrochemical deposition by the combination of any of the following three processes:

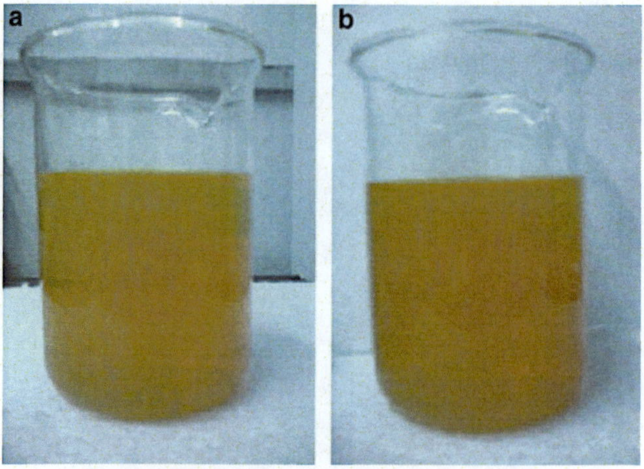

Fig. 8.8 CdS nanoparticles colloid solution prepared by the sonochemical method: freshly prepared (**a**); kept in air for 1 month (**b**) [63]

(1) reducing the mass transfer resistance (2) continuous cleaning of electrode surface (3) increase in the reaction rate with the generation of intense bubble collapse.

Doping is important for semiconductors in order to tune their optical and electrical properties for the potential applications in biotechnology and solar cells [65]. Ag-doped hexagonal CdS nanoparticles were successfully obtained by an ultrasound-assisted microwave synthesis method. Here, the doping of Ag in to CdS nanoparticles induced the evolution of crystal structure from cubic to hexagonal. Further support from photocatalytic experiment also clearly indicates the doping of Ag clusters into the CdS matrix.

8.6.4 PbS

PbS has attracted much attention due to its special direct band gap energy (0.4 eV) and a relatively large exciton Bohr radius (18 nm) and their nanoclusters have potential applications in electroluminescent devices such as light-emitting diodes. PbS nanocrystals with rod like structures with diameters of 20–60 nm and lengths of 1–2 μm have been obtained using the sonochemical method and by using PEG-6000 [66]. Addition of PEG and the time of sonication have been found to play a key role in the formation of these rods.

8.6.5 MoS_2

Mdleleni et al. [67] have reported a simple sonochemical route to generate nanostructured molybdenum sulfide (MoS_2) using molybdenum hexacarbonyl and sulfur.

Interestingly, as compared to the conventional thermal method which results in with the surface area of 32 m^2/g, the sonochemical method results in with 55 m^2/g. Also, under the same experimental conditions, the sonochemically prepared MoS_2 showed a higher catalytic activity (three to five times as compared to the catalysts prepared by conventional methods) towards thiophene hydrodesulphurization as compared to the conventional counterparts. Highly dispersed MoS_2 in aqueous solution has been prepared by the sonolysis of a simple inorganic precursor, $(NH_4)_6Mo_7O_{24}.4H_2O$. Through TEM studies, a hollow morphology has been observed, which the authors believe that these reproduce or replicates the shape of the cavitation bubbles [68]. In addition, they have observed much shorter and less ordered slabs of MoS_2 by using the sonochemical method.

8.6.6 In_2S_3

Indium sulfide (In_2S_3) was obtained by sonicating an aqueous solution of $InCl_3$ with thioacetamide. Interestingly, in this reaction, depending on the reaction temperature, different products have been observed. For example, at 0°C, In_2O_3 was the major product. Whereas, In_2S_3 was obtained as the major product at room temperature [69].

8.6.7 Bi_2S_3

Bismuth sulfide (Bi_2S_3) nanorods with different diameters and lengths have been successfully prepared by the sonochemical method using an aqueous solution of bismuth nitrate and sodium thiosulfate in the presence of complexing agents, for example, ethylenediaminetetraacetic acid (EDTA), triethanolamine, and sodium tartrate. Thioacetamide has also been used as a sulfur source. Also, by using N,N-dimethylformamide (20%) as a solvent, a higher yield as well as smaller sizes of Bi_2S_3 nanorods have been noted [70].

8.6.8 NbS_2

An alternative route to obtain NbS_2-sheathed carbon nanotubes (CNT) has been proposed by Zhu et al. [71] with this sonochemical method. In this study, CNTs act as templates to produce the uniform and well-crystallized NbS_2 nanotubes and the formation of such nanotubes has been explained by means of multi-point nuclei site growth mechanism.

8.6.9 $AgBiS_2$

Pejova et al. [72] (2008) have obtained three-dimensional arrays of close-packed semiconducting $AgBiS_2$ quantum dots with an average QD radius of 4.2 nm using sonochemical method which was twice as small as compared to the QD solid obtained without ultrasonic irradiation.

8.7 Conclusions

Above discussion/results clearly demonstrate that ultrasound or the sonochemical method has greater potential in the generation of a variety of oxides and sulfides as well as to control the size and morphology of the particles generated. Moreover, already research has been directed towards coupling the advantages of ultrasound with other green technologies (sonochemistry assisted methods), for example, with ionic liquids, or with photochemistry, or with electrochemistry or along with microwaves to obtain these materials. Variety of materials with novel properties have been obtained with these methods. It is very important to note that as special conditions are not needed, for example high temperature, high pressure, surfactants, longer reaction time etc., such sonochemical reactions could very well be scaled-up to an industrial scale in the near future.

References

1. McNamara WB, Didenko YT, Suslick KS (1999) Sonoluminescence temperatures during multibubble cavitation. Nature 401:772–775
2. Suslick KS, Price GJ (1999) Applications of ultrasound to materials chemistry. Annu Rev Mater Sci 29:295–326
3. Suslick KS (1998) Kirk-Othmer encyclopedia of chemical technology, 4th edn, vol 26, pp 517–541. Wiley: New York.
4. Suslick KS, Hyeon T, Fang M (1996) Nanostructured materials generated by high-intensity ultrasound: Sonochemical synthesis and catalytic studies. Chem Mater 8(8):2172–2179
5. Gedanken A (2004) Using sonochemistry for the fabrication of nanomaterials. Ultrason Sonochem 11(2):47–55
6. Majetich SA, Jin Y (1999) Magnetisation directions of individual nanoparticles. Science 284 (5413):470–473
7. Zarur AJ, Ying JY (2000) Reverse microemulsion synthesis of nanostructured complex oxides for catalytic combustion. Nature 403:65–67
8. Wang ZL (2004) Nanostructures of zinc oxide. Mater Today 7(6):26–33
9. Sivakumar M, Towata A, Yasui K, Tuziuti T, Iida Y (2005) Ultrasonic cavitational activation: a simple and feasible route for the direct conversion of zinc acetate to highly monodispersed ZnO. Chem Lett 35(1):60–61
10. Xiao Q, Huangb S, Zhang J, Xiao C, Tan X (2008) Sonochemical synthesis of ZnO nanosheet. J Alloys Compounds 459:L18–L22

11. Bhattacharyya S, Gedanken A (2008) A template-free, sonochemical route to porous ZnO nano-disks. Microp Mesop Mater 110:553–559
12. Mishra P, Yadav RS, Pandey AC (2009) Starch assisted sonochemical synthesis of flower-like ZnO nanostructure. Digest J Nanomaterials Biostructures 4(1):193–198
13. Pal U, Kim CW, Jadhav NA, Kang YS (2009) Ultrasound-assisted synthesis of mesoporous ZnO nanostructures of different porosities. J Phys Chem C 113(33):14676–14680
14. Jia X, Fan H, Zhang F, Qin L (2010) Using sonochemistry for the fabrication of hollow ZnO microspheres. Ultrason Sonochem 17:284–287
15. Hou X, Zhou F, Sun Y, Liu W (2007) Ultrasound-assisted synthesis of dentritic ZnO nanostructure in ionic liquid. Mater Lett 61:1789–1792
16. Goharshadi EK, Ding Y, Jorabchi MN, Nancarrow P (2009) Ultrasound-assisted green synthesis of nanocrystalline ZnO in the ionic liquid [hmim][NTf_2]. Ultrason Sonochem 16:120–123
17. Alammar T, Mudring AV (2009) Facile ultrasound-assisted synthesis of ZnO nanorods in an ionic liquid. Mater Lett 63:732–735
18. Arefian NA, Shokuhfar A, Vaezi MR, Kandjani AE, Tabriz MF (2008) Sonochemical synthesis of SnO/ZnO nano-Composite: the effects of temperature and sonication power. In: Öchsner A, Murch GE (eds) Defect and diffusion forum, vol 273–276, Diffusion in solids and liquids III., pp 34–39
19. Xiong HM, Shchukin DG, Möhwald H, Xu Y, Xia YY (2009) Sonochemical synthesis of hghly luminescent zinc oxide nanoparticles doped with magnesium (II). Angew Chem Int Ed 48(15):2727–2731
20. Pinkas J, Reichlova V, Zboril R, Moravec Z, Bezdicka P, Matejkova J (2008) Sonochemical synthesis of amorphous nanoscopic iron (III) oxide from $Fe(acac)_3$. Ultrason Sonochem 15:257–264
21. Raya I, Chakraborty S, Chowdhury A, Majumdar S, Prakash A, Pyare R, Sena A (2008) Room temperature synthesis of γ-Fe_2O_3 by sonochemical route and its response towards butane. Sens Actu B 130:882–888
22. Stengl V, Bakardjieva S, Marikova M, Bezdicka P, Subrt J (2003) Magnesium oxide nanoparticles prepared by ultrasound enhanced hydrolysis of Mg-alkoxides. Mater Lett 57:3998–4003
23. Karami H, Karimi MA, Haghdar S, Sadeghi A, Mir-Ghasemi R, Mahdi-Khani S (2008) Synthesis of lead oxide nanoparticles by sonochemical method and its application as cathode and anode of lead-acid batteries. Mater Chem Phys 108:337–344
24. Ghasemi S, Mousavi MF, Shamsipur M, Karami H (2008) Sonochemical-assisted synthesis of nano-structured lead dioxide. Ultrason Sonochem 15:448–455
25. Majumdar S, Chakraborty S, Sujatha Devi P, Sen A (2008) Room temperature synthesis of nanocrystalline SnO through sonochemical route. Mater Lett 62:1249–1251
26. Wang HC, Li Y, Yang MJ (2006) Fast response thin film SnO_2 gas sensors operating at room temperature. Sens Actu B 119:380–383
27. Mosadegh Sedghi S, Mortazavi Y, Khodadadi A (2009) Low temperature CO and CH_4 dual selective gas sensor using SnO_2 quantum dots prepared by sonochemical method. Sens Actu B (In Press)
28. Pol VG, Palchik O, Gedanken A, Felner I (2002) Synthesis of europium oxide nanorods by ultrasound irradiation. J Phys Chem B 106(38):9737–9743
29. Askarinejad A, Morsali A (2009) Synthesis and characterization of mercury oxide unusual nanostructures by ultrasonic method. Chem Eng J 153:183–186
30. Bourlinos AB, Karakassides MA, Petridis D (2001) Synthesis and characterisation of hollow clay microspheres through a resin template approach. Chem Commun 16:1518–1519
31. Hubert DHW, Jung M, Frederick PM, Bomans PHH, Meuldijk J, German AL (2000) Vesicle-directed growth of silica. Adv Mater 12(17):1286–1290
32. Rana RK, Mastai Y, Gedanken A (2002) Acoustic cavitation leading to the morphosynthesis of mesoporous silica vesicles. Adv Mater 14(19):1414–1418

33. Fan W, Gao L (2006) Synthesis of silica hollow spheres assisted by ultrasound. J Colloid Interface Sci 297:157–160
34. Taufiq-Yap YH, Wong YC, Zainal Z, Hussein MZ (2009) Synthesis of self-assembled nanorod vanadium oxide bundles by sonochemical treatment. J Nat Gas Chem 18:312–318
35. Gonzalez-Reyes L, Hernandez-Perez I, Robles Hernandez FC, Rosales HD, Arce-Estrada EM (2008) Sonochemical synthesis of nanostructured anatase and study of the kinetics among phase transformation and coarsening as a function of heat treatment conditions. J Eur Ceramic Soc 28:1585–1594
36. Gabashvili A, Major DT, Perkas N, Gedanken A (2010) The sonochemical synthesis and characterization of mesoporous chiral titania using a chiral inorganic precursor. Ultrason Sonochem 17:605–609
37. Ohayon E, Gedanken A (2010) The application of ultrasound radiation to the synthesis of nanocrystalline metal oxide in a non-aqueous solvent. Ultrason Sonochem 17:173–178
38. Murray EP, Tsai T, Barnett SA (1999) A direct-methane fuel cell with a ceria-based anode. Nature 400:649–651
39. Wang SB, Murata K, Hayakawa T, Hamakawa S, Suzuki K (1999) Excellent performance of lithium doped sulphated zirconia in oxidative dehydrogenation of ethane. Chem Commun 103–104.
40. Liang J, Jiang X, Liu G, Deng Z, Zhuang J, Li F, Li Y (2003) Characterisation and synthesis of pure ZrO_2 nanopowders via sonochemical method. Mater Res Bull 38:161–168
41. Thongtem T, Phuruangrat A, Thongtem S (2008) Sonochemical synthesis of $MMoO_4$ (M = Ca, Sr and Ba) nanocrystals. J Cer Proc Res 9(2):189–191
42. Dutta DP, Ghildiyal R, Tyagi AK (2009) Luminescent properties of doped zinc aluminate and zinc gallate white light emitting nanophosphors prepared via sonochemical method. J Phys Chem C 113(39):16954–16961
43. Kim KH, Kim KB (2008) Ultrasound assisted synthesis of nanosized lithium cobalt oxide. Ultrason Sonochem 15:1019–1025
44. Nagarajan R, Tomar N (2009) Ultrasound assisted ambient temperature synthesis of ternary oxide $AgMO_2$ (M = Fe, Ga). J Solid State Chem 182:1283–1290
45. Kang K, Meng YS, Breger J, Grey CP, Ceder G (2006) Electrodes with high power and high capacity for rechargeable lithium batteries. Science 311:977–80
46. Park JP, Kim SK, Park JY, Hwang CH, Choi MH, Kim JE, Ok KM, Kwak HY, Shim IW (2009) Syntheses of Mn_3O_4 and $LiMn_2O_4$ nanoparticles by a simple sonochemical method. Mater Lett 63:2201–2204
47. Das N, Bhattacharya D, Sen A, Maiti HS (2009) Sonochemical synthesis of $LaMnO_3$ nano-powder. Ceram Int 35:21–24
48. Baudin C, Martinez R, Pena P (1995) High temperature mechanical behaviour of stoichiometric magnesium spinel. J Am Ceram Soc 78(7):1857–1862
49. Troia A, Pavese M, Geobaldo F (2009) Sonochemical preparation of high surface area $MgAl_2O_4$ spinel. Ultrason Sonochem 16(1):136–140
50. Pirola C, Bianchi CL, Michele AD, Diodati P, Boffito D, Ragaini V (2010) Ultrasound and microwave assisted synthesis of high loading Fe-supported Fischer–Tropsch catalysts. Ultrason Sonochem 17:610–616
51. Sun L, Li J, Wang C, Li S, Lai Y, Chen H, Lin C (2009) Ultrasound aided photochemical synthesis of Ag loaded TiO_2 nanotube arrays to enhance photocatalytic activity. J Hazar Mater 171:1045–1050
52. Reisse J, Francois H, Vandercammen J, Fabre O, Kirsch-de Mesmaeker A, Maerschalk C, Delplancke JL (1994) Sonoelectrochemistry in aqueous electrolyte: a new type of sonoelectroreactor. Electrochim Acta 39(1):37–39
53. Mancier V, Daltin AL, Leclercq D (2008) Synthesis and characterization of copper oxide (I) nanoparticles produced by pulsed sonoelectrochemistry. Ultrason Sonochem 15:157–163
54. Hassan ML, Ali AF (2008) Synthesis of nanostructured cadmium and zinc sulfides in aqueous solutions of hyperbranched polyethyleneimine. J Crys Growth 310:5252–5258

55. Dhas NA, Zaban A, Gedanken A (1999) Surface synthesis of zinc sulfide nanoparticles on silica microspheres: sonochemical preparation, characterization, and optical properties. Chem Mater 11(3):806–813
56. Rana RK, Zhang L, Yu JC, Mastai Y, Gedanken A (2003) Mesoporous structures from supramolecular assembly of in situ generated ZnS nanoparticles. Langmuir 19(14):5904–5911
57. Zhou H, Fan T, Zhang D, Guo Q, Ogawa H (2007) Novel bacteria-templated sonochemical route for the in situ one-step synthesis of ZnS hollow nanostructures. Chem Mater 19(9):2144–2146
58. Elbaum R, Vega S, Hodes G (2001) Preparation and surface structure of nanocrystalline cadmium sulfide (sulfoselenide) precipitated from dimethyl sulfoxide solutions. Chem Mater 13(7):2272–2280
59. Li HL, Zhu YC, Chen SG, Palchik O, Xiong JP, Koltypin Y, Gofer Y, Gedanken A (2003) A novel ultrasound-assisted approach to the synthesis of CdSe and CdS nanoparticles. J Solid State Chem 172:102–110
60. Tao C, Zheng S, Möhwald H, Li J (2003) CdS crystal growth of lamellar morphology within templates of polyelectrolyte/surfactant complex. Langmuir 19(21):9039–9042
61. Jian D, Gao Q (2006) Synthesis of CdS nanocrystals and Au/CdS nanocomposites through ultrasound activation liquid–liquid two-phase approach at room temperature. Chem Eng J 121:9–16
62. Yadav RS, Mishra P, Mishra R, Kumar M, Pandey AC (2010) Growth mechanism and optical property of CdS nanoparticles synthesised using amino-acid histidine as chelating agent under sonochemical process. Ultrason Sonochem 17:116–122
63. Wu YD, Wang LS, Xiao MW, Huang XJ (2008) A novel sonochemical synthesis and nanostructured assembly of polyvinylpyrrolidone-capped CdS colloidal nanoparticles. J Non-Cryst Solid 354(26):2993–3000
64. Singh KV, Martinez-Morales AA, Senthil Andavan GT, Bozhilov KN, Ozkan M (2007) A simple way of synthesizing single-crystalline semiconducting copper sulfide nanorods by using ultrasonication during template-assisted electrodeposition. Chem Mater 19(10):2446–2454
65. Ma J, Tai G, Guo W (2010) Ultrasound-assisted microwave preparation of Ag-doped CdS nanoparticles. Ultrason Sonochem 17:534–540
66. Xiu Z, Liu S, Yu J, Xu F, Yu W, Feng G (2008) Sonochemical synthesis of PbS nanorods. J Alloys Compounds 457:L9–L11
67. Mdleleni MM, Hyeon T, Suslick KS (1998) Sonochemical synthesis of nanostructured molybdenum sulfide. J Am Chem Soc 120:6189–6190
68. Uzcanga I, Bezverkhyy I, Afanasiev P, Scott C, Vrinat M (2005) Sonochemical preparation of MoS_2 in aqueous solution: replication of the cavitation bubbles in an inorganic material morphology. Chem Mater 17(14):3575–3577
69. Avivi (Levi) A, Palchik O, Palchik V, Slifkin MA, Weiss AM, Gedanken A (2001) Sonochemical synthesis of nanophase indium sulphide. Chem Mater 13(6):2195–2200
70. Wang H, Zhu JJ, Zhu JM, Chen HY (2002) Sonochemical method for the preparation of bismuth sulfide nanorods. J Phys Chem B 106(15):3848–3854
71. Zhu YQ, Hsu WK, Kroto HW, Walton DRM (2002) An alternative route to NbS_2 nanotubes. J Phys Chem B 106(31):7623–7626
72. Pejova B, Grozdanov I, Nesheva D, Petrova A (2008) Size-dependent properties of sonochemically synthesized three-dimensional arrays of close-packed semiconducting $AgBiS_2$ quantum dots. Chem Mater 20(7):2551–2565

Chapter 9
Aqueous Inorganic Sonochemistry

Pankaj

Abstract This chapter discusses the effect of ultrasound propagation in water and aqueous solutions, in the atmosphere of inert and reactive gases. Sonochemical studies of aqueous solutions of divalent and trivalent metal ions and their salts have been reviewed and the precipitation behaviour of hydroxides of metal ions has been discussed. Synthesis of nanoparticles of many metals using ultrasound and in aqueous solutions has also been discussed briefly. Besides, the nephelometric and conductometric studies of sonicated solutions of these metal ions have been reported.

9.1 Introduction

The objective of this chapter is to compile work related to the beginning of sonochemical research and its extension to the aqueous solutions of metal ions. Ultrasound propagation in aqueous salt solutions leads to the hydrolysis, reduction, complexation, decomplexation and crystallization. Such works from different laboratories, along with the effect of dissolved gases on the production of free radicals in water and aqueous solutions upon sonication has been reviewed in this chapter. The generation of turbidity, due to the formation of metal hydroxides and changes in the conductivity of these aqueous solutions, carried out in this laboratory, has also been reported, to give firsthand information of the ultrasound interaction of these solutions.

Pankaj (✉)
Department of Chemistry, Dayalbagh Educational Institute, Agra 282 110, India
e-mail: pankaj2@sancharnet.in

Pankaj and M. Ashokkumar (eds.), *Theoretical and Experimental Sonochemistry Involving Inorganic Systems*, DOI 10.1007/978-90-481-3887-6_9,

The first effect of cavitation on metals was noticed by Thorneycroft [1] through the damage of the propellers of naval vessel HMS Daring, and explained later by Lord Rayleigh [2] way back in 1917, as due to cavitation. The same kind of turbulence in liquids was later found to be generated through the propagation of ultrasound, when the laboratory generation of ultrasound was possible [3]. However, a systematic scientific study of chemical effects of ultrasound was made much later, when Loomis [4] reported the effects of high frequency sound waves. The later years were devoted to the application of ultrasound for underwater acoustics mainly projected to save the defense and commercial naval fleets of the allied countries from the surprise attacks of the German U-boats in high seas. After the end of First World War, Langevin [3] played a very important role in these international research activities leading to the generation and detection of ultrasound waves, which eventually resulted in the detection of underwater objects and the discovery of many other applications of ultrasound in the diversified field of science and technology [5], as we know today.

There was a lull in such studies for quite some time, before however, Arnim Henglein initiated a series of experiments and reported some fundamental findings in aqueous solutions containing inorganic species and explained in terms of free radical formation due to sonolytic dissociation of water and resulting into the formation of reactive species such as; H_2O_2, O_3 and reduction of several metal cations or their precipitation as corresponding hydroxides. Besides, the collective effort of leading sonochemists like: TJ Mason, MA Margulis, KS Suslick, JL Luche, R Verrall, P Boudjouk, J Lindley, P Riesz, P Kruus, J Einhorn and JP Lorimer, who after the first R.S.C. Sonochemistry Symposium at Warwick University, UK, in 1986, initiated a series of systematic and well defined research, unfolding several new applications of ultrasound in the areas involving organic syntheses, polymerization, degradation, waste water treatment, metal catalysis, extraction processes of natural products, cell disruption, food technology, industrial scale up operations, photocatalysis, electrosonolytic and several such combinatorial techniques to name just few of these activities. Increase in the scientific activities had a similar reflection on the number of publication involving such studies in the form of various monographs [6], books [7] and reviews [8], which appeared during the past 2 decades.

Of the nearly a dozen of books published in this area, broadly covering various topics related to sonochemistry, ultrasound generation, industrial and chemical applications, acceleration of reaction rate or reduction in the drastic conditions in several reactions for a number of syntheses, no attention seem to have been paid to compile the work of inorganic species, especially in aqueous solutions, which is the most conventional as well as convenient way of teaching and learning sonochemistry. It is with this intention and objective the present effort has been made with the belief that it would provide researchers of this field a comprehensive idea of the work done so far and the future role in all such activities in chemical advances – especially involving inorganic species and systems. Besides, the physical effects of ultrasound are equally important. Therefore, the last segment of this chapter concludes with the process such as sterilization and crystallization.

9.2 Chemical Effects of Ultrasound

The initial study of the sonochemical oxidation of inorganic species such as iodide ions in the aqueous solutions of KI, leading to the formation of I_3^- ions even in the absence of O_2 was reported by Richards and Loomis [9] and Weissler et al. [10]. Hart and Henglein [11] later reported the simultaneous production of H_2O_2 and iodine upon the sonolysis of aqueous potassium iodide and sodium formate solutions. Gutierrez et al. reported [12] the similar oxidation for Br^- and N_3^- as well, with Br_2 and N_2 being the end products of sonolysis of potassium bromide and sodium/ceasium azide (Na/CsN_3) solutions. However, since the azide anion was also reactive towards the hydrogen atom, additional nitrogen was formed at higher azide concentrations due to scavenging of hydrogen atoms. They [12] explained the process of N_3^- reactions as under;

$$H + N_3^- \longrightarrow NH^- + N_2 \quad (9.1)$$

$$NH^- + H_2O \longrightarrow NH_2 + OH^- \quad (9.2)$$

$$NH_2 + N_3^- \longrightarrow NH_2^- + N_3 \quad (9.3)$$

$$N_3 \longrightarrow 1.5\,N_2 \quad (9.4)$$

$$NH_2^- + H_2O \longrightarrow NH_3 + OH^- \quad (9.5)$$

$$H + 2N_3^- + 2H_2O \longrightarrow NH_3 + 2.5N_2 + 2OH^- \quad (9.6)$$

Thus a lot of additional nitrogen was formed. If N_3^- scavenged H atoms, which would otherwise yield H_2, then 2 mol of NH_3 and 5 mol of N_2 molecules would be formed for each H_2 molecule not formed. When Hart and Henglein [11] sonicated aqueous solutions of potassium iodide and sodium formate with 300 kHz ultrasound and in atmospheres of argon, oxygen and Ar–O_2 mixture of several compositions, besides in pure water and water containing ozone, they reported the liberation of I_2 and H_2O_2 with maximum yield being in the atmosphere of 70% argon and 30% oxygen. The products of similar experiments with sodium formate were H_2, CO_2, H_2O_2 and oxalate in the absence of oxygen but CO_2 and H_2O_2 in the presence of oxygen. They explained the formation of above products upon sonication on the basis of $H^\bullet$ and $OH^\bullet$ free radical formation in the imploding gas bubble. The role of gas in the atmosphere of experimental set up was crucial. They concluded that in the absence of O_2, H atoms formed H_2 and the $OH^\bullet$ radicals formed H_2O_2 or reacted with nonvolatile substrates in the interfacial region. Besides, the formation of superoxide radical anion, O_2^-, with a half life of 3 min was also detected in the oxygenated formate solution of pH 14. Hart and Henglein [13] further reported the sonolysis of ozone in aqueous solution and found extremely rapid decomposition of O_3 due mainly to the thermal instability of O_3 molecules.

Hart and Henglein [14] also reported the sonolytic decomposition of nitrous oxide in aqueous solutions under pure argon, pure N_2O and the mixture of the two gases and reported the formation of species such as N_2, O_2, NO_2^- and NO_3^- with the maximum yield being in the Ar/N_2O mixture in the vol% ratio of 85:15. Although H_2O is thermodynamically much more stable than N_2O but they postulated that all H_2O and N_2O molecules in an argon bubble were converted into free radicals in the short time of adiabatic compression phase of the bubble. They proposed a series of free radical reactions for the formation of all these species in aqueous solutions.

$$N_2O \longrightarrow N_2 + O^\bullet \tag{9.7}$$

$$O^\bullet + N_2O \longrightarrow 2NO \tag{9.8}$$

$$NO + OH^\bullet \longrightarrow HNO_2 \tag{9.9}$$

$$NO + NO_2 \longrightarrow N_2O_3 \tag{9.10}$$

$$NO_2 + OH^\bullet \longrightarrow HNO_3 \tag{9.11}$$

They further reported the isotopic exchange in the sonolysis of aqueous solutions containing $^{14.14}N_2$ and $^{15.15}N_2$ [15], H/D isotopic exchange in HD–H_2O system [16], H/D isotopic exchange in D_2–H_2O system [17], isotopic exchange in aqueous solution containing D_2–CH_4 [18] isotopic distribution of $^{18.18}O_2$ with H_2O_2 [19], combustion of H_2–O_2 in cavitation bubbles [20], pyrolysis of C_2H_2 in cavitational bubbles [21] in a series of studies elaborating the role of cavitation. They explained the mechanism of reaction through the initiation by $H^\bullet$ and $OH^\bullet$ free radicals formed due to sonolysis of water and later on switching over to isotopic combinations.

In an atmosphere of HD or a mixture of Ar/HD, when water was irradiated with 300 kHz ultrasound, H_2 and D_2 were formed [16]. The final products revolved around a four centered reaction, without any involvement of free radicals.

$$2HD \longrightarrow H_2 + D_2 \tag{9.12}$$

Sonication of water in the atmosphere of argon and molecular nitrogen (1:1 mixture of $^{14.14}N_2$ and $^{15.15}N_2$), [15] yielded nitrogen containing product; $^{14.15}N_2$, NO_2^-, NO_3^-, NH_3 and N_2O from N_2 concentration between 40% and 60% in the gas atmosphere. $^{14.15}N_2$, NO_2^- and NO_3^- were produced almost in similar amounts whereas the yields for NH_3 and N_2O concentrations were comparatively small. They also observed that the yield of H_2O_2, H_2 and O_2 were maximum in the absence of N_2, which, however, decreased with increasing concentrations of N_2. Similar sonication of $^{18.18}O_2$ yielded all possible isotopic combinations of O_2 and H_2O_2 molecules with the consumption of $^{18.18}O_2$ being much faster than the production of H_2O_2 and ^{16}O products. They viewed all these products as simple combustion chemistry inside hot gas bubbles in which the isotopic identities of radicals, $^{16}OH_2$ and $^{18.18}O_2$ was practically lost. They also postulated the formation of O_2 and H_2O_2 in the cooler interfacial regions of the propagating ultrasound [19].

$$H_2{}^{16}O \xrightarrow{)))} H^\bullet + {}^{16}OH^\bullet \tag{9.13}$$

$$^{16}OH^\bullet + {}^{16}OH^\bullet \longrightarrow {}^{16.16}O_2H_2 \tag{9.14}$$

$$H^\bullet + {}^{18.18}O_2 \longrightarrow {}^{18.18}O_2H \tag{9.15}$$

$$^{18.18}O_2H + {}^{18.18}O_2H \longrightarrow {}^{18.18}O_2H_2 + {}^{18.18}O_2 \tag{9.16}$$

$$^{18.18}O_2 + 2\,{}^{16}OH_2 \longrightarrow {}^{16.16}O_2H_2 + {}^{18.18}O_2H_2 \tag{9.17}$$

$$^{18.18}O_2 + 2\,{}^{16}OH_2 \longrightarrow {}^{16.18}O_2H_2 + {}^{16.18}O_2H_2 \tag{9.18}$$

$$^{18.18}O_2 + 2\,{}^{16}OH_2 \longrightarrow {}^{16.18}O_2 + {}^{18}OH_2 + {}^{16}OH_2 \tag{9.19}$$

$$^{18.18}O_2 + 2\,{}^{16}OH_2 \longrightarrow {}^{16.16}O_2 + {}^{18}OH_2 + {}^{18}OH_2 \tag{9.20}$$

The combustion of H_2 and O_2 in the cavitation bubble [20] and the consumption of these gases, when in the ratio of 1:3 and 2:1, favoured the maximum formation of H_2O_2. Since the conditions of gases in a cavitating bubble were different from those under flame, the mechanism could not precisely be the same. The intermediates of chain reactions rapidly reached cooler interfacial region of the bubble and therefore did not allow the chain length to go beyond 10 steps. Fe^{2+} ions present in the solution were oxidized to Fe^{3+} ions in various steps;

$$Fe^{2+} + OH^\bullet \longrightarrow Fe^{3+} + OH^- \tag{9.21}$$

$$Fe^{2+} + H_2O_2 \longrightarrow Fe^{3+} + OH^- + OH \tag{9.22}$$

$$Fe^{2+} + HO_2 \xrightarrow{H+} Fe^{3+} + H_2O_2 \tag{9.23}$$

Thus yield of Fe^{3+} was: $Y[Fe^{3+}] = Y[OH] + 2Y[H_2O_2] + 3Y[HO_2]$

The yield for final product in cavitation bubble rich in oxygen was therefore different

$$H + O_2 \begin{cases} \longrightarrow HO_2 & \text{(Cooler region)} \\ \longrightarrow OH^\bullet + O & \text{(Hotter region)} \end{cases} \tag{9.24}$$

On the experimental lines as above, Hart et al. [21] reported the pyrolysis of acetylene in sonolytic cavitational bubble in aqueous solution by irradiating through 1 MHz ultrasound of about 2 W/cm^2 intensity and confirmed products as H_2, CO, CH_4, along with a great number of hydrocarbons containing two to about eight carbon atoms, such as formic and acetic acids, formaldehyde and acetaldehyde and insoluble soot (carbon particles). Besides, the formation of larger molecules such as benzene, isomers of benzene, phenylacetylene, styrene and naphthalene were also confirmed. Surprisingly the products were similar to those as observed in the pyrolysis and combustion of acetylene under flame conditions. They further observed that all products were formed in single cavitation event, even at times in the 10-s range and not through gradual multistep formation of molecules. However, at higher acetylene concentration, the direct pyrolysis of acetylene resulted in the formation of precursor species [21, 22], which produce higher carbon atom number products. The similarity of the products with those reported in combustion studies provided excellent evidence that sonolysis of acetylene generated compounds that were similar, if not identical, with those formed in flames and shock tubes.

$$C_6H_6 \xrightarrow{))))} C_6H_5^{\bullet} + H^{\bullet} \quad (9.25)$$

$$C_6H_6 + H^{\bullet} \longrightarrow C_6H_5^{\bullet} + H_2 \quad (9.26)$$

$$C_6H_5^{\bullet} \longrightarrow C_4H_3^{\bullet} + C_2H_2 \quad (9.27)$$

$$C_4H_3^{\bullet} \longrightarrow C_4H_2 + H^{\bullet} \quad (9.28)$$

Miller [23] after his study of the oxidation of Fe^{2+} to Fe^{3+} ions concluded that the effect of oxidation was in fact indirectly related to the production of $OH^{\bullet}$ radicals, which achieved an optimum concentration after sonication of 10 min in the presence of ultrasound. Riesz et al. [24–26] proved the formation of above free radicals using spin trap technique in conjunction with E.S.R. Henglein [27] later reported that in the absence of dissolved oxygen and other chemical species, 80% of the primary radicals recombine to form water, although the other data indicated the formation of high energy species, such as solvated electrons [28] and hydroperoxy [29] radicals. The presence of a particular gas in aqueous medium could affect the outcome of a sonochemical reaction. When hydrogen is present, oxidation reactions were completely suppressed due to the reducing nature of H atoms. Henglein [30] suggested two possible ways (Eqs. (9.29) and (9.30)) in which H atoms could be formed when water was sonicated in the presence of H_2 gas.

$$H_2 \xrightarrow{))))} H^{\bullet} + H^{\bullet} \quad (9.29)$$

$$H_2O \xrightarrow{))))} H^{\bullet} + OH^{\bullet} \quad (9.30)$$

$$H_2 + OH^{\bullet} \longrightarrow H_2O + H^{\bullet} \quad (9.31)$$

The formation of H_2O_2 was commonly observed when ultrasonically irradiated water contained oxygen. Relatively low amount of H_2O_2, seen on sonolysis in the absence of air, was due to the fact that $H^\bullet$ and $OH^\bullet$ radicals readily recombined [30]. Thus, in the presence of O_2 the recombination of $OH^\bullet$ and $H^\bullet$ is slowed down by the following reaction:

$$H^\bullet + O_2 \longrightarrow HOO^\bullet \tag{9.32}$$

The production of $HOO^\bullet$ radicals increased oxidation processes due to further formation of H_2O_2 by a peroxy radical combination reaction.

$$HOO^\bullet + HOO^\bullet \longrightarrow H_2O_2 + O_2 \tag{9.33}$$

H_2O_2 was also formed if the sonolysed water contained a mixture of hydrogen and oxygen gas.

$$H^\bullet + O_2 \longrightarrow HOO^\bullet \tag{9.34}$$

$$HO^\bullet + H_2 \longrightarrow H_2O + H^\bullet \tag{9.35}$$

$$H^\bullet + O_2 \longrightarrow HOO^\bullet \tag{9.36}$$

$$2HOO^\bullet \longrightarrow H_2O_2 + O_2 \tag{9.37}$$

In this case, the peroxy radical reacted more readily with $HO^\bullet$ than with the H_2 molecules present in solution and so formation of $H^\bullet$ via Eq. (9.35) was less favoured than the reaction of the peroxy radicals with the $OH^\bullet$ radical.

$$HO^\bullet + HOO^\bullet \longrightarrow H_2O + O_2 \tag{9.38}$$

Furthermore, oxygen molecules decomposed directly in the cavitation bubble to form ozone [31, 32]

$$O_2 \xrightarrow{))))} O^\bullet + O^\bullet \tag{9.39}$$

$$O^\bullet + O_2 \longrightarrow O_3 \tag{9.40}$$

When nitrogen and oxygen gases were present then it was possible to get fixation of nitrogen with the formation of nitrogen oxides. Molecules of nitrogen and oxygen dissociated in the cavitation bubbles to form initially $NO^\bullet$ radicals [33, 34].

$$\left.\begin{array}{l} N_2 \xrightarrow{))))} N^\bullet + N^\bullet \\ O_2 \xrightarrow{))))} O^\bullet + O^\bullet \end{array}\right\} \longrightarrow NO^\bullet \tag{9.41}$$

If the irradiated water was, however, saturated with hydrogen or carbon monoxide instead of oxygen then no oxides were formed [35]. Under these conditions reduction of nitrogen occurred to afford ammonia [36]. Interestingly, HCN is also observed when carbon monoxide is replaced by methane [37].Thus the result of a sonochemical reaction, in water, depends largely on the nature of the dissolved gases and their concentration.

Similarly irradiated water–methanol mixtures by 1MHz ultrasound under Ar and O_2 yielded [38] typical decomposition products such as H_2, CO, CH_2O, CH_4, C_2H_4 and C_2H_6.

$$CH_3OH \longrightarrow CH_2O + H_2 \quad (9.42)$$

$$CH_3OH \longrightarrow CO + 2H_2 \quad (9.43)$$

$$2CH_3OH \longrightarrow CH_4 + CO + H_2 + H_2O \quad (9.44)$$

$$CH_3OH \longrightarrow {}^{\bullet}CH_3 + OH^{\bullet} \quad (9.45)$$

$$OH^{\bullet} + CH_3OH \longrightarrow H_2O + {}^{\bullet}CH_2OH \quad (9.46)$$

$${}^{\bullet}CH_3 + CH_3OH \longrightarrow CH_4 + {}^{\bullet}CH_2OH \quad (9.47)$$

$${}^{\bullet}CH_2OH + M \longrightarrow H^{\bullet} + HCO^{\bullet} + M \quad (9.48)$$

$$HCO^{\bullet} + M \longrightarrow CO + H^{\bullet} + M \quad (9.49)$$

$$H^{\bullet} + CH_3OH \longrightarrow H_2 + {}^{\bullet}CH_2OH \quad (9.50)$$

The generation of these free radicals and excited molecules of solvents and dissolved solutes set in a series of reactions which are either faster than those reactions that are carried out in the absence of ultrasound or sometimes result in end products, which are entirely different and unexpected.

Lay and Low [7(iii)] reported few initial studies of inorganic species, as shown below, but without any detailed studies;

$$Fe^{2+} \longrightarrow Fe^{3+} \quad [39] \quad (9.51)$$

$$Br^{-}, Cl^{-} \longrightarrow Br_2, Cl_2 \quad [40] \quad (9.52)$$

$$CS_2 \longrightarrow H_2S + \text{sulphides} \quad [41] \quad (9.53)$$

$$HgCl_2 \longrightarrow Hg_2Cl_2 \quad [42] \quad (9.54)$$

$$KMnO_4 \longrightarrow Mn(OH)_2 \quad [42] \quad (9.55)$$

$$Ce(IV) \longrightarrow Ce(III) \quad [43] \tag{9.56}$$

$$Ti(II) \longrightarrow Ti(I) \quad [43] \tag{9.57}$$

Gutierrez et al. [44] studied the trapping of hydrogen atoms by 5,5–dimethyl-pyrroline N–oxide (DMPO) and its scavenging by inorganic species such as Br_2, I_2, MnO_4^-, $AuCl_4^-$ and Ag^+ ions in an aqueous solution irradiated with 1 MHz ultrasound and in the environment of H_2/Ar mixture. They reported that DMPO could trap only a small amount of hydrogen atoms because these were substantially scavenged by other solutes, besides, the yield of H atoms decreased with increasing concentration of H_2 in the gas mixture and the reduction of other solutes passed through a maximum at about 20 vol% H_2. They also found that thallium ions were reduced in solution of higher pH, which was due to hydrated electrons formed in the reaction of the H–atoms with OH^- anions. Based on these observations, they finally concluded that the formation of hydrated electrons was not primarily due to sonochemical process. Since solutes could not enter the gas phase, these were either oxidized or reduced in solutions only, although at a rate which were 10–100 times lower than the reactions occurring in gas phase. Br_2 and I_2 were reduced to Br_3^- and I_3^- and sonication facilitated the evaporation of these polyhalide ions into the gas phase at an appreciable rate, similar to decomposition of N_2O [45]

9.2.1 Study of Chemical Reactions of Metal Ions in Water

After understanding the generation of various free radicals as a result of ultrasonic propagation and cavitation in pure water as well as in the presence of different gases, we can perhaps understand the aqueous chemical reactions better. However, for the sake of simplicity, all reactions involving inorganic species, have been broadly classified into following sections and would be taken up one by one

1. Study of the solutions of monovalent ions [Anions, Argentous and Mercurous]
2. Study of the solutions of divalent cations [Pb^{2+}, Hg^{2+}, Cu^{2+}, Cd^{2+}, Sn^{2+}, Ni^{2+}, Zn^{2+}, Mg^{2+}, Ca^{2+}, Ba^{2+} and $Pt^{2+/4+}$]
3. Study of the solutions of trivalent cations [As^{3+}, Bi^{3+}, Sb^{3+}, Al^{3+} and Au^{3+}]

The objective of this monograph is to include all major studies of metal ions in their aqueous solutions as well as some other important studies in their zerovalent metallic state or in alloys, since the nanoparticles of many of these metals have become too important. Besides, the study of the precipitation of metal ions in aqueous solutions, upon sonication, which has been carried out in our laboratory, would also be discussed. Some of such data include unpublished work. The sequence of metallic ions in this chapter are as they come in the sequence of wet chemical analysis of basic radicals, besides the cationic charge has been kept in mind to make groups and sequences that follow the detailed description.

9.2.2 Study of Monovalent Ions

The sonochemical study of alkali metals is not possible due to highly exothermic reaction of alkali metals with water producing H_2 gas. However, Matyjaszewski et al. [46] used the reducing nature of these metals for the synthesis of polysilylenes. Reductive coupling of methylphenyldichlorosilane and di-n-hexyldichlorosilane in toluene in the presence of alkali metals and ultrasound produced monomodel polysilylenes with relatively narrow molecular weight distribution ($M_w/M_n < 1.5$) with a limiting value of $M_n \sim 50{,}000$. Concentration of alkali metals, solvent dilution, silicon atoms and the temperature were found to affect the polymerization process. However, due to higher reactivity of potassium than sodium, leading to side reaction, the polymerization of methylphenyldichlorosilane in toluene did not produce polysilene [46]. Lepoint et al. [47] reported the evidence for the emission of "alkali metal–noble gas" van der Waals molecules from cavitation bubble by recoding the visible emission spectra in the vicinity of resonance lines of alkali metals in acoustically cavitating aqueous and 1-octanol solutions with solutes as Ar (or Kr), NaCl, RbCl. They found that the emission from the alkali metals was primarily from the gas phase of bubbles, the blue satellite and line distortions were induced by the transitions of "alkali metal/rare gas" van der Waals molecules and excitation and de-excitation mechanisms were essentially chemiluminescent. Henglein et al. [44] found that the sonication of aqueous solutions of Br_2, I_2, MnO_4^-, $AuCl_4^-$ and Ag^+ could scavenge hydrogen atoms to greater extent, as measured through ESR signals. The greater scavenging effect of solutes at the liquid–gas interface was attributed to the hydrated electrons which are formed in the reaction between H atoms and OH^- ions.

$$H + OH^- \longrightarrow H_2O + e^-(aq) \tag{9.58}$$

They [48] further found that the Hg^{2+} ions could be reduced by the aqueous silver sol, forming a layer of mono or diatomic thickness around silver particles. Chemisorbed SH^- and I^- ions damp the plasmon absorption band of silver and silver–Hg particles. However, the collective excitation of the electron gas in the silver particles still strongly contributes to the optical absorption, even when they carry a mercury shell. Ramesh Kuppa and Vijayanand S Moholkar [49] reported the enhancement of $KMnO_4$ initiated oxidation of organic compounds in heterogeneous aqueous media. They treated the cavitation effect of ultrasound as a physical effect and the production of free radicals as a chemical effect of ultrasound and did experiments to differentiate between the two processes and believed such systems essentially to be a liquid–liquid heterogeneous system which were limited by the mass transfer characteristics. They further found that the free radical conserver such as Fenton's reagent assisted in the deeper penetration of radicals in aqueous medium and diffusion towards interface, thereby markedly enhancing the oxidation yield. Soudagar and Samant [50] reported the oxidation of arylalkanes to corresponding arylcarboxylic acids by aqueous potassium permanganate under heterogeneous and ultrasonic conditions in toluene/water system and found a considerable enhancement of reaction at ambient temperature.

9.2.2.1 Silver, Ag^+

Salkar et al. [51] reported the formation of amorphous silver nanoparticles of approximately 20 nm size by the sonochemical reduction of an aqueous solution of silver nitrate in an atmosphere of argon–hydrogen [in the ratio of 95:5] at 10°C. Formation of silver nanoparticles, according to them was through the generation of radical species – as a primary reaction.

$$Ag^+ + H^\bullet \longrightarrow Ag^\circ + H^+ \tag{9.59}$$

$$n\,Ag^\circ \longrightarrow Ag_n(\text{aggregate}) \tag{9.60}$$

whereas in the second step, the H_2O_2 produced in the solution could oxidize Ag° to Ag_2O

$$2Ag^\circ + H_2O_2 \longrightarrow Ag_2O + H_2O \tag{9.61}$$

In the presence of an argon and hydrogen atmosphere, the formation of H_2O_2 could be prevented by the scavenging of $OH^\bullet$ radical by hydrogen, thereby yielding pure silver nanoparticles. The Ar/H_2 mixture produced more H radicals than air, thereby enhancing the reduction of Ag^+ ions under sonochemical condition. Jhu et al. [52] reported the formation of silver nanoparticles of various shapes such as spheres, rods and dendrites, upon sonication of an aqueous solution of $AgNO_3$ in the presence of nitrilotriacetate. The shapes formed depend upon the concentration of $AgNO_3$ and nitrilotriacetate. Pol et al. [53] reported the deposition of silver nanoparticle (~5nm) on the surface of silica microspheres by irradiating a solution of $AgNO_3$ and ammonia in aqueous slurry containing preformed silica for 90 min in an atmosphere of Ar/H_2 (95:5). They could uniformly coat silica spheres with metallic silver layers through this technique. Pol et al. [54] further reported a similar coating of silver and other noble metals on polystyrene sphere, where a solution of $AgNO_3$ in a 10 wt% dispersed in water were purged with Ar/H_2 gas (95:5) and ultrasound (20 kHz, 40 W/Cm^2) for 45 min. The mechanism proposed for coating was on the basis of bursting and collapse of microjets onto the surface of polystyrene spheres. Kumar et al. [55] reported the synthesis of nanospherical Ag_2S/PVA in an ultrasonically irradiated solution of 10% ethylenediamine–water with elemental sulphur and silver nitrate and reported its band gap as 1.05 eV. Salim et al. [56] reported the complete and safer leaching of silver from solid waste using ultrasound assisted thiourea method, rather than the conventional cyanidation process. Wang [57] reported the synthesis of dendritic silver nanostructures by means of ultrasonic irradiation of an aqueous solution of silver nitrate, containing isopropanol (reducing agent) and PEG 400 (disperser) for 2 h. They further reported [58] the sonochemical synthesis of 4–7 μm long and about 100 nm diameter silver nanorods by irradiating an aqueous solution of silver nitrate, methenamine and polyvinyl pyrrolidone for 60 min. Perkas et al. [59] reported the sonochemical deposition of magnetite on silver nanocrystals by irradiating an aqueous solution of ferrous acetate and silver nanopowder. The Ag–Fe_3O_4 nanocomposite demonstrated the behavior of superparamagnetic material. The strong anchoring of magnetite to the nanosilver surface was explained as a result of local melting of silver when the magnetite nucleus was thrown at the silver surface by high speed sonochemical microjets.

In our laboratory, the sonication (20 kHz, 250 W, Air atm) of an aqueous solution of 0.1 M $AgNO_3$ for 10, 20 and 30 min produced turbidity as given in Table 9.1. Simultaneous measurement of rise in temperature and conductance was also recorded. The data of Table 9.1, however, are only qualitative and the author does not confirm to their quantitative figures. The same is true for similar tables in the subsequent sections of the chapter.

The significant increases in the turbidity of $AgNO_3$ solution could be due to the formation of colloidal solution of silver as also reported earlier by other researchers [60]. As evident from our data (Table 9.1), the increase in the turbidity could either be due to the formation of AgOH from Ag^+ or possibly Ag^o particles. Upon the addition of NH_3, the formation of $[Ag(NH_3)_2]^+$ in water itself could also be possible.

There was no appreciable increase, as should be [61], in the electrical conductance of the solution. The initial conductivity of $AgNO_3$ solution was because of

Table 9.1 Change in temperature, turbidity and conductance of $AgNO_3$ solution upon sonication for 10, 20 and 30 min

Sonication time (min)	Temperature (°C)	Turbidity (NTU)	Conductivity (ms)
First set of experiment			
0	31.8	0	1.285
10	33.4	15.0	1.288
20	34.1	20.8	1.300
30	36.6	27.5	1.398
Second set of experiment			
0	31.5	1.0	1.227
10	33.6	11.7	1.330
20	35.6	19.5	1.354
30	36.1	25.7	1.381
Third set of experiment			
0	31.0	0.6	1.194
10	33.6	17.7	1.202
20	34.2	23.1	1.219
30	36.4	26.9	1.229
Fourth set of experiment			
0	36.4	0.6	1.204
10	38.0	9.2	1.210
20	40.2	13.4	1.229
30	41.2	15.8	1.249

Ag^+ and NO_3^- ions in the solution, but when the ultrasound wave passed through the solution, Ag^+ ions were gradually removed from the system through their change into non-conducting AgOH or Ag° particles. At the same time, however, rise in temperature, due to propagating ultrasound waves in the system increased the ionic activity of residual ions. These two opposing factors acted simultaneously and cancelled each other. Therefore, there was no substantial increase in the electrical conductivity of the solution in spite of the rise in temperature of the solution by about 5°. Nevertheless, a slight increase in conductance was probably due to the formation of HNO_3 molecules, which dissociate immediately into H^+ and NO_3^- ions. The solvated protons, H_3O^+, being more active and mobile compared to Ag^+ ions, contribute towards the marginal increase in electrical conductance. In unsonicated solution, however, the Ag^+ and NO_3^- ions are held electrostatically balanced, but the equilibrium of which was disturbed due to systematic removal of Ag^+ ions in the form of Ag° and $[Ag\,(OH)_2]^-$. Change in the chemical states of ions in the solution of Ag^+ could therefore be suggested as follows:

$$Ag^+ + H_2O \longrightarrow Ag^\circ + OH^\bullet + H^+ \quad (9.62)$$

$$Ag^\circ + OH^\bullet \longrightarrow AgOH \quad (9.63)$$

$$Ag^+ \longrightarrow [Ag^+]^* \quad (9.64)$$

$$[Ag^+] + 2OH \longrightarrow [Ag(OH)_2]^- \quad (9.65)$$

Astrics (*) indicate activated species.

9.2.2.2 Mercurous ion, Hg_2^{2+}

Sonication of 0.05 M $Hg_2(NO_3)_2$ solution for 10, 20 and 30 min and the simultaneous measurements of conductivity, temperature change and turbidity (Table 9.2) indicated a rise in the turbidity due to the formation of an insoluble precipitate. This could probably be due to the formation of $Hg_2(OH)_2$, as a consequence of hydrolysis, along with $Hg^\bullet$ free radical and Hg^o particles which could be responsible for increase in the turbidity after sonication. The turbidity increased further with time. Mobility of NO_3^- ions was more or less restricted due to resonance in this ion, which helped, in the smooth and uniform distribution of charge density over NO_3^- ion surface. Hence the contribution of NO_3^- ion towards the electrical conductance was perhaps much too less than the conduction of cationic species with which it was associated in the molecular (compound) form. Since in case of $Hg_2(NO_3)_2$, $Hg_2(OH)_2$ species were being formed which also destroyed the cationic nature of Hg_2^{2+}, therefore a decrease in the electrical conductance of solution could be predicted. The simultaneous passivity of its anionic part did not increase the conductivity due to rise in temperature as anticipated and could be seen through the Table 9.2. These observations could now be summarized in reaction steps as under:

$$Hg_2(NO_3)_2 \longrightarrow Hg_2^{2+} + 2NO_3^- \tag{9.66}$$

$$Hg_2^{2+} + 2H\bullet \longrightarrow 2Hg^o\downarrow + 2H^+ \tag{9.67}$$

$$Hg_2^{2+} + 2H_2O^* \longrightarrow Hg_2(OH)_2\downarrow + 2H^+ \tag{9.68}$$

Astrics (*) indicate activated species.

9.2.3 Study of Divalent Ions

For the sonochemical studies of divalent ions especially in the aqueous solution, not many references are available in the literature. Nevertheless, in an attempt to discuss the nature of the metal and their metal ions in aqueous chemical reactions, under ultrasonic field, the available references have been reported to confirm the understanding about the behavior of such cations.

9.2.3.1 Lead, Pb^{2+}

The sonochemical studies of lead in aqueous solutions are limited due to many insoluble or sparingly soluble lead salts in water, yet few attempts to prepare

Table 9.2 Change in turbidity, conductance and temperature of 0.05 M $Hg_2(NO_3)_2$ solution after 10, 20 and 30 min of sonication

Time (min)	Temperature (°C)	Turbidity (NTU)	Conductance (mS)
First set of experiment			
0	19.5	0.1	28.4
10	24.2	2.8	28.9
20	34.0	3.0	28.8
30	33.0	5.8	28.6
Second set of experiment			
0	17.5	0.2	27.9
10	25.4	2.5	28.5
20	34.2	3.2	28.7
30	34.4	5.5	28.7
Third set of experiment			
0	17.0	0.2	27.7
10	25.2	2.3	28.2
20	34.0	3.2	28.5
30	35.1	5.7	28.6
Fourth set of experiment			
0	18.0	0.2	28.1
10	27.2	2.5	28.6
20	34.9	3.2	28.9
30	36.1	5.5	28.9

nanoparticles of lead could be found in the literature. The nanorods of lead telluride and selenides were made by Zhu et al. [62] who reported the synthesis of PbSe nanoparticles (12 nm) by the pulse sonochemical technique from an aqueous solution of sodium selenosulphate and lead acetate. They [63] further reported the sonochemical method for the preparation of monodispersed spherical and rectangular lead selenide nanoparticles from an aqueous solution of lead acetate and sodium selenosulphate in the presence of complexing agent under ambient air. The use of trisodium citrate as complexing agent yielded spherical nanoparticles (~8 nm), whereas the use of potassium nitrilotriacetate yielded rectangular nanoparticles with an average size of ~25 nm. Purkayastha et al. [64] reported sequential organic and inorganic templating in the synthesis of lead telluride by mixing an aqueous solution of $PbCl_2$ with a Te nanotube solution, already made through organic reagents and TeO_2 and cooled to 100°C, to which hydrazine monohydrate was being added drop-by-drop. Hydrazine reduced $PbCl_2$ and transformed the Te nanotubes to a single crystal PbTe nanorod. Fard-Jahromi and Morsali [65] reported the sonochemical synthesis of nanoscale mixed ligands lead (II) co-ordination polymers as precursors for the preparation of $Pb_2(SO_4)O$ and PbO nanoparticles with two dimensional polymeric units. The same group further reported the formation of nanocrystalline PbO and PbBr(OH) [66].

Sonication of 25 ml of 0.25 M aqueous solution of $Pb(NO_3)_2$ increased the turbidity and conductivity of these solutions significantly along with the rise in temperature (Table 9.3). The increase in turbidity of the solution could be attributed

Table 9.3 Change in turbidity, conductance and temperature of 0.25 M $Pb(NO_3)_2$ solution after 10, 20 and 30 min of sonication

Time (min)	Temperature (°C)	Turbidity (NTU)	Conductance (mS)
First set of experiment			
0	31.2	3.4	17.10
10	33.4	10.0	17.50
20	35.6	22.4	18.00
30	36.4	24.2	18.91
Second set of experiment			
0	31.6	3.4	16.57
10	34.7	8.5	18.12
20	36.3	14.6	18.64
30	38.3	23.3	18.91
Third set of experiment			
0	30.3	1.0	16.67
10	33.8	13.6	17.07
20	36.0	18.6	17.32
30	39.9	22.6	18.05
Fourth set of experiment			
0	34.3	2.7	18.69
10	36.6	9.4	18.97
20	39.0	15.8	19.01
30	41.2	20.1	19.12

to the formation of $Pb(OH)_2$ as in case of Hg_2^{2+} and Ag^+. The precipitation of $Pb(OH)_2$ took place via the formation of $[Pb(H_2O)_4]^{2+}$, through an aquo complex in combination with activated water molecule and was evident by the appearance of a peak at 936 nm. The aqua complex, being unstable, decomposed into $Pb(OH)_2$ which was responsible for the increase in the turbidity of the solution.

$$Pb(NO_3)_2 \longrightarrow Pb^{2+} + 2NO_3^- \quad (9.69)$$

$$Pb^{++} + 4H_2O^* \longrightarrow [Pb(H_2O)_4]^{2+} \quad (9.70)$$

$$[Pb(H_2O)_4]^{2+} \longrightarrow Pb(OH)_2 \downarrow 2H_2O + 2H^+ \quad (9.71)$$

Although, there was a net increase in the conductance of solution with increasing duration of sonication, but two opposing factors contributed significantly as under:

1. Increase in conductance due to rise in temperature with sonication
2. Increase in conductance due to formation of larger number of H^+ ions upon the hydrolysis and decomposition of $[Pb(H_2O)_4]^{2+}$
3. Decrease in conductance due to the removal of ionic species such as Pb^{2+} or $[Pb(H_2O)_4]^{2+}$ and the formation of non-conducting species $Pb(OH)_2$
4. Decrease in conductance due to the solvation of resonance stabilized NO_3^- ions by charged water molecules, rendering NO_3^- ions immobile

In another experiment, the solubility of $PbCl_2$ was examined. The $PbCl_2$ being sparingly soluble dissolved upon sonication as the temperature of the system rose, however, when the solution cooled down to the room temperature, the amount of $PbCl_2$ precipitated as before.

9.2.3.2 Mercury(II), Hg^{2+}

Ziki et al. [67] reported the sonolytic desorption of mercury from aluminium hydroxide at pH 4, which was otherwise an efficient sink for Hg(II) ions. The higher input power increased the desorption of mercury. They further reported [68] the dissolution of cinnabar (α-HgS) into the water as a result of sonication with a significant reduction in the particle size of cinnabar and the oxidation of sulphur content to SO_4^{2-}. Gil et al. [69] reported the removal of Hg(II) in trace quantities from waste water through the reduction and subsequent volatilization of mercury and the degassing effect of ultrasound. Formic acid induced reduction of Hg(II) to Hg^o through the formation of reducing gases such as H_2 and CO, which were also eliminated. At the 100 ng/mL Hg concentration, the removal was 90% within 30 min. Michele et al. [70] reported the formation of mercury acetate in a system containing metallic mercury and acetic acid in water (7% by volume) and found a drastic reduction in the crystallization time. Kristl and Drofenik [71] reported a single step sonochemical synthesis of nanocrystalline mercury chalcogenides

HgE (E = S, Se, Te), using less toxic reagents, used in earlier syntheses, by mixing mercury nitrate (dissolved in 0.1 M ethylenediamine tetraacetic acid) and the elemental chalcogenes (dissolved in aqueous NaOH solution). Wang and Zhu [72] reported the selective synthesis of α and β forms of cinnabar (α-HgS and β-HgS) in aqueous solution containing sodium thiosulphate and thiourea as the sulfur source. The band gap of α-HgS (12 nm size) was measured (2.8 eV) by the absorption spectra, whereas β-HgS (13 nm size) nanoparticles were found to absorb in the UV–Visible region. The sonochemistry of mercury could, therefore, be seen through its both Hg(I) and Hg(II) ionic states. When 0.05 M $HgCl_2$ solution was sonicated for 10, 20, 30 and 60 min, the turbidity was due to the formation of Hg_2Cl_2, which was tested qualitatively, using NH_3, KI and NaOH. These reagents gave black, green and black precipitates respectively as under:

$$Hg_2Cl_2 + 2NH_3 \longrightarrow \underset{\text{(Black ppt)}}{Hg \downarrow + Hg(NH_3)_2Cl \downarrow} + Cl^- \tag{9.72}$$

$$Hg_2^{2+} + 2I^- \longrightarrow \underset{\text{(Green ppt)}}{Hg_2I_2 \downarrow} \tag{9.73}$$

$$Hg_2^{2+} + 2OH^- \longrightarrow \underset{\text{(Black ppt)}}{Hg_2O \downarrow} + H_2O \tag{9.74}$$

The unconverted Hg^{2+} ions were also estimated quantitatively using copper ethylene di-amine mercuric iodide $[Cu(en_2)][HgI_4]$ [73]. Simultaneous turbidometric and conductometric measurements, after different sonication periods (Table 9.4), gave a good correlation. The turbidity increased as the duration of sonication in $HgCl_2$ solution increased, while conductivity of the solution decreased in spite of an increase in temperature. This was an indication of the reaction and conversion mechanism operating in the ultrasonically agitated solution. Thus it was quite obvious that this change in both the conductivity and turbidity was due to the formation of Hg_2Cl_2, which increased turbidity due to its insoluble character. The chloride ions, which were initially helping in electrical conductance, were now systematically and slowly removed from the system, hence not available for transporting electrical charge in the solution. The mechanism of this reaction could be suggested as under:

$$HgCl_2 \longrightarrow Hg^{2+} + 2Cl^- \tag{9.75}$$

$$2Hg^{2+} + 2H^{\bullet} \longrightarrow 2Hg_2^{2+} + 2H^+ \tag{9.76}$$

$$Hg_2^{2+} + 2Cl^- \longrightarrow Hg_2Cl_2 \downarrow \tag{9.77}$$

A very low percentage of reduction (Table 9.4) could be due to the dependence of the reduction on the generation and concentration of free radicals formed during sonication. This was also the reason for increase in the percentage reduction with increasing sonication period from 10 to 60 min.

Table 9.4 Turbidity, conductance, temperature and percentage conversion of 0.05 M $Hg_2(Cl_2)$ solution at different durations of sonication

Duration of sonication (min)	Temperature (°C)	Turbidity (NTU)	Conductance (μS)	Percentage (%) conversion of $Hg^{2+} \rightarrow Hg_2^{2+}$
0	23.8	0.6	112.3	
10	27.5	3.3	106.1	0.54
20	29.9	3.8	104.8	0.62
30	30.1	9.2	99.2	0.89
60	34.1	15.0	95.8	1.013

9.2.3.3 Copper, Cu^{2+}

The behavior of $Cu(II)_{(aq)}$ is relatively more understood than other metal ions. Haas and Gedanken [74] found only a partial reduction of Cu^{2+} ions to Cu^{+} (95%) instead of metallic copper (5.1%) in the presence of cetyltrimethylammonium bromide in an ultrasonic field and thus obtained CuBr particles instead of Cu. Nevertheless, when polymers such as poly(N-vinyl 2-pyrrolidone) or poly(vinyl alcohol) were used, the end product was metallic copper particles, as expected. They have proposed the reduction of Cu^{2+} ions to copper as the first stage, however, in the second stage Cu reacted with OH radicals or H_2O_2, formed by sonolysis of water to produce Cu^{+} and OH^{-} ions as under:

$$Cu^{\circ} + OH^{\bullet} \longrightarrow Cu^{+} + OH^{-} \tag{9.78}$$

They have further confirmed the role of sonication and consequently the production of OH/H_2O_2 species to facilitate the formation of CuBr by decreasing the ultrasound power from 50 to 35 and finally to 15 W and found an increase in the amount of Cu particles, clearly enough due to poor concentration of active species. The mechanism for the action of H_2O_2 with Cu^{2+} ions in aqueous solution could therefore be proposed as under:

$$Cu^{2+} + 2H^{\bullet} \longrightarrow Cu^{\circ} + 2H^{+} \tag{9.79}$$

$$nCu^{\circ} \longrightarrow (Cu^{\circ})_{n(\text{aggregate})} \tag{9.80}$$

Finally H_2O_2 produced through sonication oxidises copper to its oxide

$$2Cu^{\circ} + H_2O_2 \longrightarrow Cu_2O + H_2O \tag{9.81}$$

However, in Ar and H_2 atmospheres, the formation of H_2O_2 was arrested due to the scavenging of $OH^{\bullet}$ radicals by the hydrogen [75] producing only pure copper nanoparticles. Thus argon and hydrogen mixture produced even more $H^{\bullet}$ radicals and reduced Cu^{2+} ions to Cu^{+} as a consequence of sonication in the aqueous solution. Yonghong et al. [76] reported the synthesis of 1D $Cu(OH)_2$ nanowires and transitions to 3D CuO microstructures under ultrasonic irradiation in simple aqueous solutions containing $CuCl_2 \cdot 2H_2O$ and NaOH. They tried to explain the mechanism of the conversion from 1D $Cu(OH)_2$ nanowires to 3D CuO microstructures through the formation of a complex $[Cu(OH)_4]^{2-}$ which later decomposed to $Cu(OH)_2$ upon sonication as under:

$$Cu^{2+} + 4OH^- \longrightarrow [Cu(OH)_4]^{2-} \quad (9.82)$$

$$[Cu(OH)_4]^{2-}(aq) \longrightarrow Cu(OH)_2 + 2OH^- \quad (9.83)$$

Since $[Cu(OH)_4]^{2-}$, a layered material [77], easily formed 1D nanostructures with the orthorhombic phase due to co-ordination self assembly of $[Cu(OH)_4]^{2-}$ ions during decomposition in highly basic conditions [78], the 1D nanowires were obtained at an early stage of sonication. However, due to H-bonding interaction [79] the nanowires first grew to nanoribbons and finally crossed over to form 3D $Cu(OH)_2$ nanostructures and finally to 3D CuO microstructures. Haas et al. [80], using pulsed sonoelectrochemical method, could control the size of copper nanoparticle in the aqueous solution of copper sulphate pentahydrate stablised by poly (N-vinylpyrrolidone). They reported that the increase in the current density from 55 to 100 mA/cm^2 reduced the particle size from 29 to 10 nm. Similarly, PVC concentration from 0.2% to 2% also decreased the size from 55 to 29 nm. However, the rise in temperature from 15 to 50°C increased the particle size from 17 to 69 nm, nevertheless, there was no effect of variation in the sonic power from 35 to 76 W and the particle size remained at 29 nm. Baioni et al. [81] synthesised copper hexacyanoferrate nanoparticles in the aqueous solution containing $CuCl_2 \times H_2O$, $K_3[Fe(CN)_6]$, HCl and KCl and used hydrochloride diallylamine as stabiliser. The nanoparticles (~10 nm diameter) were immobilised onto transparent indium tin oxide electrodes through electrostatic layer by layer deposition and showed interesting electrochromic properties, changing the coloration during the redox process from brown to orange when oxidised. Salkar et al. [82] reported the synthesis of elongated copper nanoparticles coated with zwitterionic surfactant, cetyltrimethylammonium p-toluene sulfonate from the aqueous solution of copper hydrazine carboxylate, which was reduced ultrasonically in an argon–hydrogen atmosphere, at room temperature, to a reddish brown powder. A monolayer coating of the surfactant was formed on the elongated copper nanoparticles. Kumar [83] reported the sonochemical synthesis of nanocrystalline copper oxide embedded with poly (vinyl alcohol) by sonicating for 3 h a solution of copper (II) acetate monohydrate in a 10% aqueous solution of water–dimethylformamide under 1.5 atm argon at room temperature. The decrease in the pH value of the reaction mixture after the experiment indicated the reaction following as under:

$$Cu^{2+}_{(aq)} + H_2O_{(l)} \longrightarrow CuO_{(s)} + 2H^+_{(aq)} \quad (9.84)$$

They also reported [84] the sonochemical synthesis of nanocrystalline Cu_4O_3 in polyaniline matrix, starting with an aqueous solution of copper (II) acetate monohydrate and aniline. This method was found suitable for the synthesis of other oxides of transition metals such as Fe_3O_4 and Cu_2O. They [55] further reported the synthesis of CuS/PVA and Ag_2S/PVA nanocomposites of the particle size 225 and 25 nm respectively by sonochemical irradiation of copper acetate or silver nitrate in a 10% solution of ethylenediamine–water containing elemental sulphur. A band gap 2.08 and 1.05 eV was estimated for CuS/PVA and Ag_2S/PVA respectively. Mahdi et al. [85] reported the ultrasonically assisted emulsification for the recovery of Cu(II) from waste water, using an emulsion liquid membrane. Mohammad H Entezari et al. reported [86] the sonochemical degradation of phenol in water and used copper sulphate and H_2O_2 to provide different oxidative systems at different ultrasonic frequencies. They found best oxidation at 35 kHz, which was about three times faster than the degradation of phenol at 500 kHz. Weizhong Lv et al. [87] synthesised copperaluminate ($CuAl_2O_4$) nanosized spinel particles (~17 nm) by sonicating a solution containing $Cu(NO_3)_2 \cdot 6H_2O$, $Al(NO_3)_3 \cdot 9H_2O$ and urea (9M, precipitating agent).

In our laboratory, when 25 ml of 0.25 M $CuSO_4 \cdot H_2O$ (E Merck, AR) solution was sonicated in the presence of HCl, the colour changed from light blue to green, but the turbidity and conductance did not vary (Table 9.5a). This experiment, however, when repeated with 25 ml of 0.25 M $CuSO_4$ solution, without HCl, did not show any change either in the colour or in the turbidity as well as the conductance (Table 9.5b). To understand this puzzling problem, another set of 25 ml of 0.25 M solution of $CuCl_2$ (Qualigens, AR), without further addition of

Table 9.5 Effect of ultrasound on the turbidity and conductance of different copper solutions

Time (min)	Temperature (°C)	Turbidity (NTU)	Conductance (μS)
a. $CuSO_4$ (0.25 M) solution with HCl			
0	19.5	0.2	15.32
10	28.5	0.2	15.37
20	28.5	0.2	15.37
30	28.0	0.2	15.40
b. $CuSO_4$ (0.25 M) solution without HCl			
0	18.9	0.4	15.25
10	28.2	0.4	15.36
20	28.5	0.4	15.37
30	28.9	0.4	15.42
c. $CuCl_2$ (0.25 M)			
0	20.0	0.0	14.60
10	31.0	0.9	15.39
20	34.8	5.9	17.28
30	39.5	12.1	17.28
After sonication was stopped			
45		29.5	
60		48.4	

HCl, was sonicated for similar durations of 10, 20 and 30 min. The blue coloured solution changed again to green, besides, the turbidity and conductance also gradually increased in sonicated solutions. The turbidity increased remarkably after about 15 min, even when sonication was stopped. (Table 9.5c).

Initially it was thought that the HCl present in the solution of $CuSO_4$ was slowly eroding the stainless steel tip of the sonicating horn and dissolving smaller fragments of iron to generate Fe^{2+} ions, but the test for Fe^{2+} ions with potassium ferricyanide was negative, hence there appeared to be another cause for the change in colour upon sonication. Since, there was no change in colour of the first experiment, performed without the addition of HCl, the colour change in the other two experiments, with HCl or with $CuCl_2$, which both contained chloride ions derived either from the HCl or $CuCl_2$ itself, the role of Cl^- in this change of colour was anticipated. To confirm the state of chloride ions in sonicated samples, the precipitated amounts of AgCl from 25 ml of 0.25 M $CuCl_2$ with 0.25 M $AgNO_3$ (BDH, AR) (added in excess volume) in sonicated and unsonicated conditions was compared. A major difference was observed. The amount of precipitated chloride ions in sonicated sample was found to be much less than in the unsonicated samples. The ultrasonic irradiation, therefore, clearly converted aquo complex into a chloro copper (II) complex. The number of chloride ions, which were taken inside the co-ordination shell of the complex, could have varied its concentration, depending upon their availability in the solution. (Table 9.5c). Another point, which was also clear from the Table 9.5c, was that there was a competition between H_2O molecule and chloride ion to act as a complexing ligand with copper (II) ions. The ultrasound seemed to have favoured the formation of chloro complex. But when the ultrasonic propagation was stopped in the solution after 30 min, the chloride ions slowly moved out of the coordination shell of Cu(II) giving way to water molecules to form tetraaquo complex of Cu(II). This, however, in the absence of any acid in the solution, hydrolysed quickly and a turbidity due to $Cu(OH)_2$ was produced. A scheme for such conversion may, therefore, be suggested as under:

Normal Reaction of Water with $CuSO_4$ in Non-hydrolysed State

$$Cu^{2+} + 4H_2O \longrightarrow \left[Cu(H_2O)_4\right]^{2+} \qquad (9.85)$$

Reaction due to Ultrasound

$$[Cu(H_2O)_4]^{2+}SO_4^{2-} \xrightarrow{)))))} [Cu(H_2O)_4]^{2+*} + SO_4^{2-*} \qquad (9.86)$$

$$[Cu(H_2O)_4]^{2+} + 2Cl^- \xrightarrow{))))} [Cu(H_2O)_4]^{2+*} + 2Cl^{-*} \qquad (9.87)$$

$$[Cu(H_2O)_4]^{2+*} + Cl^{-*} \longrightarrow [Cu(H_2O)_3Cl]^{+*} + H_2O \qquad (9.88)$$

$$[Cu(H_2O)_3Cl]^{+*} + Cl^{-*} \longrightarrow [Cu(H_2O)_2Cl_2]^{0*} + H_2O \qquad (9.89)$$

$$[Cu(H_2O)_2Cl_2]^{0*} + Cl^{-*} \longrightarrow [Cu(H_2O)Cl_3]^{-*} + H_2O \qquad (9.90)$$

$$[Cu(H_2O)Cl_3]^{-*} + Cl^{-} \longrightarrow [CuCl_4]^{2-*} + H_2O \qquad (9.91)$$

Reaction After the Ultrasonic Source Was Stopped (Slowly Reverting to the Original Composition)

$$\begin{array}{lcl} [CuCl_4]^{2-*} \quad + \quad H_2O & \longrightarrow & [CuCl_3{\cdot}H_2O]^{-} + Cl^{-} \\ \vdots \qquad\qquad\quad \vdots & & \\ [CuCl(H_2O)_3]^{+} + \quad H_2O & \longrightarrow & [Cu(H_2O)_4]^{2+} + Cl^{-} \end{array} \qquad (9.92)$$

Astrics (*) indicate ionic species in activated states.

Reactions (9.88–9.91) proceeded in steps and depended largely on the number of activated chloride ions available in the solution. Formation of such activated chlorine radicals was also reported elsewhere during the degradation of CCl_4 in aqueous solution by ultrasound [88].

There was no change in the conductance, without HCl, in spite of an increase in the temperature of about 10°C due to sonication (Table 9.5b). The ultrasound seemed to have increased the molecular solvation, therefore the activity of SO_4^{2-} ions did not increase but remained more and more solvated due to complex ionic composition and charge distribution on the SO_4^{2-} ions (Fig. 9.1).

Later, the conductivity also did not change in spite of the presence of Cl^- ions, which ought to be less solvated due to much smaller size compared to sulphate ions, were still not available for conducting electrical current as these were systematically removed from the solution and were taken inside the co-ordination shell of the Cu (II) complex. Therefore, the total number of active chloride ions was much less and hence the electrical conductivity did not rise in spite of an increase in the temperature of the solution by about 10°C (Table 9.5a). The increase in the electrical conductance in the last experiment, with $CuCl_2$ (Table 9.5c), however, showed an increased chloride ion activity with rise in temperature, in spite

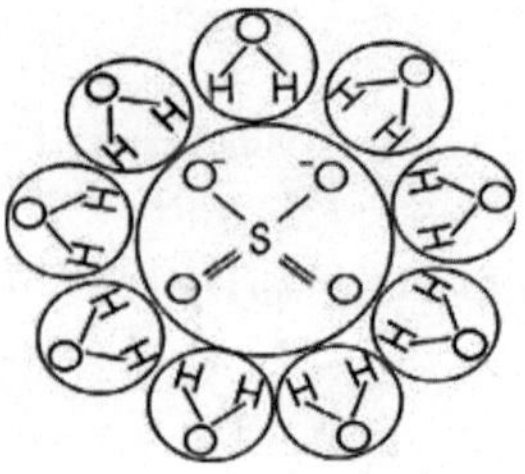

Fig. 9.1 Water solvated SO_4^{2-} ion

of the fact that the chloride ions had been removed from the solution and taken into the co-ordination shell of the Cu(II). The increase in the turbidity as seen in the Table 9.5c also suggested that the reaction did not stop here. Since the solution was not acidic, the activated water molecules attacked the $CuCl_4^{2-}$ complex and replaced Cl^- ions from the coordination shell, only to generate hydroxy derivatives of the complex through one or more of such substitutions. This was the reason why turbidity increased drastically after the sonication was stopped. The $Cu(OH)_2$ precipitated quickly and made the solution turbid as under:

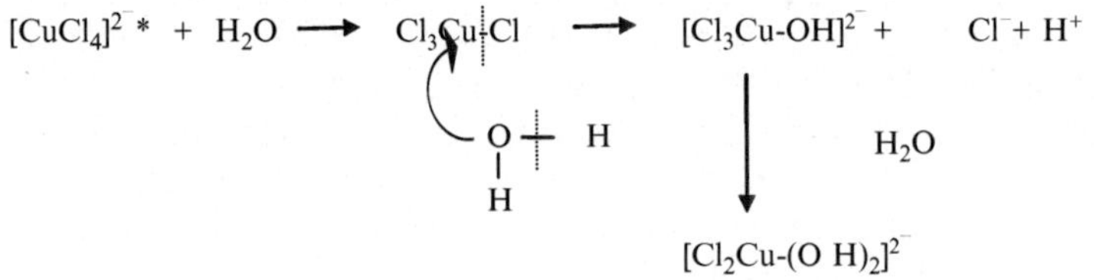

(9.93)

Therefore, chloride ions were again available in the solution to conduct electricity, hence an increase in conductance was observed.

9.2.3.4 Cadmium, Cd^{2+}

The earlier studies of aqueous solution of Cd^{2+} containing mercaptan were found to produce colloidal CdS by sonication [89] under nitrogen atmosphere. The H atoms produced during sonication attacked on molecules possessing a thiol moiety and lead to either H atoms extraction or H_2S formation. The proportion of H_2 to H_2S was highly dependent on the type of mercaptan. Possible reaction mechanism was reported as follows [90],

$$Cd^{2+} + H_2S \longrightarrow CdS + 2H^+ \quad (9.94)$$

$$n\,CdS \longrightarrow (CdS)_n \quad (9.95)$$

Absorption spectra of CdS colloid indicate the formation of quantum sized CdS particles. The particle size increased upon sonication, indicated by the red shift in the onset of absorption. The particle size was highly dependent on the mercaptan used, because of the absorption of the mercaptan on the particle acting as a 'capping agent' and the rate of H_2S produced. Study of mercaptan systems revealed that there was also a thermal process responsible for CdS formation. 25% of the total CdS produced sonochemically was formed via a thermal mechanism presumably in the hot shell around the compressed bubble. CdS colloid could be dissolved quite readily by sonicating solutions under air saturated conditions [89] by the following reaction,

$$CdS + 2OH^{\bullet} \longrightarrow Cd^{2+} + 2OH^- + S \quad (9.96)$$

The dissolution reaction could, however be inhibited by the addition of Na_2S to the colloidal solutions. Yadav et al. [91] tried to explain the growth mechanism of the nanosized CdS particles upon the use of amino acid, histidine. They suggested that the imidazole ring of histidine could capture the cadmium ions from the solution and prevented the growth of CdS nanoparticles. They further suggested that the growth of the nanosized CdS particles proceeded via two processes - either through the diffusion of reactants or by reactions on the surface of the crystallites. In one of very interesting studies on the effect of Cd^{2+} on the yield of $H^{\bullet}$ during the sonolysis of water, Misik and Riesz reported [92] against the formation of hydrated electrons, a view in support of the results reported by Henglein [75], but postulated and predicted earlier by Margulis [7viii]. Mastai et al. [93] reported the synthesis of cadmium selenide by sonicating and electrodepositing an aqueous solution of $CdSO_4$ complexed with potassium nitriloacetate and sodium seleno-sulphate at a pH greater than 7. The nanoparticles formed were amorphous because the rapid cooling in the cavitating bubbles did not allow the precipitate to crystallize before their electrodeposition.

25 ml of an aqueous solution of $CdCl_2$ (0.25 M), when sonicated for 10, 20 and 30 min, in our laboratory, showed an increase in the turbidity and conductivity (Table 9.6) with increasing sonication time. The increase in the turbidity of the solution could be attributed to the formation of $Cd(OH)_2$ [Cf. Ag^+, Cu^{2+}, Hg^{2+} ions] and the increase in the conductance of solution was because of the formation of more mobile H^+ ions. Although the Cd^{2+} ions were removed from the system through the formation of $Cd(OH)_2$, the removal of Cd^{2+} provided H^+ ions to the system resulting in the increased conductance of the system in spite of the formation of hydroxide as under:

$$CdCl_2 \longrightarrow Cd^{2+} + 2Cl^- \quad (9.97)$$

$$Cd^{2+} + 4H_2O^* \longrightarrow [Cd(H_2O)_4]^{2+} \quad (9.98)$$

$$[Cd(H_2O)_4]^{2+} \longrightarrow Cd(OH)_2 + 2H^+ + 2H_2O \quad (9.99)$$

9.2.3.5 Tin, Sn^{2+}

Zhu et al. [94] reported the synthesis of SnO_2 semiconductor nanoparticles by ultrasonic irradiation of an aqueous solution of $SnCl_4$ and azodicarbonamide under ambient air. They found that the sonochemically synthesized SnO_2 nanoparticles improved remarkably the performance of Li ion batteries such that there was about threefold increase (from 300 to 800 mAh/g) in the reversible capacity in the first lithiation to delithiation cycles. Similarly the irreversible capacity also increased by about 70% (from 800 to 1400 mA h/g). Wang et al. [95] reported the synthesis of positively charged tin porphyrin adsorbed onto the surface of silica and used as photochemically active templates to synthesise platinum and palladium shell and

Table 9.6 Change in turbidity, conductance and temperature with different durations of sonication in 0.25 M $CdCl_2$ solutions

Time (min)	Temperature (ºC)	Turbidity (NTU)	Conductance (mS)
First set of experiment			
0	21.9	0.1	14.64
10	44.1	3.0	17.76
20	45.5	7.1	21.0
30	48.0	11.0	22.2
Second set of experiment			
0	21.0	0.1	13.05
10	44.9	3.7	18.65
20	45.2	7.0	21.4
30	48.0	11.2	22.1
Third set of experiment			
0	21.5	0.1	14.54
10	44.1	3.2	17.76
20	45.7	7.4	21.1
30	47.0	11.0	21.9
Fourth set of experiment			
0	22.0	0.1	14.84
10	44.2	3.2	17.78
20	45.7	7.4	21.1
30	47.8‘	11.1	22.0

core–shell nanostructures. The cationic tin porphyrin served a dual function of reducing the metal photocatalytically and nucleate growth sites and also as surface modifier that promoted binding of platinum metal to the surface of the nanospheres.

When 25 ml of 0.25 M $SnCl_2$ solution was sonicated for 10, 20 and 30 min, in our laboratory, the changes in nephelometric and conductometric values were found as reported in Table 9.7. The turbidity of the solution increased considerably due perhaps to the formation of hydroxide of Sn(II) as $Sn(OH)_2$ which agglomerated and settled rapidly under the influence of ultrasonic agitation. The conductance increased rapidly only for a solution sonicated for 10 min, but did not increase significantly even after further sonication and so was the temperature of the system. It appeared that the system attained equilibrium in dissipating the heat equivalent to the heat generated due to sonication after 10 min. Since there was no rise in temperature the conductance did not increase further. To see the reproducibility of the experimental observations, the experiment was repeated 5–6 times. However, surprisingly in the later two experiments the colour of the solution, after sonicating for 30 min, changed from colourless to green, with few black particles settled in the bottom of one reaction beaker. Reaction between the stainless steel tip of the sonicator with the highly acidic experimental solution was anticipated. Iron present in these particles dissolved to produce Fe^{2+} ions and was confirmed through a deep blue colour (prussian blue) developed with $K_3[Fe(CN)_6)]$ in one drop of sonicated solution. The conversion of Fe° to Fe^{2+} ion in the presence of Sn^{2+} was normally

Table 9.7 Change in turbidity, conductance and temperature with different durations of sonication in 0.25 M $SnCl_2$ solutions

Time (min)	Temperature (°C)	Turbidity (NTU)	Conductance (mS)
First set of experiment			
0	20.8	1.2	141.4
10	38.0	18.3	166.1
20	38.4	21.1	168.0
30	38.7	28.1	165.1
Second set of experiment			
0	19.1	1.0	138.1
10	38.9	17,9	166.9
20	39.2	22.1	169.9
30	39.4	29.0	166.2
Third set of experiment			
0	20.0	1.0	140.2
10	38.0	18.1	166.2
20	38.8	23.0	169.0
30	39.0	29.4	167.2
Fourth set of experiment			
0	21.0	1.0	142.4
10	40.1	18.3	170.1
20	41.0	25.0	172.5
30	41.0	30.0	169.9

difficult but since Sn^{2+} was itself unstable and a fraction of it oxidised to Sn^{4+} in the presence of atmospheric oxygen, the formation of Sn^{4+} in the solutions was possible. These Sn^{4+} ions oxidised iron particles to produce Fe^{2+} ions and got themselves reduced to Sn^{2+} again in the presence of acid and ultrasound. This was the reason for the change in the colour of $SnCl_2$ solutions in later experiments.

$$Sn^{2+} + H^{\bullet} \xrightarrow{O_2} Sn^{4+} + 2OH^{-} \tag{9.100}$$

$$Sn^{4+} + Fe \longrightarrow Sn^{2+} + Fe^{2+} \tag{9.101}$$

Nevertheless, when the above experiment was repeated again with a new polished ultrasonic tip, there was no change in the colour of the sonicated solution even after 30 min of sonication. This was an interesting observation. It was then noticed that the first stainless steel tip of the ultrasonic horn got severely corroded in this experiment and deep cracks had developed. The reason was understandable. The solution of $SnCl_2$, which was strong reducing and highly unstable, was prepared in 8% HCl with few pieces of Sn metal, which was still reacting with conc. HCl to produce H_2 gas. The experimental solution was filtered quickly before use. Strong acid in the solution eroded the ultrasonic tip very fast and developed deep cracks. Some of these small particles of steel could have fallen into the test solutions of the later experiments. The reason for the change in colour of the experimental solution under ultrasonic agitation was therefore not chemical but mechanical. The reaction mechanism for the precipitation of Sn(II) as indicated in nephelometric observations could therefore be suggested as under:

$$SnCl_2 \longrightarrow Sn^{2+} + 2Cl^- \tag{9.102}$$

$$Sn^{2+} + 4H_2O \longrightarrow [Sn(H_2O)_4]^{2+} \tag{9.103}$$

$$[Sn(H_2O)_4]^{2+} \longrightarrow Sn(OH)_2 \downarrow + 2H_2O + 2H^+ \tag{9.104}$$

9.2.3.6 Nickel, Ni^{2+}

Although additional oxidation states of +3 and +4 are known for Ni but in most of its aqueous reactions, Ni forms compound in +2 oxidation state. In a study of electroless plating of Ni at low temperature by applying ultrasonic waves, the Ni deposit had been found to have higher P content and hardness and possessed more compact and uniform microstructure than conventional electroless Ni deposit [96]. Ultrasound also extended the leaching rate of nickel and the maximum extraction of 95% nickel was obtained in 14 days, in contrast to 24.9% nickel with conventional in situ bioleaching [97]. Porous nickel and cobalt oxides were prepared using their alkoxides as inorganic precursors. The stabilization of the mesostructure was especially critical for divalent elements such as Ni and Co, which did not form any network structure, like silicates. A relatively better surface area had been obtained for the Co oxide, but in Ni the surface area was not found to be good [98]. Ramesh et al. [99] reported the deposition of nanophasic amorphous clusters of elemental nickel (10–15 nm) on submicrospheres of amorphous silica by the sonication of a suspension containing nickeltetracarbonyl and silica submicrospheres. Shafi et al. [100] synthesised nanosized amorphous $NiFe_2O_4$ powder by the sonochemical decomposition of a solution of $Fe(CO)_5$ and $Ni(CO)_4$ in decalin at 273 K under and oxygen pressure of 100–150 kPa. As-prepared ultrafine particles were superparamagnetic which crystallised to a sextet pattern with A (tetrahedral) and B (octahedral) sublattices respectively of the inverse spinel $NiFe_2O_4$ with Curie temperatures of 440°C and 560°C respectively for the amorphous and crystalline forms. Zhong et al. reported [101] the coating of nickel on the amorphous and crystallised alumina by first preparing γ-Al_2O_3 through the hydrolysis of aluminium sec-butoxide and then treating sonochemically with nickel tetracarbonyl in argon atmosphere. The amorphous form of Ni-coated alumina had more active sites due to the presence of hydroxyl groups which were apparently absent in the crystallised alumina coated nickel sample. Koltypin et al. [102] further reported the formation of relatively larger sized amorphous nickel nanoparticles (200 nm) through the use of different precursor $Ni(cyclooctadiene)_2$ with carbon particles also being present within the lattice. From the same laboratory, later Jeevanandam et al. [103]

reported the synthesis of nanosized α-nickel hydroxide with an interlayer spacing of 7.2 Å through a rather simpler sonochemical method and better product from the point of its application in the secondary alkaline batteries.

In the study of effects of ultrasound on the aqueous reactions of nickel, we found some interesting results, for example, the colloidal formation of Ni-DMG complex and degassing of NH_3 during different experiments. When 25 ml of 0.001 M $NiSO_4$ solution was complexed with 5 ml of 1% dimethyl glyoxime (DMG) in faintly alkaline ammonia medium and sonicated for 30 minutes and compared with another set of 25 ml of complexed solution which was stirred mechanically, a colloidal solution of Ni-DMG complex was formed in sonicated condition. Particles of Ni-DMG complex did not settle even after keeping 3–4 h because of their smaller size, in sonicated solution, whereas in the unsonicated condition large particles of Ni-DMG complex settled down immediately.

Similarly in another experiment, 25 ml of 0.01 M $NiSO_4$ solution was sonicated for 20 min, after the addition of different volumes (0.2, 0.3, 0.5, 1.0, 2.0, 4.0, 6.0 and 8.0 ml) of 5 M ammonia solution and compared with control sample, stirred mechanically for 20 min. Turbidity of sonicated and control samples were measured after each addition of ammonia.

Precipitation of $NiSO_4$ to $Ni(OH)_2$ by liquid NH_3 gave a very interesting result in sonicated and unsonicated condition. The turbidity of control samples increased to maxima before decreasing to the zero gradually. But in the sonicated sample, the turbidity increased initially before decreasing gradually but never reached the zero value (Table 9.8).

In yet another experiment, three set of 25 ml of 100 ppm $NiSO_4$ solutions were complexed with 0.1 ml of 1% 1-Nitroso-2-Naphthol and sonicated for 10, 20 and 30 min and compared with a control sample which was stirred mechanically. Turbidity of these solutions was measured. From the Table 9.9, it is clear that the turbidity of the solution increased as the time of sonication increased.

In the last experiment, the rate of crystallization of $NiSO_4$, $NiCl_2$ and $Ni(SCN)_2$ was compared for solutions under normal condition and those agitated by ultrasound for 1 hr. As could be seen in Table 9.10 the ultrasound could reduced the

Table 9.8 Turbidity in $NiSO_4$ solution in sonicated and unsonicated condition

Volume of 5 M ammonia (ml)	Time of sonication/ agitation (min)	Sonicated	Unsonicated
		Turbidity (NTU)	
0.0	20	00.4	00.4
0.2	20	120.7	134.7
0.3	20	129.0	135.0
0.5	20	131.0	136.0
1.0	20	122.0	138.0
2.0	20	112.0	108.0
4.0	20	114.0	19.7
6.0	20	114.7	2.5
8.0	20	113.0	0.0

Table 9.9 Turbidity in $NiSO_4$ – 1-Nitroso-2-Naphthol after sonication

Volume of sample (ml)	Time of sonication (min)	Turbidity (NTU)
25	0.0	4.0
25	10	5.1
25	20	6.3
25	30	7.6

Table 9.10 Effect of ultrasound on the starting time of crystallization

Time of sonication (min)	Solution	First observable crystallite formation (hours)	
		Unsonicated	Sonicated
60	$NiSO_4$	144	144
60	$NiCl_2$	96	48
60	Ni $(SCN)_2$	120	96

crystallisation time. Crystals of controlled samples were light green in colour and the edges were smooth, but the crystals in sonicated samples were dark green and their edges were multilayered and needle shaped.

From the above, it could be observed that ultrasound broke down the agglomerated precipitate of Ni-DMG complex to very fine particles, resulting in the formation of colloidal solution of Ni-DMG complex in the presence of an ultrasonic field. The settlement of particles was slow in the sonicated sample than in control sample. However, in the normal reaction, green precipitate of nickel (II) hydroxide dissolved in the excess NH_3 solution and formed deep blue solution of hexamminenickel(II) ions, as under:

$$Ni^{2+} + 2NH_3 + 2H_2O \quad \rightarrow \quad Ni(OH)_2 + 2NH_4^+ \tag{9.105}$$

$$Ni(OH)_2 + 6NH_3 \quad \rightarrow \quad [Ni(NH_3)_6]^{2+} + 2OH^- \tag{9.106}$$

But in sonicated condition hexamminenickel(II) ions did not form, because of the degassing effect of the ultrasound leading to the removal of ammonia gas during the sonication process. Because of this reason the turbidity in sonicated sample did not come to the zero value (Table 9.8). Ultrasound was also effective in the reduction of crystallization time by 50% in Ni-chloro and by 25% in Ni-thiocyanato complex, which, however, remained the same in $NiSO_4$ solution as shown in Table 9.10.

The conclusions derived from all above experiments could now be summarized as:

1. Ultrasound formed colloidal solution of Ni-DMG complex, due to the promotion of molecular fragmentation.
2. Ultrasound accelerated the degassing phenomenon and removed NH_3 from the solution, therefore nickel remained in its hydrolysed form only.
3. Ultrasound reduced the crystallisation time of Ni-thiocyanate and Ni-chloro complex but did not affect the crystallization of $NiSO_4$.

9.2.3.7 Zinc, Zn^{2+}

Zinc, in spite of being a transition metal, does not show the characteristic features because of its d^{10} configuration. Zinc behaves like a bivalent cation. Earlier work [104] on zinc metal indicated an acceleration of the liberation of H_2 gas in an ultrasonic field. Corrosion passivation mechanism of zinc in sonicated NaOH solutions (0.1 and 1 M) was investigated using electrochemical techniques, to determine corrosion and passivation kinetics parameters [105]. A study of the effect of 20 and 500 kHz ultrasound on the corrosion of zinc precoated steel in Cl^-, SO_4^{2-}, HCO_3^- and H_2O_2 electrolytes revealed an intensive corrosive effect of ultrasound in spite of the fact that the reaction sequence remained unchanged in the ultrasonic field compared to the normal silent condition. [106]. Suslick et al. [107] reported in detail the sonochemistry of zinc powder and explained the mechanism of heterogeneous reactions involving zinc metal and many other transition metal catalysts in both organic as well as inorganic chemical reactions and attributed the intensification of reactivity in the ultrasonic field mainly to the interparticle collisions, as a result of the turbulent flow and shock waves produced by the ultrasonic field. Dhas et al. [108] reported the formation of zinc sulphide nanoparticles on silica microspheres by sonicating for 3 h, a slurry of amorphous silica microspheres with zinc acetate and thioacetamide in an aqueous medium. They observed either a uniform coating of ZnS or clustered spots on the surface of silica microspheres. They could further fabricate supramolecular assembly of zinc sulphide nanoparticles by using dodecylamine as a structure directing agent [109]. On similar lines Zhu et al. [110] further synthesised zinc selenide nanoparticles of about 3 nm size by sonicating an aqueous solution containing selenourea and zinc acetate under argon and observed that many other selenides could be synthesised sonochemically. Seung-Ho Jung et al. [111] synthesised sonochemically highly uniform zinc phosphate hexagonal bipyramid crystals under ambient conditions with smooth facets and average size (from pole to pole) of 1.437 μm. Similarly Baranchikov et al. [112] reported the microstructural evaluation of Fe_2O_3 and $ZnFe_2O_4$ during high temperature (600–800°C) sonochemical synthesis of zinc ferrite and interpreted in terms of a qualitative model for ultrasonically activated solid state reactions. They further discussed the kinetics of the formation of zinc ferrite in an ultrasonic field [113].

Keeping in mind the above work, experiments were carried out to examine the effects of ultrasound, on the dissolution of zinc metal in an alkaline medium and the decomposition of zinc-dithizone complex in the presence of an ultrasonic field. To examine the effect of power ultrasound on the dissolution of zinc metal in alkaline media, 0.0480 g zinc metal was treated with 10 ml of 5 M NaOH solution. Two samples of this solution were exposed to ultrasound for 15 and 30 min, while, control samples were also kept in the similar condition and for the same duration. To compare their spectra and concentration of dissolved zinc in sonicated and control conditions, zinc-dithizone complex was formed by adding 0.5 ml of 0.005% dithizone solution. The red coloured complex, thus obtained, was extracted in chloroform and made upto to the mark in 25 ml volumetric flask with chloroform.

Table 9.11 Effect of ultrasound on the dissolution of zinc metal in alkaline medium

Time of sonication/exposure (min)	Concentration of Zn^{2+} ions	
	Sonicated	Control
15	0.02627	0.01115
30	0.03265	0.01124

Table 9.12 Nephelometric and conductometric analysis of zinc ions in alkaline medium

Time of sonication (min)	Turbidity (NTU)	Conductivity (ms)
0.0	0.5	63.0
10	10.1	68.9
20	12.0	74.8
30	15.2	80.7

UV spectra of these solutions were recorded and the concentration of zinc was measured from the absorbance value and the results obtained have been shown in Table 9.11. In another experiment, the nephelometric and conductometric analysis were carried out but only after diluting the solutions from 5 to 1 M of NaOH, since the conductivity of solutions above this concentration was beyond the scale of the conductivity meter.

As is obvious from the Table 9.11, the concentration of Zn^{2+} ions in alkaline solution increased as the time of sonication increased. In particular, the increase in sonicated solutions was much more than in the unsonicated solution. Turbidity and conductivity of the solutions also increased gradually as time of sonication increased (Table 9.12).

To examine the effect of ultrasound on the decomposition of Zn–dithizone complex, 0.2264 g Zn metal was treated with 10 ml of 5 M NaOH and sonicated for 30 min for maximum dissolution of Zn metal. After treating with ultrasound, 0.5 ml of 1% dithizone was added to form Zn-dithizone red coloured complex. This red colour Zn-dithizone complex was extracted in chloroform and made upto to the mark in 50 ml volumetric flask with chloroform. 10 ml of this complex was sonicated for different duration of time (10, 20 and 30 min) and UV-vis spectrophotometric analysis was carried out.

As could be seen in the Fig. 9.2, the peak of Zn-dithizone complex gradually decreased as the time of sonication increased (10–30 min). It could be due to the dissociation of Zn-dithizone complex in the presence of an ultrasonic field.

It may therefore be concluded that the pure zinc metal dissolved very slowly in alkaline medium, in the form of tetrahydroxozincate(II) ions as under:

$$Zn + 2OH^- + 2H_2O \rightarrow [Zn(OH)_4]^{2-} + H_2 \qquad (9.107)$$

Ultrasound enhanced the rate of dissolution of zinc metal in alkaline media, 2–3 times in sonicated condition rather than the unsonicated condition, due to the cleaning of metal surface in ultrasonic field. Decomposition of Zn-dithizone complex was also observed in ultrasonic field. In sonicated samples, the peak of

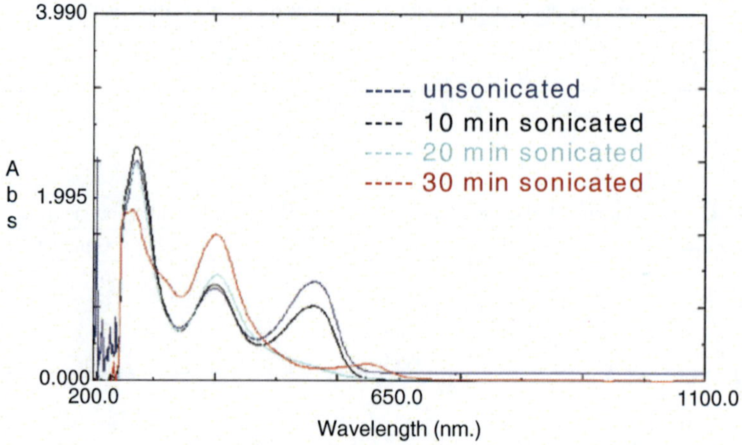

Fig. 9.2 Decomposition of Zn-dithizone complex

Zn-dithizone complex gradually decreased (Fig. 9.2). Mechanism for the decomposition of this complex could thus be explained that the complex was broken into zinc ions and dithizone before dithizone molecules were oxidised to S=C (N=NC$_6$H$_5$)$_2$, which, however, did not react further with metal ions [114] and therefore, the Zn-dithizone complex was decomposed. The decomposition of Zn-dithizone complex could thus be attributed to the oxidation of dithizone molecule in ultrasonic field.

9.2.3.8 Alkaline Earth Metals (Mg^{2+}, Ca^{2+}, Sr^{2+} and Ba^{2+})

The ultrasound enhanced the reactivity of calcium, 10 times, in the reduction of aromatic hydrocarbon [115]. The influence of ultrasonic compaction conditions of the (Ba, Sr)TiO_3 nanopowder on the pore distribution in sintered ceramics have been reported. The distribution regularities of small and large pores depended on sample manufacture conditions [116]. In the Weissler reaction, the yield of I_3^- was determined by irradiating aqueous solutions of NaI containing CCl_4 and $MgCl_2$ at various concentration, under air with 1 MHz ultrasound [117]. Shafi et al. reported [118] the olympic ring formation of amorphous $BaFe_{12}O_{19}$ from the sonochemical decomposition of barium hexaferrite. Rings of smaller dimension trapped inside the larger ones besides the intersection of two rings was in contradiction to the proposed mechanism of the ring formation. The creation of such amazing feature was attributed to the interplay of magnetic forces with the regular particle–substrate interaction. They further reported the sonochemical preparation of strontium hexaferrite nanoparticle for its use in the body lotion/ cosmetic creams [119].

Out of these above four alkaline earth metals in which the sonochemical examinations were carried out in this laboratory, the hydroxide of magnesium

Table 9.13 Turbidity and conductivity in the $MgCl_2$, $CaCl_2$, $SrCl_2$ and $BaCl_2$ solutions

Time of sonication	$MgCl_2$		$CaCl_2$		$SrCl_2$	$BaCl_2$
	Turbidity (NTU)	Conductivity (ms)	Turbidity (NTU)	Conductivity (ms)	Turbidity (NTU)	Turbidity (NTU)
0.0	0.6	46.2	0.6	48.7	0.6	0.6
10	13.7	47.7	2.0	50.2	1.8	2.0
20	15.1	48.8	4.0	54.3	3.0	5.0
30	20.1	51.6	6.0	65.5	4.0	5.5
60	56.5	55.4	8.7	74.7	6.0	6.6

was extremely insoluble in water, but the other hydroxides were relatively soluble. The solubility increased in the order; Ca < Sr < Ba as the basic character increased. Neutral solutions of the chlorides of these metals did not show any precipitation except in case of $MgCl_2$, in which some turbidity was observed as could be seen from Table 9.13. In another experiment, however, the precipitation of bicarbonates of Mg and Ca was observed, which helped reducing the hardness in underground water. This is discussed later in this chapter under the physical effects of ultrasound.

9.2.3.9 Platinum, $Pt^{2+/4+}$

$[PtCl_6]^{2-}$ forms colloidal particles when sonicated with 20 kHz frequency under acidic conditions [120]. Reduction [121] of Pt (II) has been reported which proceeded by reduction of secondary reducing radicals formed by hydrogen abstraction from organic additions with OH radicals and H atoms. Pt (IV) has been reported to reduce to zero-valent metal in the presence of various kinds of surfactants [122]. The reduction occurs in two steps;

$$\mathrm{Pt(IV)} \longrightarrow \mathrm{Pt(II)} \tag{9.108}$$

$$\mathrm{Pt(II)} \longrightarrow \mathrm{Pt}^{\circ} \tag{9.109}$$

In the sonolysis of aqueous surfactant solution, two kinds of organic reducing radicals were proposed to contribute. One R_{ab} was formed from the reaction of surfactant with primary radical such as $OH^{\bullet}$ or $H^{\bullet}$ and other radical R_{py} was formed from direct thermal decomposition of surfactant in the interfacial region between the collapsing cavities and bulk water. R_{ab} was effective for both reduction steps whereas R_{py} is involved only in Ist step. Pt particles, with an average diameter of 1 nm, was prepared from the non-ionic surfactant system of polyethylene glycol monostearate, which was smaller than those obtained from anionic surfactants such as sodium dodecylsulfate (3 nm) and sodium dodecylbenzenesulphonate (3 nm). The sonochemical synthesis of Pt, Au and Pd nanoparticles, immobilized onto the surface of titania has been reported with higher rate of hydrogen production from an aqueous solution of ethanol [123]. Smaller Pt nanoparticles effectively restricted recombination of electrons and holes and provided H_2 at a higher rate.

Torok et al. [124] reported the sonochemical enantioselective hydrogenation of ethyl pyruvate over various platinum catalyst and different solvents. Pt/SiO_2 and Pt/K-10 catalysts showed best enantioselectivity. The reactions took place in quantitative yield and with complete chemoselectivity and the hydrogenation rates increased with one order of magnitude even in very mild (atmospheric hydrogen pressure, room temperature) reaction conditions. They further reported that the ultrasonic pretreatment of the supported platinum was very beneficial, improving both the catalytic activity and selectivity [125]. The enantioselectivity of the hydrogenation of ethyl pyruvate increased upto 97.1%, besides, in case of cinnamaldehyde hydrogenation, the selective preparation of cinnamyl alcohol became possible. However, similar effects on other unsaturated aldehydes and alcohols were not very encouraging [126]. Similar pronouncement of the sonochemically induced photocatalytic efficiencies of platinised rutile titania has been reported by Sivakumar et al. [127] for the enhanced photodegradation (~2 times) of methyl orange.

However, in our experiments the complex of platinum (IV) with KI, $K_2[PtI_6]$, was not affected by 1 h of sonication, probably because of the low concentration and inert nature of the metal ion involved and the relatively weak power of the ultrasound (20 kHz, 6 W/cm^2).

9.2.4 *Study of Trivalent Ions*

9.2.4.1 Arsenic, As^{3+}

Ashokkumar et al. [128] have reported the sonochemical conversion of As(III) to As(V) in an aqueous solution as a process for the removal of arsenic in contaminated water. A very interesting pH independent sonochemical conversion of arsenic (III) to arsenic (V), besides its precipitation as sulphide in aqueous solutions, in the pH range ~5 to ~9, and subsequent adsorption on coagulants such as $Fe(OH)_3$ and $Al(OH)_3$ has also been reported [129] from this laboratory.

In our experiments, the stock solution (1,032 ppm) of As^{3+} was prepared by dissolving As_2O_3 (LOBA, AR) in the minimum volume of 1 M NaOH solution, which was neutralized with dilute HCl solution before making up to the calibration mark of the volumetric flask. Fine variation in the pH was adjusted later to prepare solutions with acidic (pH 3.4–5.1), neutral (pH 6.9–7.2) and basic (pH 8.8) solutions of arsenic using NaOH or HCl and the pH meter. Aqueous solutions of arsenic in different concentrations were treated with H_2S in neutral, faintly acidic and basic media and precipitation was observed. H_2S was passed into 25 ml of the solution until its saturation and sonicated for 60 min, using 12 mm tip which was dipped 20 mm in the solution. H_2S gas was purged moderately during the process of sonication to avoid depletion of the gas due to degassing property of ultrasound. Precipitate or colloidal suspension of As_2S_3 was filtered, washed with 8N HCl (saturated with H_2S), C_2H_5OH, CS_2 and again with C_2H_5OH and dried at 105°C before weighing. Since arsenic could be precipitated irrespective of the pH of the test medium, its precipitation along with CdS was also carried out. H_2S was passed to a mixture of Cd^{2+} and As^{3+} in 0.25 M HCl. The acid soluble fraction was tested for Cd^{2+} qualitatively using NaOH and NH_3, while alkali hydroxide soluble fraction was tested for As^{3+}. Surprisingly, both were found to have precipitated in the same solution under ultrasound.

Removal of Arsenic Using a Coagulant

Keeping in mind the toxic effects of arsenic and the severe health hazards that it poses through contaminated underground drinking water, removal of arsenic under ultrasonic field using a coagulant was also carried out. Solutions of different concentrations were prepared by diluting the initially prepared solution of As^{3+} of approximately 1,000 ppm (1,032 ppm) and the pH of subsequent solutions were adjusted by adding 2 M NaOH. The initial pH (5.1) of the arsenic stock solution was increased by adding 2 M NaOH solution until it settled to 7.0 and 8.0 respectively for neutral and partly alkaline solutions of arsenic. Three identical streams of H_2S gas were drawn simultaneously from a single source to maintain uniform supply of the gas to a set of three solutions (50 ml each) of arsenic. In the first beaker, only H_2S gas was passed in the arsenic solution and this was treated as blank. In the second solution, small amount of solid potash alum (0.0105 g) was added. In the third solution, the potash alum was added in the similar quantity, besides ultrasonic agitation was also carried out. The same process was repeated for arsenic solutions of 516 and 103 ppm with 0.005 and 0.0008 g of alum respectively and also at pH 5.1, 7.0 and 8.0. Only yellowing of the first solution at all pH values and all concentrations of arsenic was observed. A white gelatinous precipitate formation was observed in the second test solution, whereas, a clear thick precipitate of arsenic sulphide in the third set of beakers at all concentrations and pH of arsenic was obtained. This precipitate was digested overnight and filtered through G-4 sintered crucible. The precipitate was washed with 8N HCl, saturated with H_2S gas, to remove As^{3+} ions adhering to the precipitate and treated with CS_2 to dissolve free sulphur particles and finally with ethanol before drying at 105°C. From different amounts of As_2S_3 precipitate, the amount of arsenic was calculated.

The precipitation of arsenic with H_2S gas in the normal condition could occur only in strongly acidic medium whereas another cation of the second group Cd(II), precipitates only in faintly acidic medium, therefore, the precipitation of both cadmium and arsenic with H_2S gas in the same solution was not easily possible. To precipitate both in the same solution, the H_2S gas is conventionally first passed into the strongly acidic original solution of basic radicals followed by its bubbling into the diluted solution. To examine the role of ultrasound on the precipitation of arsenic in faintly acidic or neutral medium, few experiments were carried out. The results obtained showed effective precipitation of arsenic even in mild reaction solutions, with their pH ranging from 5.1 to 8.8. under ultrasonic field. Hence Cd^{2+} and $As^{3+/5+}$ both could be precipitated in the same solution at low pH under the

Table 9.14 Precipitation of Arsenic (as sulphide) at different concentration and pH for 1 h sonication

As in stock solution (g/l)	pH	Precipitation of arsenic by sonication as sulphide (%)
Acidic medium		
7.4	3.4	5.6
0.7490	5.0	8.01
0.3745	5.0	24.03
0.2996	5.1	18.35
0.2496	5.1	20.03
Alkaline medium		
7.4	8.8	1.28
0.7490	8.8	9.3
0.3745	8.8	17.3
Neutral medium		
7.4	6.9	1.35
0.7490	7.2	11.34
0.3745	6.9	14.6

ultrasonic effect. The precipitation of As (III and V), under normal interaction with H_2S gas, is highly pH dependent. No precipitation occurred unless the solution was strongly acidic (9N HCl) and only a yellow colouration was visible because of the formation of colloidal arsenic sulphide at mild acidic pH. However, when ultrasound was introduced to this yellow solution, precipitation began irrespective of the pH of the solution. Precipitation of the order of 24% was observed by the ultrasound alone (Table 9.14) which however increased further upon the addition of a coagulant such as alum and went upto 87% (Table 9.15) even in very dilute acidic solutions of arsenic (in ppm range). Sonication helped in coagulation and agglomeration of finer suspended arsenic sulphide particles spread all over the solution as colloidal particles, thus turning solution of arsenic sulphide turbid. With increasing time of sonication, turbidity increased and the particles did not settle even after keeping the solution overnight. The time required for the turbidity to appear after sonication was inversely proportional to the concentration of arsenic in the solution. When this experiment was repeated with potash alum, turbidity appeared almost instantaneously upon sonication and also settled within 30 min. In order to explain this effect, the nature of the arsenic sulphide sol needed to be understood. Arsenic sulphide sol is hydrophobic and negatively charged because of the adsorption of S^{2-} ions on the surface of sol particles, furnished by ionisation of H_2S gas. The sol gets precipitated upon sonication due to the fact that ultrasound helps in desorption phenomenon [130]. Stresses generated during ultrasound propagation, cavitational collapse, microjet formation helped S^{2-} to get desorbed from arsenic sulphide particles. Further, during the tearing of water molecular chains in the aqueous system, the separation of charged suspended particles as suggested by Margulis [131] facilitated agglomeration by withdrawing charges of the sol particles which result in the formation of neutral (charge less) arsenic sulphide particles. These particles then collided at a very high speed, as result of jet formation by collapsing

Table 9.15 Precipitation of Arsenic (as sulphide) using Potash alum at different concentrations and pH for 1-h sonication

As present in parent solution (g/l)	Potash Alum (g/l)	pH	Precipitate of Arsenic (g/l)	Removal of Arsenic (%)
Acidic medium				
0.1032	0.032	5.9	0.828	80.26
	0.22		0.828	80.26
0.515	0.032	5.7	0.4360	84.51
	0.22		0.4336	84.04
1.032	0.032	5.1	0.8867	85.9
	0.22		0.8989	87.1
	0.42		0.8989	87.1
Neutral medium				
0.1032	0.032	7.0	0.0803	77.9
	0.22		0.0803	77.9
0.516	0.032	7.0	0.4263	82.62
	0.22		0.4239	82.15
1.032	0.032	7.0	0.8770	84.98
	0.22		0.8867	85.92
	0.42		0.8867	85.92
Alkaline medium				
0.1032	0.032	8.0	0.0803	77.9
	0.22		0.0803	77.9
0.516	0.032	8.0	0.4190	81.2
	0.22		0.4190	81.2
1.032	0.032	8.0	0.8526	82.62
	0.22		0.8697	84.27
	0.42		0.8697	84.27

bubble, making the smaller particles to cling together to form larger particles. The increase in the precipitation of arsenic sulphide upon the addition of alum was, thus, only a secondary process of co-precipitation along with the precipitation of Al^{3+} ions as $Al(OH)_3$ due to hydrolysis of the later. $Al(OH)_3$ acted as a carrier and tiny particles of arsenic sulphide got adsorbed/stuck to the surface of $Al(OH)_3$ and agglomerated still quicker.

A scheme could thus be proposed for the precipitation and removal of arsenic in the ultrasonic field in the ppm and ppb range at normal pH as provided in Fig. 9.3.

9.2.4.2 Bismuth, Bi^{3+}

We could not find any study of Bi(III) ions in aqueous solutions except that Wang et al. [132] obtained nanorods of bismuth sulphide by sonicating an aqueous solution of bismuth nitrate and sodium thiosulphate in the presence of complexing agents such as ethylenediamine tetraacetic acid, triethanolamine and sodium tartarate. Similar results were found when thioacetamide was used in place of sodium thiosulphate as a source of sulfur. However, the results improved with higher yield

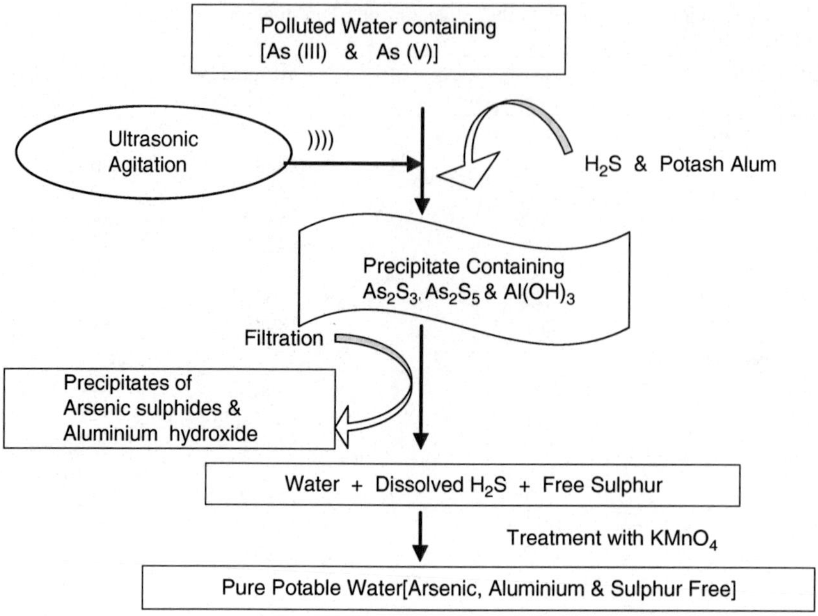

Fig. 9.3 Schematic diagram of purification of underground water contaminated with arsenic

and smaller size Bi_2S_3 particles were obtained when sonication was carried out in 20% N, N-dimethylformamide as a solvent.

In our experiments, the solutions of $Bi(NO_3)_3$ (E-Merck, AR), in different concentrations (0.071, 0.078, 0.087, 0.095, 0.103 and 0.111 M), were prepared and sonicated for 30 min, keeping in mind the relation between pH and solubility after which the salt got hydrolysed. Precipitate thus obtained after the sonication was found to be of $BiO(NO_3)$ which was soluble in dilute HCl and HNO_3 but insoluble in tartaric acid.

$$Bi^{3+} + NO_3^- + H_2O \longrightarrow BiO(NO_3)\downarrow + 2H^+ \qquad (9.110)$$

In order to estimate% conversion, precipitate was filtered, washed with C_2H_5OH and dried at 60°C before weighed. Different amounts of bismuth precipitated in solutions of bismuth nitrate, $Bi(NO_3)_3$, of the concentration 0.071–0.111 M between the pH range 0.9–0.5 respectively, were estimated gravimetrically and given in the Table 9.16. These Bi^{3+} solutions were sonicated above autohydrolytic concentration and pH. Hydrolysis began in the form of precipitate due to the formation of $BiO(NO_3)$. At lower concentration [0.071 M of $Bi(NO_3)_3$], precipitation started immediately upon sonication, whereas, the time for the appearance of turbidity increased as the concentration increased to 0.111 M. This was perhaps because of the availability of more water molecules for interaction with Bi^{3+} ions in dilute solutions. Ultrasonically induced hydrolysis was therefore

Table 9.16 Percentage precipitation of Bi^{3+} as BiO^+ at different concentrations and pH

Concentration of $Bi(NO_3)_3$ (M)	Bi^{3+} in solution (g/l)	Bi in ppt. of $BiONO_3$ (g/l)	pH	Precipitation of Bi (%)
0.071	15.04	3.084	0.90	20.5
0.078	16.72	1.636	0.87	9.78
0.087	18.39	1.596	0.79	8.67
0.095	20.06	1.004	0.67	5.0
0.103	21.73	0.904	0.60	4.16
0.111	23.40	0.876	0.57	3.7

inversely proportional to the concentration of the bismuth salt in the solution. Under normal conditions, hydrolysis of the Bi^{3+} ion took place as:

$$Bi^{3+} + H_2O \longrightarrow BiO^+ + 2H^+ \tag{9.111}$$

However, if there was sufficient amount of H^+ ions in the solution, equilibrium shifted towards the left and bismuth remained in the solution as Bi^{3+}. Therefore, during sonication activated water molecules were formed which reacted with Bi^{3+} ions. But instead of forming $Bi(OH)_3$, they formed bismuthyl ions, BiO^+, due to the fact that $Bi(OH)_3$ was a weaker base, therefore hydrolysed readily to generate bismuthyl ion, BiO^+. These steps could be summarised as under:

$$Bi(NO_3)_3 \longrightarrow Bi^{3+} + 3NO_3^- \tag{9.112}$$

$$Bi^{3+} + H_2O^* \longrightarrow BiO^+ + 2H^+ \tag{9.113}$$

$$BiO^+ + NO_3^- \longrightarrow BiONO_3 \downarrow \tag{9.114}$$

The amount of precipitated bismuth decreased as the concentration of bismuth salt increased (Table 9.16) and the duration of sonication required to bring about hydrolysis also increased. The initial reaction was spontaneous as per Eq. (9.111), which, however, seemed to be facilitated by ultrasonic cavitation at high concentration of bismuth. Since the H^+ ions were also produced during the formation of bismuthyl ion, at the point where the sum of concentration of H^+ ions present initially and formed by Eq. (9.110) was equal to the concentration required to shift the equilibrium of Eq. (9.111) towards left side, the hydrolysis did not occur even after sonication.

9.2.4.3 Antimony, Sb^{3+}

The only references of antimony available in the literature were the reactions in ethanolic solutions. Nowak et al. [133] have reported the sonochemical synthesis of SbSI gel by irradiating an ethanolic solution containing elemental antimony, sulfur and iodine for 2 h by the ultrasound of 35 kHz and 2 W/cm^2 at 50°C. They also reported the sonochemical preparation of antimony subiodide (Sb_3I) by irradiating sonochemically an ethanolic solution containing elemental antimony and iodine at 50°C [134]. However, the reported XRD pattern was found to be identical to the one reported earlier for $Sb_4O_5I_2$. They further reported the synthesis of antimony-sulfur-selenium-iodine nanowires of the unknown composition, $SbS_{1-x}Se_xI$ by sonicating elemental antimony, selenium, sulfur and iodine in ethanol by ultrasound of 35 kHz and 2 W/cm^2 at 50°C for 2 h [135]. The nanowires with about 10–50 nm diameter and length reaching upto several micrometers were of single crystalline nature. Wang et al. [136] reported the synthesis of regular stibnite (Sb_2S_3) nanorods of the diameter of 20–40 nm and length 200–350 nm by sonicating an ethanolic solution of antimony chloride ($SbCl_3$) and thioacetamide under ambient air. The high intensity ultrasound affected the synthesis in four steps, starting from the ultrasound induced decomposition of the precursor leading to the formation of amorphous Sb_2S_3 nanospheres. In the second step, ultrasound induced crystallization of these amorphous Sb_2S_3 nanoparticles. The third step involved the sonochemical acceleration in the formation and growth of crystal size and finally, in the last step, the surface erosion and the fragmentation were observed.

In our experiments, the solutions of $SbCl_3$ (E-Merck, AR) were prepared in the concentrations viz., 0.111, 0.119, 0.127 and 0.135 M, which were above the autohydrolytic concentration (0.111 M) at 0.2 pH, and sonicated for 30 min. A white precipitate of SbOCl, due to hydrolysis, was formed which was soluble in concentrated HCl and tartaric acid solution. To estimate the percentage precipitation, the precipitate was filtered, washed with ethyl alcohol and estimated gravimetrically after drying at 60°C.

The amounts of SbOCl precipitated upon sonication of 30 min duration are given in Table 9.17. The hydrolysis of the salt took place at low pH (less than1.0) as well. Solutions of all concentrations were sonicated for a period of 30 min. The rate of ultrasonically induced hydrolysis was found to be inversely proportional to the concentration in the solution and seemed to have been confirmed by observing the delay in the appearance of turbidity along with increasing concentration. This was possibly because at lower concentration more water molecules were available for interaction, resulting in the subsequent hydrolysis of antimony ion.

Since both Bi^{3+} and Sb^{3+} cations belong to the same group of the periodic table, the formation of antimonyl ion from antimony(III) ion upon sonication seemed to be similar to that as suggested in case of bismuth earlier, following the similar sequences of steps as under;

$$SbCl_3 \longrightarrow Sb^{3+} + 3Cl^- \quad (9.115)$$

$$Sb^{3+} + H_2O^* \longrightarrow SbO^+ + 2H^+ \quad (9.116)$$

$$SbO^+ + Cl^- \longrightarrow SbOCl \downarrow \quad (9.117)$$

Another similarity between antimony and bismuth, in the hydrolysis behaviour upon sonication, was that the hydrolysis due to sonication did not begin as long as

Table 9.17 Percent precipitation of Antimony at different concentrations and pH for the duration of 30 min

Molar concentration of $SbCl_3$ (M)	Antimony (g/l)	Precipitated Antimony in SbOCl (g/l)	pH	Precipitate of Sb (%)
0.111	3.62	8.82	0.3	64.75
0.119	14.59	8.67	0.2	59.42
0.127	5.57	7.81	0.2	50.16
0.135	16.54	7.16	0.2	43.28

the sum of the concentration of H^+ ions, already present in the solution and formed in Eq. (9.116), was above the concentration required to keep antimony in the solution as Sb^{3+} ion, whereas any decrease in H^+ ion concentration initiated the process of hydrolysis of both Sb^{3+} and Bi^{3+} and subsequent precipitation in the solution. Nevertheless, antimony showed a relatively higher degree of precipitation compared to bismuth. This could perhaps be due to the relatively smaller size but higher charge density on antimony ion with respect to bismuth. The antimony therefore attracts higher number of water molecules around itself. However, accommodating more water molecules on the relatively smaller size ionic surface was difficult. Therefore, hydrolysis occurred, resulting in the formation of antimonyl ion.

9.2.4.4 Aluminium, Al^{3+}

Although, there is no reported sonochemical study of aluminium salts in aqueous medium, there are few references in the literature for ultrasonic studies of Al, as a metal [137], slurry [138], alloy [139, 140] and alum [141], which cover partly some of the sonochemical behavior of aluminium. Ultrasound waves decreased the supersaturation limits and modified the morphology of the crystals of potash alum, besides having abrasive effect on crystals at high power inputs of ultrasound [141]. Agglomerates produced during the sonication of a series of mixed metal powders [138] in decane indicated fused metal particles of mixed metal systems such as Ni/Co, Al/Co, Al/Ni, Cu/Mo with substantially different tribological characteristics. Coatings were generated by both adhesive wear and direct impact. Low frequency treatment of Al-Sc, Al-Ti and Al-Zr alloy melts contributed to the fineness of the structure and intermetallics, increase in density and electrical resistance, decrease in gas content and higher strength characteristics of cast products [140]. Brazing of aluminium alloy parts had been observed due to ultrasound, which also destroyed the oxide film on the bonding surface and generated heat for bonding [139]. Morris and Brett [137] examined the influence of power ultrasound on the corrosion of aluminium and high speed steel in chloride medium. Cavitation led to the formation of pits, erosion and increased the mass transport. Ultrasound destroyed the oxide film on aluminium and the corrosion products from the surface of high speed steel due to increased solution mass transport. Aluminium, however, owing to its impervious oxide film formation tendency recovered considerably. The nucleation rate of cavities in super plastic aluminium composites was found to be low because stress concentrations were relieved by the presence of a liquid phase [142]. Ultrasound brazed alumina to copper using zinc-aluminium-copper fillers containing Al-Cu content$\leq$30 mass% where the ratio of Al/Cu was fixed as 3/2 [143]. The increase in application time of ultrasound and metalizing temperature increased the joining strength of Al_2O_3/Zn-Al-Cu joints. It may therefore, be seen that there is no sonochemical study of aluminium salt in aqueous medium. An attempt has therefore been made to examine the effect of ultrasound on the chemical reactivity of aluminium salts in aqueous medium. However, the poor tendency of the Al^{3+} for complexation and strong tendency for hydrolysis of dilute aqueous solutions made such a study rather limited. Al^{3+} salts in aqueous medium and their reactions with other reagent seemed to be transparent to the passage of ultrasound or in other words to cavitation. Even strong cavitation did not show any change on direct interactions involving aluminium, therefore, an attempt was made to examine the effect of ultrasound on its adsorption complex with aluminon reagent and precipitation with sodium acetate.

To examine the cavitational effect of ultrasound on bright-red complex of aluminon adsorbed on $Al(OH)_3$, in our experiments, 25 ml of 0.005 M aluminium sulphate was treated with 5 ml of 1% aluminon (triammonium aurine-tricarboxylate). This adsorption complex was sonicated for 10, 20 and 30 min, while the control sample was agitated for the same duration with a magnetic stirrer. The turbidity in sonicated sample increased, as time of sonication increased compared to the unsonicated condition (Table 9.18 and Fig. 9.4). In sonicated sample, the colour of the adsorption complex was dark compared to the control sample and the settlement of the adsorption complex was also slower due to the smaller size particle of the complex.

Ultrasound could play a dual role of creating higher interfacial area as well as facilitating the process of interfacial transport. This phenomenon seemed to be responsible for the increase in the colour intensity of the solution of Al^{3+}-aluminon adsorption complex and could be explained as follows.

The $Al(OH)_3$ precipitate in the aqueous solution floated as a fluffy and loafy mass with little agglomeration. The addition of aluminon reagent in the solution initiated adsorption on the active sites of this mass, developing red colour. When ultrasound was propagated in the solution, the bursting of cavitating bubbles pushed the organic molecules of aluminon deep inside the fluffy mass with of $Al(OH)_3$, creating greater precipitate to aluminon linkage. However, further propagation of ultrasound could dissociate the feebly interlinked precipitate to a uniformly dispersed heterogeneous solution of the adsorption complex. This could have also created new active sites for the adsorption of aluminon reagent. Some increase in

Table 9.18 Turbidity in Al^{3+} – aluminon complex after sonication/agitation

Time (min) of sonication/agitation	Turbidity (NTU)	
	Sonicated	Agitated
0.0	1.4	1.4
10	3.7	1.2
20	4.2	1.2
30	4.8	1.3

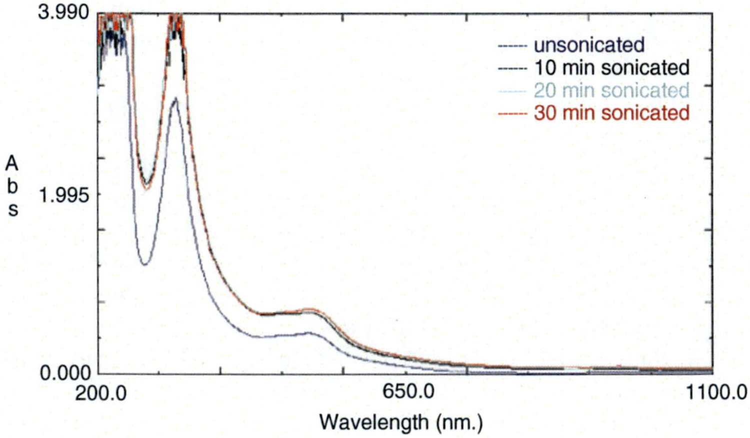

Fig. 9.4 Absorbance of Al^{3+} – aluminon complex after sonication

Table 9.19 Turbidity in Al^{3+} solutions by CH_3COONa

Volume of 1,000 ppm $Al_2(SO_4)_3$ (ml)	Volume of 4,000 ppm CH_3COONa (ml)	Turbidity(NTU)		
		Unsonicated	Sonicated	Boiled
10	2.5	40.2	58.7	137
10	5.0	117	147	145
10	10	128	153	191

the turbidity could be seen from the measurements recorded in Table 9.18. This might have deepened the colour of the precipitate (solution) as well.

In another experiment, different volumes (2.5, 5.0 and 10.0 ml) of 4,000 ppm solution of sodium acetate was added to 10.0 ml of 1,000 ppm solution of aluminium sulphate. In the normal reaction, no precipitate was obtained in cold and neutral solutions, but on boiling with excess reagent, a voluminous precipitate of basic aluminium acetate, $Al(OH)_2COOCH_3$, was formed.

$$Al^{3+} + 3CH_3COO^- + 2H_2O \rightarrow Al(OH)_2COOCH_3 + 2CH_3COOH \qquad (9.118)$$

Turbidity of the sonicated (30 min), control and boiled samples were measured and given in Table 9.19. As could be seen, the turbidity increased gradually, as the volume of solution or of CH_3COONa increased in all solutions (control, sonicated and boiled). When 20 ml of Al^{3+} was sonicated for 30 min with 5 ml of CH_3COONa, turbidity did not appear, but when a solution containing, 10 ml of Al^{3+} and 2.5 ml of CH_3COONa was sonicated for the same duration, the turbidity appeared, indicating the role of ultrasonic power which decreased to about half as the volume of the solution increased.

The turbidity gradually increased as shown in Table 9.19, with an increase in the volume of solution or amount of CH_3COONa in all samples (control, sonicated and boiled). But the increase in precipitation was not in proportion to the reagent added to the $Al_2(SO_4)_3$ solution. The turbidity in the first solution for unsonicated, sonicated and boiled solution was roughly in the ratio 1:1½:3. Turbidity in sonicated and boiled samples was almost equal in solution containing 10 ml of aluminium sulphate and 5.0 ml of sodium acetate. Turbidity in sonicated samples (10 + 10) increased marginally compared to the boiled sample. This trend of sonicated samples, as seen in Table 9.19, indicated the role of ultrasonic power to be important. When 20 ml of $Al_2(SO_4)_3$ solution was mixed with 5 ml of sodium acetate and sonicated, the amount of precipitate formed was negligible compared to the precipitation in a mixture of 10 ml of $Al_2(SO_4)_3$ and 2.5 ml of sodium acetate, where the intensity of the ultrasonic power was almost double.

9.2.4.5 Gold, Au^{3+}

Henglein et al. [144] prepared gold colloid in an aqueous solution of $AuCl_4^-$, under argon saturated atmosphere. The reduction of $AuCl_4^-$ complex was suggested to be due to $H^\bullet$ radical produced by ultrasonic dissociation of water. The gold complex was much more efficient in scavenging H atoms than the Ag^+ because of the ability of solute to accumulate at bubble water interface, which played a major role in radical scavenging process. Yeung et al. [145] reported that colloidal gold formation could be enhanced greatly if $H^\bullet/OH^\bullet$ radical scavengers, such as short chained alcohols, were used. The amount $AuCl_4^-$ reduced was highly dependent on the type and concentration of alcohols. The effect was due to the ability of different alcohols to get accumulated at bubble-solution interface. At any given concentration, the more hydrophobic was any alcohol, the greater was its concentration at the interface. It was further reported that more hydrophobic alcohol form smaller sized particles due to greater adsorption on particles thereby limiting the particle growth. Grieser et al. [146] found that the gold nanoparticles synthesized from the reduction of gold chloride under the effect of microgravity were much smaller (~10 nm) than those synthesized by the same method but under normal gravity (80 nm) in the laboratory. They further found that the rate of sonochemical reduction of Au(III) to produce nanoparticles in aqueous solutions containing 1-propanol was highly dependent upon the ultrasound frequency. The size and distribution of the Au nanoparticles produced could also be correlated with the rate of Au(III) reduction, which in turn was influenced by the applied frequency. The rate of Au(III) reduction as well as the size distribution of Au particles was governed more by the chemical effects of cavitation rather than its physical effect [147]. Park et al. [148] found that the gold nanoparticles of multiple and controlled shapes could be synthesized from the aqueous solution of Au(III) by changing the concentration of stabilizer (sodium dodecylsulphate) and the ultrasonic irradiation power. Pradhan et al. [149] reported that the morphology of original magnetite nanoparticles was maintained in the sonochemically synthesised gold–magnetite nanoparticles, when made in aqueous solution of dissolved tetrachloroauric acid and preformed magnetite nanoparticles. They controlled the loading of gold particles by adding different amounts of modifiers such as methanol, diethyl glycol and oleic acid.

In our experiment, the study of the effect of ultrasound on the reduction of $AuCl_3$ by stannous chloride in aqueous medium and also the formation of gold sols due to its reduction in an ultrasonic field was carried out. In 25, 50 and 75 ppm solution, only a purple colouration was visible upon the addition of $SnCl_2$ whereas in 100 and 1,000 ppm solutions a purple precipitate was obtained, which however, dissolved upon sonication forming a pink coloured solution. The pink colour, which was formed after 60 min of sonication, was the characteristic of gold sols having particle size less than 20 nm [146]. Here the gold particles, adsorbed on $Sn(OH)_2$ and indicated in Fig. 9.5 through a peak at 311 nm, seemed to have got desorbed because of cavitation. Stresses generated by the collapse of cavitation bubbles deformed, fractured and detached particles from the surface of nearby solids [150, 151]. These stresses helped in the desorption of gold. The precise morphological changes, which occurred in the solid/solid particle during sonication, were dependent upon the size of the particle and the nature of solid - bubble interface [152]. Cavitational collapse, which occurred near a surface that was larger than the diameter of a bubble immediately prior to the collapse (~150 nm at 20 kHz) could generate high speed microjets directed towards the solid surface which led to particle fracture [153]. Formation of these microjets helped gold to detach from $Sn(OH)_2$ through particle fracture. Upon sonication, the small particles were strongly shaken and collided head on or tangentially [154]. Therefore, there was a competition between collision and microjets [155], as a result, the particles were broken and smoothened. This involved a large decrease of grain size and consequently an increase in the number of particles. Conclusively, it could be said that when a particle of gold adsorbed on $Sn(OH)_2$ was exposed to ultrasound, the adsorption bond between $Sn(OH)_2$ and Au had broken due to cavitational collapse,

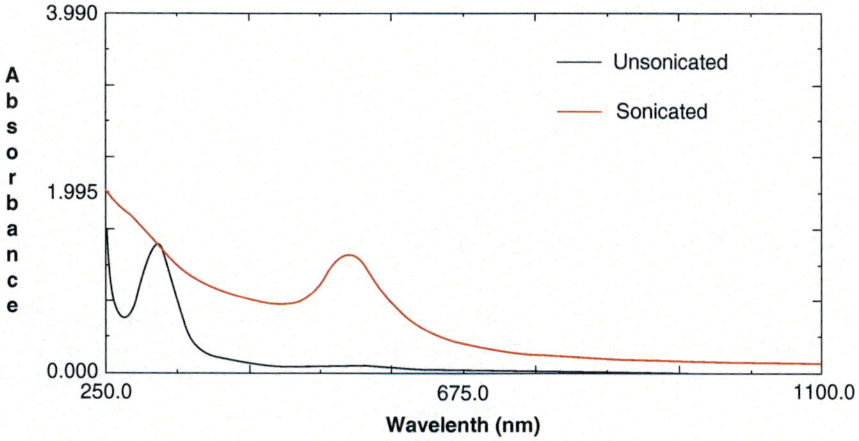

Fig. 9.5 Absorption spectra of 100 ppm Au^{3+} with $SnCl_2$

shock waves and particle collision, releasing Au particles into the solution. Various steps of the chemical changes for the above reaction could, therefore, be suggested as under:

$$2H^{\bullet}(OH^{\bullet}) + SnCl_2 \longrightarrow Sn(OH)_2 + 2HCl \quad (9.119)$$

$$AuCl_3 + 3H^{\bullet} \longrightarrow Au^{\circ} + 3HCl \quad (9.120)$$

$$Sn(OH)_2 + Au^{o} \longrightarrow [Sn(OH)_2]Au \quad (9.121)$$

$$nAu \longrightarrow Au_n \quad (9.122)$$

As reported in the literature by Heard et al. [156], the colour of the gold sol depended strongly on the size of the particles. The blue sol corresponded to 20–30 nm elongated agglomerated particles whereas pink sols corresponded to <20 nm particle size.

In the second experiment, the reaction of $AuCl_3$ with H_2O_2 in the presence of NaOH, was not very distinct. However, upon sonication the solution turned blue. This blue colour was the typical of gold sols when agglomeration of sol particles had occurred appreciably. In such sols cluster formation of somewhat elongated particles (generated from particle–particle fusion) of the range 20–30 nm diameter had taken place. Contrary to our observation with $SnCl_2$, where precipitated mass of $Sn(OH)_2$ was heterogeneously distributed over the solution system, the solid surface helped in the deposition of gold particles immediately after their generation, as if, these particles were captured and adsorbed over the floating solid surface. The phenomenon of desorption, when operated upon sonication, removed these embedded smaller particles into the solution and the colour of the solution was, therefore, red. However, in the second experiment, since there was no solid surface, which could have removed gold particles just after its formation, the size of the particle, therefore, grew relatively to attain a size around 20–30 nm and thus the solution looked blue. Now it could be easily concluded that ultrasound helped in generating Au particles of specific size depending on whether the solution phase was homogenous or heterogeneous and size of the solid surface present, through either by particle fusion or particle fracture.

9.2.5 Hardness Mitigation and Bacterial Decay

The biological contamination and the hardness of underground water are the two very serious problems of water. Dissolved salts deteriorate water quality and may cause diseases related to joints and bones, while infected water may cause many water-borne diseases such as cholera, dysentery, typhoid etc. Ultrasound may disinfect the potable water by blasting off micro organisms such as zooplanktons, phytoplanktons, pathogenic bacteria and produce germ-free water in few minutes of

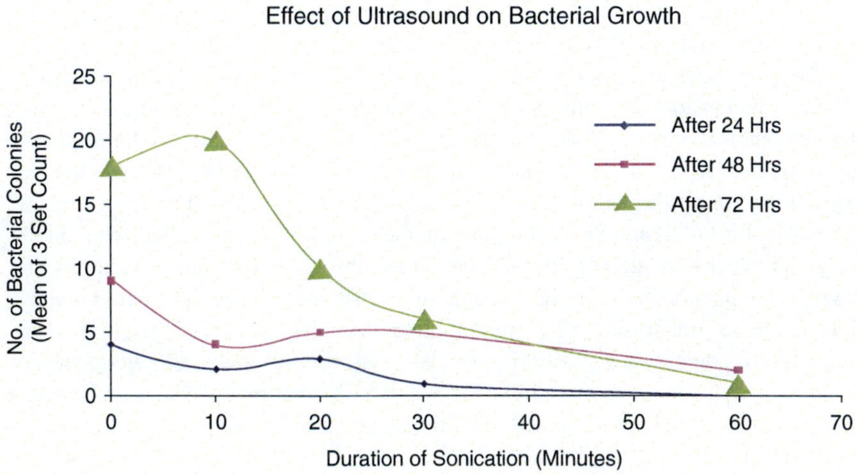

Fig. 9.6 Decrease in the bacterial colonies with time and sonication period

sonication without the addition of chemicals [157]. In laboratory experiments, sonication reduced bacterial colonies (*E. Coli*) of the water samples, which were treated with ultrasound for 60 min and remained clean and safe even upto 72 h after sonication. Figure 9.6 shows the decreasing number of bacterial colonies with increasing duration of sonication. These results were found to be in good agreement with the results reported elsewhere [158–160]. Similarly, sonication also converted water soluble bicarbonates of calcium and magnesium to their respective water insoluble carbonates. The precipitated carbonates of calcium and magnesium could then be filtered off. Sonication could efficiently remove nearly 14% of dissolved salts present in water, as verified experimentally.

9.2.6 Ultrasound Initiated Crystallization

A number of methods to grow crystals are reported in the literature [161]. The size and physical appearances of the crystals grown under ultrasonic medium has, however, attracted the attention recently [162] due to its application in the synthesis and growth of nanometer to centimeter sized crystals on an industrial scale [163]. Sonochemical method for crystallization, with controllable conditions is rapid besides, being able to produce crystals of uniform size, high purity and narrow size distribution [164]. The acoustic cavitation process offers significant potential for modifying and improving products in terms of their crystal purity and morphological forms [165], even in ambient conditions [162]. The electrical [166] and optical properties [167], which depend upon the size and shape of inorganic nanocrystals, can be controlled better using ultrasound. Thus, the crystals of desired molecular shape, symmetry and intermolecular forces can be synthesized [168] through the variation in the concentration of precursors and sonication time [169].

The synthesis and modification of crystals of $PbCl_2$, PbI_2, Au_2S, PtS_2 and $PdO \cdot nH_2O$, grown in aqueous solutions both in normal as well as in ultrasonic medium at the frequency 20 kHz and 6 W/cm^3 power has been described below to highlight the apparent difference in the shape, morphology and crystallinity of salts in their crystals, as found through scanning electron microscope. An increase in the size of crystals for $PbCl_2$ and PbI_2, blistering on the surface of Au_2S, surface cleaning in PtS_2 and ideal crystallinity in PdO has been observed. Heating, impaction and cavitational phenomenon of the propagating ultrasound has been attributed to these drastic effects on crystals. The precipitated flocculent or crystalline mass was subjected to sonication for 30 min at 6 W of ultrasonic power and frequency of 20 kHz in 25 ml aqueous medium. From the SEM pictures of salts, synthesized under normal as well as sonicated conditions and shown in Fig. 9.7, the following interesting inferences may be drawn.

$PdO.nH_2O$ does not show any sign of crystallization in the precipitate of $PdO \cdot nH_2O$, under normal condition even at $\times 1{,}500$ magnification. However after sonicating the samples for 30 min the crystals begin to form agglomerates and big rhomboidal crystals with very clear edges and faces were obtained even at $\times 1{,}000$ magnification. The size of a single crystal was larger than 10 μm and there was no sign of pitting or physical breaking of crystals. This could have perhaps been due to the hardness of the crystalline form of $PdO \cdot nH_2O$. This effect of ultrasound is similar to our observation in case of PbI_2.

Long, smooth and fragile crystals of lead chloride, $PbCl_2$ with rounded ends, virtually got crushed after the ultrasonic irradiation for 30 min. However, the size of the crystals grew 3–4 times than those of the parent crystals. These crystals seemed to first dissolve into the solution as a result of the rise in temperature followed by the reappearance of the second crop of crystals and quick formation of $PbCl_2$ lattices due to the vapourisation of the solvent (water) as a consequence of cavitation. Incessant cavitation in the solution accelerated the process of deposition of $PbCl_2$ molecules further on the crystal nucleation sites as the vapour formation in the cavitating bubbles continued. New nucleation sites could also be seen on the surface of already grown crystals even under $\times 430$ magnification.

Drastic changes were apparently visible in the crystals of PbI_2, under $\times 1{,}500$ and $\times 2{,}000$ magnification. PbI_2 crystals were initially spherical and very small (<10 μm) in size, however, after the 30 min of ultrasonic irradiation, changed to larger size and layered crystals with bright reflecting surfaces. Impaction and inter particle collision resulted into the formation of bigger agglomerates, which were further smoothened due to shock waves generated by the ultrasound.

Gold sulphide had a featureless non-crystalline plain surface with very sharp and smooth edges of the dried mass. However, the smoothness of the surface was destroyed completely with blisters appearing as a result of cavitation, after 30 min of ultrasound irradiation. A closer look at the surface at $\times 7{,}500$ magnification reveals extensive surface erosion as a result of the micro-jets streaming.

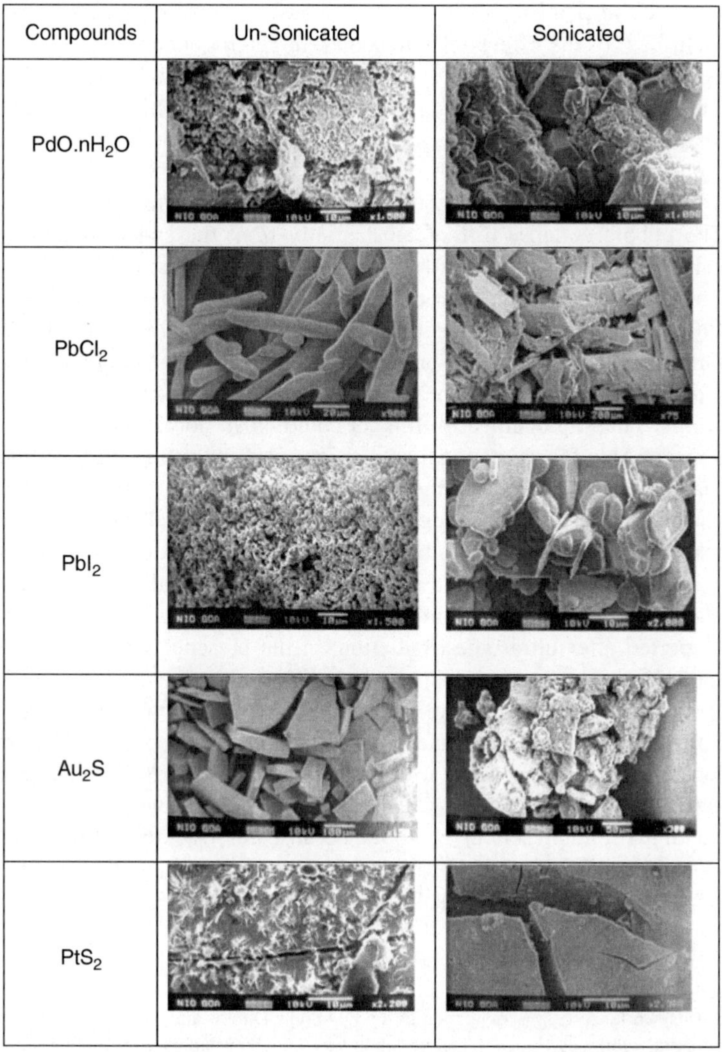

Fig. 9.7 SEM pictures of sonicated and unsonicated crystals of few inorganic compounds

Flower shaped crystalline deposit on the surface of the solid non-crystalline mass of platinum sulphide was probably due to the precipitation of elemental sulphur, which deposited as a floral growth on the non-crystalline platinum sulphide precipitate. Ultrasonic irradiation seemed to have broken tender sulphur flakes and cleaned the surface. The free sulphur, however, did not deposit further. This was probably due to the formation of other compounds of sulphur such as H_2S, SO_2, etc. which could have been removed from the solution due to the phenomenon of degassing.

Although the finer mechanistic detail of acoustic cavitation created a complex picture, but most of the effects of ultrasound, in heterogonous systems, had been attributed to this process. There was little doubt that the collapse of cavitation bubbles in the vicinity of a solid surface could cause considerable turbulence due to shock waves and high velocity solvent micro-jets, which on interaction with the surface, may lead to considerable changes in surface morphology. In addition, the stirring effect, caused by acoustic streaming, also affected crystallization [170]. The bubble collapse differs substantially from the spherically symmetric implosion [171], observed normally in homogenous liquids, compared to when cavitation occurred in a liquid near an extended solid surface. The solid surface in a solution would effectively distort the pressure of sound field, affecting the force of cavitational collapse near the surface [172]. This could generate a moving jet of liquid directed at the surface. The speed of the jet had been reported to be greater than 100 m/s. The impingement of this jet could create localised surface damage and was responsible for cavitational erosion and surface pitting. Second contribution to cavitational erosion was the impact of shockwave generated by bubble collapse, which could be large enough to deform metals, crystals or solid substrates. The shock waves created in liquids by ultrasound leads to particle collisions, which could be so intense that they could cause localised melting at the point of impact with metal powders [172]. A significant decrease in the particle size of the host solid is reported after ultrasonic irradiation, in the presence or absence of guest compounds [173]. The particle diameters could be reduced to 15 μm after 10 min of sonication, starting with solids of 60–90 μm diameters. However, further irradiation, may not change the particle size. In all reported cases here, ultrasonic irradiation breaks particles down to only some finite size. At 20 kHz, this size seems to be 1–10 μm in diameter. Change in surface morphology largely depends on the size of the solid present in the liquid, because micro-jet formation requires an extended surface [174]. At 20 kHz, the bubble size before the collapse is between 100–200 μm [174] in diameter. In solid particles, having this size or smaller, bubble deformation during cavitational collapse will not occur.

Acknowledgements Author is grateful to DST, AICTE, BRNS-DAE and SAP-UGC research grants for this work. A very special thanks to Dr. Sonu Dubey and Dr. Manju Chauhan for the experimental work, carried out in our laboratory and reported in this monograph and to Mr. Mayank Verma (JRF-UGC) and Ms. Shikha Goyal (JRF-CSIR) for their referral assistance.

References

1. Thorneycroft J, Barnaby SW (1895) Torpedo-boat destroyers (including appendix and plate at back of volume). Inst C E 122:51–69
2. Rayleigh L (1917) On the pressure developed during the collapse of a spherical cavity. Philos Mag Ser 6(34):94–98

3. Langevin MP and Chilowsky MC (1918) Procédés et appareils pour la production de signaux sous-marins diriges et pour la localization á distance d'obstacles sous marins, French Patent No. 502913
4. Wood RW, Loomis AL (1927) The physical and biological effects of high-frequency sound-waves of great intensity. Philos Mag Ser 7:417–436
5. i. Mason TJ, Lorimer JP (1989) Sonochemistry – the theory, applications and uses of ultrasound in chemistry, Ellis Horwood, Chichester. ii. Pankaj, Sonu Misra (1999) Sonochemistry – An introduction, Chem Educ Rev 14(3): 21–24
6. i. Yeager E, Hovorka F (1953) Ultrasonic waves and electrochemistry. I. A survey of the electrochemical applications of ultrasonic waves. J Acoust Soc Am 25(3): 443–455. ii. Yeager E, Bugosh J, Hovorka F, McCarthy J (1949) The application of ultrasonics to the study of electrolyte solutions. II. The detection of Debye effect J Chem Phys 17(4): 411–415. iii. Yeager E, Hovorka F (1949) The application of ultrasonics to the study of electrolyte solutions. III. The effect of acoustical waves on the hydrogen electrode. J Chem Phys 17(4): 416–417. iv. Compton RG, Eklund JC, Marken F (1997) Sonoelectrochemical processes: a review. Electroanalysis 9: 509–522
7. i. Blandamer MJ (1973) Introduction to chemical ultrasonics. Academic Press, New York. ii. Suslick KS (1988) Ultrasound: its chemical, physical and biological effects. VHS Verlag, , New York. iii. Ley Steven V, Low CMR (1990) Ultrasound in synthesis. Springer Verleg, New York. iv. Mason TJ (1990) Sonochemistry: the uses of ultrasound in chemistry. Royal Society of Chemistry, Cambridge. v. Mason TJ (1990) Chemistry with ultrasound, Springer, Berlin. vi. Timothy J Mason (1991) Practical sonochemistry: user's guide to application in chemistry and chemical engineering. Ellis Horwood, Chichester. vii. Gareth J Price (1992) Current trends in sonochemistry. Springer Verleg, New York. viii. Margulis MA (1995) Sonochemistry & cavitation. Taylor & Francis, London. ix. Jean-Louis Luche (1998) Synthetic organic sonochemistry. Springer, London. x. Timothy J Mason (1999) Sonochemistry. Oxford University Press (Premier Series no. 70), Oxford. xi. Timothy J Mason, John P Lorimer (2002) Applied sonochemistry: uses of power ultrasound in chemistry. Wiley-VCH, Weinheim. xii. Timothy J Mason (2002) Practical sonochemistry: power ultrasound uses and application. Horwood Publishing, Chichester.
8. i. Suslick KS, Price GJ (1999) Applications of ultrasound to material chemistry. Annu Rev Mater Sci 29: 295–326. ii. Thompson LH, nd Doraiswamy LK (1999) Sonochemistry: science and engineering. Ind Eng Chem Res 38: 1215–1249. iii. Chen D, He Z, Weavers LK, Chin Y-P, Walker HW, Hatcher PG (2004) Sonochemical reactions of dissolved organic matter. Res Chem Intermed 30(7–8): 735–753
9. William RT, Alfred LL (1927) The chemical effects of high frequency sound waves 1. A Preliminary Survey. J Am Chem Soc 49:3086–3100
10. Weissler A, Herbert Cooper W, Snyder Stuart (1950) Chemical effects of ultrasonic waves: oxidation of potassium iodide solution by carbon tetrachloride. J Am Chem Soc 72:1769–1775
11. Edwin Hart J, Henglein Arnim (1985) Free radical and free atom reactions in the sonolysis of aqueous iodide and formate solutions. J Phys Chem 89:4342–4347
12. Martiza G, Arnim H, Fernando I (1991) Radical scavenging in the sonolysis of aqueous solutions of I^-, Br^- and N_3^-. J Phys Chem 95:6044–6047
13. Hart Edwin J, Henglein Arnim (1986) Sonolysis of ozone in aqueous solution. J Phys Chem 90:3061–3062
14. Hart Edwin J, Henglein Arnim (1986) Sonolytic decomposition of nitrous oxide in aqueous solution. J Phys Chem 90:5992–5995
15. Hart Edwin J, Fischer Ch-Herbert, Henglein Arnim (1986) Isotopic exchange in the sonolysis of aqueous solutions containing $^{14.14}N_2$ and $^{15.15}N_2$. J Phys Chem 90:5989–5991
16. Fischer C-H, Hart Edwin J, Henglein Arnim (1986) H/D isotopic exchange in the HD – H_2O system under the influence of ultrasound. J Phys Chem 90:3059–3060

17. Fischer Christian-H, Hart Edwin J, Henglein Arnim (1986) H/D isotopic exchange in the $D_2 - H_2+$ system under the influence of ultrasound. J Phys Chem 90:222–224
18. Hart Edwin J, Fischer Ch-H, Henglein A (1987) Isotopic exchange in the sonolysis of aqueous solutions containing D_2 and CH_4. J Phys Chem 91:4166–4169
19. Fischer Christian-H, Hart Edwin J, Henglein Arnim (1986) Ultrasonic irradiation of water in the presence of $^{18.18}O_2$: Isotope exchange and isotopic distribution of H_2O_2. J Phys Chem 90:1954–1956
20. Hart Edwin J, Henglein A (1987) Sonochemistry of aqueous solutions: H_2–O_2 combustion in cavitation bubbles. J Phys Chem 91:3654–3656
21. Hart Edwin J, Fischer C-H, Henglein A (1990) Pyrolysis of acetylene in sonolytic cavitation bubbles in aqueous solution. J Phys Chem 94:284–290
22. P Christian, Dominick C (2001) The sonochemical degradation of aromatic and chloroaromatic contaminants. In: Mason TJ and Tiehm A (eds) Advances in sonochemistry: ultrasound in environmental protection, Elsevier 6: 102–103
23. Miller N (1950) Chemical actions of sound waves on aqueous solutions. J Chem Soc Faraday Trans 46:546–550
24. Makino K, Mosobba MM, Riesz P (1982) Chemical effects of ultrasound in aqueous solutions. Evidence for OH and H by spin-trapping. J Am Chem Soc 104:3537–3539
25. Rosentel IK, Mosobba MM, Riesz P (1981) Sonolysis of perhalomethane as studied by EPR and spin trapping. J Magn Reson 45:359–361
26. Makino K, Mosobba MM, Riesz P (1983) Formation of .OH and .H in aqueous solutions by ultrasound using clinical equipment. Radiat Res 96(2):416–421
27. Henglein A (1987) Sonochemistry: historical developments and modern aspects. Ultrasonics 25:6–16
28. Margulis MA, Mal'tsev (1968) Estimation of energetic yield of chemical reactions initiated by ultrasonic waves. II. Initial chemical acoustic yields of products of water decomposition by ultrasound. Zh Fiz Khim 42:2660–2663, Russ J Phys Chem (42): 1447–1451
29. Lippitt B, McCord JM, Fridovitch IJ (1972) The sonochemical reduction of cytochrome c and its inhibition by superoxide dismutase. Biol Chem 247:4688–4690
30. Henglein A (1956) The acceleration of chemical reactions of ultrasound in solutions of oxygen-rare gas mixtures. Naturwissenschaften (in German) 43(12):277–278
31. DelDuca M, Yeager E, Davies MO, Havorka F (1958) Isotopic technique in the study of the sonochemical formation of hydrogen peroxide. J Acoust Soc Am 30(4):301–307
32. Henglein A (1957) The formation of hydrogen peroxide through ultrasound in aqueous solutions of hydrogen in presence of argon and oxygen (in German). Naturwissenschaften 44(6):179
33. Virtanen AJ, Ellfolk NJ (1950) Nitrogen fixation in an ultrasonic field. J Am Chem Soc 72:1046–1047
34. Virtanen AJ, Ellfolk NJ (1950) Oxidative nitrogen fixation in ultrasonic field. Acta Chem Scand 4:93–102
35. Virtanen AJ, Ellfolk NJ (1952) Inhibition of oxidative nitrogen fixation in ultrasonic field by volatile substances. Acta Chem Scand 6(5):660–666
36. Sokol'skaya, El'piner IE (1957) The halogen treatments of coals under the action of ultrasonic waves. Dokl Akad Nauk SSSR 114(2):372–374
37. Sokol'skaya, Elpiner IE (1958) Chemical synthesis under the action of supersonic waves in a water saturated with gases of a reducing atmosphere. Dokl Akad Nauk USSR 119(6): 1180–1182
38. Buttner J, Gutiérrez M, Henglein A (1991) Sonolysis of water – methanol mixtures. J Phys Chem 95:1528–1530
39. Lindstrom O (1955) Physico-chemical aspects of chemically active ultrasonic cavitation in aqueous solutions. J Acoust Soc Am 27(4):654–671
40. Gueguen H (1963) Recherches sur les effets chimiques suscites par les ultrasons dans les solutions aqueuses de quelques halogenures. Ann Chim 8(11–12):667–713

41. Prudhomme RO, Grabar P (1949) De l'action Chimique des ultrasons sur certain solutions aqueuses. J Chim Phys 46:323–331
42. Beuthe U (1933) Concerning the influence of ultrasonic waves on chemical processes. Z Phys Chem 163(3/4):161–171
43. Srivastava SC (1958) Chemical reactions initiated by ultrasonic waves. Nature 182:47
44. Gutierrez M, Henglein A, Dohrmann Jurgen K (1987) H atom reactions in the sonolysis of aqueous solutions. J Phys Chem 91:6687–6690
45. Hart EJ, Henglein A (1986) Sonolytic decomposition of nitrous oxide in aqueous solution. J Phys Chem 90:5992–5995
46. Matyjaszewski Krzysztof, Dorota Gresta, Hrkach Jeffrey S, Hwan Kyu Kim (1995) Sonochemical synthesis of polysilylenes by reductive coupling of disubstituted dichlorosilanes with alkali metals. Macromolecules 28:59–72
47. Francoise L-M, Voglet N, Thierry L, Rudi A (2001) Evidence for the emission of 'alkali-metal–noble-gas' van der Waals molecules from cavitation bubbles. Ultrason Sonochem 8(2):151–158
48. Lynne K, Maritza G, Arnim H (1996) Bimetallic colloids: Silver and Mercury. J Phys Chem 100(27):11203–11206
49. Kuppa R, Moholkar VS (2010) Physical features of ultrasound-enhanced heterogeneous permanganate oxidation. Ultrason Sonochem 17(1):123–131
50. Soudagar SR, Samant SD (1995) Investigation of ultrasound catalyzed oxidation of arylalkanes using aqueous potassium permanganate. Ultrason Sonochem 2(1):S15–S18
51. Salkar RA, Jeevanandam P, Aruna ST, Yuri K, Gedanken A (1999) The sonochemical preparation of silver nanoparticles. J Mater Chem 9:1333–1335
52. Jhu J, Suwen Liu O, Palchik YL, Gedanken A (2000) Shape-controlled synthesis of silver nanoparticles by pulse sonoelectrochemical methods. Langmiur 16(16):6396–6399
53. Pol VG, Srivastava DN, Palchik O, Palchik V, Slifkin MA, Weiss AM, Gedanken A (2000) Sonochemical deposition of silver nanoparticles on silica spheres. Langmuir 18(8): 3352–3357
54. Pol VG, Grisaru H, Gedanken A (2005) Coating noble metal nanocrystals (Ag, Au, Pd and Pt) on polystyrene spheres via ultrasound irradiation. Langmuir 21:3635–3640
55. Vijaya Kumar R, Palchik O, Koltypin Yu, Diamant Y, Gedanken A (2002) Sonochemical synthesis and characterization of Ag_2S/PVA and CuS/PVA nanocomposite. Ultrason Sonochem 9(2):65–70
56. Salim OM, Mahir I, Mahmut Bayramo glu (2005) Leaching of silver from solid waste using ultrasound assisted thiourea method. Ultrason Sonochem 12(3):237–242
57. Wang Wang Xi-Kui, Li S, Wei-Lin G, Jin_Gang Wang, Yu-ping Zhu, Chen Wang (2009) Synthesis of dendritic silver nanostructures by means of ultrasonic irradiation. Ultrason Sonochem 16(6):747–751
58. Yu-ping Z, Xi-Kui W, Wei-lin G, Jin-gang W, Chen W (2010) Sonochemical synthesis of silver nanorods by reduction of sliver nitrate in aqueous solution. Ultrason Sonochem 17:675–679
59. Nina P, Galina A, Claudio R, de la Fernando Vega, Aharon Gedanken (2009) Sonochemical deposition of magnetite on silver nanocrystals. Ultrason Sonochem 16(1):132–135
60. Nagata Y, Watananbe S, Fujita S, Dohmaru T and Taniguchi (1992) Formation of colloidal silver in water by ultrasonic irradiation. J Chem Soc Commun 1620–1622
61. Eckfeldt Edgar L, Eynon James U (1964) Conductivity (electrical) measurements. In: Hampel CA (ed) Encyclopedia of electrochemistry. Krieges, Huntington, NY, pp 232–236
62. Zhu JJ, Aruna ST, Yuri K, Gedanken A (2000) A novel method for the preparation of lead selenide: pulse sonoelectrochemical synthesis of lead selenide nanoparticles. Chem Mater 12:143–147
63. Zhu Jun-Jie, Wang Hui, Xu Shu, Chen Hong-Yuan Chen (2002) Sonochemical method for the preparation of monodisperse spherical and rectangular lead selenide nanoparticles. Langmuir 18:3306–3310

64. Purkayastha A, Qingyu Y, Gandhi Darshan D, Li H, Gyana P, Theodorian B-T, Ravishankar N, Ramanath G (2008) Sequential organic–inorganic templating and thermoelectric properties of high-aspect-ratio single-crystal lead telluride nanorods. Chem Mater 20(15):4791–4793
65. Mohammad Jafar Soltanian Fard-Jahromi and Ali Morsali (2010) Sonochemical synthesis of nanoscale mixed-ligands lead(II) coordination polymers as precursors for preparation of $Pb_2(SO_4)O$ and PbO nanoparticles; thermal, structural and X-ray powder diffraction studies. Ultrason Sonochem 17(2):435–440
66. Sadeghzadeh H, Morsali A, Yilmaz VY, Buyukgungor O (2010) Sonochemical synthesis of nano-scale mixed-ligands lead(II) coordination polymers as precursors for preparation of PbO and PbBr(OH) nano-structures; thermal, structural and X-ray powder diffraction studies. Ultrason Sonochem 17(3):592–597
67. Ziki H, Samual TJ, Jerry BM, Linda WK (2005) Sonolytic desorption of mercury from aluminium oxide. Environ Sci Technol 39(4):1037–1044
68. Ziki H, Samual TJ, Linda WK (2007) Sonochemical dissolution of cinnabar (α-HgS). Environ Sci Technol 41(3):773–778
69. Gil S, Lavilla I, Bendicho C (2008) Mercury removal from contaminated water by ultrasound-promoted reduction/vaporization in a microscale reactor. Ultrason Sonochem 15(3):212–216
70. Di Michele A, Diodati P, Morresi A, Sassi P (2009) Mercury acetate produced by metallic mercury subjected to acoustic cavitation in a solution of acetic acid in water. Ultrason Sonochem 16(1):141–144
71. Kristl M, Drofenik M (2008) Sonochemical synthesis of nanocrystalline mercury sulfide, selenide and telluride in aqueous solutions. Ultrason Sonochem 15(5):695–699
72. Wang H, Zhu J-J (2004) A sonochemical method for the selective synthesis of α-HgS and β-HgS nanoparticles. Ultrason Sonochem 11(5):293–300
73. Vogel AI (1995) A text book of quantitative inorganic analysis. ELBS, London, 133
74. Haas I, Gedanken A (2006) Sonoelectrochemistry of Cu^{2+} in the presence of cetyltrimethylammonium bromide: obtaining CuBr instead of copper. Chem Mater 18(5):1184–1189
75. Gutterez M, Henglein A, Dohrmann JK (1987) Hydrogen atom reactions in the sonolysis of aqueous solutions. J Phys Chem 91(27):6687–6690
76. Yonghong N, li Hua, Jin Lina, Hong Jianming (2009) Synthesis of ID $Cu(OH)_2$ nanowires and transition to 3D CuO microstructures under ultrasonic irradiation, and their electrochemical property. Cryst Growth Des 9(9):3868–3873
77. Wen XG, Zhang WX, Yang SH (2002) Solution phase synthesis of $Cu(OH)_2$ nanoribbons by coordination self-assembly using Cu_2S nanowires as precursors. Nano Lett 2(12):1397–1401
78. Lu CH, Qi LM, Yang JH, Jhang DY, Wu NZ, Ma JM (2004) Simple template-free solution route for the controlled synthesis of $Cu(OH)_2$ and CuO nanostructures. J Phys Chem B 108: 17825–17831
79. Zhang W, Wen X, Yang S, Yolande B, Wang ZL (2003) Single-crystalline scroll-type nanotube arrays of copper hydroxide synthesized at room temperature. Adv Mater 15(10): 822–825
80. Haas I, Shanmugam S, Gedanken A (2006) Pulsed sonoelectrochemical synthesis of size-controlled copper nanoparticles under protection of poly (N-vinylpyrrolidone). J Phys Chem B 110:16947–16952
81. Baioni AP, Vidotti M, Fiorito PA, Ponzio EA, Cordoba de Torresi SI (2007) Synthesis and characterization of copper hexacyanoferrate nanoparticles for building up long-term stability electrochromic electrodes. Langmuir 23(12):6796–6800
82. Salkar RA, Jeevanandam P, Kataby G, Aruna ST, Koltipin yuri, Palchik O, Gedanken A (2000) Elongated copper nanoparticles coated with a zwitterionic surfactant. J Phys Chem B 104:893–897
83. Vijay Kumar R, Elgamiel R, Diamant Y, Gedanken A (2001) Sonochemical preparation and characterization of nanocrystalline copper oxide embedded in poly(vinyl alcohol) and Its effect on crystal growth of copper oxide. Langmuir 17(5):1406–1410

84. Vijay Kumar R, Mastai Y, Gedanken A (2000) Sonochemical synthesis and characterisation of nanocrystalline paramelaconite in polyaniline matrix. Chem Mater 12:3892–3895
85. Mahdi C, Oualid H, Fatiha A, Christian P (2010) Study on ultrasonically assisted emulsification and recovery of copper(II) from wastewater using an emulsion liquid membrane process. Ultrason Sonochem 17(2):318–325
86. Entezari Mohammad H, Christian P, Pierre D (2003) Degradation of azo dyes by hybrid ultrasound-Fenton reagent. Ultrason Sonochem 10(2):103–108
87. Lv W, Luo Z, Yang H, Liu B, Weng W, Liu J (2010) Effect of processing conditions on sonochemical synthesis of nanosized copper aluminate powders. Ultrason Sonochem 17(2):344–351
88. Francony A, Petrier C (1996) Sonochemical degradation of carbon tetrachloride in aqueous solution at two frequencies: 20 kHz and 500 kHz. Ultrason Sonochem 3(2):S77
89. Hobson RA, Mulvaney P, Grieser F (1994) Formation of Q-state CdS colloids using ultrasound. J Chem Soc Chem Commun 7:823–824
90. Swayambunathan V, Hayes D, Schmidt KH, Liao YX, Meisel D (1990) Thiol surface complexation on growing cadmium sulfide clusters. J Am Chem Soc 112:3831–3837
91. Yadav Raghvendra S, Priya M, Rupali M, Kumar M, Pandey Avinash C (2010) Growth mechanism and optical property of CdS nanoparticles synthesized using amino-acid histidine as chelating agent under sonochemical process. Ultrason Sonochem 17(1):116–122
92. Misik V, Riesz P (1997) Effect of Cd^{2+} on the •H atom yield in the sonolysis of water. Evidence against the formation of hydrated electrons. J Phys Chem A 101(8):1441–1444
93. Mastai Y, Polsky R, Koltypin Yu, Gedanklen A, Hodes G (1999) Pulsed electrochemical synthesis of cadmium selenide nanoparticles. J Am Chem Soc 121:10047–10052
94. Zhu Junjie Lu, Zhonghua AST, Doron A, Aharon G (2000) Sonochemical synthesis of SnO_2 nanoparticles and their preliminary study as Li insertion electrodes. Chem Mater 12:2557–2566
95. Wang H, Song Y, Wang Z, Medforth Craig J, James ME, Lindsay E, Peng L, John SA (2008) Silica-Metal core-shells and metal shells synthesized by porphyrin-assisted photocatalysis. Chem Mater 20:7434–7439
96. Li H, Liu J, Tusi DY (1998) Guangzhaushi Erginggongye Yanjiuso, 21(ch) 17(3): 1–4
97. Swamy KM, Sukla LB, Narayana KL, Kar KN, Panchanadikar VV (1995) Use of ultrasound in microbial leaching of nickel from laterites. Ultrason Sonochem 2(1):S5–S9
98. Srivastava DN, Perkas N, Seisen Bueva GA, Koltypin Y, Kessler VP, Gedanken A (2003) Preparation of porous cobalt and nickel oxides from corresponding alkoxides using a sonochemical technique and its application as a catalyst in the oxidation of hydrocarbons. Ultrason Sonochem 10(1):1–9
99. Ramesh S, Yuri K, Ruslan P, Aharon G (1997) Sonochemical deposition and characterisation of nanophasic amorphous nickel on silica microspheres. Chem Mater 9:546–551
100. Shafi KVPM, Yuri K, Aharon G, Raslan P, Balogh LJ, Israel F (1997) Sonochemical preparation of nanosized amorphous $NiFe_2O_4$ particles. J Phys Chem B 101:6409–6414
101. Zhong Z, Yitzhak M, Yuri K, Yanming Z, Aharon G (1999) Sonochemical coating of nanosized nickel on alumina submicrospheres and interaction between the nickel and nickel oxide with substrate. Chem Mater 11:2350–2359
102. Koltypin Y, Fernandez A, Rojas CT, Campora J, Palma P, Prozorov R, Gedanken A (1999) Encapsulation of nickel nanoparticles in carbon obtained by the sonochemical decomposition of $Ni(C_8H_{12})_2$. Chem Mater 11:1331–1335
103. Jeevanandam P, Koltypin Yu, Gedanken A (2006) Synthesis of nanosized α-Nickel hydroxide by sonochemical method. J Mater Sci 41:5598–5601
104. Gandhi KS (1995) Analysis of ultrasonically enhanced hydrogen evolution for Zn–$NiCl_2$ system. Chem Eng Sci 50(15):2409
105. Doche M-L, Hihn J-Y, Touyeras F, Lorimer JP, Mason TJ, Plattes M (2001) Electrochemical behaviour of zinc in 20 kHz sonicated NaOH electrolytes. Ultrason Sonochem 8(3):291–298

106. Ligier V, Hihn JY, Wery M, Tachez M (2001) The effect of 20 kHz and 500 kHz ultrasound on the corrosion of zinc precoated steel in [Cl^-] [SO_4^{2-}] [HCO_3^-] [H_2O_2] electrolytes. J Appl Electrochem 31:213–322
107. Suslick KS, Doktyez SJ (1989) The sonochemistry of Zn powder. Am Chem Soc 111: 2342–2344. Prfozorov T, Prozorov R, Suslick KS (2004) High velocity interparticle collisions driven by ultrasound. J Am Chem Soc 126: 13890–13891
108. Dhas Arul N, Zaban A, Gedanken A (1999) Surface synthesis of zinc sulphide nanoparticles in silica microspheres: sonochemical preparation, characterisation and optical properties. Chem Mater 11:806–813
109. Kumar RR, Zhang Lizhi Yu, Jimmy C, Yitzhak M, Aharon G (2003) Mesoporous structures from supramolecular assembly of *in situ* generated ZnS nanoparticles. Langmuir 19: 5904–5911
110. Zho JJ, Yuri K, Gedanken A (2000) General sonochemical method for the preparation of nanophasic selenide: synthesis of ZnSe nanoparticles. Chem Mater 12:73–78
111. Jung Seung-Ho, Eugene Oh, Hanna L, Dae-Seob S, Seungho C, Kun-Hong L, Soo-Hwan J (2009) Shape selective fabrication of zinc phosphate hexagonal bipyramids via a disodium phosphate-assisted sonochemical route. Cryst Growth Des 9(8):3544–3547
112. Baranchikiv AE, Ivanov VK, Oleinikov NN, Tretyakov DYu (2004) Microstructural evaluation of Fe_2O_3 and $ZnFe_2O_4$ during sonochemical synthesis of zinc ferrite. Inorg Mater 40(10):1091–1094
113. Baranchikiv AE, Ivanov VK, Muraveva GP, Oleinikov NN, Tretyakov Yu D (2004) Kinetics of the formation of zinc ferrite in an ultrasonic field. Dokl Chem 397(Part I):146–148
114. Vogel AI (1961) Quantitative inorganic analysis, 3rd edn. ELBS, London, 900
115. Peter C, Suslick KS (2000) Ultrasound-enhanced reactivity of calcium in the reduction of aromatic hydrocarbons. Ultrason Sonochem 7:53–61
116. Khasanov OL, Hahn SR, Dvilis ES, Han MS, Lim SS, Sokolov VM, Milovanova (2001) Proc. Korus 2001, The Korean–Russia international symposium on science and technology 2:251–254
117. Gutierrez M, Henglein A, Moeckel H (1995) Observations on the role of $MgCl_2$ in the Weissler reaction. Ultrason Sonochem 2(2):S111–S113
118. Shafi KVPM, Felner I, Mastai Y, Gedanken A (1999) Ólympic ring formation from newly prepared barium hexaferrite nanoparticle suspension. J Phys Chem B 103:3358–3360
119. Perelshtein I, Perkas N, Magdassi Sh, Zioni T, Royz M, Maor Z, Gedanken A (2008) Últrasound assisted dispersion of $SrFe_{12}O_{19}$ nanoparticles in organic solvents and the use of the dispersion as magnetic cosmetics. J Nanopart Res 10:191–195
120. YeungAu S, Hobson RA, Biggs S, Grieser F (1994) 8th international conference on surface and colloid science, Adelaide, South Australia
121. Okitsu K, Mizukoshi Y, Bandow H, Maeda Y, Yamamoto T, Nagata Y (1996) Formation of noble metal particles by ultrasonic irradiation. Ultrason Sonochem 3:S249–S251
122. Mizukoshi Y, Takagi E, Okuno H, Oshima R, Maeda Y, Nagata Y (2001) Preparation of platinum nanoparticles by sonochemical reduction of the Pt(IV) ions: role of surfactants. Ultrason Sonochem 8(1):1–6
123. Yoshiteru M, Yoji M, Shuto Tatsuya Hu, Jinwei TA, Sayoko S, Shuji T (2007) Immobilization of noble metal nanoparticles on the surface of TiO_2 by the sonochemical method: photocatalytic production of hydrogen from an aqueous solution of ethanol. Ultrason Sonochem 14(3):387–392
124. Torok B, Karoly F, Gerda S, Mihaly B (1997) Sonochemical enantioselective hydrogenation of ethyl pyruvate over platinum catalysts. Ultrason Sonochem 4(4):301–304
125. Torok B, Gyorgy S, Katalin B, Karoly F, Istvan K, Mihaly B (1999) Ultrasonics in heterogeneous metal catalysis: sonochemical chemo- and enantioselective hydrogenations over supported platinum catalysts. Ultrason Sonochem 6(1–2):97–103

126. Gyorgy S, Istvan K, Bela T, Mihaly B (2000) Ultrasonics in chemoselective heterogeneous metal catalysis. Sonochemical hydrogenation of unsaturated carbonyl compounds over platinum catalysts. Ultrason Sonochem 7(4):173–176
127. Sivakumar M, Atsuya T, Kyuichi Y, Toru T, Teruyuki K, Yasuo I (2010) Dependence of sonochemical parameters on the platinization of rutile titania – an observation of a pronounced increase in photocatalytic efficiencies. Ultrason Sonochem 17(3):621–627
128. Neppolian B, Doronila A, Grieser F, Ashokkumar M (2009) Simple and efficient sonochemical method for the oxidation of arsenic (III) to arsenic (V). Environ Sci Technol 43: 6793–6798
129. Pankaj, Manju C (2006) pH independent precipitation of arsenic as sulphide under ultrasonic field. J Ind Council Chem 23(1):41–43
130. Breitbach M, Bathen D (2001) Influence of ultrasound on adsorption processes. Ultrason Sonochem 8(3):277–283
131. Margulis MA (1985) Sonoluminescence from non-aqueous liquids. Ultrasonics 23:157
132. Wang H, Jun-Jie J, Jian-Min Z (2002) Sonochemical method for the preparation of bismuth sulfide nanorods, J Phys Chem B 106(15):3848–3854
133. Nowak M, Szperlich P, Bober L, Szala J, Moskal G, Stroz D (2008) Sonochemical preparation of SbSI gel. Ultrason Sonochem 15(5):709–716
134. Nowak M, Szperlich P, Tilak E, Szala J, Rzychon T, Stroz D, Nowrot A, Solecka B (2010) Sonochemical preparation of antimony subiodide. Ultrason Sonochem 17(1): 219–227
135. Nowak M, Kauch B, Szperlich P, Stroz D, Szala J, Rzychon T, Bober L, Toron B, Nowrot A (2010) Sonochemical preparation of $SbS_{1-x}Se_xI$ nanowires. Ultrason Sonochem 17(2): 487–493
136. Wang H, Yi-Nong Lu, Jun-Jie J, Hong-Yuan C (2003) Sonochemical fabrication and characterization of stibnite nanorods. Inorg Chem 42:6404–6411
137. De Morais NLPA, Brett CMA (2002) Influence of power ultrasound on the corrosion of aluminium and high speed steel. J Appl Electrochem 32:653–660
138. Sweet JD, Casadonte DJ Jr (1994) Sonochemical formation of intermetallic coatings. Chem Mater 6(11):2113–2117
139. Watanabe T, Sato S (1993) Jpn. Kokai Tokkyo Koho JP 05 69, 120
140. Kim HC, Ahn TH, So CH, Ma Y, Zhao X, Langdon TG (1992) A 1st report on the use of nondestructive technique to investigate cavitation in a superplastic alluminium-alloy. Scr Metall Mater 26(3):423–428
141. Amara N, Ratsimba B, Marie A, Wilhelm, Delmas Henri (2001) Crystallization of potash alum: effect of power ultrasound. Ultrason Sonochem 8(3):265–270
142. Iwasaki H, Mabuchi M, Higashi K, Langdon TG (1996) The development of cavitation in superplastic aluminum composites reinforced with Si_3N_4. Mater Sci Eng A 208(1): 116–121
143. Naka M, Maeda M (1991) Application on ultrasound on joining of ceramics to metals. Eng Fract Mech 40(4–5):951–956
144. Gutierrez M, Henglein A, Dohrmann JK (1987) Hydrogen atom reactions in the sonolysis of aqueous solutions. J Phys Chem 91:6687
145. Yeung S Au, Hobson R, Biggs S and Grieser F (1993) Formation of gold sols using ultrasound. J Chem Soc Chem Commun 4:378–379
146. Reed Justin A, Andrew C, Halaas HJ, Paul P, Alex R, Thomas MJ, Grieser F (2003) The effects of microgravity on nanoparticle size distributions generated by the ultrasonic reduction of an aqueous gold-chloride solution. Ultrason Sonochem 10(4–5):285–289
147. Okitsu K, Ashokkumar M, Grieser F (2005) Sonochemical synthesis of gold nanoparticles: effect of ultrasound frequency. J Phys Chem B 109:20673–20675
148. Park Jong Eun, Mahito A, Toshio F (2006) Synthesis of multiple shapes of gold nanoparticles with controlled sizes in aqueous solution using ultrasound. Ultrason Sonochem 13(3): 237–241

149. Pradhan A, Jones Robin C, Daniela C, O'Connor Charles J, Tarr Matthew A (2008) Gold-magnetite nanocomposite materials formed via sonochemical methods. Ultrason Sonochem 15(5):891–897
150. Preece CM (1979) In: Preece CM (ed) Treatise on material advances and technology: erosion, vol 16. Academic, New York, p 249
151. Karimi A, Martin JL (1986) Cavitation erosion of material. Int Met Rev 31:1–26
152. Doktycz SJ, Suslick KS (1990) Interparticle collisions driven by ultrasound. Science 247:1067–1069
153. Prosperetti A (1984) Bubble phenomena in sound fields: part two. Ultrasonics 22:115
154. Gasgnier M (2000) Ultrasound effects on metallic (Fe and Cr); iron sesquioxide (α-, γ-Fe_2O_3): calcite; copper, lead and manganese oxides as powders. Ultrason Sonochem 7:25–39
155. Alex CCT, Goh NN, Chia LS (1995) Effects of particle size morphology on ultrasound induced cavitational mechanism in heterogeneous systems. J Chem Soc Chem Commun 2:201–201
156. Heard SM, Grieser F, Barraclough CG, Sanders JV (1983) The characterization of Ag sols by electron microscopy, optical absorption and electrophoresis. J Colloid Interface Sci 93(2):545–555
157. Pankaj DS, Manju C (2006) Sonochemical removal of hardness and sterilization of underground potable water. J Ind Council Chem 23(1):38–40
158. Kruger O, Schulz Th L, Peters D (1999) Sonochemical treatment of natural ground water at different high frequencies: preliminary results. Ultrason Sonochem 6(1):123–128
159. Phull SS, Newman AP, Lorimer JP, Pollet B, Mason TJ (1997) The development and evaluation of ultrasound in the biocidal treatment of water. Ultrason Sonochem 4(2): 157–164
160. Nagata Y, Nakagawa M, Okuno H, Mizukoshi Y, Yim B, Maeda Y (2000) Sonochemical degradation of chlorophenols in water. Ultrason Sonochem 7:115–120
161. (i) Carvajal JJ, Aznar A, Sole R, Gavalda J, Massons J, Solans X, Aguilo M, Diaz F (2003) Growth and structural characterization of $Rb_2Ti_{1.01}Er_{0.99}(PO_4)_3$. Chem Mater 15: 204–211. (ii) Leite, ER, Vila C, Bettini J, Longo E (2006) Synthesis of niobia nanocrystals with controlled morphology. J Phys Chem B 110: 18088–18090. (iii) Sankar R, Raghvan CM, Jayavel R (2007) Bulk growth and characterization of semi-organic nonlinear optical bis thiourea bismuth chloride single crystals. Cryst Growth Des 7(3): 501–505. (iv) Vijayan N, Rajasekaran S, Bhagvannarayana G, Babu RR, Gopalkrishnan R, Palanichamy M, Ramasamy P (2006) Growth and characterization of nonlinear optical amino acid single crystal: l-alanine. Cryst Growth Des 6(11): 2441–2445. (v) Wu D, Ugurlu O, Chumbley LS, Kramer MJ, Lograsso TA (2006) Synthesis and characterization of hexagonal $Cd_{51}Yb_{14}$ single crystals. Philos Mag 86(3–5): 381–387. (vi) Nitsch K, Nikl M, Ganschow S, Reiche P, Uecker R (1996) Growth of lead tungstate single crystal scintillators. J Cryst Growth 165: 163–165. (vii) Barre M, Crosnier-Lopez MP, Le Berre F, Emery J, Suard E, Fourquet J-L (2005) Room temperature crystal structure of $La_{1/3}Zr_2(PO_4)_3$, NASICON-type compound. Chem Mater 17: 6605–6610. (viii) Arora SK, Chudasama B (2007) Flux growth and optoelectronic study of $PbWO_4$ single crystals, Cryst Growth Des 7(2): 296–299. (viii) Prabukanthan P, Dhanasekaran R (2007) Growth of $CuGaS_2$ single crystals by chemical vapor transport and characterization. Cryst Growth Des 7(4): 618–623. (ix) Parreu I, Carvajal JJ, Solans X, Diaz F, Aguilo, M (2006) Crystal structure and optical characterization of pure and Nd-substituted type III $KGd(PO_3)_4$, Chem Mater 18: 221–228
162. Suslick KS, Price GJ (1999) Applications of ultrasound to material chemistry. Annu Rev Mater Sci 29:295–326
163. Cains PW, Martin PD, Price CJ (1998) The use of ultrasound in industrial chemical synthesis and crystallization. Part1. Applications to synthetic chemistry. Org Process Res Dev 2:34
164. Zhu L, Liu X, Meng J, Cao X (2007) Facile sonochemical synthesis of single-crystalline europium fluorine with novel nanostructure. Cryst Growth Des 7(12):2505–2511

165. McCausland LJ, Cains PW (2002) Sonocrystallization – ultrasonically promoted crystallization for the optimal Isolation of drug actives, Drug Delivery Systems & Sciences 2(2):47
166. Lieber CM (1998) One-dimensional nanostructures: chemistry, physics and applications. Solid State Commun 107:607
167. Alivisatos AP (1996) Semiconductor clusters, nanocrystals, and quantum dots. Science 271:933–937
168. Sharma KCV (2002) Crystal engineering – where do we go from here? Cryst Growth Des 2(6):465–474
169. Jung S-H, Oh E, Lee K-H, Park CG, Park W, Jeong S-H (2008) Sonochemical preparation of shape-selective ZnO nanostructures. Cryst Growth Des 8(1):265–269
170. Lindley J (1992) Sonochemical effects on syntheses involving solid and supported catalysts. Ultrasonics 30(3):163–167
171. Plesset MS, Chapman RB (1971) Collapse of an initially spherical vapor cavity in the neighborhood of a solid boundary. J Fluid Mech 47:283–290
172. Crum LA (1982) Acoustic cavitation. Proc Ultrason Symp 1:1–11
173. Suslick KS, Doktycz SJ (1990) In: TJ Mason (ed) Advances in sonochemistry, vol 1, JAI Press, London, pp 204
174. Neppirass EA (1980) Acoustic cavitation. Phy Rep 61:159–251

Chapter 10
Sonochemical Study on Multivalent Cations (Fe, Cr, and Mn)

Pankaj and Manju Chauhan

Abstract The behaviour of many metal ions which are stable in more than one oxidation states in their aqueous solutions has been studied under sonochemical reaction conditions. Fe(II) is oxidized to Fe(III) and Fe(III) is reduced to Fe(II) with equal ease under sonochemical conditions. Besides, the oxidizing power of $Cr_2O_7^{2-}$ is found to be less than the MnO_4^- ions, therefore, in a system containing both species, Cr^{3+} is susceptible to oxidation to Cr^{6+} and the MnO_4^- to reduction to Mn^{2+}.

10.1 Introduction

A number of transition metals exist in more than one stable oxidation states in aqueous solutions, therefore, the inter-conversion of these oxidation states in the presence of an oxidizing or reducing agent is quite possible. Some of such ions, Fe^{2+}/Fe^{3+}, Cr^{3+}/Cr^{6+}, Mn^{2+}/Mn^{7+} and Co^{2+}/Co^{3+}, are frequently encountered in day-to-day inorganic sonochemistry. Ultrasound in water generates H• and •OH free radicals which upon recombination produces a number of secondary active redox species such as H_2, O_3, H_2O_2, •OOH etc. Therefore, even in the absence of any oxidising or reducing agent, redox reactions or inter-conversions may occur under sonication conditions. A brief description of such studies reported in the literature is reviewed in this chapter.

Manganese in its various oxidation states, such as, Mn^{+7}, Mn^{+4} and Mn^{+3}, had been exploited in several studies. An electron transfer mechanism for the inter-molecular arylation of malonate esters in the presence of manganese (III) triacetate

Pankaj (✉)
Department of Chemistry, Dayalbagh Educational Institute, 282 110 Agra, India
e-mail: pankaj2@sancharnet.in

Pankaj and M. Ashokkumar (eds.), *Theoretical and Experimental Sonochemistry Involving Inorganic Systems*, DOI 10.1007/978-90-481-3887-6_10,

has been reported [1]. Similarly, upon sonication at 90°C, an aqueous solution of Mn(III) acetylacetonate and LiOH produced nanoparticles through a strongly pH dependent reaction [2]. In tetravalent state Mn^{+4}, MnO_2 produced gold sol by the reduction of $AuCl_4^-$ in aqueous solution using 20 kHz ultrasound. The reduction process increased significantly in the presence of hydrophobic aliphatic alcohol [3]. Similarly, the sonochemical degradation of an azo dye, acid red B, and its decolourisation in the presence of MnO_2 was studied. The effect was found to be highly pH dependent and the efficiency increased with increase in the pH. The O_2 favoured mineralisation. The presence of SO_4^{2-} and NO_3^- had a negative effect in the mineralisation process, whereas argon gas was relatively less effective than O_2 [4]. In the presence of ultrasound, the Mn^{+7} ($KMnO_4$) oxidised benzyl alcohol and alkylarenes to the corresponding carbonyl compounds and reduced the reaction time as well as the temperature of the reaction. The $KMnO_4$ supported copper sulphate pentahydrate could also selectively oxidise benzoyl alcohol to benzaldehyde [5]. The role of sonochemically generated H_2O_2, H_2 and H atoms was found to be crucial for the reduction reaction MnO_4^- to MnO_2. The pH value of the reaction solution also increased as the sonochemical reduction of MnO_4^- proceeded [6].

Surprisingly, under different sonochemical reaction conditions, the products and final outcome of the reaction were unexpectedly diverse. In one such example, the formation of •NO was detected by the Electron Paramagnetic Resonance spectroscopy in an aqueous nitrogen solution sonicated in a 50 kHz ultrasonic bath. By trapping •NO with sodium *N*-methyl-d-glucamine dithiocarbamate iron(II) complex $\{(MGD)_2Fe^{2+}\}$ or by measuring the conversion of the nitronyl nitroxide, 2-(4-carboxyphenyl)-4,4,5,5-tetramethylimidazoline-3-oxide-1-oxyl (carboxy-PTIO), to the imino nitroxide, 2-(4-carboxyphenyl)-4,4,5,5-tetramethylimidazoline-1-oxyl, the rate of •NO production was determined to be ~0.5 μM/min. The presence of dissolved oxygen was not essential for the production of •NO and the highest yield of •NO (~1.2 μM •NO/min) was found in an aqueous solution containing 40% N_2 and 60% argon. The formation of •NO by ultrasound in aqueous solutions could be understood in terms of combustion chemistry-type reactions occurring inside the "hot" collapsing cavitation bubbles [7]. However, using ultrasound and zerovalent nanoparticles of iron (NZVI), dispersed in the solution, the reduction and subsequent removal of nitrites from the aqueous solution was achieved. Factors, such as, the length of ultrasonication time, the dosage of NZVI, the initial nitrite concentration, the temperature, and the solution pH were crucial in this process. The reaction was found to follow a pseudo-first-order reaction kinetics with respect to the concentration of nitrite in the given experimental conditions [8]. In another similar example, the thermally unstable prussian blue, $Fe_4[Fe(CN)_6]_3$ [9] was produced in an ultrasonic field [10]. Single-crystalline nanocubes of different sizes by the direct dissociation of the precursor, $K_4Fe(CN)_6$ was achieved. Here also the temperature, concentration of reactants and the ultrasonic conditions greatly influenced the final results. Numerous such reactions of complexation – decomplexation, polymerisation – depolymerisation, agglomeration – fragmentation could also be found in the literature.

Metal specific reactions of these kinds are quite common, where metals can be oxidised or metal ions can be reduced in aqueous solutions with equal ease. This is what makes sonochemical studies unique and interesting. Nanoparticles of iron may be oxidised to higher oxidation state as well as the reduction of Fe^{3+} to Fe^{2+} (Fenton reagent) is also known. Such behaviour of metal and their salts is often used for the degradation of various organic molecules and pollutants. Enhanced rate of the sonolytic degradation of CCl_4 in the presence of Fe^0 in argon saturated solution was attributed to the continuous cleaning and chemical activation of the Fe^0 particle surface by the physical effects of acoustic cavitation and the acceleration in the mass transport rates of reactants to the Fe^0 particle surface [11]. In a similar enhancement of the sonochemical degradation of methylene blue in aqueous solution in the presence of Fe(III), the effect was attributed to a sonochemically induced Fenton reaction, where both reactants (Fe(II) and H_2O_2) were sonochemically produced. Hydroperoxide/superoxide, generated upon sonochemical processes in aerated solution, were the key species involved in both Fe(III) reduction to Fe(II) and in the production of H_2O_2. The Fenton reaction between Fe(II) and H_2O_2 then produced hydroxyl radicals, enhancing the degradation of methylene blue. The addition of external H_2O_2 was found to increase the degradation process in the presence of Fe (III) but the H_2O_2 alone, in the absence of Fe(III), had a very limited effect on the sonochemical degradation rate [12]. A recent review on the sonochemical degradation of chlorinated aromatic hydrocarbons concluded that the application of ultrasound in combination with Fenton's reagent was a better advanced oxidation system rather than the either methods [13]. The degradation and mineralization of bisphenol A (BPA), an organic compound largely used in the plastic industry as a monomer for the production of epoxy resins and polycarbonate, could only be degraded to the extent of 50%, after ~90 min of sonication, using a 300 kHz frequency and 80 W electrical power and the rest remained recalcitrant until the addition of Fe^{2+} (100 $\mu mol\ L^{-1}$) and/or UV radiation (254 nm). Both UV and Fe^{2+} induced hydrogen peroxide dissociation, leading to additional •OH radical formation, thus could successfully lead to the mineralization of micropollutants like BPA, which was otherwise not possible under ultrasonic action alone [14]. Similar degradations of 2,4-dichlorophenol [15] and pentachlorophenol [16] in their aqueous solutions and in the presence of Fenton's reagent and $Fe@Fe_2O_3$ core-shell nanowires as iron reagent, respectively have been reported. However, while the first method could only effectively decompose (77.6%) of dichlorophenol, the nanowires, in the second method, could degrade the recalcitrant pentachlorophenol completely. Giancarlo et al [17] recently reported the oxidative degradation of chlorophenol derivatives both in microwave and power ultrasound conditions and discussed their mechanisms. In a similar study, an efficient degradation (99.9%) of acid red 97 dye in aqueous solution by the combination of ultrasound and Fenton reagent has been reported. Fenton like other reagents, such as, Cu^{2+}/H_2O_2, Co^{2+}/H_2O_2 and Mn^{2+}/H_2O_2 also produced similar effects, however, out of Cl^-, SO_4^{2-} and $H_2PO_4^-$ anions, except for SO_4^{2-}, others had a negative effect on the degradation [18].

Sonochemical reduction of aqueous solutions containing ammonium dichromate and potassium permanganate produced ultrafine powders of Cr_2O_3 and

Mn_2O_3, respectively. The yields of both these oxides were found to increase by raising the temperature or using an aqueous solution of ethanol (0.1M). Powders of Cr_2O_3 and Mn_2O_3, upon heating to 600 and 900 K, respectively changed to crystalline forms [19]. Similar preparations of iron and ferrous alloy nanoparticles, including chromium, through the reduction of their metal halides with lithium triethyl borohydride have been reported [20]. Reduction of hexavalent chromium in soil and ground water using zero-valent iron and ultrasound conditions in several physico-chemical states of iron, such as Fe (II) sulphate solution, Fe^0 granulated elemental iron, non-stablised zerovalent iron and colloidal zerovalent iron was found to be effective to the extent of 98.5% reduction and the colloidal zero-valent iron were reported to be the most effective [21].

In a very interesting and innovative study recently, the ultrasound-assisted microbial reduction of chromium [22], Mathur et al. reported the reduction of hexavalent chromium using B*acillus sp,* isolated from tannery effluent contaminated site. The optimum reduction was found at pH 7 and 37°C. The percent reduction increased with an increase in biomass concentration and decreased with an increase in the initial concentration of hexavalent chromium.

Ultrasound dramatically enhances the reactivity of transition metals dispersed in the solution, principally due to micro-streaming [23] and acoustic cavitation [24–26]. The examination of the reactivity of transition metals with carbon monoxide during sonication revealed the reduction of a variety of transition metal salts to active forms that would react even in low pressures of CO to form respective simple carbonylate anions [27]. The manganese and iron reactions with CO were quite efficient but nickel clusters were formed only in small numbers. Ultrasound, however, oxidised the solution of $Fe(OH)_2$ to FeOOH (goethite) or Fe_3O_4 (magnetite) through the process involving dissolution of $Fe(OH)_2$, oxidation of Fe^{2+} to Fe^{3+} and the precipitation of the end product [28]. Ultrasound decreased the particle size in the slurries of chromium and iron (as metal) and calcite, copper, lead, manganese (as powdered oxide) in water, dodecane or dilute acetic acid (5%) besides prompting the formation of unexpected and unknown compounds [29] on mesoscopic scale such as Cu_2O, PbO, Pb_3O_4 and Mn_3O_4. The dissolution of MnO_2 colloids in ultrasonic field in the presence or absence of aliphatic alcohols at pH 4.5 revealed an enhancement in the amount and rate of dissolution of MnO_2 in the presence of alcohols [30]. Ultrasound was found to be effective in both the rate and the degree of completion of the removal of cobalt ions from solution at temperature greater than 50°C in the reaction of zinc powder with aqueous cobaltous solution [31]. The physical effects induced by ultrasound are also equally interesting. The removal of iron coating by ultrasonic irradiation, from the surface of silica sand used in glass making was found to be an improvement in the electrostatic separation [32]. The synthesis [33] of magnesium ferrite ($MgFe_2O_4$) from Fe_2O_3 and MgO for catalytic applications has been reported by the use of ultrasound of 20 kHz and power 100 W/cm^2. Short range ordering of the alloy of Ni-(0.2–10.3) atomic percent in Cr solid could be affected by 17.5 Hz sound wave at liquid helium temperature (40 K) [34].

10.2 Experimental

Keeping in mind the above studies of multivalent cations (Fe, Cr and Mn) in aqueous medium, some experiments involving redox or complexometric reactions of these metal ions have been carried out , using ultrasound (20 kHz) and its effect on the precipitation, oxidation, reduction and decomposition of complex have been evaluated. An Ultrasonic Processor model P2 with a titanium tip of diameter 12 mm and 250 watts power was used. In the subsequent sections details of some of the interesting experiments, carried out in aqueous solutions of salts of Fe, Cr and Mn in their different oxidation states, have been discussed.

10.2.1 Iron, Fe

Preliminary observations of the reactions of Fe(II) and Fe(III) in an ultrasonic field revealed conversion of Fe(II) to Fe(III) and vice-versa, even without any external oxidant/reductant, which were subsequently confirmed experimentally and sequentially through the following four qualitative conversions. The details are available in the literature [35].

10.2.1.1 Reduction of Fe^{3+} to Fe^{2+}

In the first experiment 10 ml of 500 ppm solutions of $FeCl_3$ were sonicated for 15, 30, 45 and 60 min. To examine the reduction of Fe(III) to Fe(II), 0.1 ml of the sonicated sample was transferred to a 10 ml volumetric flask and mixed with 0.5 ml of 2,000 ppm $K_3[Fe(CN)_6]$ solution before making up to the mark with distilled water. Final concentration of this sonicated sample in 10 ml of volumetric flask was 5 ppm. UV-vis absorbance at λ_{max} 795 was recorded. Sonication reduced Fe^{3+} to Fe^{2+}, which reacted with $K_3[Fe(CN)_6]$, already added in the solution, to form blue colour due to prussian blue. Continuous sonication gradually increased the concentration and intensity of prussian blue complex, as is clear from the Table 10.1.

Since there were no Fe^{2+} ions in the solution initially, the increase in the intensity of prussian blue colour and the increase in the absorbance at λ_{max} 795 nm could be

Table. 10.1 Increase in absorbance with sonication due to reduction of Fe(III) to Fe(II)

Sonication (min)	Fe^{3+} ion concentration in diluted solution (ppm)	After sonication		Reduction of Fe^{3+} to Fe^{2+} (%)
		Concentration of Fe^{2+} ions (ppm)	Absorbance at λ_{max} 795 nm	
15	5	0.113	0.039	2.3
30	5	0.327	0.058	6.5
45	5	0.975	0.114	19.5
60	5	2.1631	0.217	43.26

attributed only to the conversion of Fe^{3+} to Fe^{2+} and the subsequent reaction of ferrous ions with $K_3[Fe(CN)_6]$ to produce prussian blue, $Fe_4[Fe(CN)_6]_3$.

10.2.1.2 Oxidation of Fe^{2+} to Fe^{3+}

To examine the oxidation of Fe^{2+} to Fe^{3+}, in the second experiment, 10 ml solution of 0.1M ferrous ammonium sulphate was taken separately in four different beakers and sonicated for 15, 30, 45 and 60 min, before transferring the solution to a 25 ml volumetric flask and adding to it 10 ml of 0.01M KSCN and making upto the mark with deionised water. The absorbance of these solutions was measured at λ_{max} 451 nm. Sonication of ferrous ammonium sulphate solutions oxidised ferrous ions to ferric ions, which in the presence of thiocyanate ions, produced an intense red coloured complex $Fe(SCN)_6^{3-}$, in proportions to the oxidation of ferrous ions to ferric ions, as could be seen in Fig. 10.1.

10.2.1.3 Decomposition of $[Fe(SCN)_6]^{3-}$ complex

As stated earlier that a paradoxical reactivity could often be encountered in sonochemical studies. The decomposition of $[Fe(SCN)_6]^{3-}$ complex had been witnessed, if the sonication of the solution was continued long. Contrary to the sonochemically assisted formation of $[Fe(SCN)_6]^{3-}$ in the preceding experiment, used for the detection of the conversion of Fe^{2+} to Fe^{3+}, here is another opposite result of the sonicated

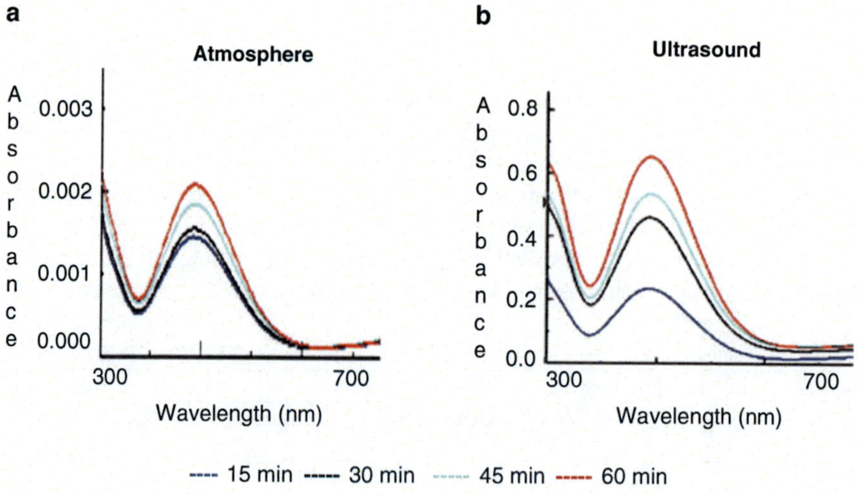

Fig. 10.1 Comparison of oxidation of Fe^{2+} to Fe^{3+} under atmospheric (**a**) and ultrasonic (**b**) conditions

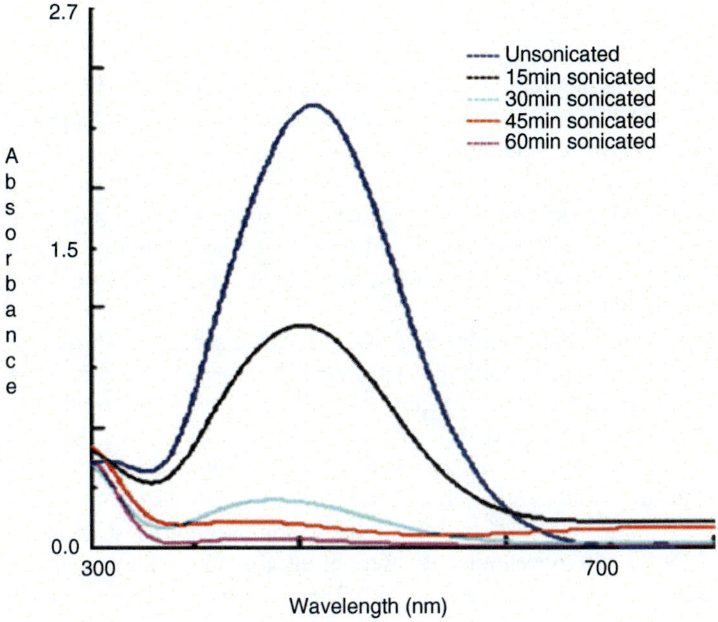

Fig. 10.2 Sonochemical Decomposition of $[Fe(SCN)_6]^{3-}$ complex

solution. When a pink coloured faintly acidic solution of $FeCl_3$ (100 ppm) and KSCN (0.0005M) was sonicated for 15, 30, 45 and 60 min, the absorption of a peak at λ_{max} 451 nm (due to $[Fe(SCN)_6]^{3-}$ complex] was found diminishing until its complete disappearance at the end of 1 h. (Fig. 10.2). However, this explicates that the results of the previous experiment about the % degradation of Fe^{2+} to Fe^{3+}, as given in Table 10.1, were only qualitative and not quantitative.

10.2.1.4 Oxidation of Cl^- and SCN^-

The possibility of the oxidation of Cl^- and SCN^- ions were also examined. 10 ml of 500 ppm neutral $FeCl_3$ solution was sonicated for 30 min, while the control sample was kept for the same duration in the atmospheric condition. These sonicated and controlled samples were treated with $AgNO_3$ to precipitate unaffected remaining amount of chloride ions as silver chloride. The amount of precipitable chloride was same in both sonicated as well as unsonicated solutions. Similarly, the reappearance of pink colour in the sonicated and decolourised solution of $[Fe(SCN)_6]^{3-}$ upon the addition of $FeCl_3$ confirmed that the SCN^- too did not oxidize during the decomposition and decolourisation of the complex but remained unaffected even after the complex had perished.

10.2.2 Chromium (Cr)

Cr^{3+} and Cr^{6+} are the most stable oxidation states of chromium, but with the only difference that while +3 oxidation state is cationic where as the +6 oxdation state is oxoanionic. However, the other oxidation states of +1, +2, +4 and +5 are also known for chromium, especially in aqueous solution at different pH. Inter-conversion of these oxidation states too is very frequent. With this view, an attempt is made here to examine the effect of ultrasound on the inter-convertibility of chromium among various oxidation states in aqueous solutions. The details of this study is reported in the literature [36].

15 mL of 0.001 M solution of $K_2Cr_2O_7$ was taken separately in four volumetric flasks of 25 ml capacity and the appropriate volumes of conc. HCl was added to make solutions of 0.4, 0.6, 0.8, and 1.0 N HCl in 25 ml of test solutions. 4.0 ml of 25% diphenyl carbazide (DPC) solution was added to each of these solutions and made upto the mark with deionised water. DPC first reduces Cr^{+6} to Cr^{+3} and then makes a water soluble red-violet complex, $[Cr(DPC)_2]^{3+}$, which has an absorption λ_{max} at 545 nm. 10 mL of these homogeneous test solutions were sonicated for 20 min to examine the effect of ultrasound on this complex. UV-vis spectra of these solutions were recorded. The same was done for K_2CrO_4 solutions as well. This complex decomposed upon sonication and the solution

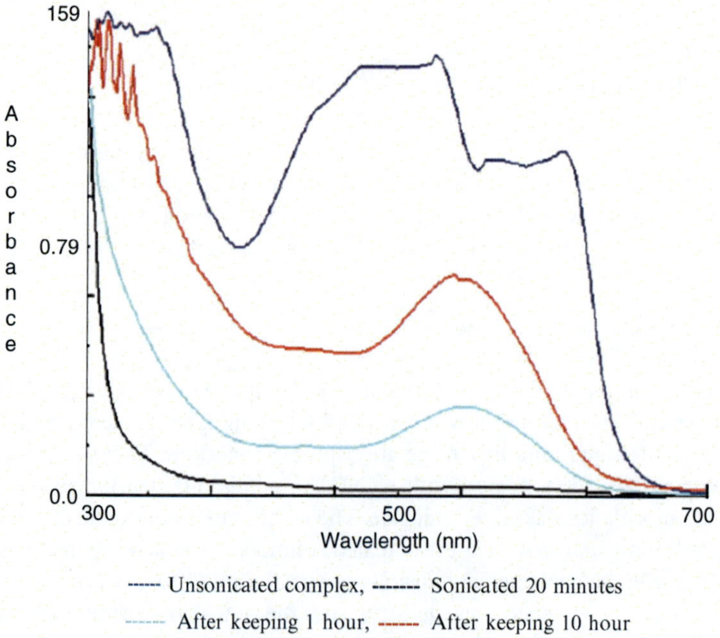

Fig. 10.3 UV Spectra of Cr-DPC complex with $K_2Cr_2O_7$

decolorized. However, after keeping the solution in air for 1 and 10 h respectively, the colour reappeared and the intensity also increased (Fig. 10.3).

The DPC first reduces Cr(VI) to Cr(III) and then forms a red-violet coloured complex as under.

Diphenylcarbazide **Diphenylcarbazone**

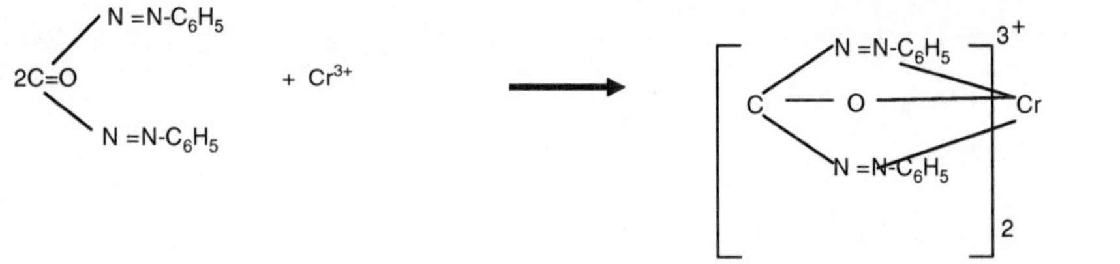

Diphenylcarbazone-Chromium(III)complex
Red-violet colour

The propagation of ultrasound in aqueous solution changes the environment and therefore new oxidation states may be expected. To verify the exact nature of action of ultrasound in this system, a series of experiments were carried out to confirm whether,

(a) The DPC molecule decomposed as a result of sonication?
(b) The Cr(III) was reduced further to Cr(II) or Cr(I) states?
(c) The decolourisation was due to the oxidation of Cr(III) to Cr(VI)?

To confirm (a) the decolourised solution of Cr-DPC complex was divided into two fractions. The first fraction was treated with DPC stock solution and was found to show no change. However, when $K_2Cr_2O_7$ was added to the sonicated and decolourised solution, the red-violet colour reappeared. This confirmed that the DPC molecule remained unaffected in the solution, in spite of the vigorous sonication.

However the second question, whether the Cr^{+3} species either underwent some chemical change so that they became inert in the solution or Cr^{+3} ions were not available to DPC for complexation from the existing dichromate ions remain to be explained. Since either oxidation (c) or reduction (b) would occur in the solution in the given set of experimental condition, another experiment was performed to ascertain the cause of decomposition of Cr-DPC complex resulting into the decolourisation. A current of N_2 gas was purged into the decolourised solution for about 10 min to remove all dissolved O_2 gas from the solution and create an oxidation free atmosphere in and above the solution in the flask. The solution was sealed and left for an hour. The colourless solution changed to feebly pinkish colour and intensified over night (about 10 h). This confirmed the restoration of chromium ions to +3 oxidation state even in the absence of atmospheric oxygen. But whether these Cr^{3+} ions were formed from Cr^{6+} or Cr^{+}/Cr^{2+} species still remained to be resolved. A keen observation of the UV spectra of sonicated solution and magnification of absorption spectra in the region 400–600 nm, did not show any absorption in the region and ruled out completely the oxidation of Cr^{3+} to CrO_4^{2-} / $Cr_2O_7^{2-}$ species. Therefore, the only possibility left now was the conversion to +1 or +2 species of Cr which were oxidised to Cr^{3+} by the small traces of H_2O_2 formed due to sonication, and was the cause of the restoration of the coloured Cr(III)-DPC complex in the solution. The mechanism for this conversion may thus be proposed as under.

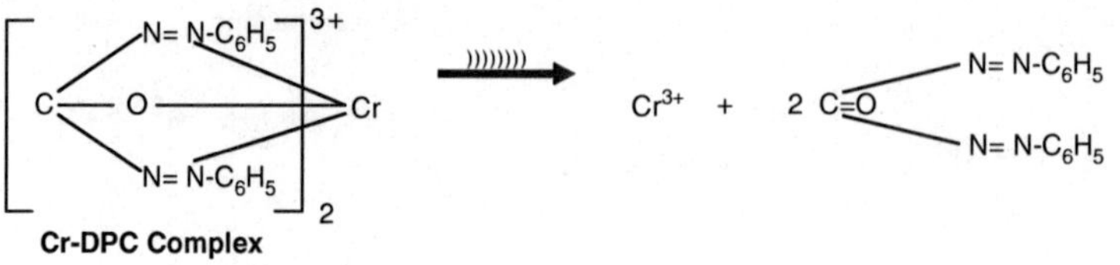

(10.1)

$$H_2O \xrightarrow{)))))))} \bullet H + \bullet OH$$

$$Cr^{3+} + \bullet H + H_2O \rightarrow Cr^{2+} + [H_3O]^{+} \qquad (10.3)$$

or

$$Cr^{2+} + \bullet H + H_2O \rightarrow Cr^{+} + [H_3O]^{+} \qquad (10.3)$$

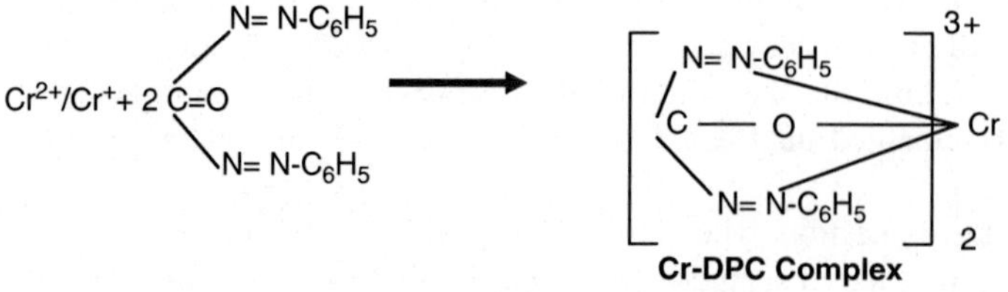

(10.4)

10.2.3 *Chromium and Manganese*

General sonochemical studies of both chromium and manganese have been reported separately due to strong oxidising nature of these ions. However, comparative strength of their oxidising powers, in an ultrasonic field, is of some technical interest. In order to study the changes in the oxidation states of the multivalent cations of Cr and Mn, the following experiment was carried out by combining these ions under different experimental conditions and examined spectrophotometrically, conductometrically as well as nephelometrically, as and where required.

10 mL of a mixture of 5 ml of 0.0008 M potassium permanganate and 5 ml of 0.002 M chromium sulphate solutions were sonicated for different durations from 5 to 20 min with an interval of 5 min, to examine the effect of ultrasound on normal reaction between Cr^{3+} and MnO_4^- ions. Equimolar concentrations (0.002 M) of Cr and Mn could not be taken as the 0.002 M $KMnO_4$ solution would precipitate out immediately upon sonication, besides, the absorption spectra at this concentration went beyond the range of the spectrophotometeric measurements, hence only a dilute solution of $KMnO_4$ [0.0008 M] was used. The duration of sonication was also, therefore, reduced, as the reaction was complete within 10 min of sonication (cf. 40 min in the normal reaction).

The solutions of $KMnO_4$, upon sonication, tend to become turbid due to the formation of hydrated manganese dioxide, hence the experiments were carried out only at lower concentration (below 0.0008 M). No such precipitation of $Cr_2(SO_4)_3$ was observed in ultrasonic field. Therefore, a mixture of 5 ml each of 0.0008 M $KMnO_4$ and 0.002 M $Cr_2(SO_4)_3$ solution was sonicated. After 10 min of sonication, the solution containing $KMnO_4$ and $Cr_2(SO_4)_3$ began changing to yellow colour. UV-vis spectra of sonicated and unsonicated solutions are shown in Figs. 10.4a and b. However when a solution containing 10 ml of 0.002 M K_2CrO_4 and 1.0 ml of $Mn(CH_3COO)_2$ was sonicated, no significant change in the turbidity and conductivity was observed. Similarly, the sonication of $K_2Cr_2O_7$ and $KMnO_4$ too did not have any significant change.

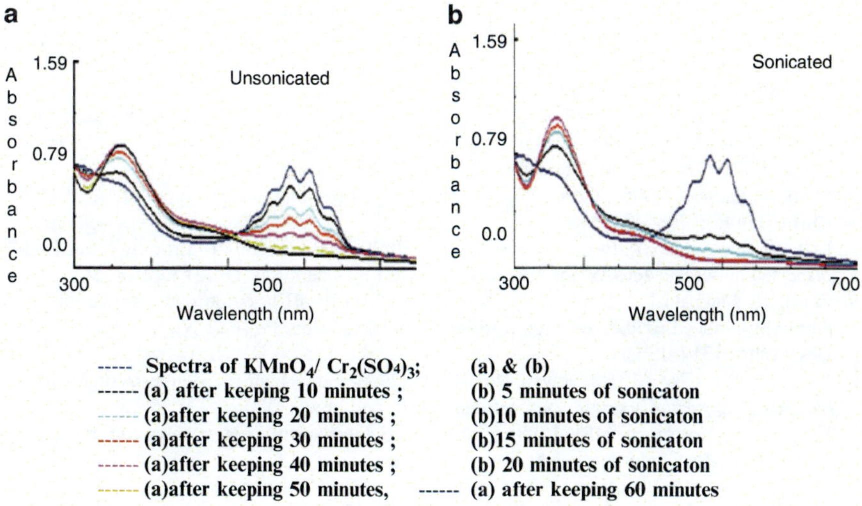

Fig. 10.4 UV Spectra of reactions between $KMnO_4$ and $Cr_2(SO_4)_3$

10.3 Conclusion

From the above experiments, we conclude that

1. $KMnO_4$ is a stronger oxidising agent than K_2CrO_4 and $K_2Cr_2O_7$.
2. Hydrolysis of Mn^{2+} ions upon sonication increased as the concentration of Mn^{2+} ions increased, in its solution.

The solution of $KMnO_4$, upon sonication, tends to become turbid due to the formation of hydrated manganese dioxide (unconfirmed), hence the experiments were to be carried out only at lower concentration (below 0.0008 M). However, no such precipitation of $Cr_2(SO_4)_3$ was observed in ultrasonic field. After 10 min of sonication, the solution of $KMnO_4$ and $Cr_2(SO_4)_3$ began changing to yellow, due to the oxidation of Cr^{+3} to CrO_4^{-2}, which was a clear indication of the reduction of MnO_4^- to Mn^{2+} ions and thus facilitating the conversion of Cr^{+3} to Cr^{+6}. Since in the unsonicated condition, the complete reduction of MnO_4^- to Mn^{2+} ions had taken 40 min, therefore, the ultrasound enhanced the rate of reduction (MnO_4^- to Mn^{2+}) or oxidation (Cr^{3+} to Cr^{6+}) to four times.

References

1. Meciarova M, Toma S, Luche J-L (2001) The sonochemical arylation of malonic esters mediated by manganese triacetate. Ultrason Sonochem 8(2):119–122
2. Kumar GV, Aurboch D, Gedanken A (2002) Influence of pH on the structure of the aqueous sonolysis products of manganese (III) acetylacetone. J Mater Res A 17(7):1706–1710
3. Grieser F, Hobson R, Sostric J (1996) Sonochemical reduction processes in aqueous colloidal systems. Ultrasonics 34(2–5):547–550
4. Ge J, Qu J (2003) Degradation of azo dye acid red B on manganese dioxide in the absence and presence of ultrasonic irradiation. J Hazard Mater 100:197–207
5. Meciarova M, Toma S, Heribanova A (2000) Ultrasound assisted heterogeneous permanganate oxidation. Tetrahedron 56(43):8561–8566
6. Sonochemical reduction of permanganate to manganese dioxide: the effect of H_2O_2 formed in the sonolysis of water on the rates of reduction: Kenji O, Masaki I, Ben Nishimura Rokura N, Yasuaki M (2009) Ultrason Sonochem 16(3):387–391
7. Mišík V, Riesz P (1996) Nitric oxide formation by ultrasound in aqueous solutions. J Phys Chem 100(45):17986–179948
8. Liang F, Fan J, Guo Y, Fan M, Wang J, Yang H (2008) Reduction of nitrite by ultrasound-dispersed nanoscale zero-valent iron particles. Ind Eng Chem Res 47(22):8550–8554
9. Zboril R, Machala L, Mashlan M, Sharma V (2004) Iron(III) oxide nanoparticles in the thermally induced oxidative decomposition of prussian blue, $Fe_4[Fe(CN)_6]_3$. Cryst Growth Design 4(6):1317–1325
10. Xinglong Wu, Cao M, Changwen Hu, He X (2006) Sonochemical synthesis of prussian blue nanocubes from a single-source precursor. Cryst Growth Design 6(1):26–28
11. Hung H-M, Hoffmann MR (1998) Kinetics and mechanism of the enhanced reductive degradation of CCl_4 by elemental iron in the presence of ultrasound. Environ Sci Technol 32(19):3011–3016
12. Minero C, Lucchiari M, Vione D, Maurino V (2005) Fe(III)-enhanced sonochemical degradation of methylene blue in aqueous solution. Environ Sci Technol 39(22):8936–8942

13. Liang J, Komarov S, Hayashi N, Kasai E (2007) Recent trends in the decomposition of chlorinated aromatic hydrocarbons by ultrasound irradiation and Fenton's reagent. J Mat Cycles Waste Manage 9(1):47–55
14. Torres RA, Pétrier C, Combet E, Moulet F, Pulgarin C (2007) Bisphenol a mineralization by integrated ultrasound-UV-iron (II) treatment. Environ Sci Technol 41(1):297–302
15. Dorathi PJ, Ranjit KP, Lee C-S (2008) Degradation of 2, 4-dichlorophenol in aqueous solution by sono-Fenton method. Korean J Chem Engi 25(1):112–117
16. Luo T, Ai Z, Zhang L (2008) Fe@Fe_2O_3 core-shell nanowires as iron reagent. 4. Sono-Fenton degradation of pentachlorophenol and the mechanism analysis. J Phys Chem C 112(23): 8675–8681
17. Cravotto G, Binello A, Di Carlo S, Orio L, Zhi-Lin Wu, Ondruschka B (2010) Oxidative degradation of chlorophenol derivatives promoted by microwaves or power ultrasound: a mechanism investigation. Environ Sci Poll Res 17(3):674–687
18. Tai Li Ji, Ya-Li S (2009) Degradation of AR 97 aqueous solution by combining ultrasound and Fenton reagent. Environ Prog Sustain Energy 29(1):101–106
19. Arul Dhas N, Koltypin Y, Gedanken A (1997) Sonochemical preparation and characterization of ultrafine chromium oxide and manganese oxide powders. Chem Mater 9(12):3159–3163
20. Gonsalves KE, Rangarajan SP, Law CC, Feng CR, Chow G.-M, Garcia-Ruiz A (1998) In: Chow G.-M, Gonsalves KE (eds) Nanotechnology: molecularly designed materials. In ACS symposium series, vol 622, chap 15, pp 220–236. Oxford University Press, USA
21. Franco DV, Da Silva LM, Jardim WF (2009) Reduction of hexavalent chromium in soil and ground water using zero-valent iron under batch and semi-batch conditions. Water Air Soil Poll 197(4):49–60
22. Kathiravan MN, Karthick R, Muthu N, Muthukumar K, Velan M (2010) Sonoassisted microbial reduction of chromium. Biochem Biotechnol 160(7):2000–2013
23. Ultrasonics: Rooney JA (1981) In Edmonds PD (ed) Methods of Experimental Physics, vol 19, pp 299–353. Academic Press, New York
24. Lauterborn W (1982) Cavitation bubble dynamics – new tools for an intricate problem. Appl Sci Res 38:165
25. Apfel RE (1981) In Edmonds PD (ed) Methods in Experimental Physics, vol 19, pp 356–413. Academic Press, New York.
26. Neppiras EA (1980) Acoustic cavitation. Phys Rep 61(3):159–251
27. Suslick KS, Johnson RE (1984) Sonochemical activation of transition metals. J Am Chem Soc 106:6856–6858
28. Enomoto N, Akagi JI, Z-l N (1996) Sonochemical powder processing of iron hydroxides. Ultrason Sonochem 3:S97–S103
29. Gasgnier M, Beaury L, Derout J (2000) Ultrasound effects on metallic (Fe and Cr); iron sesquioxides (α-, γ-Fe2O3); calcite; copper, lead and manganese oxides as powders. Ultrason Sonochem 7:25–33
30. Sostaric JZ, Mulvaney P, Grieser F (1995) Sonochemical dissolution of MnO_2 colloids. J Chem Soc Faraday Trans 91:2843–2846
31. Kruss P, Robertson DA, Mcmillen LA (1991) Effects of ultrasound on the cementation of cobalt on zinc. Ultrasonics 29:370–375
32. Farmer AD, Collings AF, Jameson GJ (2000) The application of power ultrasound to the surface cleaning of silica and heavy mineral sands. Ultrason Sonochem 7(4):243–247
33. Kuznetsov VM, Baranov AN, Oleinikov NN (1997) Sonochemical synthesis of Magnesium Ferrite. Dokl Akad Nauk 352(3):355–357
34. Belostotskii VF, Bemkin VM (1988) Ordre dans les solutions solides au cours de la déformation par les ultrasons aux températures de l'hélium liquide Ordering in solid solution during ultrasonic deformation at liquid helium temperatures. Metallokfizika 10(6):99–101
35. Pankaj, Manju C (2004) Effect of ultrasound on the redox reactions of iron (II) and (III). Ind J Chem 43(A):2098–2101
36. Pankaj, Chauhan (2004) Sonochemical studies of aqueous solutions of chromium and manganese in their cationic and oxoanionic states. Ind J Chem 43(A):1206–1209

Chapter 11
Sonochemical Degradation of Phenol in the Presence of Inorganic Catalytic Materials

Pankaj and Mayank Verma

Abstract Degradation of phenol in its aqueous solutions, using various techniques, including ultrasound, have been examined and discussed to better understand the mechanisms involved therein and the advantages as well as the disadvantages associated with the use of inorganic catalytic materials.

11.1 Introduction

Sonochemical degradation of organic molecules [1], employing cavitational effect is not new. Many simple [2] and complex molecules [3] from modest to very high toxicity levels, polluting the aquatic life and the environment have been successfully reduced to relatively non-toxic or lesser harmful species, if not degraded completely by the ultrasound. Although the degree of degradation and the duration of sonication varied considerably, depending upon the nature of moieties, but the sonochemical degradation undergoes either without the addition of extra chemicals or involves the minimum amounts of non-toxic chemicals such as photocatalyst TiO_2, Fenton's reagent etc., hence can be termed as a green technology. The purpose of this article is to review the sonochemical degradation of phenol, under different experimental conditions and in the presence of several other active chemicals to understand better the mechanisms involved in all such processes.

Phenol is an important chemical, from the point of its industrial and commercial applications, but is of great environmental apprehension due to its toxicity. Its chemical name is hydroxybenzene which is more acidic compared to aliphatic alcohols hence is also known as carbolic acid [4]. At ambient temperature and pressure it is a hygroscopic crystalline solid, which is colourless when pure [5] but

Pankaj (✉)
Department of Chemistry, Dayalbagh Educational Institute, Agra 282 110, India
e-mail: pankaj2@sancharnet.in

Pankaj and M. Ashokkumar (eds.), *Theoretical and Experimental Sonochemistry Involving Inorganic Systems*, DOI 10.1007/978-90-481-3887-6_11,

mostly light pink in color due to impurities [6]. Some of its physical and chemical properties are available in the literature [5, 7–9].

Phenol and its derivatives, due to their high volatility and poor water solubility, generally contaminate the environment through their natural, anthropogenic and endogenous sources. Some phenols enter the environment through natural processes such as by decomposition of organic matter or synthesis of chlorinated phenols by fungi and plants [10], whereas many industries release their effluents directly into the water, river or soil and contaminate them with phenolic compounds.

Due to its high chemical reactivity and the ease of handling phenol is used in industries as preservative for paints, leather and textile goods, in the production of phenolic and/or formaldehyde resins, pesticides, disinfectants, medicine, caprolactam and bisphenol A [11], as reagent in chemical analysis [12]. It is also used in electroplating, timber products processing, metal finishing, paving and roofing, paint formulating, metal molding and casting, and in the formation of electrical and electronic components [13]. Phenol serves as intermediates in industrial syntheses of products like adhesives and antiseptics [14]. Besides, it is used in perfumes, tanning agents and lubricating oils [15], explosives [16] and as precursor for nylon 6 and other man-made fibers as well [17]. Therefore, Phenol is released as effluents from steel, dyestuff, coal gasification and liquefaction industries and oil refineries [18] and surface runoff from coal mines [19] as well. Phenol is found in surface water, ground water, soil and sediments and is released to air and water as a result of wood burning, auto exhaust [20, 21] as well. It is also released through waste water from different industries, e.g., phenol and pesticides manufacturing plants, pharmaceutical and wood processing industries, petroleum and coal refineries, tannery, pulp and paper mills etc. Besides, phenol is released from various sources due to its commercial use as well as many phenol containing products, e.g., slimicides, disinfectants, medicinal ointments, ear and nose drops, cold sore lotion, mouthwashes, gargles, toothaches, antiseptic lotions, analgesic rubs, etc.

Phenol is very stable when present alone [22] and produces undesirable taste, color and odor to water [23]. Workers may be exposed to phenol during their work in industries related to phenol or phenol containing products. The use of smoked food with high phenol content may also cause toxic effects [24]. It is most widely used worldwide as its production at the rate of about six million ton/year has been reported [25]. The presence of phenol in potable and irrigation water presents a serious health problem not only to humans but also to animals, plants and microorganisms as well [26–28]. Phenol creates digestive disturbances as well as vomiting, difficulty in swallowing, ptyalism (hypersalivation), diarrhea, anorexia, headache, fainting, mental disturbances, muscle pain, enlarged liver and high serum enzymes (LDH, GOT and GPT) characteristic of liver damage [9, 29–32]. Therefore phenol has been considered as priority pollutant by Environmental Protection Agency (EPA) since 1976 due to its high toxicity [33]. According to WHO, the concentration upto 1.0 μg/l of phenol is prescribed safe for drinking water [34]. The main path of contact is principally by ingestion, inhalation or dermal absorption and inhalation [35] leading to anorexia, weight loss, headache, vertigo, salivation and dark urine whereas ingestion [24, 34–37] shows adverse cardiovascular effects, respiratory distress, metabolic acidosis, renal failure, neurological effects, shock, coma and death, necrosis of the skin and mucous membranes of the throat, abdominal pain, gastrointestinal irritation, nausea, vomiting, sweating and diarrhea, altering in respiratory and pulse rate. The visible symptoms for the dermal absorption [34, 35, 38] are dermal inflammation, erythma, necrosis of the skin, severe burns, cardiovascular shock, cardiac arrhythmias, bradycardia and metabolic acidosis.

Phenols are carcinogenic [39–42] and mutagenic thus affect the central nervous system. Long term contact to phenol may even paralyze the body and damage liver, kidneys [41] and heart [43]. Phenol and its vapour are corrosive to the eyes, skin and respiratory tract [44]. Renal failure and pulmonary toxicity has been reported with overdose of 89% injectable phenol solution [45]. According to Central Pollution Control Board (CPCB) the discharge limit of phenol in inland water should be lower than 1 mg/l [46].

From the above discussion, it is clear that while phenol is essential for the well being and advancement of the humans, it also brings in innumerable hazards and toxic effects, if the concentration limits are exceeded. Therefore, it is essential that the excess amount of phenol is degraded and brought under the permissible limits before it is discharged into the environment. Several such efforts have already been attempted with partial to complete success, which needs a review, before some new effort of this laboratory is discussed at the end of the chapter.

11.2 Remediation Methods of Phenol

Several efforts to degrade phenol have been undertaken using various techniques but some important methods involve the application of electromagnetic [47, 48], laser [49, 50] photocatalytic [51–54], mechanochemical [55], electro-catalytic [56–59], photo-electrocatalytic [60], biodegradation [61–71], advanced oxidation process [72] and glow discharge plasma [73–75] processes. However, in the following section, various sonochemical methods, used for the degradation of phenol in aqueous solutions, would be discussed.

11.2.1 Sonochemical Methods

Unlike conventional chemical reactions, the altered reactivity of chemical reactions undergoing ultrasonic irradiation is principally due to acoustic cavitation which essentially involves the free radical formation. The ultrasound produces highly reactive free radical species like H• and OH• radicals from the homolytic cleavage of water. Further they may react with any of other free radicals present or with neutral molecules like O_2 and O_3 to produce peroxy species, superoxide, hydrogen peroxide or hydrogen. When the aqueous solution is saturated with O_2, extra reactions exist due to reaction of molecular oxygen and hydrogen atom and thermal decomposition of oxygen [11]. Similarly, on mixing O_3, some other reactions occur, augmenting the production of OH• radicals in aqueous and gas phase reaction [11]. Such reactions have already been discussed in detail in Chapter 10. We may, therefore, switch over to other sonochemical studies related to phenol degradation in water.

Li et al. [76] confirmed that efficacy of phenol degradation depends on micro-bubble formation. In their experiments, they observed no change in phenol concentration if micro-bubble formation was stopped. The phenol decomposition rate was found maximum in the case when O_2 was passed in the solution due to highest micro-bubble formation followed by air and N_2 respectively.

When ultrasound of different frequencies [77] was applied to degrade phenol, the degradation efficiency seemed increased at higher frequencies. The reason had been explained on the basis of formation of OH• radicals under acoustic cavitation. The more the production of the OH• radicals during cavitation the more was the chance to attack phenol molecules leading to its degradation. Petrier et al. [78] found 200 kHz frequency to be most efficient in phenol degradation when compared with 20, 200 and 500 kHz. Perhaps the 200 kHz frequency facilitated the formation of radicals. Gogate et al. [79] examined the efficiency of acoustic cavitation with different frequency using a novel triple frequency flow cell with a maximum capacity of 7.5 l and compared with the conventional set up containing single frequency transducer to explore efficiency of sonoreactors on industrial scale. The advantage of the novel reactor was its ability to operate at multiple frequencies of 20, 30 and 50 kHz individually and in combination as well. Initial concentration (100–500 ppm) of phenol at constant temperature of 30°C ± 1°C, power supply of 300 W per frequency of irradiation with total irradiation time of 90 min was examined. Since the region of maximum energy created by ultrasonic horn was restricted to near the horn surface, the destruction occurring in the vicinity of the transducer surface was too low to be felt as an overall degradation due to poor mixing conditions existing in the reactor. To overcome the drawback of conventional reactor (ultrasonic horn with single transducer) this failed to induce destruction of phenol on industrial scale. Gogate's group used multiple frequency irradiations and found maximum percentage degradation of phenol when triple frequency in combination was operated followed by double frequency and single frequency operations. The frequency order was found to be in the sequence $20 < 30 < 50$ (single frequency operation) $<20 + 30 <20 + 50 < 30 + 50$ (double frequency operation) $<20 + 30 + 50$ (triple frequency operation).

Wu et al. [12] used both microwave (MW) and ultrasound (US) methods individually and in combination to examine the combined effect. The rapid thermal effect of MW could be seen on polar chemicals and more OH radicals were produced due to US. Microwave irradiations have shown enhanced degradation effect when applied with sonication in absence of additional catalyst though the rate increased more in presence of H_2O_2. The rate order was found to be MW-US > MW > US.

When the ultrasound was combined with the photocatalytic method, the combination method [80, 81] proved to be more effective in the degradation of phenol compared to both the method applied separately. These experiments have also shown that lower pH, high concentration of dissolved O_2 and ultrasonic power density also enhanced the phenol degradation. Kubo et al. [82] examined the ultrasonically induced phenol degradation with TiO_2 in complete dark and found the phenol degradation rate to be higher in the presence the TiO_2 compared to that of sonication alone. The rate also enhanced linearly with the catalyst amount. As the phenol degradation did not proceed in the presence of TiO_2 without ultrasound, the study suggested that catalytic property of TiO_2 was activated only by the ultrasound which provided the initial energy to activate TiO_2.

In photocatalysis, the photon energy is more important than the band gap of the semiconductor, TiO_2 produces OH• radicals similar to ultrasonic waves inside the cavitation bubble. Therefore, there is a general idea that by combining these two methods more OH• radicals may be produced to attack phenol molecules. The thermal energy produced during ultrasonic irradiation is also thought to generate positive holes near TiO_2 surface [82]. These positive holes also react with water to produce additional OH• radicals as shown by Eq. 11.2 which react with phenol further by following reactions;

$$\text{PhO}\bullet + \text{OH}\bullet \longrightarrow \text{non-radical products} \tag{11.1}$$

$$2\text{OH}\bullet \longrightarrow H_2O_2 \tag{11.2}$$

Many workers have since tried ultrasound induced Fenton's reagent methods to degrade phenol. Papadaki et al. [83] reported the lower efficiency of Fenton's reagent for phenol degradation due to competition of both Fe^{2+} and ultrasound for H_2O_2, resulting in the reduction of concentration of Fenton's reagent in solution.

In sono-Fenton methods, the addition of oxidant species [84] like H_2O_2 and $CuSO_4$ have been examined at different ultrasonic frequencies, 20, 35 and 500 kHz. Iron-mesoporous silica (SBA-15) composite [85–87] has been reported to increase the degradation. The rate was found linearly dependent upon the frequencies but when oxidant species like H_2O_2 and $CuSO_4$ were added to the medium, the most efficient degradation was found at 35 kHz. The degradation followed the frequency order 35 > 500 > 20 kHz. This result signifies the role not only of frequency of the ultrasound but also of intensity, surface area of the sonicator and the geometry of the reactor. Bremner et al. [86] examined the influence of different ultrasonic frequencies (382, 584 and 1,142 kHz) on the efficiency of the US/Fe_2O_3/SBA-15/H_2O_2 system. They reported a better total organic carbon removal per ultrasound power applied at 584 kHz frequency possibly due to the promotion of the number and distribution of cavitations instead of formation of microjets and violent local cavitation at 20 kHz. Sonication of 30 min was reported sufficient to completely destroy phenol and aromatic byproducts. Molina et al. [85] investigated phenol oxidation using catalytic wet oxidation assisted with ultrasound irradiation.

The catalyst, composite of crystalline hematite particles embedded into a mesoporous silica SBA-15 matrix was used. They accounted the following reactions for ultrasound coupled with Fenton-like reagent:

Direct pyrolysis (Reactions 11.3 and 11.4)

$$RH + US \longrightarrow \text{Degradation products} \quad (11.3)$$

$$H_2O + US \longrightarrow H\bullet + OH\bullet \quad (11.4)$$

Indirect oxidation by attack of OH• radicals formed in bulk solution or interface between the collapsing bubbles (Reactions 11.5 and 11.6)

$$RH + HO \longrightarrow \text{Degradation products} \quad (11.5)$$

$$H_2O_2 + US \longrightarrow 2OH\bullet \quad (11.6)$$

reaction with H_2O_2 (Reaction 11.7)

$$\Xi Fe^{3+} + H_2O_2 \longrightarrow \Xi Fe(OOH)^{2+} + H^{+} \longleftrightarrow \Xi Fe^{2+} + HO_2 \quad (11.7)$$

complex redox reactions of Fe^{2+} and Fe^{3+} (Reactions 11.8–11.10)

$$\Xi Fe(OOH)^{2+} + US \longrightarrow \Xi Fe^{2+} + HO_2 \quad (11.8)$$

$$\Xi Fe^{2+} + H_2O_2 \longrightarrow \Xi Fe^{3+} + HO^{-} + HO \quad (11.9)$$

$$\Xi Fe^{3+} + HO_2 \longrightarrow \Xi Fe^{2+} + H + O_2 \quad (11.10)$$

Chand et al. [88] degraded phenol under sonication at different frequencies with H_2O_2 and O_3 in presence of zero valent iron and copper. The effectiveness of 300 kHz sonochemical reactor was found to be the maximum for the generation of OH• radicals among 20, 300 and 520 kHz. The complete removal of phenol was achieved when sonicated with 300 kHz frequency with zero valent iron compared to zero valent copper. In their mechanism it was supposed that at first iron metal was corroded in the presence of hydrogen peroxide, under acidic conditions, oxidizing Fe^0 to Fe^{2+}, which on further reaction with H_2O_2 produce OH• radicals and Fe^{3+}. The Fe^0 then reduces the Fe^{3+} back to Fe^{2+} and the chain was maintained as

$$Fe^0 + 2H^+ \longrightarrow Fe^{2+} + H_2 \quad (11.11)$$

$$Fe^{2+} + H_2O_2 \longrightarrow Fe^{3+} + OH^{-} + OH\bullet \quad (11.12)$$

$$H_2O_2 \longrightarrow 2OH\bullet \quad (11.13)$$

$$H_2O_2 + OH\bullet \longrightarrow H_2O + O_2H\bullet \quad (11.14)$$

$$Fe\bullet + 2Fe^{3+} \longrightarrow 3Fe^{2+} \quad (11.15)$$

$$Fe^{3+} + HO_2 \longrightarrow Fe^{2+} + H^+ + O_2 \quad (11.16)$$

And the mechanism for ozone oxidation is explained in following steps (Reactions 11.17–11.21)

$$H_2O + O_3 \longrightarrow O_2H\bullet + O\bullet + OH\bullet \quad (11.17)$$

$$O_2^+H\bullet + \bullet H \longrightarrow H_2O_2 \quad (11.18)$$

$$H_2O_2 + O_3 \longrightarrow O_2H\bullet + OH\bullet + O_2 \quad (11.19)$$

$$O_2H\bullet + O_3 \longrightarrow OH\bullet + 2O_2 \quad (11.20)$$

$$OH\bullet + O_3 \longrightarrow O_2H\bullet + O_2 \quad (11.21)$$

In the experiment involving oxidative enzyme HRP (EC 1.11.1.7, RZ 1.9, 240 purpuro gallin (units/mg)) [89] for the enzymatic treatment and ultrasonic waves of 423 kHz and 5.5 W, the phenol degradation rate was found to increase. The ultrasound assisted biodegradation method has been found to be more efficient method than the sonolysis and enzyme treatment when operated individually.

Sonawane et al. [90] investigated the affect of ultrasound and nanoclay for the adsorption of phenol. Three types of nanoclay tetrabutyl ammonium chloride (TBAC), N-acetyl-N,N,N trimethyl ammonium bromide (CTAB) and hexadecyl trimethyl ammonium chloride (HDTMA), modified under sonication, were synthesized which showed healthier adsorption of phenol within only 10 min in waste water. The interlamellar spacing of all the three clay increased due to incorporation of long chain quaternary salts under cavitational effect.

Unlike conventional chemical reactions, the sonochemical reactions are dependent upon the rate of scavenging [91] of the radicals. The scavenging means the reactions of different types of radicals to produce other oxidative species. Thus higher the concentration of the scavenging radicals in the system, the lower is the recombination and higher the degradation. Therefore, not only the production of the free radicals is necessary, an enhancement of the scavenging of the radicals is also equally important to enhance the degradation. The radicals formed outside the bubbles are extremely reactive as compared to formed inside the bubbles. They react very rapidly just after their formation. So if the concentration of phenol near the cavitation bubble is lower, they recombine and discourage the phenol degradation. Sivasankar et al. [91] used both the molecular species (O_2) and ionic species (Fe^{2+}) for scavenging of radicals and reported more efficacy of the ionic species Fe^{2+} even at lower concentration levels as compared to that of O_2. It can be attributed due to faster reactivity of Fe^{2+} ions and uniform concentration of the Fe^{2+} ions during sonication. They anticipated the mechanisms for both the species O_2 and Fe^{2+}. When O_2 is supplied to the solution, during cavitation it may be inside bubble in the form of vapor or gas and in bulk solution in liquid or dissolved form. During implosion the oxygen molecules decompose homolytically to produce oxygen atom which may be inside and outside the bubble due to scavenging. Thus oxygen molecule and its atom react with different molecules like H_2O and its fragmented free radicals H• and OH• in different ways to produce different oxidative free radical species i.e., OH•, O_2H• etc. The reactions are explained as under (Reactions 11.22–11.24).

$$O_2 +))) \longrightarrow 2O\bullet \quad (11.22)$$

$$O\bullet + H_2O \longrightarrow 2OH\bullet \quad (11.23)$$

The reaction of H• with O_2 are

$$H\bullet + O_2 \longleftrightarrow O_2H\bullet \quad (11.24)$$

The reaction of OH• with O_2

Unlike the oxygen, Fe^{2+} scavenging process is due to ionic nature which is limited to bulk medium only. Fe^{2+} reacts mainly with H_2O_2, produced by recombination of OH• radicals and liberated in the bulk system with transient collapse of the bubble. Experimentally [91] the average concentration of Fe^{2+} in the bulk medium was found to be almost constant possibly due to continuous regeneration, providing an effective radicals scavenging.

Zheng et al. [92] investigated the effect of CCl_4, perfluorohexane (volatile) and iodate (non-volatile) hydrogen atom scavenger for improving the phenol degradation. They reported the enhancing effect with the volatile H atom scavenger. Therefore the use of H atom scavenger indirectly helped in more production of OH• radicals and thereby enhanced degradation rate of phenol.

Bapat et al. [93] introduced the interfacial phenomenon in explaining sonochemical degradation of phenol. They analyzed theoretically, the degradation of phenol and its chloro-derivatives produced by the reaction of CCl_4 and phenol. They concluded quantitatively that the o- and p-chloro derivatives of phenol had greater possibility to gather at the cavity–liquid interface than that of phenol. Therefore chlorophenols generated during the reaction of CCl_4 with phenol degraded at faster rate than the phenol itself.

After the review of literature, we report here the results of the degradation of phenol, carried out in our laboratory in the presence of ultrasound, TiO_2, rare earths and transition metal ions to highlight our interpretation of the mechanism. Various transition metal salts are known for their catalytic properties due to partly filled d-orbital of the metal atom. Mesoporous transition metal oxides are used not only as catalyst but also as highly functional materials with shape-specific and/or quantum effects. Besides, rare earth oxides, mixed oxides and other rare earth doped materials have also been synthesized and reported as materials of luminescent, catalytic, electric, magnetic, transport and thermodynamic behavior, due to their 4-f electrons [94–106]. However, the relative catalytic behavior of transition metal ions is many times more than the inner transition metal ions due to their relatively shielded f-electrons, which are not as freely available as the d-electrons of transition metals. Photocatalytic performance of lanthanum doped TiO_2 has been examined by varying the rare earth oxide content, calcinating temperature and calcinating time [107]. Catalytic activity of lanthanum-cobalt metal has also been reported for methane combustion [108]. La^{3+} and N co-doped TiO_2 have been used for photo-catalytic decomposition of acetaldehyde [109]. The effect of samarium doping on photocatalytic activity of TiO_2 for methylene blue has also been studied [110]. However, there seem to be no study on the synthesis of new solid composites of common transition metal salts with rare earths under ultrasonic field and their study as a catalyst in the degradation of organic molecules. Among lanthanides, Dy^{3+} ion is found best suited for the yellow laser applications due to its confinement of characteristic fluorescence spectra consisting of $^4F_{9/2} \rightarrow {}^6H_{13/2}$ transition[2]. Besides, Dy^{3+} doped chloride crystals were found to amplify 1.3 μm signal used in telecommunications network [111]. Magnetic properties [112] of dysprosium single crystals and photoelectronic properties [113] of its complexes with acetylacetonate and meso-tetraalkyltetrabenzoporphyrin are also known. Similarly, in the recent past, ceria, Ce ($4f^2\ 5d^0\ 6s^2$), dispersed with Pt, Au and Cu, and in the form of mixed oxides such as CeO_2-ZrO_2, have been used for the catalytic oxidation of carbon monoxide, combustion of ethyl acetate and volatile organic compounds and in water-gas shift reactions [114–118]. Besides, cerium promoted copper oxide on alumina and Pd have also been used as catalyst in the oxidation of CO and methane [119, 120]. Cerium oxide has potential application as a dielectric in silicon-based microelectronic devices [121]. Due to optical absorption which is associated with the 4f-5d electronic transition of Ce^{3+} ions, sulfides of Ce^{3+} ($4f^1\ 5d^0$) ions such as γ-Ce_2S_3 and its alkali-metal derivatives γ- $Ce_{2\text{-}X}A_{3X}S_3$ (A = alkali) are found important substitutes, for the hazardous red and orange Cd pigments conventionally used in plastic industry [122]. Therefore we carried out similar such degradation studies of phenol using Cu(II), Co(II) and Mn(II) salts doped with cerium(III) [123] and latter extended to Dy and subsequently to other rare earths such as La (III), Sm(III) and Gd(III) doped with transition metal ions Cu(II) and Co(II). Tetra-n-butylorthotitanate was used as a source of TiO_2 for the reason already explained in the preceding chapter. Besides, the propagation of ultrasound prevents agglomeration of TiO_2 resulting into finer particles, thereby increasing the activated surface area of the catalyst many fold. [124]. A systematic study of the same is being reported here to give readers of this chapter the precise details of the experiments and the results obtained thereafter for such combinatorial studies.

11.3 Experimental

11.3.1 Synthesis of Catalyst

The rare earth oxides of lanthanum, samarium and gadolinium were converted into soluble nitrate salts by dissolving them in the minimum amount of concentrated nitric acid. Then two sets were prepared by adding 2.0 ml of aqueous solution of $La(NO_3)_3.6H_2O$ [0.2 M] and 0.01 ml of $(n\text{-}BuO)_4Ti$ to 25 ml of aqueous solution of $Cu(NO_3)_2$ [1.0 M]. Similarly, two sets were prepared with $Co(NO_3)_3$. Same procedures were followed for $Sm(NO_3)_3$ [0.2 M] and $Gd(NO_3)_3$ [0.2 M]. One set of all these solutions were sonicated under ultrasonic bath (Model – Meltronics, 20 kHz, 250 W) for half an hour. The solutions prepared in normal and sonicated conditions were kept in muffle furnace (Model – Deluxe Zenith) first at 100°C for 2 h and then the temperature of the furnace was raised up to 900°C and calcined for 2 h. The solid composites prepared were then cooled to room temperature and treated as catalyst for phenol degradation.

Similarly for the synthesis of catalysts consisting of rare earths (Ce and Dy) with transition metals (Cu, Co and Mn) were also prepared as following. For dysprosium catalyst synthesis, 10 ml of $DyCl_3$ (0.1 M) was mixed with two aliquots of aqueous solution of 25 ml of $CuSO_4.5H_2O$ (1.4 M). One of these solutions was kept for crystallization through evaporation whereas the other sample was sonicated with ultrasonic bath until the crystallization. Similar procedure was followed for $MnCl_2.4H_2O$ (1.0M) and $CoCl_2.6H_2O$ (2.0M). The reaction vessel (100 ml beaker) was dipped in the ultrasonic bath and kept 15.5 cm above to the point of origin of sonication. The device operated at a fixed frequency of 20 ± 2 kHz and power of 6 W/cm^2. The synthesis of cerium catalyst was also done in the same manner as for dysprosium with the only difference that in this case the synthesis of sonicated composites was performed with ultrasonic probe only for half an hour before the crystallization. All the sets were covered with Sartorius filter paper (Germany, Grade-293) with three holes of 0.5 cm diameter on surface. In normal condition the solutions were gently heated for 5 min to speed up the crystallization. Crystals synthesized under normal and sonicated conditions were isolated under ambient conditions, washed with distilled acetone and allowed to dry under vacuum, before being used as catalysts. Besides, these crystals were also not calcined as in case of the crystals doped with La, Sm and Gd.

The elemental analysis of La, Sm, Gd, Dy, Cu, Co and Mn was done with atomic absorption spectrophotometer (Analytik Zena ZEEnit 700) with graphite furnace attached to the auto sampler. The calibration curve was plotted for all the metals by taking the standard solution of AAS analysis grade before estimating the metals present in solid sample. However, Ce was estimated spectrophotometrically [125] (Shimadzu, UV-VIS-1601) at λ_{max} 500 nm, using 1-(2-pyridylazo)-2-naphthol as complexing agent, after dissolving the catalyst composite in Conc. HCl. The concentrations of metals present in the solid sample are given in the Table 11.1.

Table 11.1 Composition of La, Sm, Gd, Cu and Co in composites synthesized in normal and sonicated conditions

S. No.	Transition metal-rare earth metal composition	Transition metal (mg/l)	Rare earth metal (mg/l)	Transition metal: Rare earth metal
1.	Cu–La composite	Cu	La	Cu : La
	normal	0.4629	0.0057	81.21 : 1.00
	sonicated	0.2983	0.0041	72.76 : 1.00
	Co–La composite	Co	La	Co : La
	normal	1.2200	0.0097	125.77 : 1.00
	sonicated	0.9141	0.0118	77.47 : 1.00
2.	Cu–Ce composite	Cu	Ce	Cu : Ce
	normal (blue)	2.578	0.373	6.91 : 1.00
	normal (white)	0.270	0.502	0.54 : 1.00
	sonicated	3.582	0.454	7.89 : 1.00
	Mn–Ce composite	Mn	Ce	Mn : Ce
	normal	4.077	0.116	35.15 : 1.00
	sonicated	4.501	0.127	35.44 : 1.00
	Co–Ce composite	Co	Ce	Co : Ce
	normal	6.841	0.139	49.2 : 1.00
	sonicated	6.651	0.158	42.1 : 1.00
3.	Cu–Sm composite	Cu	Sm	Cu : Sm
	normal	0.3599	0.0078	46.14 : 1.00
	sonicated	0.8912	0.0067	133.01 : 1.00
	Co–Sm composite	Co	Sm	Co : Sm
	normal	1.4330	0.0103	139.13 : 1.00
	sonicated	1.0612	0.0118	89.93 : 1.00
4.	Cu–Gd composite	Cu	Gd	Cu : Gd
	normal	0.5528	0.0037	149.41 : 1.00
	sonicated	0.3916	0.0032	122.37 : 1.00
	Co–Gd composite	Co	Gd	Co : Gd
	normal	1.0210	0.0097	105.28 : 1.00
	sonicated	0.9165	0.0194	47.24 : 1.00
5.	Cu–Dy composite	Cu	Dy	Cu : Dy
	normal	1.349	0.196	6.88 : 1.00
	sonicated	1.517	0.382	3.97 : 1.00
	Mn–Dy composite	Mn	Dy	Mn : Dy
	normal	3.543	0.766	4.63 : 1.00
	sonicated	3.494	0.360	9.71 : 1.00
	Co–Dy composite	Co	Dy	Co : Dy
	Normal	6.822	0.235	29.03 : 1.00
	sonicated	7.156	0.213	33.60 : 1.00

From AAS estimations, the ratio of rare earth metal to transition metal was found to be quite different in solid composites synthesized in sonication conditions compared to normal conditions. In all the cases of rare earth-transition metal-TiO_2 composite the concentration of both transition metals (Cu and Co) and rare earths (La, Sm and Gd) was found changed due to sonication. The composition of transition metals (Cu and Co) was found to be lower compared to rare earths (La, Sm and Gd) respectively in solid composites synthesized in sonicated conditions than those solid composites synthesized in normal conditions (Table 11.1). In other words we can say that the concentration of rare earths to transition metal is higher as is clear from the ratio of transition metal-rare earth ratio given in Table 11.1. The only difference was found in case of Sm–Cu–TiO_2 composite ratio where the concentration of copper was slightly higher in sonicated conditions.

In case of crystals of Cu–Dy composite formed under sonication, the concentration of dysprosium increased while in case of the crystals of Mn–Dy and Co–Dy composites, the concentration of dopant, Dy, decreased indicating a strong attraction of Dy for Cu compared to its weak interaction for Mn and Co ions. Nevertheless, the possibility of some of the Dy having been ejected out due to forceful cavitational effect of the ultrasound from the lattice of Mn and Co cannot be ruled out. Higher percentage of Cu, Mn, and Ce in case of Cu–Ce, Co–Ce and Mn–Ce composites, synthesized under sonication compared to normal crystals, could be attributed to the change in the composition of the lattice pattern due to the mechanical impact of ultrasound, whereas, such an effect has not been found in Co salts. These can be seen in Table 11.1.

11.3.2 Sonophotocatalytic Degradation of Phenol

Degradation of phenol in aqueous solution (0.5 mg/50 ml) was monitored spectrophotometrically [126] under several experimental conditions and different duration of time such as, (i) under sonicated condition (ii) in the presence of $(n\text{-BuO})_4$ Ti, (iii) in presence of $(n\text{-BuO})_4$ Ti and ultrasound (iv) in the presence of La-, Sm- and Gd-doped transition metal oxides such as (a) La–Cu, (b) La–Co, (c) Sm–Cu, (d) Sm–Co, (e) Gd–Cu, (f) Gd–Co. 0.01 ml of $(n\text{-BuO})_4$ Ti was taken in case of sonocatalytic and photocatalytic degradation of phenol (condition (2) and (3)), whereas in cases of (4) and (5) 10 mg of rare earths (La/Sm/Gd) – transition metals (Cu/Co)-TiO_2 composite was mixed to 0.5 mg/50 ml aqueous solution of phenol as shown in Figs. 11.1–11.6. The same process was followed for the dysprosium and cerium catalysts with the difference that in these cases tetra-n-butylorthotitanate, $(n\text{-BuO})_4$Ti, was used at the time of phenol degradation in all conditions and not used in catalyst synthesis (Figs. 11.8–11.10). The ultrasonic equipment consisting of ultrasonic sonicator with titanium horn (model Vibronics Ultrasonic Processor) with a tip diameter of 1.2 cm was used and immersed up to 2.0 cm in the liquid. The device operated at a fixed frequency of 20 ± 2 kHz and power of 6 W/cm^2. Phenol was degraded for 30, 60, 90 and 120 min rspectively. Experiments were run in triplicate and results reported were the average value. However, before the UV–Vis spectrophotometric analysis of phenol, all the samples were filtered using filtration assembly, fitted with filter discs (Sartorius, Germany, Grade-393) to remove turbidity generated due to sonication of solid catalyst and $(n\text{-BuO})_4$ Ti in water. The percentage degradation of phenol in presence of different rare earth doped transition metals and under different conditions for different durations has been given in Tables 11.2–11.7 and graphically shown through Fig. 11.1–11.6.

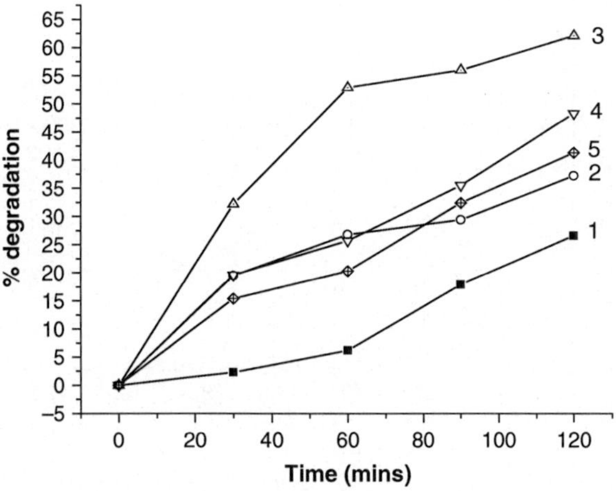

Fig. 11.1 Degradation of phenol with La – Cu composite. *1* – US; *2* – TiO_2; *3* – US + TiO_2; *4* – US + La–Cu–TiO_2 (normal solid composite); *5* – US + La–Cu–TiO_2 (sonicated solid composite)

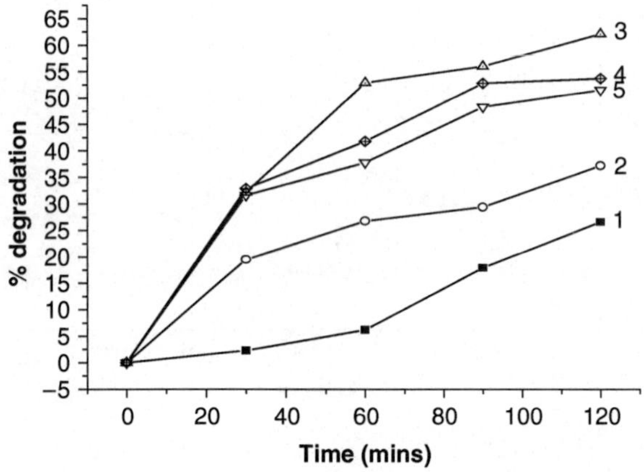

Fig. 11.2 Degradation of phenol with La – Co composite. *1* – US; *2* – TiO_2; *3* – US + TiO_2; *4* – US + La–Co–TiO_2 (normal solid composite); *5* – US + La–Co–TiO_2 (sonicated solid composite)

Percentage degradation of phenol in the presence of ultrasound and TiO_2 photocatalyst individually, in combination and in the presence of additives like La–Cu–TiO_2, La–Co–TiO_2, Sm–Cu–TiO_2, Sm–Cu–TiO_2, Gd–Cu–TiO_2 and Gd–Co–TiO_2 at ambient conditions were examined. The percentage degradation of phenol after 2 h was found to be 26.6% and 37.2% in the presence of ultrasound and TiO_2 respectively and 62.1% when the two techniques were combined. Kubo et al. [82] also reported higher degradation in the presence of TiO_2 compared to sonication alone. The results showed the synergistic effect of ultrasound and TiO_2 for the phenol degradation.

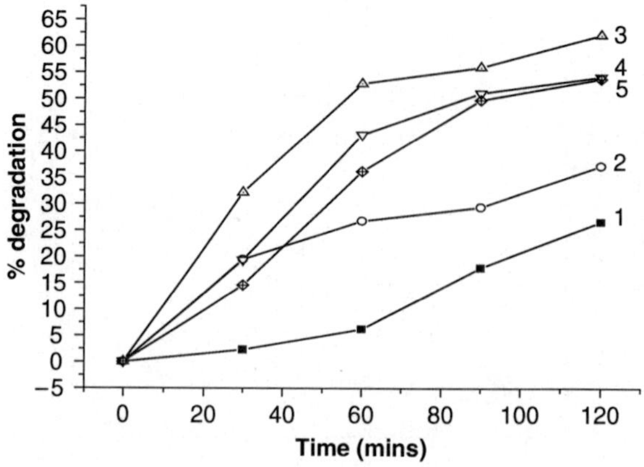

Fig. 11.3 Degradation of phenol with Sm – Cu composite. *1* – US; *2* – TiO_2; *3* – US + TiO_2; *4* – US + Sm–Cu–TiO_2 (normal solid composite); *5*. US + Sm–Cu–TiO_2 (sonicated solid composite)

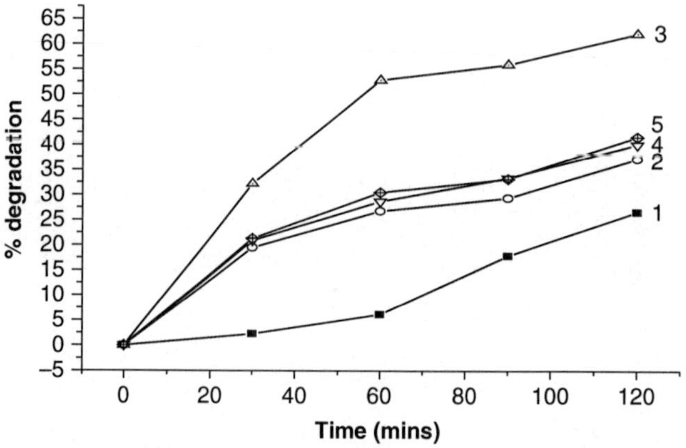

Fig. 11.4 Degradation of phenol with Sm – Co composite. *1* – US; *2* – TiO_2; *3* – US + TiO_2; *4* – US + Sm–Co–TiO_2 (normal solid composite); *5* – US + Sm–Co–TiO_2 (sonicated solid composite)

The addition of different rare earth-transition metal-TiO_2 solid composites showed higher percentage of phenol degradation than that of pure TiO_2 but lower than that of sono-photocatalytic method after 2 h. The percentage degradation after 2 h in different conditions and presence of different additives is given in Fig. 11.7. The lower percentage degradation of phenol in presence of these additives compared to degradation in combined sono-photocatalytic method could be attributed to slower rate of formation of OH° radicals in cases of additives possibly due to partial conversion of TiO_2 from anatase to rutile phase during sintering, deficiency of scavenging radicals perhaps due to formation of some complex compounds of rare earth-transition metal-TiO_2, and/or low availability of dissolved O_2 in the system due to degassing.

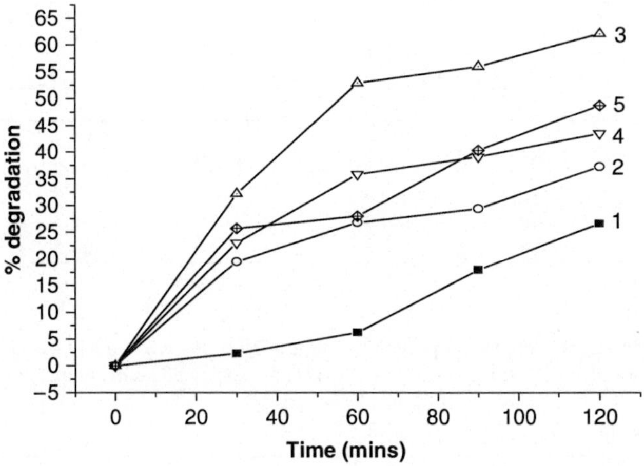

Fig. 11.5 Degradation of phenol with Gd – Cu composite. *1* – US; *2* – TiO_2; *3* – US + TiO_2; *4* – US + Gd–Cu–TiO_2 (normal solid composite); *5* – US + Gd–Cu–TiO_2 (sonicated solid composite)

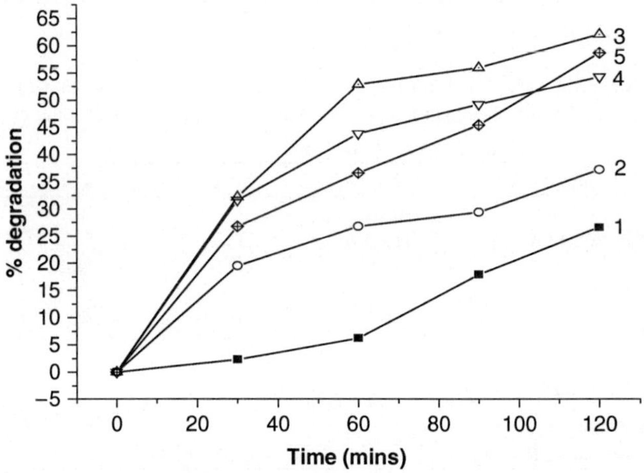

Fig. 11.6 Degradation of phenol with Gd – Co composite. *1* – US; *2* – TiO_2; *3* – US + TiO_2; *4* – US + Gd–Cu–TiO_2 (normal solid composite); *5* – US + Gd–Cu–TiO_2 (sonicated solid composite)

Table 11.2 Lanthanum–Copper composite (*1* – US; *2* – TiO_2; *3* – US + TiO_2; *4* – US + La–Cu–TiO_2 (normal solid composite); *5* – US + La-Cu-TiO_2 (sonicated solid composite))

Time (min)	Percentage degradation of phenol				
	1	2	3	4	5
0	0	0	0	0	0
30	2.3	19.5	32.2	19.6	15.4
60	6.2	26.8	52.9	25.7	20.3
90	17.9	29.4	56	35.5	32.4
120	26.6	37.2	62.1	48.3	41.3

Table 11.3 Lanthanum–Cobalt composite (*1* – US; *2* – TiO_2; *3* – US + TiO_2; *4* – US + La–Co–TiO_2 (normal solid composite); *5* – US + La–Co–TiO_2 (sonicated solid composite))

Time (min)	Percentage degradation of phenol				
	1	2	3	4	5
0	0	0	0	0	0
30	2.3	19.5	32.2	31.6	32.9
60	6.2	26.8	52.9	37.8	41.8
90	17.9	29.4	56	48.3	52.8
120	26.6	37.2	62.1	51.5	53.1

Table 11.4 Samarium–Copper composite (*1* – US; *2* – TiO_2; *3* – US + TiO_2; *4* – US + Sm–Cu–TiO_2 (normal solid composite); *5* – US + Sm–Cu–TiO_2 (sonicated solid composite))

Time (min)	Percentage degradation of phenol				
	1	2	3	4	5
0	0	0	0	0	0
30	2.3	19.5	32.2	19.4	14.6
60	6.2	26.8	52.9	43.2	36.2
90	17.9	29.4	56	51.1	49.8
120	26.6	37.2	62.1	54.2	53.8

Table 11.5 Samarium–Cobalt composite (*1* – US; *2* – TiO_2; *3* – US + TiO_2; *4* – US + Sm–Co–TiO_2 (normal solid composite); *5* – US + Sm–Co–TiO_2 (sonicated solid composite))

Time (min)	Percentage degradation of phenol				
	1	2	3	4	5
0	0	0	0	0	0
30	2.3	19.5	32.2	20.9	21.3
60	6.2	26.8	52.9	28.7	30.5
90	17.9	29.4	56	33.4	33.2
120	26.6	37.2	62.1	40.1	41.6

Table 11.6 Gadolinium–Copper composite. (*1* – US; *2* – TiO_2; *3* – US + TiO_2; *4* – US + Gd–Cu–TiO_2 (normal solid composite); *5* – US + Gd–Cu–TiO_2 (sonicated solid composite))

Time (min)	Percentage degradation of phenol				
	1	2	3	4	5
0	0	0	0	0	0
30	2.3	19.5	32.2	23.0	25.7
60	6.2	26.8	52.9	35.8	28.0
90	17.9	29.4	56	39.1	40.3
120	26.6	37.2	62.1	43.4	48.7

TiO_2 is instantly produced and suspended in the system by the hydrolysis of $(n\text{-}BuO)_4Ti$ in water when dissolved and hence more formation and availability of OH• radicals to attack phenol molecules. It was observed that the percentage degradation of phenol was lower in the cases where additives were used compared

Table 11.7 Gd–Co composite. *1* – US; *2* – TiO_2; *3* – US + TiO_2; *4* – US + Gd–Co–TiO_2 (normal solid composite); *5* – US + Gd–Co–TiO_2 (sonicated solid composite))

Time (min)	Percentage degradation of phenol				
	1	2	3	4	5
0	0	0	0	0	0
30	2.3	19.5	32.2	31.6	26.8
60	6.2	26.8	52.9	43.8	36.6
90	17.9	29.4	56	49.2	45.4
120	26.6	37.2	62.1	54.3	58.7

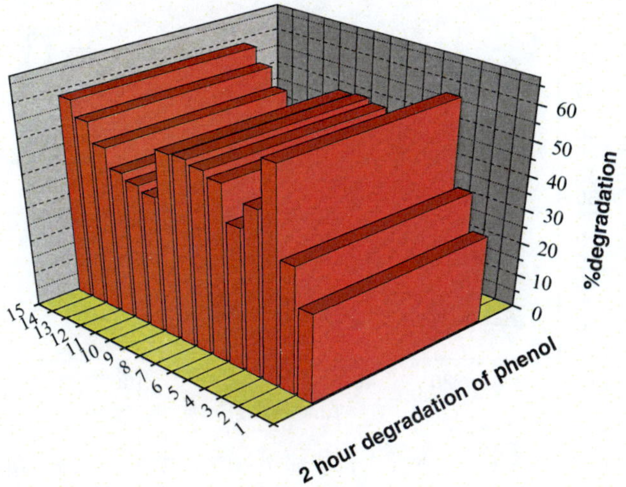

Fig. 11.7 Two hour degradation of phenol in different conditions and in presence of different rare earth-transition meatl–TiO_2 composite. *1* – US; *2* – TiO_2; *3* – US + TiO_2; *4* – US + La-Cu-TiO_2 (normal solid composite); *5* – US + La-Cu-TiO_2 (sonicated solid composite); *6* – US + La-Co-TiO_2 (normal solid composite); *7* – US + La–Co–TiO_2 (sonicated solid composite); *8* – US + Sm-Cu-TiO_2 (normal solid composite); *9* – US + Sm–Cu–TiO_2 (sonicated solid composite); *10* – US + Sm–Co–TiO_2 (normal solid composite); *11* –US + Sm–Co–TiO_2 (sonicated solid composite); *12* – US + Gd–Cu–TiO_2 (normal solid composite); *13* – US + Gd–Cu–TiO_2 (sonicated solid composite); *14* – US + Gd–Co–TiO_2 (normal solid composite); *15* – US + Gd–Co–TiO_2 (sonicated solid composite)

to that in combined sono-photocatalytic method but comparing the percentage degradation of phenol using different additives of rare earths synthesized in normal and sonicated conditions, phenol degradation rate was found to be dependent on the amount of rare earths as it could be seen through Table 11.2–11.7. The only exceptions were in case of La–Co–TiO_2 (Fig. 11.2) and Gd–Cu–TiO_2 (Fig. 11.5) solid composites where the percent degradation of phenol did not depend on the amount of rare earth (Table 11.1).

Figures 11.8–11.10 and Tables 11.8–11.10 explain the sono-photochemical degradation of phenol under sonicated and normal conditions and in the presence of photocatalyst $(n\text{-BuO})_4Ti$ with and without Cu–Dy, Mn–Dy and Co–Dy composites respectively. These Figs. 11.8–11.10 show the percentage degradation of phenol during a period of 2 h through an interval of 30 min under sonicated, photocatalytic, sonophotocatalytic and crystal induced sono-photo-catalytic conditions.

In case of Cu–Dy composite the degradation of phenol is in the following order:

sonicated < photo-catalyzed < sono-photo-catalyzed < sono-photo-catalyzed with normal crystals < sono-photo-catalyzed with sonicated crystals.

While in case of Mn–Dy and Co–Dy composites the degradation of phenol is in following order:

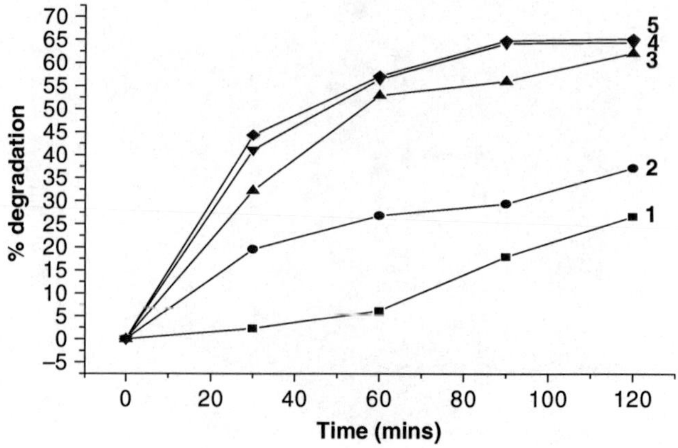

Fig. 11.8 Degradation of Phenol with Cu–Dy crystal. *1* – US; *2* – TiO_2; *3* – US + TiO_2; *4* – US + TiO_2 + Cu–Dy (normal solid composite); *5* – US + TiO_2 + Cu–Dy (sonicated solid composite)

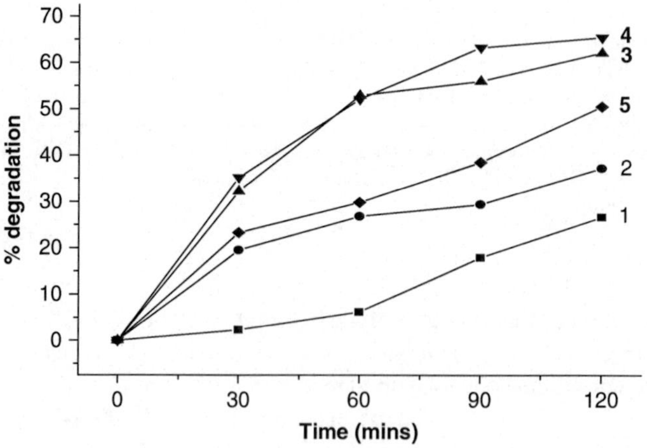

Fig. 11.9 Degradation of Phenol with Mn–Dy crystals. *1* – US; *2* – TiO_2; *3* – US + TiO_2; *4* – US + TiO_2 + Mn–Dy (normal solid composite); *5* – US + TiO_2 + Mn–Dy (sonicated solid composite)

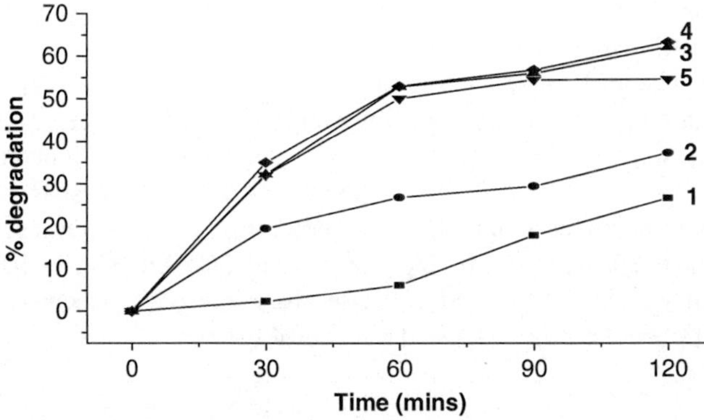

Fig. 11.10 Degradation of Phenol with Co–Dy crystals. *1* – US; *2* – TiO_2; *3* – US + TiO_2; *4* – US + TiO_2 + Co–Dy (normal solid composite); *5* – US + TiO_2 + Co–Dy (sonicated solid composite)

Table 11.8 Cu–Dy composite (*1* – US; *2* – TiO_2; *3* – US + TiO_2; *4* – US + TiO_2 + Cu–Dy (normal solid composite); *5* – US + TiO_2 + Cu–Dy (sonicated solid composite))

Time (min)	Percentage degradation of phenol				
	1	2	3	4	5
30	2.3	19.5	32.2	41.1	44.3
60	6.2	26.8	52.9	56.4	57.1
90	17.9	29.4	56.0	64.2	64.8
120	26.6	37.2	62.1	64.4	65.2

Table 11.9 Mn–Dy composite (*1* – US; *2* – TiO_2; *3* – US + TiO_2; *4* – US + TiO_2 + Mn–Dy (normal solid composite); *5* – US + TiO_2 + Mn–Dy (sonicated solid composite))

Time (min)	Percentage degradation of phenol				
	1	2	3	4	5
30	2.3	19.5	32.2	35.2	23.3
60	6.2	26.8	52.9	52.1	29.8
90	17.9	29.4	56.0	63.6	38.5
120	26.6	37.2	62.1	65.6	50.5

Table 11.10 Co–Dy composite (*1* – US; *2* – TiO_2; *3* – US + TiO_2; *4* – US + TiO_2 + Co–Dy (normal solid composite); *5* – US + TiO_2 + Co–Dy (sonicated solid composite))

Time (min)	Percentage degradation of phenol				
	1	2	3	4	5
30	2.3	19.5	32.2	32.0	35.0
60	6.2	26.8	52.9	50.1	53.0
90	17.9	29.4	56.0	54.5	56.8
120	26.6	37.2	62.1	54.6	63.3

sonicated < photo-catalyzed < sono-photo-catalyzed with sonicated crystals < sono-photo-catalyzed < sono-photo-catalyzed normal crystals.

In case of Cu–Dy composites (Table 11.1) dysprosium concentration is higher in sonicated condition in comparison to normal condition, whereas, in case of Mn–Dy and Co–Dy composites the dysprosium concentration is lower in sonicated condition and therefore the degradation of phenol was lower. The sonication adds to the degradation of phenol in aqueous solution but it is less than the decomposition of phenol under photocatalytic condition. However, there is still enhanced degradation when the process is carried out under sonication in combination with photocatalysis. Besides, degradation is slightly higher in solutions containing crystals synthesized under sonic condition in case of Cu–Dy composite (Fig. 11.8). Nevertheless there is not much difference in cases of Mn–Dy (Fig. 11.9) and Co–Dy (Fig. 11.10) composite under similar conditions. Based on these observations the mechanism for the degradation of phenol can be suggested as detailed below.

11.4 Mechanism

Faster degradation of chloro (o- and p-chlorophenols) and hydroxyl (catechol, resorcinol and hydroxoquinone) derivatives of phenol is attributed to multicentered electron delocalization due to chlorine and oxygen atoms, which are both more electronegative than carbon atom and facilitate delocalized electrons of the phenyl ring to drain out and concentrate outside the ring. Therefore their degradation in the fields of ultrasound and microwave as well as in the presence of photocatalysts (TiO_2) increases many fold. All these processes follow free radical mechanism and is explained [123]. The formation of phenoxide ion, which is resonance stabilized, is further facilitated by the ultrasound. But in the presence of photocatalyst the phenoxide ion seem to be attracted towards the alkyl chain of the n-Bu group through lyophilic interaction and transfers the electron density to the vacant d orbital of Ti and decomposes. Besides, the four oxygen atom surrounding Ti ion attract electronic shell of the Ti and leave a δ^+ charge on the Ti, making it more attractive for the phenoxide ions to attack the titanium atom and transfer their electron density to the induced polarized titanium ion. At this stage the phenoxide ions, due to the transfer of their electron density to titanium ion, are no more resonance stabilized and therefore, decompose far more easily due to the cavitation effect of the ultrasound. Figure 11.11 makes this mechanism more explicit. Similarly in the presence of transition metals such as Cu(II), Co(II), and Mn(II) which all have unpaired electrons in their valence shell, the attraction for the phenoxide ion is facilitated and the transfer of electron density from phenoxide ring to metal ion results.

The same reason may be applied to Ce(III) with $4f^1$ electron, where electron may be accommodated but since the loss of the $5f^1$ electron of the Ce(III) makes it more stable, the same is lost to H^+ ions of the system generating free $H^{\cdot}$ or even to any electron acceptor.

$$Ce^{3+} \overset{e^-}{\curvearrowright} H^+ \longrightarrow H^{\cdot} + Ce^{4+}$$

Now the possible explanation for the greater degradation of phenol by the cerium doped copper crystals than those of either cobalt or manganese ions could

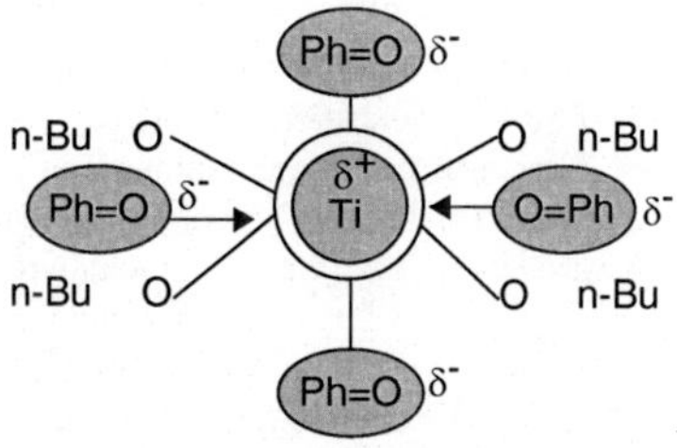

Fig. 11.11 Mechanism showing the attraction of phenoxide ion towards Ti ion

be easily understood. Since the cupric ion is a $3d^9$ electronic system and requires just one electron to fill its valence shell to become stable, the 4f electron of Ce^{3+} is donated readily to Cu^{2+} ion making both copper and cerium stable. However, the $3d^5$ and $3d^7$ electronic configurations of Mn^{2+} and Co^{2+} ions, even after accepting 4f electron from cerium, do not acquire stability. Rather, Mn^{2+} may even become slightly unstable due to possible $3d^6$ electronic configuration and therefore these ions do not show any preference. This difference in the electronic configuration of copper from those of cobalt and manganese make copper a better catalyst than the other two in combination with cerium.

The higher degradation of phenol in the presence of copper and cerium salts can now be explained better. Since transition metal ions (and the inner transition metal ions as well) acquire extra stability, when these ions have completely vacant, half filled or completed d (or f) shells, there is an effort by these ions to invariably go to such electronic states either by the loss or gain of electrons with the help of surrounding species. Thus, cupric ion attains $3d^{10}$ electronic configuration by accepting one electron and Ce^{3+} ion looses one electron to acquire completely vacant $4f^0$ electronic configuration. Therefore, while cupric ion accepts electron from the resonance stabilised phenoxide ion to destabilise and finally decompose phenol, the cerous ion looses electron to a proton resulting in the formation of more hydrogen free radicals, which also degrade phenol through enhanced free radical reactions in the aqueous medium.

Thus to conclude, it can be inferred that the ultrasound alone is less effective in degrading phenol compared to the photocatalyst, TiO_2. However, when the two techniques are combined, the result is most effective. Nevertheless, the role of transition and inner transition metal ions is still not clear although their effect could be seen – may be the lower concentrations of these ions could not play any effective role as of now until a further study is carried out.

References

1. (a) Kaur S and Singh V (2007) Visible light induced sonophotocatalytic degradation of reactive Red dye 198 using dye sensitized TiO_2. Ultrason Sonochem 14: 531–537. (b) Tangestaninejad S, Moghadam M, Mirkhani V, Mohammadpoor-Baltork I, Salavati H (2008) Sonochemical and visible light induced photochemical and sonophotochemical degradation of dyes catalyzed by recoverable vanadium-containing polyphosphomolybdate immobilized on TiO_2 nanoparticles. Ultrason Sonochem 15(5):815–822. (c) Song Y-L, Li J-T (2009) Degradation of C.I. Direct Black 168 from aqueous solution by fly ash/H_2O_2 combining ultrasound. Ultrason Sonochem 16(4):440–444

2. (a) Zourab SM, Ezzo EM, El-Aila HJ, Salem JKJ (2005) Study of kinetics of oxidation of amines by potassium ferricyanide in the presence of N,N-dimethyldodecylamine N-oxide. J Surfactants Detergents 8(1) 83–89. (b) Calza P, Pelizzetti E (2001) Photocatalytic transformation of organic compounds in the presence of inorganic ions. Pure Appl Chem 73(12): 1839. (c) Kado Y, Atobe M and Nonaka T (2001) Ultrasonic effects on electroorganic processes – Part 20. Photocatalytic oxidation of aliphatic alcohols in aqueous suspension of TiO_2 powder. Ultrason Sonochem 8(2):69
3. (a) Kim T-K, Kim M-K, Lim Y-J, Son Y-A (2005) Degradation of the disazo acid dye by the sulfur-containing amino acids of wool fibers. Dyes Pigments 67(2): 127–132. (b) Knoevenagel K, Himmelreich R (1976) Degradation of compounds containing carbon atoms by photooxidation in the presence of water. Arch Environ Contamin. Toxicol 4(1): 324–333. (c) Park S-H, Wei S, Mizaikoff B, Taylor AE, Favero C, and Huang C-H (2009) Degradation of amine-based water treatment polymers during chloramination as N-nitrosodimethylamine (NDMA) precursors. Environ Sci Technol 43 (5):1360–1366
4. Gardner W, Cooke EI, Cooke RWI (1978) Handbook of chemical synonyms and trade names. CRC Press, Boca Raton, FL
5. Budavari S, O'Neil MJ, Smith A, Heckelmen PE (1989) The Merck index. Merck, Whitehouse station, NJ, p 1150
6. HSDB (1998) Hazardous substances data bank. National Library of Medicine, National Toxicology Information Program, Bethesda, MD
7. Lide DR (1993) CRC handbook of chemistry and physics. CRC Press, Boca Raton, FL
8. Amoore JE, Hautala E (1983) Odors as an aid to chemical safety: odor threshold limit values and volatilities for 214 industrial chemicals in air and water dilution. Appl Toxicol 3:272–290
9. Baker EL, Landrigan PJ, Bertozzi PE, Field PH, Basteyns BJ, Skinner HG (1978) Phenol poisoning due to contaminated drinking water. Arch Environ Health 33:89–94
10. Swarts M, Verhage NF, Field J, Wijn Berg J (1998) Trichlorinated phenols from Hypholoma elongatum. Phytochemistry 49(1):203–206
11. Kidak R, Ince NH (2006) Ultrasonic destruction of phenol and substituted phenols: a review of current research. Ultrason Sonochem 13:195–199
12. Wu Z-L, Dondruschka B, Cravotto G (2008) Degradation of phenol under combined irradiation of microwaves and ultrasound. Environ Sci Technol 42:8083–8087
13. Meister JJ (ed) (2000) Polymer modification: principles techniques and applications. CRC Press, 936
14. Alnaizy R, Akgerman A (2000) Advanced oxidation of phenolic compounds. Adv Environ Res 4(3):233–244
15. Lurascu B, Siminiceanu I, Vione D, Vicente MA, Gil A (2009) Phenol degradation in water through a heterogeneous photo-Fenton process catalyzed by Fe-treated laponite. Water Res 43:1313–1322
16. Bu Davari S (2001) The Merck index, 13th edn. Merck, Whitehouse station, NJ, pp 1299–1367
17. Idris A, Saed K (2002) A precursor for nylon 6 and other manmade fibres. Global Nest Int J 4 (2–3):139–144
18. Leonardo SA, Rocha-Filho RC, Bocchi N, Biaggio SR, Garcia-Garcia JIV, Montiel V (2008) Degradation of phenol using Co- and Co, F-doped PbO_2 anodes in electrochemical filter-press cells. J Hazordous Mater 153:252–260
19. Saravanan P, Pakshirajan K, Saha P (2009) Degradation of phenol by TiO_2-based heterogeneous photocatalysts in presence of sunlight. J Hydroenviron Res 3:45–50
20. International Programme on Chemical Safety (IPCS) (1994) Environmental Health Criteria 161. Phenol. World Health Organisation, Geneva
21. International Programme on Chemical Safety (IPCS) (1994) Phenol. Health and safety guide no 88. WHO, Geneva
22. Chaliha S, Bhattacharyya KG (2008) Using Mn(II) – MCM41 as an environment – friendly catalyst to oxidize phenol, 2-chlorophenol and 2-nitrophenol in aqueous solution. Ind Eng Chem Res 47:1370–1379

23. Klibanov AM (1982) Enzymatic removal of hazardous pollutants from industrial aqueous effluents. Enzym Eng 6:319–323
24. Del Fino F, Dube D (1976) Persistent contamination of ground water by phenol. J Environ Sci Health 43:345
25. Jordan W, van Barneveld H, Gerlich O, Kleine-Boymann M, Ullrich J (2002) Phenol in: Ullmann's encyclopedia of industrial chemistry. Weinheim, Wiley-VCH Verlag
26. Shailubhai K (1986) Treatment of Petroleum industry oil sludge in soil, Tibtech. Elsevier Science Publishers B.V., Amsterdam, pp 202–206
27. Salonen M, Middeldorp P, Briglia M, Valo R, Haggblom M and McBain A Kamely D, Chakrabarty A and Omenn, GS (eds) (1989) Cleanup of old industrial sites. In: Biotechnology and biodegradation. Portfolio Publishing, The Woodlands, TX, pp 347–365
28. Sharma H, Barber JT, Ensley HE, Polito MA (1997) A comparison of the toxicity of phenol and chlorinated phenols by Lemna gibba with reference to 2, 4, 5-trichloorophenol. Environ Toxicol Chem 16:346–350
29. Tuah PBM (2006) The performance of phenol biodegradation by candida tropicalis Retl-Cr-1 using batch and fed-batch fermentation techniques. Ph.D. Thesis, Universiti Teknologi Malaysia
30. Bruce RN, Santodonato J, Neal MW (1987) Summary review of the health effects associated with phenol. Toxicol Ind Health 3:535–568
31. Gosslin RE, Smith RP, Hodge HC, Braddock JE (1984) Phenol in: clinical toxicology of commercial products, vol 3. Williams and Wilkins, Baltimore, MD, pp 345–346
32. Merliss RR (1972) Phenol marasmus. J Occu Environ Med 14:55–56
33. Keith H, Telliard WA (1979) Priority pollutants I. A perspective view. Environ Sci Technol 13:416–423
34. World Health Organization (WHO) (1994) Phenol, environmental health criteria-EHC 161. WHO, Geneva
35. International Programme on Chemical Safety (IPCS) (1999) Phenol poisons information monograph. PIM 412. (http://www.inchem.org/documents/pims/chemical/pim412.htm)
36. International Programme on Chemical Safety (IPCS) (1994) Phenol. Health and safety guide no 88. WHO. Geneva. Printed by Wissenschsftliche Verlagsgesellschaft, Stuttgart. http://www.inchem.org/documents/hsg/hsg/hsg88_e.htm
37. Barker EL, Peter EB, Petrecia HF, Grant SK (1978) Phenol poisoning due to contaminated drinking water. Arch Environ Health 33:89–94
38. Agency for Toxic Substances and Disease Registry (ATSDR) (1998) Toxicological Profile for Phenol. US Department of Health and Human Services. Atlanta, US. http://www.epa.gov/ttn/atw/hlthef/phenol.html
39. Kim JH, Oh KK, Lee ST, Kim SW, Hong SI (2002) Biodegradation of phenol and chlorophenols with defined mixed culture in shake-flasks and a packed-bed reactor. Process Biochem 37(12):1367–1373
40. US Environmental Protection Agency (2002) Toxicological review of phenol (CAS No. 108-95-2) EPA/635/R-02/006, In support of summary information on the integrated risk information system (IRIS), Washington DC
41. Iurascu B, Siminiceanu I, Vione D, Vicente MA, Gil A (2009) Phenol degradation in water through a heterogeneous photo-Fenton process catalyzed by Fe-treated laponite. Water Res 43:1313–1322
42. American conference of Governmental Industrial Hygienists standards (ACGIH) Manual, 2005, and United States Environmental Protection Agency, EPA, 816-F-01-007, 2006
43. Warner MA, Harper JV (1985) Cardiac dysrhythmias associated with chemical peeling with phenol. Anesthesiology 62:366–7
44. Budavari S (ed) (1996) The Merck index: an encyclopedia of chemical, drugs, and biologicals. Merck, Whitehouse Station, NJ
45. Gupta S, Ashrith G, Chandra D, Gupta AK, Finkel KW, Guntupalli JS (2008) Acute phenol poisoning: a life – threatening hazard of chronic pain relief. Clin Toxicol 46:250–253

46. Lathasree S, Nageswara Rao A, Sivasankar B, Sadasivam V, Rengaraj K (2004) Heterogeneous photocatalytic mineralization of phenols in aqueous solutions. J Mol Cat A Chem 223:101–105
47. Sato K, Takimoto K, Tsuda S (1978) Degradation of aqueous phenol solution by gamma irradiation. J Am Chem Soc 12(9):1043–1046
48. Lai T-L, Lee C-C, Wu K-S, Shu Y-Y, Wang C-B (2006) Microwave-enhanced catalytic degradation of phenol over nickel oxide. Appl Catal B 68:147–153
49. Gondal MA, Seddiqi Z (2006) Laser-induced photo-catalytic removal of phenol using n-type WO_3 semiconductor catalyst. Chem Phys Lett 417:124–127
50. Gondal MA, Sayeed MN, Seddiqi Z (2008) Laser enhanced photo-catalytic removal of phenol from water using p-type NiO semiconductor catalyst. J Hazard Mater 155:83–89
51. Matthews RW, McEvoy SR (1992) Destruction of phenol in water with sun, sand and photocatalysis. Sol Energy 49(6):507–513
52. Guo Z, Ma R, Li G (2006) Degradation of phenol by nanomaterial TiO_2 in wastewater. Chem Eng J 119:55–59
53. Karunakaran C, Dhanalakshmi R (2008) Semiconductor-catalyzed degradation of phenols with sunlight. Sol Energy Mater Sol Cells 92:1315–1321
54. Karunakaran C, Dhanalakshmi R (2009) Phenol degradation on Pr_6O_{11} surface under UV-A light. Synergistic photocatalysis by semiconductors. Radiat Phys Chem 78:8–12
55. Cotto MC, Emiliano A, Nieto S, Duconge J, Roque-Malherbe R (2009) Degradation of phenol by mechanical activation of a rutile catalyst. J Colloid Interface Sci 339:133–139
56. (a) Feng YJ, Li XY (2003) Electro-catalytic oxidation of phenol on several metal-oxide electrodes in aqueous solution, Water Res 37:2399–2407. (b) Szpyrkowicz L, Kaul SN, Neti RN, Satyanarayan S (2005) Influence of anode material on electrochemical oxidation for the treatment of tannery wastewater. Water Res 39: 1601–1613. (c) Li XY, Cui YH, Feng YJ, Xie ZM Dong JD (2005) Reaction pathways and mechanisms of the electrochemical degradation of phenol on different electrodes. Water Res 39: 1972–1981
57. Wang Y-Q, Gu B, Xu W-L (2009) Electro-catalytic degradation of phenol on several metal-oxide anodes. J Hazard Mater 162:1159–1164
58. Torres RA, Torres W, Peringer P, Pulgarin C (2003) Electrochemical degradation of p-substituted phenols of industrial interest on Pt electrodes. Attempt of a structure–reactivity relationship assessment. Chemosphere 50:97–104
59. Trabelsi F, Ait-Lyazidi H, Ratsimba B, Wilhelm AM, Delmas H, Fabre P-L, Berlan J (1996) Oxidation of phenol in waste water by sonoelectrochemistry. Chem Eng Sci 51(10):1857–1865
60. Liu Z, Zhang X, Nishimoto S, Jin M, Tryk DA, Murakami T, Fujishima A (2008) Highly ordered TiO_2 nanotube arrays with controllable length for photoelectrocatalytic degradation of phenol. J Phys Chem C 112:253–259
61. Kryst K, Karamanev DG (2001) Aerobic phenol biodegradation in an inverse fluidized-bed biofilm reactor. Ind Eng Chem Res 40:5436–5439
62. Jiang H-L, Tay J-H, Maszenan AM, Tay ST-L (2006) Enhanced phenol biodegradation and aerobic granulation by two coaggregating bacterial strains. Environ Sci Technol 40:6137–6142
63. Mordocco A, Jenkins CKR (1999) Continuous degradation of phenol at low concentration using immobilized Pseudomonas putida. Enzyme Microb Technol 25:530–536
64. Pai S-L, Hsu Y-L, Chong N-M, Sheu C-S, Chen C-H (1995) Continuous degradation of phenol by Rhodococcus SR immobilized on granular activated carbon and in calcium alginate. Biores Technol 51:37–42
65. Santos VL, Linardi VR (2004) Biodegradation of phenol by a filamentous fungi isolated from industrial effluents – identification and degradation potential. Process Biochem 39:1001–1006
66. Perron N, Welander U (2004) Degradation of phenol and cresols at low temperatures using a suspended-carrier biofilm process. Chemosphere 55:45–50
67. Fanga HHP, Lianga DW, Zhanga T, Liu Y (2006) Anaerobic treatment of phenol in wastewater under thermophilic condition. Water Res 40:427–434
68. Karlsson A, Ejlertsson J, Nezirevic D, Svensson BH (1999) Degradation of phenol under meso- and thermophillic, anaerobic conditions. Anaerobe 5(1):25–35

69. Akaya G, Erhan E, Keskinler B, Algur OF (2002) Removal of phenol from wastewater using membrane-immobilized enzymes Part II. Cross-flow filtration. J Membr Sci 206:61–68
70. Luo H, Liu G, Zhang R, Jin S (2009) Phenol degradation in microbial fuel cells. Chem Eng J 147:259–264
71. Moussavi G, Mahmoudi M, Barikbin B (2009) Biological removal of phenol from strong wastewaters using a novel MSBR. Water Res 43:1295–1302
72. Esplugas S, Gim!enez J, Contreras S, Pascual E, Rodríguez M (2002) Comparison of different advanced oxidation processes for phenol degradation. Water Res 36:1034–1042
73. Sano N, Yamamoto T, Takemori I, Kim S-I, Eiad-ua A, Yamamoto D, Nakaiwa M (2006) Degradation of phenol by simultaneous use of gas-phase corona discharge and catalyst-supported mesoporous carbon gels. Ind Eng Chem Res 45:2897–2900
74. Li J, Sato M, Ohshima T (2007) Degradation of phenol in water using a gas–liquid phase pulsed discharge plasma reactor. Thin Solid Films 515:4283–4288
75. Liu YJ, Xuan Zhen Jian G (2005) Phenol degradation by a nonpulsed diaphragm glow discharge in an aqueous solution. Environ Sci Technol 39:8512–8517
76. Li P, Takahashi M, Chiba K (2009) Degradation of phenol by the collapse of microbubbles. Chemosphere 75:1371–1375
77. Petrier C, Lamy M-F, Francony A, Benahcene A, David B (1994) Sonochemical degradation of phenol in dilute aqueous solutions: comparison of the reaction rates at 20 and 487 kHz. J Phys Chem 98:10514–10520
78. Petrier C, Francony A (1997) Ultrasonic waste-water treatment: incidence of ultrasonic frequency on the rate of phenol and carbon tetrachloride degradation. Ultrason Sonochem 4(4):295–300
79. Gogate PR, Mujumdar S, Thampi J, Wilhelm AM, Pandit AB (2004) Destruction of phenol using sonochemical reactors: scale up aspects and comparison of novel configuration with conventional reactors. Sep Purif Technol 34:25–34
80. Chen Y-C, Smirniotis P (2002) Enhancement of photocatalytic degradation of phenol and chlorophenols by ultrasound. Ind Eng Chem Res 41:5958–5965
81. Wu C, Liu X, Wei D, Fan J, Wang L (2001) Photosonochemical degradation of phenol in water. Water Res 35(16):3927–3933
82. Kubo M, Matsuoka K, Takahashi A, Shibasaki-Kitakawa N, Yonemoto T (2005) Kinetics of ultrasonic degradation of phenol in the presence of TiO_2 particles. Ultrason Sonochem 12:263–269
83. Papadaki M, Emery RJ, Abu-Hassan MA, Bustos AD, Metcalfe IS (2004) Sonocatalytic oxidation processes for the removal of contaminants containing aromatic rings from aqueous effluents. Sep Purif Technol 34(1–3):35–42
84. Entezari MH, Petrier C, Devidal P (2003) Sonochemical degradation of phenol in water: a comparison of classical equipment with a new cylindrical reactor. Ultrason Sonochem 10:103–108
85. Molina R, Martinez F, Melero JA, Bremner DH, Chakinala AG (2006) Mineralization of phenol by a heterogeneous ultrasound/Fe-SBA-15/H_2O_2 process: multivariate study by factorial design of experiments. Appl Catal B Environ 66:198–207
86. Bremner DH, Molina R, Martnez F, Melero JA, Segura Y (2009) Degradation of phenolic aqueous solutions by high frequency sono-Fenton systems (US–Fe_2O_3/SBA-15–H_2O_2). Appl Catal B 90:380–388
87. Segura Y, Molina R, Martínez F, Melero JA (2009) Integrated heterogeneous sono–photo Fenton processes for the degradation of phenolic aqueous solutions. Ultrason Sonochem 16:417–424
88. Chand R, Ince NH, Gogate PR, Bremner DH (2009) Phenol degradation using 20, 300 and 520 kHz ultrasonic reactors with hydrogen peroxide, ozone and zero valent metals. Sep Purif Technol 67:103–109
89. Entezari MH, Petrier C (2003) A combination of ultrasound and oxidative enzyme: sono-biodegradation of substituted phenols. Ultrason Sonochem 10:241–246

90. Sonawane S, Chaudhari P, Ghodke S, Ambade S, Gulig S, Mirikar A, Bane A (2008) Combined effect of ultrasound and nanoclay on adsorption of phenol. Ultrason Sonochem 15:1033–1037
91. Sivasankar T, Moholkar VS (2009) Mechanistic approach to intensification of sonochemical degradation of phenol. Chem Eng J 149:57–69
92. Zheng W, Maurin M, Tarr MA (2005) Enhancement of sonochemical degradation of phenol using hydrogen atom scavengers. Ultrason Sonochem 12:313–317
93. Bapat PS, Gogate PR, Pandit AB (2008) Theoretical analysis of sonochemical degradation of phenol and its chloro-derivatives. Ultrason Sonochem 15:564–570
94. Wang Y, Yin L, Gedanken A (2002) Sonochemical synthesis of mesoporous transition metal and rare earth oxides. Ultrason Sonochem 9(6):285–290
95. Rozenberg E, Gorodetsky G, Felner I, Sominski E, Gedanken A, Mukovskii YM (2006) Magnetic properties of crystalline $La_{0.9}Ca_{0.1}MnO_3$: Comparison of bulk and nanometer-sized samples. J Appl Phys 99:08Q305
96. Raebiger J, Miller W (2002) Magnetic ordering in the rare earth molecule-based magnets, Ln (TCNE)3 (Ln = Gd, Dy; TCNE = tetracyanoethylene). J S Inorg Chem 41(12):3308–3312
97. Weang ZC, Wang LS (1997) Thermodynamic properties of the rare earth element vapor complexes $LnAl_3Cl_{12}$ from Ln = La to Lu. Inorg Chem 36(8):1536–1540
98. Viswanathan B (1984) Solid state and catalytic properties of rare earth orthocobaltites – a new generation catalysts. J Sci Ind Res 44:66–74
99. Zou Z, Ye J, Arakawa H (2002) Role of R in Bi_2RNbO_7 (R = Y, Rare earths); Effect on band structure and photocatalytic properties. J Phys Chem B 106(3):517–520
100. Cussen EJ, Lynham DR, Rogers J (2006) Magnetic order arising from structural distortion: structure and magnetic properties of Ba_2LnMoO_6. J Chem Mater 18:2855–2866
101. Huang YH, Fjellvag H, Karppinen M, Hauback BC, Yamuchi H, Goodenough JB (2006) Crystal and magnetic structure of the orthorhombic perovskite $YbMnO_3$. Chem Mater 18:2130–2134
102. Ryaznov M, Keinle L, Simon A, Mattausch HU (2005) New synthesis route to and physical properties of lanthanum monoiodide. Inorg Chem 45:2068–2074
103. Mills AM, Ruck M (2005) $Ce_{53}Fe_{12}S_{90}X_3$ (X = Cl, Br, I): The first rare-earth transition-metal sulphide halides. Inorg Chem 45:5172–5178
104. Colina JZ, Nix RM, Weiss H (2005) Growth, structure, and stability of ceria films on Si (111) and the application of CaF_2 buffer layers. J Phys Chem B 109:10978–10985
105. Gauthier G, Jobic S, Evain M, Koo HJ, Whangbo MH, Fouassier C, Brec R (2003) Synthesis, structures and optical properties of yellow Ce_2SiS_5, $Ce_6Si_4S_{17}$, and $Ce_4Si_3S_{12}$ materials. Chem Mater 15:828–837
106. Bernot K, Bogani L, Caneschi A, Gatteschi D, Sessoli R (2006) A family of rare-earth based single chain magnets: playing with anisotropy. J Am Chem Soc 128:7947–7956
107. Lin-hai Y, Miao S, Wang-Liang Z, Zhu-De X (2001) Photocatalytic activity of lanthanum doping TiO_2. J Zhejiang University (Sci) 2(3):271–274
108. O'Connell M, Norman AK, Huttermann CF, Morris MA (1999) Catalytic oxidation over lanthanum-transition metal perovskite materials. Catal Today 47:123–132
109. Sakatani Y, Nunoshige J, Ando H, Okusako K, Koike H, Takata T, Kondo JN, Hara M, Domen K (2003) Photocatalytic decomposition of acetaldhyde under visible light irradiation over La^{3+} and N Co-doped TiO_2. Chem Lett 32(12):1156–1157
110. Xiao Q, Si Z, Zhang J, Xiao C, Yu Z, Qiu G (2007) Effects of samarium dopant on photocatalytic activity of TiO_2 nanocrystallite for methylene blue degradation. J Mater Sci 42:9194–9199
111. Page RH, Schaffers KI, Payne SA, Krupke WF (1997) Dy-doped chlorides as gain media for 1.3 μm telecommunications amplifiers. J Lightwave Technol 15(5):786–793
112. Behrendt DR, Legvold S, Spedding FHC (1958) Magnetic properties of dysprosium single crystals. Phys Rev 109:1544–1547

113. Qi M-H, Liu G-F (2004) Synthesis and photoelectronic properties on a series of lanthanide dysprosium(III) complexes with acetylacetonate and meso-tetraalkyltetrabenzoporphyrin. Solid State Sci 6(3):287–294
114. Perkas N, Roffer H, Vradman L, Landau MV, Gedanken A (2006) Sonochemically prepared Pt/CeO_2 and its application as a catalyst in ethyl acetate combustion. Langmuir 22:7072–7077
115. Fu Q, Deng W, Saltsburg H, Flytzani-Stephanopoulos M (2005) Activity and stability of low-content gold-cerium oxide catalysts for the water-gas shift reaction. Appl Catal B 56:57–68
116. Thammachart M, Meeyoo V, Risksomboon T, Osuwan S (2001) Catalytic activity of CeO_2-ZrO_2 mixed oxide catalysts prepared via sol-gel technique: CO oxidation. Catal Today 68 (1):53–61
117. Scire S, Minico S, Crisafulli C, Satriano C, Pistone (2003) Catalytic combustion of volatile organic compounds on gold/cerium oxide catalysts. Appl Catal B Environ 40(1–8):43–49
118. Zheng X-C, Wu S-H, Wang S-P, Wang S-R, Zheng S-M, Huang W-P (2005) The preparation and catalytic behavior of copper–cerium oxide catalysts for low-temperature carbon monoxide oxidation. Appl Catal A 283(1–2):217–223
119. Park PW, Ledford JS (1998) The influence of surface structure on the catalytic activity of cerium promoted copper oxide catalysts on alumina: oxidation of carbon monoxide and methane. Catal Lett 50(1–2):41–48
120. Thevenin PO, Pettersson AALJ, Jaras SG, Fierro JLG (2003) Catalytic combustion of methane over cerium-doped palladium catalysts. J Catal 215:78–86
121. Zarraga-Colina J, Nix RM, Weiss H (2005) Growth, structure, and stability of ceria films on Si(111) and the application of CaF_2 buffer layers. J Phys Chem B 109(21):10978–10985
122. Gauthier G, Jobic S, Evain M, Koo H-J, Whangbo M-H, Fouassier C, Brec R (2003) Syntheses, structures, and optical properties of yellow Ce_2SiS_5, $Ce_6Si_4S_{17}$, and $Ce_4Si_3S_{12}$ materials. Chem Mater 15(4):828–837
123. Pankaj, Verma M (2009) Sonophotocatalytic behavior of cerium doped salts of Cu(II), Co(II) and Mn(II) in the degradation of phenol. Ind J Chem 48A:367–371
124. Juengsuwattananon K, Jaroenworaluck A, Panyathanmaporn T, Jinawath S, Supothina S (2007) Effect of water and hydrolysis catalyst on the crystal structure of nanocrystalline TiO_2 powders prepared by sol-gel method. Physica status solidi a 204(6):1751–1756
125. Ali Z, Zuhri A (1984) Spectrophotometric studies and analytical application of Ce(III) chelates with1-(2-Pyridylazo)-2-naphthol (PAN). Monatsh Chem 115(1):57–58
126. Franson MAH (1985) Standard methods for the examination of water and wastewater. American Public Health Association, Washington DC, pp 560–561

Chapter 12
Sonophotocatalytic Degradation of Amines in Water

Mayank Verma and Pankaj

Abstract Hazardous effects of various amines, produced in the environment from the partial degradation of azo dyes and amino acids, adversely affect the quality of human life through water, soil and air pollution and therefore needed to be degraded. A number of such studies are already available in the literature, with or without the use of ultrasound, which have been summarized briefly. The sonochemical degradation of amines and in the combination with a photocatalyst, TiO_2 has also been discussed. Similar such degradation studies for ethylamine (EA), aniline (A), diphenylamine (DPA) and naphthylamine (NA) in the presence of ultrasound, TiO_2 and rare earths (REs); La, Pr, Nd, Sm and Gd, in aqueous solutions at 20 kHz and 250 W power have been carried out and reported, to examine the combinatorial efficacy of ultrasound in the presence of a photocatalyst and rare earth ions with reactive f-electrons.

12.1 Introduction

The degradation of amines is a matter of great concern. These have toxic and hazardous effects to human and aquatic life, but are required to produce thousands of dyes with more than 6,00,000 tons of which are released into the environment every year. Azo dyes constitute more than 80% of all dyes [1]. Degradation of dyes with a single treatment technology is less efficient besides, may also lead to the generation of carcinogenic and mutagenic fragments of amines upon aerobic treatment. Hence their proper degradation and mineralization is needed to protect the environment.

Pankaj (✉)
Department of Chemistry, Faculty of Science, Dayalbagh Educational Institute, Agra 282 110 Uttar Pradesh, India
e-mail: pankaj2@sancharnet.in; executive_editor_jicc@yahoo.com

Pankaj and M. Ashokkumar (eds.), *Theoretical and Experimental Sonochemistry Involving Inorganic Systems*, DOI 10.1007/978-90-481-3887-6_12,

Amines are valuable synthetic tool. Aliphatic amines are used in the manufacturing of fatty acids, drug [2], antifreeze solutions [3], lubricating oils [4], cosmetics and toiletries industries [5], whereas aromatic amines are used in different industries e.g. dye, plastic, paint, pigments, herbicides, pharmaceutical preparation and rubber [6–11], for the generation of gold thread [12], carbazoles [13], synthesis of microemulsions [14], high yielding urea [15], amine-stabilized iron oxide nanoparticles [16], cyclometallated Ru(II), Rh(III) and Ir(III) catalyst with high chemo- and regio-selectivity [17], α-aminonitriles intermediates [18], polysubstituted pyrroles [19], N- substituted hydrazines [20] and disubstituted oxamidines [21]. Amines are also used for direct aldol addition [22], in removal of acid gases like CO_2 and H_2S from natural gases [23], in modification of water repellent properties of paints [24], in decarboxylation of oxaloacetic acid [25], in activation of high-Tc ferromagnetism in Mn^{2+} doped ZnO [26], in degradation of peptides [27], and in intercalation [28]. Similarly the heterocyclic amines are used in corrosion inhibitors, rubber additives, solvents, chemical intermediates for the synthesis of optical brightners, drugs, herbicides, pesticides, and photographic developers [29]. Intercalated compounds of non-aromatic heterocyclic amines e.g. piperazine, piperidine and morpholine into the layer region of zirconium glycine-N,N-dimethylphosphonate are extensively used in catalysis, photochemistry, medicine and for environmental protection purposes [28]. The oxidative cyanation of tertiary amines by Ru catalyst with molecular O_2 or H_2O_2 and NaCN [18] has been used to synthesize a range of nitrogen compounds and photoreaction of aromatic tertiary amines with 60-fullerene [30]. Sulfonated amines, soluble in water due to presence of sulfonated groups, find their place in river and surface water as O'Neill et al. [31] could fix only 50–90% of total sulfonated reactive dyes.

Besides, the industrial production, all over the world, a variety of biological activities also produces amines by the breakdown of amino acids. Therefore both the biogenic and anthropogenic amines enter into natural water through domestic as well as industrial effluents. The aromatic amines formed during the degradation of pesticides [32,33], nitroaromatics [34,35] and azo dyes [36,37] find their place in aquifers and in soil respectively due to polar nature of aromatic amines [38] and sorption efficiency of amino group with solid particles through covalent bonding [39]. The presence of monomethylamine (MA), dimethylamine (DMA) and diethylamine (DEA) in significant concentrations at the Cape Verde Atmospheric Observatory on the island Sao Vicnte [40] and nitrosamine in municipal sludge [41] have been reported. Thermal degradation of polyurethane (PUR) in car repair shops produces amines during welding [42]. Use of plastic utensils made up of nylon (polyamide) produces primary aromatic amines [43]. Lower alkyl amines are soluble in water although the solubility decreases with increase in number of carbon atom in molecule [44].

Cytotoxicity of methylamine [45], kidney liver and myocardial damage due to ethylamine [46], hepatosplenomegaly and eosinophilia due to aniline [47], euphoria, dyspnea, teratogenicity, renal failure, hematuria, proteinurea, anorexia and methanoglobinemia due to α-naphthylamine and diphenylamine have been reported in the literature [48–54]. Therefore the remediation and mineralization of amines is imperative to sustain a pollution free environment. Since amines can be categorized into different classes e.g. aliphatic amines, aryl amines, heterocyclic amines, inorganic amines, hydrazines and hydroxylamines on the basis of presence of different functional groups, these require different single method or a combination of such degradation methods for their total degradation and mineralisation.

12.2 Remediation Methods

A number of methods for the degradation of various amines, involving techniques both with and without ultrasound are reported in the literature. Some of these important degradation processes, which do not involve ultrasound are redox [55–64] catalytic [65], pyrolysis [66], photochemical [67–73], photocatalytic [74–81], electrochemical [82–87], oxidative [88–93], microbial [94–108], biological activated carbon systems [109] and nanostructured and amorphous metals, alloys, used as catalysts [110–116]. The use of TiO_2, both in the presence and absence of ultrasound have been found to be an efficient method [117]. Besides, the sonocatalytic degradation in presence of various dopants such as transition metal and rare earths have also been found to be very effective in the removal of amines. The doped and undoped TiO_2 catalysis can be explained by three mechanisms namely, sonoluminescence, hot-spot and oxygen atom escape [117]. The photocatalytic activity sturdily depends on the better adsorption of organic substrate and the improvement of the interfacial charge transfer reactions [118,119]. Pre-adsorption of the donor and / or acceptor and its amount, before the photocatalytic reaction, is necessary for interfacial charge transfer. The factors leading to the enhanced adsorption capacity should involve the change of the physical or chemical properties of the catalysts owing to doping. The transitions of 4f electrons of rare earth lead to the enforcement of the adsorption of catalysts and favour the separation of photo-generated electron hole pairs. In addition, the red shift of the adsorption edge of TiO_2 by rare earth ion doping is also helpful to the improvement of visible light photocatalytic activity of TiO_2. Therefore, the transitions of 4f electrons of rare earth and red shifts of the optical adsorption edge of TiO_2 by rare earth ion doping enhances visible light photocatalytic activity of TiO_2 [120]. Similar sonochemical methods, although not numerous, have also been successfully employed to degrade those azo dyes which are resistant to biodegradation. The work reported in the literature, using sonochemical methods is, therefore, being described briefly, before discussing the work carried out in this laboratory.

For the sonochemical mineralization of reactive dye CI Reactive Black 5 with 20, 279 and 817 kHz irradiation, the discoloration and radical formation both are directly dependent upon ultrasonic frequency, acoustic power and irradiation time and indirectly on the number of free radicals thus generated, as their suppression decreased the discoloration rate due to radical scavenging effect. Although ultrasound alone is capable of decolorizing Reactive Black 5 but inefficient in mineralization as only 50% degradation was observed after 6 h of ultrasonic irradiation [121]. The sonochemical discoloration of 40 μM of CI Acid Orange 7 and CI Acid Orange 8 in their hydrazone form in dye solution is through the cleavage of either N-N bond in nitrosoaryl intermediates or the C-N bond with the generation of benzene [122].

Ultrasound has been found to be a potent method in enhancing the photocatalytic removal of various organic pollutants [123,124] but the utilization of high frequency ultrasonic equipments is costlier than those of lower frequencies, e.g. 20 and 40 kHz, available commercially. For the sonocatalyzed photodegradation of orange II (Org II), ethyl orange (EO) and Acid Red G (ARG) using Au-loaded TiO_2 with low-frequency ultrasonic transducer (40 kHz), the rate of both discoloration and total organic (TOC) removal of azo dyes by Au/TiO_2 is more effective compared to TiO_2 alone. In the argon/air atmosphere both OH• referred oxidation and H• mediated reduction played important role in the decomposition and mineralization of non-volatile and hydrophilic azo dyes due to higher hydrogen production by Au/TiO_2 [125]. The discoloration rate of arsenazo I solutions under ultrasound is not as effective as when applied in combination, the sono-ozonation treatment [1]. Both ozonation and sono-ozonation treatment followed the same destruction mechanism through the breakdown of –N=N- bonds, the conversion of benzene ring to carboxylic acid, and $–HSO_3$ bonds to H_2SO_4. Mineralization of some industrial azo dyes namely, Acid orange 5 & 52, Direct Blue 71, Reactive Black 5 and Reactive orange 16 & 107 to non-toxic end products was found to be complete within 3–15 h at 90 W and 1–4 h at 120 W respectively. The mass spectra confirmed the hydroxyl radical attack on azo dyes by simultaneous azo bond scission, oxidation of nitrogen atoms and hydroxylation of aromatic ring structures [126]. However, in the sono-degradation of C-I Reactive red 22 and methyl orange, using 200 kHz and 1.25 W/cm^2 ultrasound, in homogeneous aqueous solution, the effective decolouration decreases only in presence of radical scavenger, t-butyl alcohol [127].

Rare earth (RE) ions particularly La^{3+}, Nd^{3+}, Sm^{3+}, Eu^{3+}, Gd^{3+} and Yb^{3+} are known for their ability to form complexes with various Lewis bases e.g. acids, amines, aldehydes, alcohols, thiols, etc. through the f-orbital. Hence these can carry organic pollutants to the active surface of TiO_2 and thus facilitate degradation. The improvement in the photocatalytic reactions due to rare earth doping in TiO_2 is also perhaps due to the increase in the stability of anatase phase and prevention of the segregation of TiO_2 [128]. Thus, incorporation of RE ions into a TiO_2 matrix could provide a means to concentrate organic substrate at the semiconductor surface. [129,130] and consequently enhance the photoactivity of titania [131]. From the literature review for the sonochemical degradation of amines and azo dyes it could be seen that although there are some studies of azo dyes but the degradation of amine, which is the basic constituents of azo dye, has been completely overlooked. Therefore we investigated the sono-photo-degradation of four structurally different amines such as; ethylamine (EA), aniline (A), diphenylamine (DPA) and 1-naphthylamine (NA) in the presence of ultrasound, TiO_2 and several rare earths. Although the combination of ultrasound and TiO_2 has already been found to be quite effective [132], but how the presence of rare earths in the same system further affects the degradation process is another aspect of the understanding the mechanism involved.

In our experiments, the rare earths, REs, were taken not as dopants to the TiO_2 particles but as co-adds in the aqueous solutions containing organic pollutants. This avoided the possible alteration in the anatase phase of TiO_2 leading sometimes to the collapse of lattice structure, when the concentration of dopants increases exorbitantly. Before the experiments, a known amount of the corresponding rare earth oxide, RE_2O_3, was dissolved in the minimum amount of conc. HNO_3 and diluted to a known concentration before being added to the experimental solutions. This helped in making rare earth ions available to TiO_2 particles either just on its surface or close enough to participate in photocatalytic activity without upsetting the lattice patterns and holes in TiO_2. The degradation increased in almost all the cases of such experiments.

Although the role of rare earth ions on the surface of TiO_2 or close to them is important from the point of electron exchange, still more important is the number of f-electrons present in the valence shell of a particular rare earth. As in case of transition metal doped semiconductor catalysts, which produce n-type WO_3 semiconductor [133] or p-type NiO semiconductor [134] catalysts and affect the overall kinetics of the reaction, the rare earth ions with just less than half filled (f^{5-6}) shell produce p-type semiconductor catalysts and with slightly more than half filled electronic configuration (f^{8-10}) would act as n-type of semiconductor catalyst. Since the half filled (f^7) state is most stable, ions with f^{5-6} electrons would accept electrons from the surface of TiO_2 and get reduced and rare earth ions with f^{8-9} electrons would tend to lose electrons to go to stabler electronic configuration of f^7. The tendency of rare earths with f^{1-3} electrons would be to lose electrons and thus behave as n-type of semiconductor catalyst to attain completely vacant f^0- shell state [135]. The valence electrons of rare earths are rather embedded deep into their inner shells (n-2), hence not available easily for chemical reactions, but the cavitational energy of ultrasound activates them to participate in the chemical reactions, therefore some of the unknown oxidation states (as Dy^{+4}) may also be seen [136,137].

The rare earths, REs, chosen for the study were La = [Xe] $5d^1$ $6s^2$, Pr = [Xe] $4f^3$ $6s^2$, Nd = [Xe] $4f^4$ $6s^2$, Sm = [Xe] $4f^6$ $6s^2$ and Gd, [Xe] $4f^7$ $5d^1$ $6s^2$, whereas the TiO_2 particles, both in anatase as well as rutile phase, were produced *in situ* from the hydrolysis of $(n\text{-}BuO)_4Ti$ [138]. 0.02 mL of $(n\text{-}BuO)_4Ti$ was used for each experimental solution, wherever, TiO_2 was needed. Similarly, 10 mg of RE_2O_3 (RE= La, Pr, Nd, Sm & Gd) was dissolved in the minimum volume of conc. HNO_3 and made up to 10.0 mL with deionized water to prepare a standard solution for influencing further the sono-photo catalyzed degradation of amines.

It would not be out of place to discuss the significance of using $(n\text{-}BuO)_4Ti$ as a source of TiO_2, since the solid TiO_2 is insoluble in water as well as all mineral acids, therefore, a fixed and known amount of TiO_2 catalyst cannot be added to every experimental solution. However, $(n\text{-}BuO)_4Ti$ is a liquid at room temperature (Liquid range: – 55°C to 312°C) and a fixed volume of this reagent, which hydrolyses to TiO_2 immediately [138,139] can be added and the different sets of experiments can be carried out with known concentration of TiO_2 nanoparticles in the aqueous phase. Besides, the n-butyl radicals produced in the solution as a result of hydrolysis of $(n\text{-}BuO)_4Ti$ create tunnels due to their lyophobic nature.

This facilitates the flow of degradable species through these tunnels onto the surface of the TiO_2 where electron could be donated to the holes of the anatase phase and the photocatalytic action in combination with the cavitational effect of the ultrasound can accelerate the fragmentations of pollutants. The details of this mechanism are however discussed at the end of chapter. Ultrasound also breaks TiO_2 particles to still smaller size and increases the active surface area manifold.

The experimental conditions for the degradation for all the four amines with different rare earths, REs, were as under;

1. Degradation only under ultrasound
2. Degradation in presence of ultrasound and $(n\text{-}BuO)_4Ti$
3. Degradation in the presence of ultrasound and RE^{3+} ions
4. Degradation in the presence of ultrasound, $(n\text{-}BuO)_4Ti$ and RE^{3+}

Ethyl amine was estimated by the method of Citron et al. [140], aniline by the method of Norwitz [141], diphenylamine by the method of DeAtley [142] and naphthyl amine by the method of Pandey et al. [143].

12.3 Degradation of Amines

12.3.1 Ethyl Amine (EA)

The degradation of ethylamine (36 ppm) was carried out in different experimental conditions for 1 h through an interval of 15 min. The percentage degradation is given in Table 12.1 and shown graphically in Fig. 12.1.

After 1 h of sonication with ultrasound, the degradation of EA is more effective than in combination with TiO_2 but improves drastically when rare earths are added. In spite of higher initial degradation of EA in the presence of TiO_2 (approximately nine times) the degradation does not accelerate at a rate with which it increases in case of ultrasound alone. This is quite understandable. All photocatalytic reactions are almost instantaneous whereas the sonochemical reactions depend upon the generation of free radicals and reactive species such as; OH•, H_2O_2, O_3, etc. to accelerate the reaction. In sonicated solutions, the concentration of these species increase with time, which is not the case with photochemical degradation.

Table 12.1 Percentage degradation of ethylamine (EA)

Time (mins)	Percentage degradation of ethylamine under different experimental conditions											
	Ultrasound)))	TiO_2 +)))	La +)))	La + TiO_2 +)))	Pr +)))	Pr + TiO_2 +)))	Nd +)))	Nd + TiO_2 +)))	Sm +)))	Sm + TiO_2 +)))	Gd +)))	Gd + TiO_2 +)))
15	3.5	36.0	57.1	81.4	53.4	68.7	81.0	58.0	45.8	77.1	66.4	71.9
30	34.7	40.6	69.1	87.0	72.3	71.8	92.0	63.0	57.2	84.6	70.8	78.3
45	58.0	47.5	89.1	97.3	77.0	72.4	93.0	80.0	62.1	87.5	74.9	85.0
60	69.6	54.0	97.7	98.9	78.2	73.8	94.0	85.0	65.7	90.4	82.6	91.2

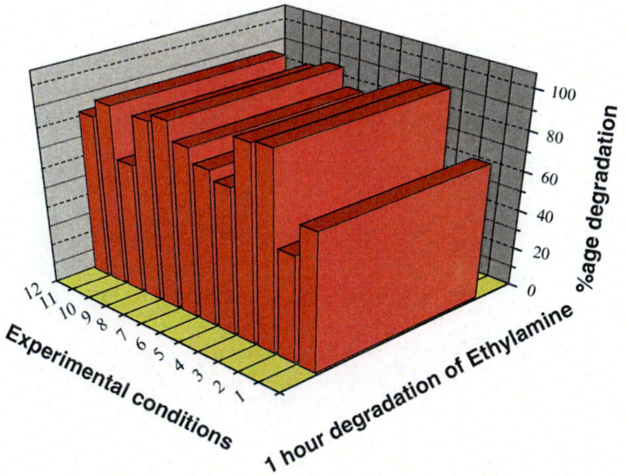

Fig. 12.1 Degradation of ethylamine (1 h) under different experimental conditions and in presence of different rare earths. [1. Ultrasound; 2. $(n\text{-BuO})_4Ti$ (ultrasound); 3. $(n\text{-BuO})_4Ti$ + La (ultrasound); 4. La (ultrasound); 5. $(n\text{-BuO})_4Ti$ + Pr (ultrasound); 6. Pr (ultrasound); 7. $(n\text{-BuO})_4Ti$ + Nd (ultrasound); 8. Nd (ultrasound); 9. $(n\text{-BuO})_4Ti$ + Sm (ultrasound); 10. Sm (ultrasound); 11. $n\text{-BuO})_4Ti$ + Gd (ultrasound); 12. Gd (ultrasound)]

Therefore, higher degradation of EA under ultrasound was possible only after some gestation period of sonication, when sufficient active species were produced in the aqueous solution.

The effect of different rare earths with ultrasound was additive compared to ultrasound alone. For the degradation of EA, the efficiencies of different rare earths by the ultrasound alone has been found to be in the order; Sm + US < US < Pr + US < Gd + US < Nd + US < La + US, whereas the order of efficiency for the same altered in the presence of ultrasound and TiO_2 as; US + TiO_2 < Pr + US + TiO_2 < Nd + US + TiO_2 < Sm + US + TiO_2 $\approx$ Gd + US + TiO_2 < La + US + TiO_2 (Fig. 12.2). The maximum degradation was, however, when the combination La + TiO_2 + US was tried, while minimum was found due to TiO_2 + US. The enhanced degradation of EA was found, when La, Sm and Gd were added one by one in the presence of TiO_2, which however, decreased in the solution containing TiO_2 and Pr or Nd (Fig. 12.2).

Taken as a whole, the degradation in all conditions followed the following sequence; TiO_2 + US < Sm + US < US < Pr + TiO_2 + US < Pr + US < Gd + US < Nd + TiO_2 + US < Sm + TiO_2 + US < Gd + TiO_2 + US < Nd + US < La + US < La + TiO_2 + US

12.3.2 Aniline (A)

The percentage degradation of aniline (4.0 ppm), in various conditions has been given in Table 12.2 and shown graphically as the final degradation after 1 h in

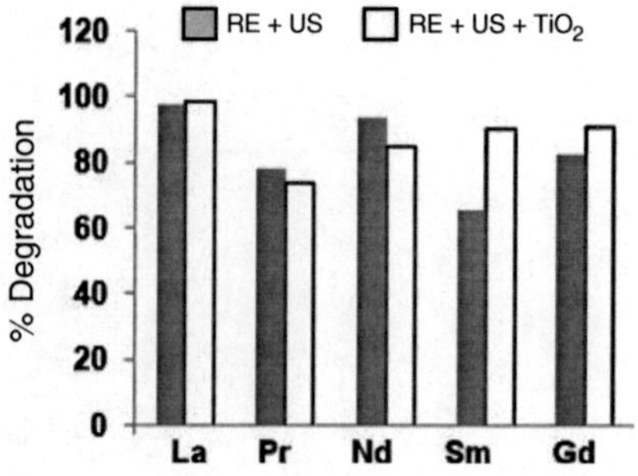

Fig. 12.2 Degradation of ethylamine with different rare earths

Table 12.2 Percentage degradation of aniline (A)

Time (mins)	Percentage degradation of aniline under different experimental conditions											
	Ultrasound)))	TiO_2 +)))	La +)))	La + TiO_2 +)))	Pr +)))	Pr + TiO_2 +)))	Nd +)))	Nd + TiO_2^+ +)))	Sm +)))	Sm + TiO_2 +)))	Gd +)))	Gd + TiO_2 +)))
15	9.8	12.1	4.9	6.7	3.9	4.9	7.0	3.0	6.2	6.6	5.2	11.0
30	16.0	36.2	5.7	9.8	4.9	5.7	15.0	6.0	7.9	8.7	8.9	13.0
45	19.5	37.9	8.8	10.2	5.2	8.8	29.0	14.0	11.8	10.1	10.7	14.7
60	20.8	48.9	10.1	17.8	11.2	10.1	30.0	25.0	12.0	11.9	11.4	18.8

Fig. 12.3. Degradation due to ultrasound alone was more (20.8%) compared to when carried out in the combination with REs (10.1–12.0%), except for Nd (30%), and maximum in combination with ultrasound and TiO_2 (48.9%). A comparison of the percentage degradation of aniline in the presence of different rare earths and ultrasound and with the further addition of TiO_2 can be had from the Fig. 12.4.

12.3.3 Diphenylamine (DPA) and Naphthyl Amine (NA)

The percentage degradation of diphenylamine (4 ppm) and 1-naphthylamine (20 ppm) has been given in Tables 12.3 and 12.4 and shown graphically as the final degradation after 1 h in Figs. 12.5 and 12.6 respectively. Like aniline, both diphenylamine and 1-naphthylamine, were degraded better with ultrasound in combination with TiO_2 compared to ultrasound alone. The relative degradation of diphenylamine in the presence of different REs with ultrasound has been found as additive (Fig. 12.7)

The degradation of DPA due to different REs and ultrasound has been found in the order e.g. US < Pr + US < La + US ~ Nd + US < Sm + US < Gd + US. However, the order of the degradation in presence of REs, ultrasound and TiO_2 has been found as; Pr + US + TiO_2 < Nd + US + TiO_2 < US + TiO_2 + US + TiO_2 < La + US + TiO_2 < Sm + US + TiO_2 < Gd + US + TiO_2 (Fig. 12.7). The highest

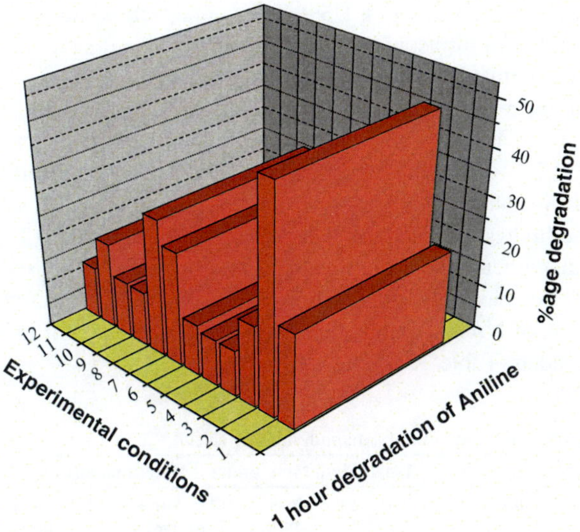

Fig. 12.3 Degradation of aniline (1 h) under different experimental conditions and in presence of different rare earths. [1. ultrasound; 2. (n-BuO)$_4$Ti (ultrasound); 3. (n-BuO)$_4$Ti + La (ultrasound); 4. La (ultrasound); 5. (n-BuO)$_4$Ti + Pr (ultrasound); 6. Pr (ultrasound); 7. (n-BuO)$_4$Ti + Nd (ultrasound); 8. Nd (ultrasound); 9. (n-BuO)$_4$Ti + Sm (ultrasound); 10. Sm (ultrasound); 11. n-BuO)$_4$Ti + Gd (ultrasound); 12. Gd (ultrasound)]

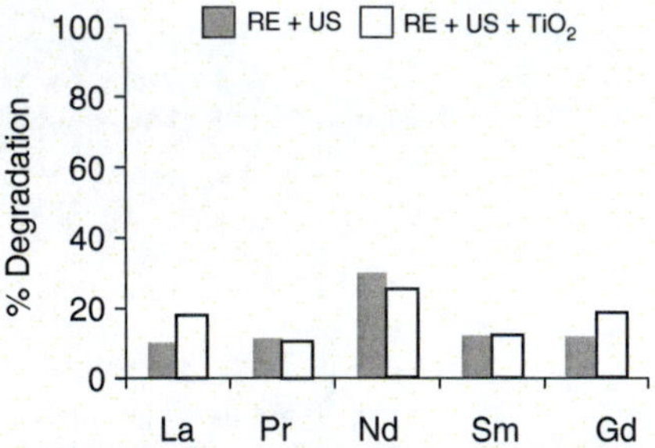

Fig. 12.4 Degradation of aniline with different rare earths

Table 12.3 Percentage degradation of diphenylamine (DPA)

Time (mins)	Percentage degradation of diphenylamine under different experimental conditions											
	Ultrasound)))	TiO_2 +)))	La +)))	La + TiO_2 +)))	Pr +)))	Pr + TiO_2 +)))	Nd +)))	Nd + TiO_2 +)))	Sm +)))	Sm + TiO_2 +)))	Gd +)))	Gd + TiO_2 +)))
15	26.8	16.4	39.1	42.9	35.0	43.7	47.0	51.0	47.8	34.2	57.7	61.3
30	30.3	29.1	40.6	45.1	35.5	45.9	52.0	53.0	52.9	40.2	60.5	62.4
45	33.2	46.4	45.7	49.1	36.6	47.3	53.0	55.0	59.6	55.6	64.5	70.4
60	34.2	56.9	53.9	59.7	42.5	49.0	54.0	56.0	61.7	65.0	67.8	75.1

degradation was observed in Gd + US + TiO_2 condition while the minimum by ultrasound alone. Under ultrasonic condition an enhanced effect of TiO_2 was found in case of all rare earths, La, Pr, Nd, Sm and Gd (Fig. 12.7).

The degradation of NA, however, in presence of ultrasound and/or TiO_2 with different rare earths has been found in the order; US < Pr + US < Sm + US < Gd + US < Nd + US < La + US while with ultrasound and TiO_2 was different (Fig. 12.8) as Pr + US + TiO_2 < Nd + US + TiO_2 ~ La + US + TiO_2 < Sm + US + TiO_2 < Gd + US + TiO_2. The highest degradation was observed in case of US+TiO_2 condition while minimum with ultrasound alone. The addition of TiO_2, exception in the presence of Gd, decreased the degradation. The role of RE, therefore, was better in the degradation of NA in combination with ultrasound and still improved when TiO_2 was also added (Table 12.3).

Table 12.4 Percentage degradation of 1-naphthylamine (NA)

Time (mins)	Percentage degradation of 1-naphthylamine under different experimental conditions											
	Ultrasound)))	TiO_2 +)))	La +)))	La + TiO_2 +)))	Pr +)))	Pr + TiO_2 +)))	Nd +)))	Nd + TiO_2 +)))	Sm +)))	Sm + TiO_2 +)))	Gd +)))	Gd + TiO_2 +)))
15	19.3	36.5	34.4	21.7	23.8	9.1	25.0	22.0	40.7	44.4	50.4	53.0
30	20.0	63.3	42.8	48.5	26.3	19.6	40.0	23.0	44.8	47.5	55.4	56.0
45	21.9	66.4	46.3	51.5	33.7	23.0	54.0	53.0	50.5	63.4	56.0	63.1
60	25.7	75.7	68.2	58.2	37.6	28.2	65.0	57.0	52.0	65.9	60.0	66.1

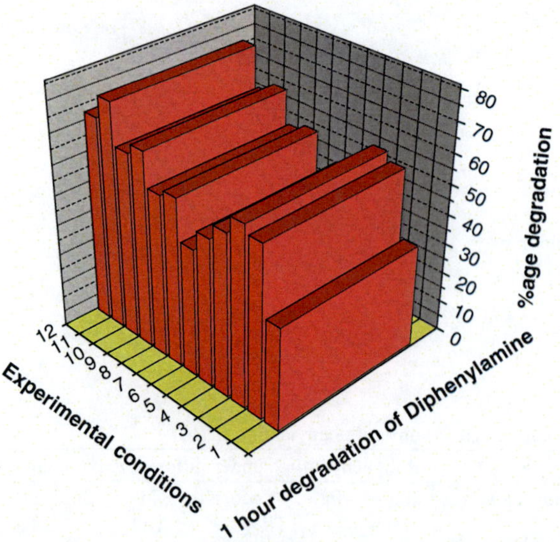

Fig. 12.5 Diphenylamine Figures 12.5 and 12.6. Degradation of diphenylamine and naphthylamine (1 h) under different experimental conditions and in presence of different rare earths [1. ultrasound; 2. $(n\text{-}BuO)_4Ti$ (ultrasound); 3. $(n\text{-}BuO)_4Ti$ + La (ultrasound); 4. La (ultrasound); 5. $(n\text{-}BuO)_4Ti$ + Pr (ultrasound); 6. Pr (ultrasound); 7. $(n\text{-}BuO)_4Ti$ + Nd (ultrasound); 8. Nd (ultrasound); 9. $(n\text{-}BuO)_4Ti$ + Sm (ultrasound); 10. Sm (ultrasound); 11. $n\text{-}BuO)_4Ti$ + Gd (ultrasound); 12. Gd (ultrasound)]

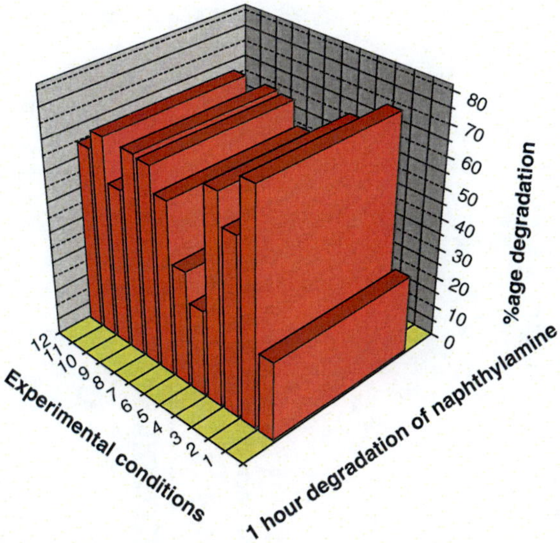

Fig. 12.6 Naphthylamine Figures 12.5 and 12.6. Degradation of diphenylamine and naphthylamine (1 h) under different experimental conditions and in presence of different rare earths [1. ultrasound; 2. (n-BuO)$_4$Ti (ultrasound); 3. (n-BuO)$_4$Ti + La (ultrasound); 4. La (ultrasound); 5. (n-BuO)$_4$Ti + Pr (ultrasound); 6. Pr (ultrasound); 7. (n-BuO)$_4$Ti + Nd (ultrasound); 8. Nd (ultrasound); 9. (n-BuO)$_4$Ti + Sm (ultrasound); 10. Sm (ultrasound); 11. n-BuO)$_4$Ti + Gd (ultrasound); 12. Gd (ultrasound)]

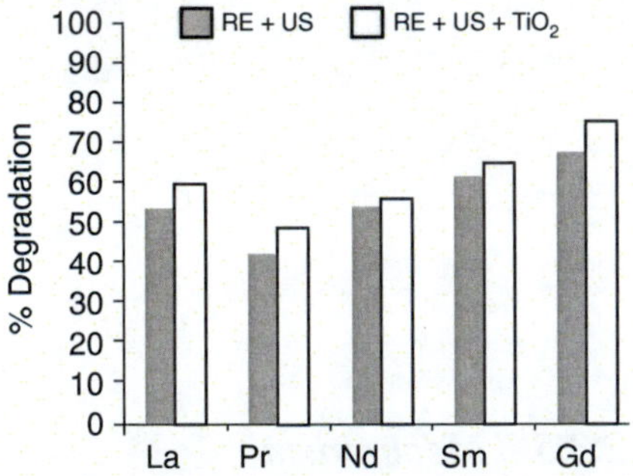

Fig. 12.7 Degradation of diphenylamine with different rare earths

12.3.4 Effect of La, Pr, Nd, Sm and Gd ions

The relative degradations of EA, A, NA and DPA, in 1 h of sonication and in the presence of REs (La, Pr, Nd, Sm and Gd) have been shown through Fig. 12.9–12.13 respectively. The degradation of EA was maximum, whereas minimum for A. Therefore the EA is the softest target for its degradation by all REs, whereas aniline has been a very hard to degrade under all experimental conditions (Tables 12.1–12.4). The overall order of degradation has been found to be EA > NA ~ DPA > A.

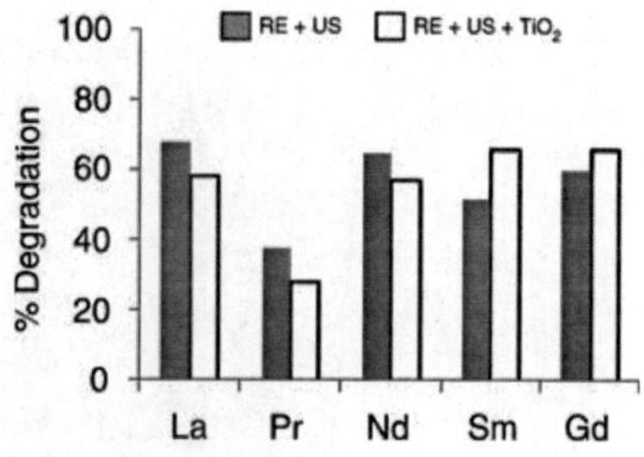

Fig. 12.8 Degradation of naphthylamine with different rare earths

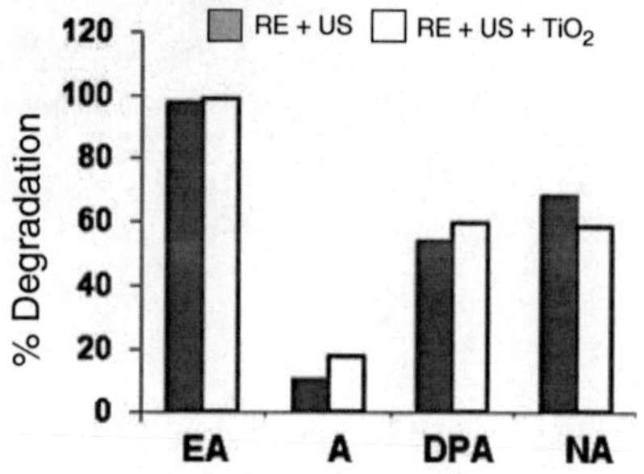

Fig. 12.9 Degradation of different amines due to La^{3+} ion

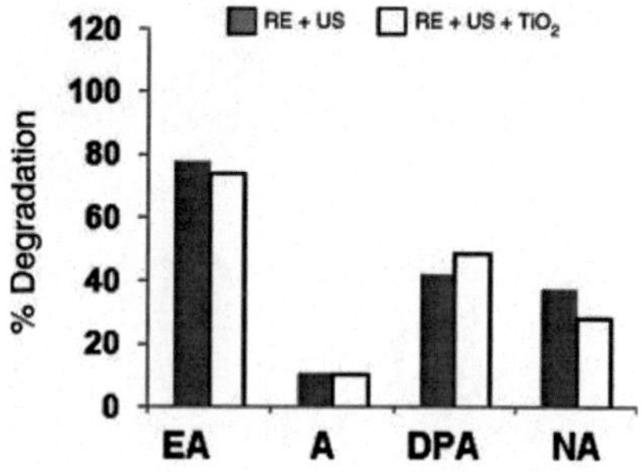

Fig. 12.10 Degradation of different amines due to Pr^{3+} ion

12.3.5 Mechanism

To understand the mechanism of degradation of amines with variables such as ultrasound, TiO_2, REs etc. and their combination, we need to discuss the delocalization of electrons in these molecules. EA does not show any delocalization of electrons on the nitrogen atom of $-NH_2$ group and is, therefore, different from the rest of the amines studied, whereas, the other three amines exhibit delocalization of the lone pair of electrons, available on their nitrogen atoms to different extents. Both DPA and NA have two benzene rings with the only difference that, in case of NA, these are fused [144] whereas in case of DPA, two independent benzene rings are connected through NH link [145], therefore the delocalization of electrons

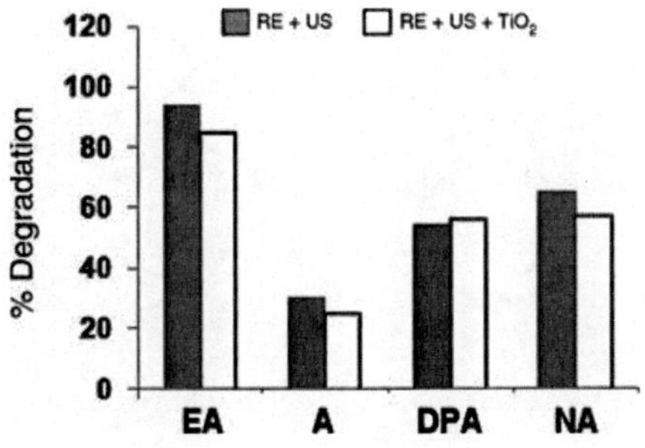

Fig. 12.11 Degradation of different amines due to Nd^{3+} ion

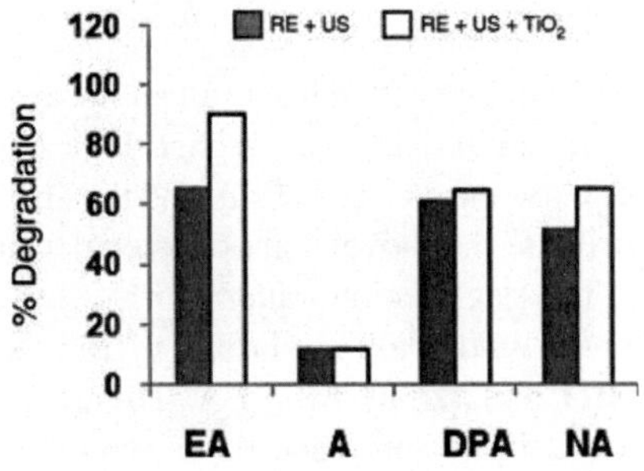

Fig. 12.12 Degradation of different amines due to Sm^{3+} ion

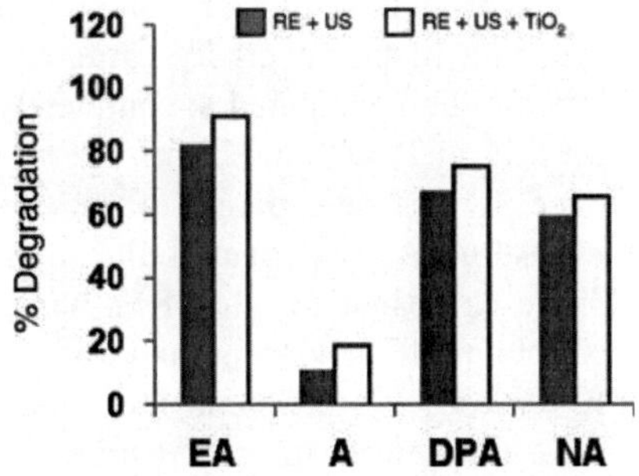

Fig. 12.13 Degradation of different amines due to Gd^{3+} ion

in these two molecules is not similar. The benzene ring in A after accepting lone pair of electrons from the nitrogen atom of NH_2 group delocalizes electron density to three positions (2 ortho and 1 para) only [146], contrary to this the NA and DPA molecules delocalize to five and six different positions respectively [147] (Fig. 12.14). Greater is the delocalization, lower is the electron density. The electron density on A would therefore be much more (almost double) compared to NA and DPA. Since amines interact with polar solvent molecules (water, H_2O) electrostatically, when migrating in the solution, the electrostatic interaction of A would be almost twice than for NA and DPA delocalized molecules. In other words, the movement of A would be hindered more than NA and DPA molecules to reach the TiO_2/RE surface. Higher charge density on the delocalized molecules would obstruct its movement. Since, the electron density is maximum in case of aniline

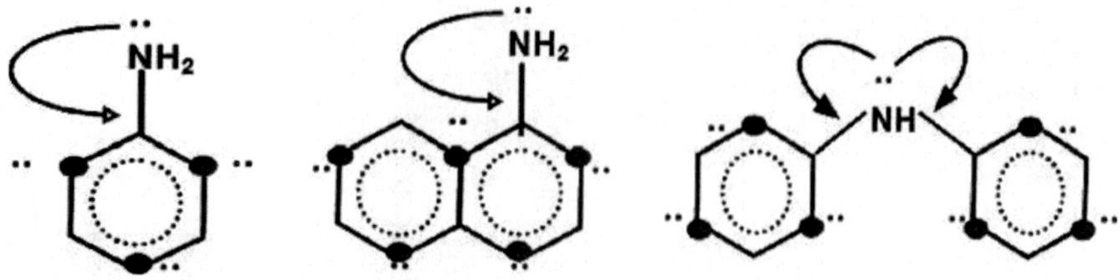

Fig. 12.14 Delocalisation of electrons in aniline, naphthylamine and diphenylamine molecules

molecules and therefore, all molecules would not reach the surface of TiO_2 for degradation. However, higher delocalization of electrons on NA and DPA would decrease the charge density and thus reduce the electrostatic interaction. This would facilitate the movement of molecules onto the surface of TiO_2 for degradation.

The degradation pattern for NA and DPA, involving REs is, however, almost similar under all conditions of the experiment. Nevertheless, these two amines differ remarkably from EA. The degradation of NA and DPA is maximum with titania in the presence of Gd compared to Sm and La and surprisingly exactly opposite to what is found in case of EA, suggesting a different kind of mechanism of degradation to be in operation in systems involving benzene rings. Gd, when co-added with titania, degrades to maximum extent because of its half filled ($4f^7$) electronic shell compared to completely vacant 4f electronic shell in La and partly vacant 4f shell of Sm. Therefore, probably the necessity of holes on the TiO_2 surface, for the degradation of molecules, with benzene rings is not required as much as for EA. The effect of ultrasound and TiO_2 in the presence of gadolinium is additive compared to other REs namely; lanthanum ($4f^0$), praseodymium ($4f^2$), neodymium ($4f^3$) and samarium ($4f^5$), inspite of the identical oxidation states. The only reason for a different photocatalytic behavior of gadolinium may thus be related to its electronic configuration ($4f^7$). This is in agreement to other similar studies which corroborate that the stability of a closed electronic shell makes the hole/electron trapping unfavorable, and metal ions as dopants in TiO_2 with a close-shell electronic configuration had little effect on the photoreactivity of TiO_2 [148,149]. The main interaction between aniline and the surface of the catalyst is, nevertheless, through the amino group, rather than via a π-interaction due to aromatic ring, that constitutes a minor, but relevant, mode of interaction.

The overall process of degradation can be explained through Fig. 12.15, where respective amines move to the surface of the titania, with their different electrical charge densities, spread over the surface of the molecules and get fragmented. n-Bu species create hydrophobic tunnels, facilitating organic pollutants to migrate to TiO_2 surface.

Nevertheless, this is only our observation and explanation for the degradation of four amines, the study of which has been undertaken and a lot of work is still

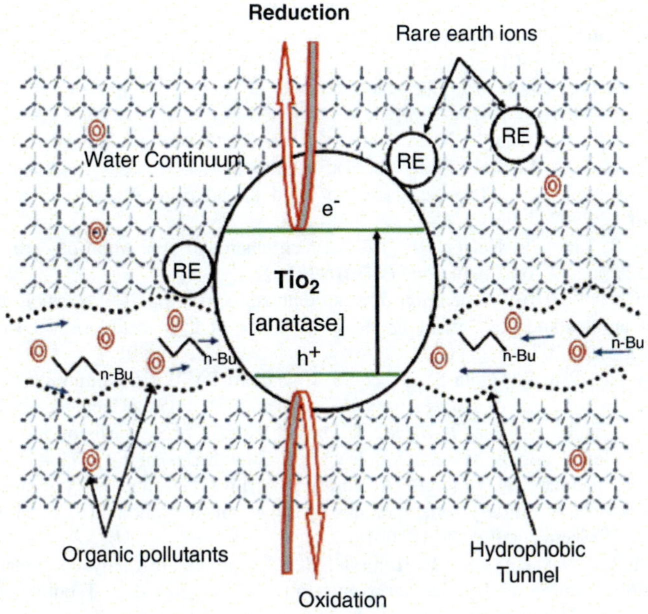

Fig. 12.15 Approach of amine molecules on the holes of TiO_2 through the hydrophobic tunnels made by -O-n-Bu chains.

required to verify and confirm the intermediate steps, suggested by us, so that this mechanism may be confirmed.

Acknowledgements Authors are grateful to Arti Bhatnagar, Shita Tyagi, Manju, and Suchita Saxena for carrying out experiments as reported in their M.Phil. and M.Sc. dissertations.

References

1. Wenrong H, Haiyan PEI (2002) Decomposition characteristics of azo dyes by ozonization with ultrasonic enhancement. Chinese Sci Bull 47(12):986–989
2. Shonle HA (July 12, 1932) U.S. Patent, 1,867,332
3. Calcott W-S, Walker HW (March 1, 1932) U.S. Patent 1,847,711
4. Adam CE (Nov 15, 1932) U.S. Patent, 1,888,023
5. Lawrence SA (2004) (ed.) Amines: Synthesis, properties and applications, Cambridge University Press, Cambridge
6. Drzyzga O (2003) Diphenylamine and derivatives in the environment: A review. Chemosphere 53(8):809–818
7. Molineaux CJ, Batzinger RP, Schmidt W, Bueding E (1980) Mutagenic activation of an antischistosomal drug by enteric Streptococcus sps. in vitro and in vivo. Teratog Carcinog Mutagen 1(2):129–139
8. Datta S, Bhattacharya PK, Verma N (2003) Removal of aniline from aqueous solution in a mixed flow reactor using emulsion liquid membrane. J Membr Sci 2269(1–2):185–201

9. Brillas E, Casado J (2002) Aniline degradation by electro-Fenton and peroxi-coagulation processes using a flow reactor for waste water treatment. Chemosphere 47(3):241–248
10. O'Neill FJ, Bromley-Challenork KCA, Grrenwood RJ, Knapp JS (2000) Bacterial growth on aniline implication for the biotreatment of industrial wastewater. Water Res 34(18):4397–4409
11. Qi-Xun H, Zhuang Yuan Y, Yuan You C, Gu Wen X (2002) Decomposition of aniline in supercritical water. J Hazard Mater 90(14):51–62
12. Kuo P-L, Chen C-C (2006) Generation of gold thread from Au(III) and triethylamine. Langmuir 22:7902–7906
13. Reynolds R, Line LL, Nelson RF (1974) Electrochemical generation of carbazoles from aromatic amines. J Am Chem Soc 96(4):1087–1092
14. Hou A, Chen S (2010) Preparation of microemulsions of the polysiloxanes modified with different amines and their effect on the color shade of dyed cellulose. J Dispersion Sci Technol 31:102–107
15. Didgikar MR, Roy D, Gupta SP, Joshi SS, Chaudhari RV (2010) Immobilized palladium nanoparticles catalyzed oxidative carbonylation of amines. Ind Eng Chem Res 49:1027–1032
16. Aslam M, Schultz EA, Sun T, Meade T, Dravid VP (2007) Synthesis of amine-stabilized aqueous colloidal iron oxide nanoparticles. Cryst Growth Design 7(3):471–475
17. Sortais J-B, Pannetier N, Holuigue A, Barloy L, Sirlin C, Pfeffer M, Kyritsakas N (2007) Cyclometalation of primary benzyl amines by ruthenium(II), rhodium(III), and iridium(III) complexes. Organometallics 26:1856–1867
18. Murahashi S-I, Nakae T, Terai H, Komiya N (2008) Ruthenium-catalyzed oxidative cyanation of tertiary amines with molecular oxygen or hydrogen peroxide and sodium cyanide: sp^3 C-H bond activation and carbon-carbon bond formation. J Am Chem Soc 130:11005–11012
19. Liu W, Jiang H, Huang L (2010) One-pot silver-catalyzed and PIDA-mediated sequential reactions: Synthesis of polysubstituted pyrroles directly from alkynoates and amines. Org Lett 12(2):312–315
20. Diamond LH, Audrieth LF (1955) Preparation of N-substituted hydrazines from amines and chloramine. J Am Chem Soc 77(11):3131
21. Woodburn HM, Morehead BA, Chih CM (1950) The reaction of cyanogens with organic compounds. II. Primary aliphatic amines. J Org Chem 15(3):535–540
22. Markert M, Mulzer M, Schetter B, Mahrwald R (2007) Amine- catalyzed direct aldol addition. J Am Chem Soc 129(23):7258–7259
23. Amiri AP, Haghighi ASLA, Ehsani MR (2007) Alkanolamines foaming tendency: Effect of amine mixing, amine degradation and gas flow rate. Farayandno 8:2–7
24. Watanabe O, Nagai K (2003) Chemical modification of lacquer tree paint using primary amines. Bull Chem Soc Jpn 76(4):799–804
25. Thalji NK, Crowe WE, Waldrop GL (2009) Kinetic mechanism and structural requirements of the amine-catalyzed decarboxylation of oxaloacetic acid. J Org Chem 74(1):144–152
26. Kittilstved KR, Gamelin DR (2005) Activation of high-T_c ferromagnetism in Mn^{+2}-doped ZnO using amines. J Am Chem Soc 127(15):5292–5293
27. Than C, Ferguson GA, Raghavachari K (2010) Quaternary amine-induced peptide degradation via cyclization. J Phys Chem A 114:481–485
28. Yang XB, Fu X-K, Zeng R-Q (2010) Intercalation of non-aromatic heterocyclic amines into layered zirconium glycine-N, N-dimethylphosphonate. Chem Paper 64(1):118–122
29. Mijos K (ed) (1978) Cyclic amines: In Kirk Othmer Encyclopedia of chemical technology, 3rd edn, vol 2, Wiley, New York, pp 295–308
30. Nakamura Y, Suzuki KO-kawa M, Konno T, Nishimura J (2005) Photoreactions between [60] Fullerene and various aromatic tertiary amines. J Org Chem 70(21):8472–8477
31. O'Neill C, Hawkes FR, Hawkes DL, Lourenco ND, Pinheiro HM, Delee W (1999) Colour in textile effluents – sources, measurements, discharge, contents and simulation: a review. J Chem Technol Biotechnol 74:1009–1018
32. Lores EM, Bristal DW, Moseman RF (1978) Determination of electro- chemical and ultraviolet detection. J Chromatogr Sci 16:358–362

33. de Kok A, Vos YJ, van Garderen C, de Jong T, van Opstal M, Frei RW, Geerdink RB, Brinkman UAT (1984) Chromatographic determination of phenylurea herbicides and their corresponding anilinedegradation products in environmental samples. J Chromatogr Sci 288:71–90
34. Spain JC (ed) (1995) Biodegradation of nitroaromatic compounds. Plenum Press, New York
35. Larson RA, Weber EJ (1994) Reaction mechanism in environmental organic chemistry. Lewis, Boca Raton, FL
36. Baughman GL, Weber E (1994) Transformation of dyes and related compounds in anoxic sediment: Kinetics and products. J Environ Sci Technol 28:267–276
37. Weber EJ, Adams RL (1995) Chemical- and sediment-mediated reduction of the azo dye disperse blue 79. Environ Sci Technol 29:1163–1170
38. Boer G, Schlett C, Tnier H-P (1993) Substituted anilines: Gas chromatographic determination and behaviour during a simulated subsoil passage. Vom Wasser 80:59–63
39. Weber EJ, Spidle DL, Thorn KA (1996) Covalent binding of aniline to humic substances. 1. Kinetic studies. Environ Sci Technol 30(9):2755–2763
40. Muller C, Iinuma Y, Karstensen J, van Pinxteren D, Lehmann S, Gnauk T, Herrmann H (2009) Seasonal variation of aliphatic amines in marine sub-micrometer particles at the cape verde islands. Atmos Chem Phys 9:9587–9597
41. Padhye L, Tezel U, Mitch WA, Pavlostathis SG, Huang C-H (2009) Occurrence and fate of nitrosamines and their precursors in municipal sludge and anaerobic digestion systems. Environ Sci Technol 43:3087–3093
42. Karlsson D, Spanne M, Dalenea M, Skarping G (2000) Airborne thermal degradation products of polyurethane coatings in car repair shops. J Environ Monit 2:462–469
43. Sendon R, Bustos J, Sanchez JJ, Paseiro P, Cirugeda ME (2010) Validation of a liquid chromatography-mass spectrometry method for determining the migration of primary aromatic amines from cooking utensils and its application to actual samples. Food Additiv Contamin 27(1):107–117
44. Morrison RT, Boyd RN (2008) Text book of organic chemistry, Pearson Education Print Ltd, Upper Saddle River, NJ, Chapter 22, p 856
45. Gilad GM, Gilad VH (1986) Cytotoxic effects of monodansylcadaverina and methylamine in primary cultures of rat cerebellar neurons. Int J Dev Neurosci 4(5):401–405
46. http://www.osha.gov/SLTC/healthguidelines/ethylamine/recognition.html
47. Tabuenca JM (1981) Toxic-allergic syndrome caused by ingestion of rapseed oil denatured with aniline. Lancet 318(8246):567–568
48. Davis KR, Schultz TW, Dumont JN (1981) Toxic and teratogenic effects of selected aromatic amines on embryos of the amphibian Xenopus laevis. Arch Environ Contamin Toxicol 10(3):371–391
49. Fouarge M, Mercier M, Poncelet F (1984) Liver, kidney and small-intestine microsomal-mediated mutagenicity of carcinogenic aromatic amines. Mutat Res 125(1):23–31
50. http://www.cdc.gov/niosh/docs/81-123/pdfs/0441.pdf
51. http://www.sciencelab.com/xMSDS-Diphenylamine-9927159
52. Thomas JO, Ribelin WE, Woodward JR, Deeds F (1967) The chronic activity of diphenylamine for dogs. Toxicol Appl Pharmacol 11:184–194
53. Drzyzga O, Jannsen S, Blotevogel KH (1995) Toxicity of diphenylamine and some of its nitrated and aminated derivatives to the luminescent bacterium vibrio fischeri. Environ Safety 31(12):149–152
54. Masubuchi Y, Yamada S, Horie T (2000) Possible mechanism of hepatocyte injury induced by diphenylamine and its structuraaly related nonsteroidal anti-inflammatory drugs. J Pharmacol Exp Ther 292(3):982–987
55. Wei MM, Stewart R (1966) The Mechanisms of permanganate oxidation. VIII. substituted benzylamines. J Am Chem Soc 88:1974–1979
56. Zamora R, Gallardo E, Hidalgo FJ (2006) Amine degradation by 4, 5-epoxy-2-decenal in model systems. J Agric Food Chem 54:2398–2404

57. Ho C-T (1996) Thermal degradation of Maillard aromas. In: Ikan R (ed) the Maillard reaction: Consequences for the chemical and life sciences. Wiley, Chichester, UK, pp 27–53
58. Hidalgo FJ, Zamera R (2004) Strecker-type degradation produced by the lipid oxidation products 4, 5 epoxy-2-alkenals. J Agri Food Chem 52:7126–7131
59. Lepaumier H, Picq D, Carrette P-L (2009) New amines for CO_2 capture. II. oxidative degradation mechanisms. Ind Eng Chem Res 48:9068–9075
60. Lepaumier H, Picq D, Carrette P-L (2009) New amines for CO_2 capture. I. Oxidative degradation mechanisms. Ind Eng Chem Res 48:9061–9067
61. Mitch WA, Schreiber IM (2008) Degradation of tertiary alkylamines during chlorination/chloramination: implications for formation of aldehydes, nitriles, halonitroalkanes, and nitrosamines. Environ Sci Technol 42:4811–4817
62. (i) Murray RW, Iyanar K, Chen J, Wearing JT (1996) Oxidation of organonitrogen compounds by the methyltrioxorhenium-hydrogen peroxide system, Tetrahedr Lett 37(6):805–808; (ii) Zhu Z, Espenson JH (1995) Kinetics and mechanism of oxidation of anilines by hydrogen peroxide as catalyzed by methylrhenium trioxide, J Org Chem 60:1326–1332; (iii) Coperet C, Adolfsson H, Khuong T-AV, Yudin AK, Sharpless KB (1998) A simple and efficient method for the preparation of pyridine N-oxides, J Org Chem 63:1740–1741
63. Marinescu L, Molbach M, Rousseau C, Bols M (2005) Supramolecular oxidation of anilines using hydrogen peroxide as stoichiometric oxidant. J Am Chem Soc 127(50):17578–17579
64. Stewart TD, Aston JG (1927) The decomposition of quaternary ammonium hydroxides. Alkoxymethyldiethyl-methylammonium hydroxides. J Am Chem Soc 49(7):1718–1728
65. Boux LJ, Milligan JR, Archer MC (1988) Base-catalyzed decomposition of N-Nitrosobis (2-oxopropyl)amine. Chem Res Toxicol 1:32–34
66. Meadows GW, Kirkland JJ (1952) Low pressure, fast flow pyrolysis of methylamines. J Phys Chem 65:2139–2143
67. Chen Y, Hu C, Hu X, Qu J (2009) Indirect photodegradation of amine drugs in aqueous solution under simulated sunlight. Environ Sci Technol 43:2760–2765
68. Cohen SG, Davis GA, Clark WDK (1972) Photoreduction of Π, Π^* triplets by amines, 2-naphthaldehyde, and 2-acetonaphthone. J Am Chem Soc 94:869–874
69. Cohen SG, Parola A, Parsons GH (1973) Photoreduction by amines. Chem Rev 73:141–161
70. Mailhot B, Morlat-Therias S, Bussiere P-O, Gardette J-L (2005) Study of the degradation of an epoxy/amine resin, kinetics and depth-profiles. Macromol Chem Phys 206:585–591
71. Emeleus HJ, Taylor HS (1931) The photochemical decomposition of amines and the photochemical interaction of amines and ethylene. J Am Chem Soc 53:3370–3377
72. Hossein HM, Majid MMS (2000) A photocatlytic method for the degradation of pyrrolidine in water. Iran J Chem Chem Eng 19(2):84–87
73. Baum AA, Karnischky LA, McLeod Jr D, Kasai PH (1973) Mercury photo-sensitized oxidation of primary and secondary aliphatic amines J Am Chem Soc 95(2):617–618
74. Augugliaro V, Baiocchi C, Prevot AB, Garcia-Lopez E, Loddo V, Malato S, Marci G, Palmisano L, Pazzi M, Pramauro E (2002) Azo-dyes photocatalytic degradation in aqueous suspension of TiO_2 under solar irradiation. Chemosphere 49:1223–1230
75. Zhu C, Wang L, Kong L, Yang X, Wang L, Zheng S, Chen F, MaiZhi F, Zong H (2000) Photocatalytic degradation of azo dyes by supported TiO_2 + UV in aqueous solution. Chemosphere 41:303–309
76. Tang WZ, An H (1995) UV/TiO_2 photocatalytic oxidation of commercial dyes in aqueous solutions. Chemosphere 31:4157–4170
77. Tang WZ, An H (1995) Photocatalytic degradation kinetics and mechanism of acid blue 40 by UV/TiO_2 in aqueous solution. Chemosphere 31:4171–4183
78. Tang WZ, Zhang Z, An H, Quintana MO, Torres DF (1996) TiO_2/UV photodegradation of azo dyes in aqueous solutions. Environ Technol 18:1–12

79. Mirkhani V, Tangestaninejad S, Habibi MH, Rostami-Vartooni A (2009) Photocatalytic degradation of azo dyes catalyzed by Ag doped TiO_2 photocatalyst. J Iran Chem Soc 6(3):578–587
80. Gultekin I, Ince H (2004) Degradation of azo dyes by UV/H_2O_2: Impact of radical scavengers. J Environ Sci Health A 39(4):1069–1081
81. Devi LG, Kumar SG, Reddy KM, Munikrishnappa C (2009) Photodegradation of methyl orange an azo dye by advanced Fenton process using zero valent metallic iron: Influence of various reaction parameters and its degradation mechanism. J Hazard Mater 164:459–467
82. Mamian M, Torres W, Larmat FE (2009) Electrochemical degradation of atrazine in aqueous solution at a platinum electrode. Portugaliae Electrochim Act 27(3):371–379
83. Zanoni MVB, Stradiotto NR (2005) Electrochemical behaviour of aromatic amines protected by nitrobenzenesulfonyl group. Electroanalysis 7(4):365–369
84. Mann CK, Barnes KK (1967) Electrochemical reactions in non-aqueous systems, Marcel Decker: New York, 1970. Chapter 9. J Electroanal Chem 13:1474
85. Santos V, Morão A, Pacheco MJ, Ciríaco L, Lopes A (2008) Electrochemical degradation of azo dyes on BDD: effect of chemical structure and operating conditions on the combustion efficiency. J Environ Eng Manage 18(3):193–204
86. Pacheco MJ, Ciriaco MLF, Lopes A, Goncalves IC, Nunes MR, Pereira MI (2006) Electrodegradation of azo dyes using the oxide $BaPb_{0.9}Sb_{0.1}O_{3-\delta}$ as anode material. Portugaliae Electrochim Acta 24:273–282
87. Kim GY, Lee K-B, Cho S-H, Shim J, Moon S-H (2005) Electroenzymatic degradation of azo dye using an immobilized peroxidase enzyme. J Hazad Mater B 126:183–188
88. Mohan SV, Prasad KK, Rao NC, Sarma PN (2005) Acid azo dye degradation by free and immobilized horseradish peroxidase (HRP) catalysed process. Chemosphere 58:1097–1105
89. Nam S, Tratnyek PG (2000) Reduction of azo dyes with zero-valent iron. Water Res 34(6):1837–1845
90. Cao J, Wei L, Huang Q, Wang L, Han S (1999) Reducing degradation of azo dye by zero-valent iron in aqueous solution. Chemosphere 38(3):565–571
91. Silva AC, Pic JS, Sant'Anna GL Jr, Dezotti M (2009) Ozonation of azo dyes (orange II and acid red 27) in saline media. J Hazard Mater 169:965–971
92. Ozen AS, Aviyente V (2004) Modeling the substituent effect on the oxidative degradation of azo dyes. J Phys Chem A 108:5990–6000
93. Petrova SN, Volodarski MV, Makarov SV, Li LZ (2008) Oxidation of azo dyes with inorganic peroxides in the presence of cationic surfactants. Russian J Appl Chem 81(9):1573–1577
94. Delnavaz M, Ayati B, Ganjidoust H (2008) Biodegradation of aromatic amine compounds using moving bed biofilm reactors. Iran J Environ Health Sci Eng 5(4):243–250
95. Scott EM, Jakoby WB (1959) Soluble gamma-aminobutyricglutamic transaminase from Pseudomonas fluorescens. J Biol Chem 234:932–936
96. Poupin P, Godon JJ, Zumstein E, Truffant N (1999) Degradation of morpholine, piperidine and pyrrolidine by microbacteria: evidence for the involvement of a cytochrome P450. Can J Microbiol 45:209–216
97. Gupta RC, Kaul SM, Shukla OP (1975) Pyrrolidine metabolism and its regulation in Arthrobacter sp. Ind J Biophys 12:263–268
98. Bae H-S, Im W-T, Suwa Y, Lee JM, Lee S-T, Chang Y-K (2009) Characterization of diverse heterocyclic amine-degrading denitrifying bacteria from various environments. Arch Microbiol 191:329–340
99. Tan NCG, van Leeuwen A, van Voorthuizen EM, Slenders P, Prenafeta-Boldu FX, Temmnik H, Lettinga G, Field A (2005) Fate and biodegradability of sulfonated aromatic amines. Biodegr 16:527–537
100. Schrader T, Schuffenhauer G, Sielaff B, Andreesen JR (2000) High morpholine degradation rates and formation of cytochrome P450 during growth on different cyclic amines by newly isolated Mycobacterium sp. Strain HE5 Microbiol 146:1091–1098

101. Chen H, Xu H, Heinze TM (2009) Decolorization of water and oil-soluble azo dyes by Lactobacillus acidophilis fermentum. J Ind Microbiol Biotechnol 36:1459–1466
102. Olukanni OD, Osuntoki AA, Gbenle GO (2009) Decolourization of azo dyes by a strain of Micrococcus isolated from a refuse dump soil. Biotechnology 8(4):442–448
103. Togo CA, Mutambanengwe CCZ, Whiteley CG (2008) Decolourization and degradation of textile dyes using a sulphate reducing bacteria (SRB) – biodigester microflora co-culture. Afr J Biotechnol 7(2):114–121
104. Ali N, Hameed A, Siddiqui MF, Ghumro P-b, Ahmed S (2009) Application of Aspergillus niger SA1 for the enhanced bioremoval of azo dyes in simulated textile effluent. Afr J Biotechnol 8(16):3839–3845
105. Senan RC, Abraham TE (2004) Bioremediation of textile azo dyes by aerobic bacterial consortium. Biodegradation 15:275–280
106. Omar HH (2008) Algal decolorization and degradation of monoazo and diazo dyes. Pakistan J Biol Sci 11(10):1310–1316
107. Martins MAM, Ferreira IC, Santos IFM, Queiroz MJ, Lima N (2001) Biodegradation of bioaccessible textile azo dyes by Phanerochaete chrysosporium. J Biotechnol 89:91–98
108. Liu L, Li F-b, Feng CH (2009) Microbial fuel cell with an azo-dye-feeding cathode. Appl Microbiol Technol 85:175–183
109. Mezohegyi G, Fabregat A, Font J, Bengoa C, Stuber F (2009) Advanced bioreduction of commercially important azo dyes: modeling and correlation with electrochemical characteristics. Ind Eng Chem Res 48:7054–7059
110. Suslick KS, Choe SB, Cichowlas AA, Grinstaff MW (1991) Sonochemical synthesis of amorphous iron. Nature 353:414–416
111. Suslick KS, Hyeon T, Fang M, Cichowlas AA, Gonsalves KE, Chow GM, Xiao TO, Cammarata RC (1994) Molecular Design Nanostruct Maters 351:201–206
112. Suslick KS, Fang M, Hyeon T, Cichowlas AA, Gonsalves KE, Chow GM, Xiao TO, Cammarata RC (1994) Nanostructured Fe-Co catalysts generated by ultrasound. Molecular Design Nanostruct Mater 351:443–448
113. Bellissent R, Galli G, Hyeon T, Magazu S, Majolino D, Migliardo P, Suslick KS (1995) Structural properties of amorphous bulk Fe, Co, and Fe-Co binary alloys. Phys Scripta 57:79–83, References and further reading may be available for this article. To view references and further read purchase this article
114. Suslick KS, Hyeon T, Fang M, Cichowlas AA (1995) Sonochemical synthesis of nanostructured catalysts. Mater Sci Eng, A 204(1–2):186–192
115. Hyeon T, Fang M, Suslick KS (1996) Nanostructured molybdenum carbide: sonochemical synthesis and catalytic properties. J Am Chem Soc 118(23):5492–5493
116. Suslick KS, Hyeon T, Fang M (1996) Nanostructured materials generated by high-intensity ultrasound: sonochemical synthesis and catalytic studies. Chem Mater 8(8):2172–2179
117. Özkan A, Özkan MH, Gürkan R, Akçay M, Sökmen M (2004) Photocatalytic degradation of a textile azo dye, Sirius Gelb GC on TiO_2 or Ag-TiO_2 particles in the absence and presence of UV irradiation and the effects of some inorganic anions on the photocatalysis. J Photochem Photobiol A Chem 163:29–35
118. Xu A-W, Gao Y, Liu H-Q (2002) The preparation, characterization, and their photocatalyticactivities of rare-earth-doped TiO_2 nanoparticles. J Catal 207:151–157
119. Colombo DP, Bowman RM (1995) Femtosecond diffuse reflectance spectroscopy of TiO_2 powder. J Phys Chem 99:11752e6
120. Liang C-H, Li F-B, Liu C-S, Lu J-L, Wang X-G (2008) The enhancement of adsorption and photocatalytic activity of rare earth ions doped TiO_2 for the degradation of orange I. Dyes Pigm 76:477–484
121. Vajnhandl S, Marechal AML (2007) Case study of the sonochemical decolouration of textile azo dye Reactive Black 5. J Hazard Mater 141:329–335

122. Ozen AS, Aviyente V, Tezcanli-Guyer G, Ince NH (2005) Experimental and modeling approach to decolorization of azo dyes by ultrasound: degradation of the hydrazone tautomer. J Phys Chem A 109:3506–3516
123. Petrier C, Lamy MF, Francony A, Benahcene A, David B, Renaudin N, Gondrexon V (1994) Sonochemical degradation of phenol in dilute aqueous solutions; comparison of the reaction rates at 20 kHz and 487 kHz. J Phys Chem 98:10514–10520
124. Hua I, Hoffmann MR (1997) Optimization of ultrasonic irradiation as an advanced oxidation technology. Environ Sci Technol 31:2237–2243
125. Wang Y, Zhao D, MA W, Chen C, Zhao J (2008) Enhanced sonocatalytic degradation of azo dyes by Au/TiO_2. Environ Sci Technol 42:6173–6178
126. Rehorek A, Tauber M, Gubitz G (2004) Application of power ultrasound for azo dye degradation. Ultrason Sonochem 11:177–182
127. Kenji O, Kazuya I, Yoshihiro Y, Hiroshi B, Rokuro N, Yasuaki M (2005) Sonochemical degradation of azo dyes in aqueous solution: a new heterogeneous kinetics model taking into account the local concentration of OH radicals and azo dyes. Ultrason Sonochem 12(4):255–262
128. El-Bahy ZM, Ismail AA, Mohamed RM (2009) Enhancement of titania by doping rare earth for photodegradation of organic dye (Direct Blue). J Hazard Mater 166:138–143
129. Ranjit KT, Willner I, Bossmann SH, Braun AM (2001) Lanthanide oxide doped titanium dioxide photocatalysts: Novel photocatalysts for the enhanced degradation of p-chlorophenoxyacetic acid. Environ Sci Technol 35:1544–1549
130. Ranjit KT, Willner I, Bossmann SH, Braun AM (2001) Lanthanide oxide doped titanium dioxide photocatalysts: Effective photocatalysts for the enhanced degradation of salicylic acid and t-cinnamic acid. J Catal 204:305–313
131. Uzunova-Bujnova M, Todorovska R, Dimitrov D, Todorovsky D (2008) Lanthanide-doped titanium dioxide layers as photocatalysts. Appl Surfac Sci 254:7296–7302
132. (i) Chen Y-C, Smirniotis P (2002) Enhancement of photocatalytic degradation of phenol and chlorophenols by ultrasound Ind Eng Chem Res 41:5958–5965; (ii) Wu C, Liu X, Wei D, Fan J, Wang L (2001) Photosonochemical degradation of phenol in water Wat Res 35(16):3927–3933
133. Gondal MA, Seddigi Z (2006) Laser-induced photo-catalytic removal of phenol using n-type WO_3 semiconductor catalyst. Chem Phys Lett 417:124–127
134. Gondal MA, Sayeed MN, Seddigi Z (2008) Laser-inhanced photo-catalytic removal of phenol from water using p-type NiO semiconductor catalyst. J Hazard Mater 155:83–89
135. Xu AW, Gao Y, Liu H-Q (2002) Preparation, characterization, and their photocatalytic activities of rare earth doped TiO_2 nanoparticles. J Catal 207:151–157
136. Sharipov GL, Gainetdinov RKh, Abdrakhmanov AM (2008) Effect of argon on the multibubble sonoluminescence of cerium, terbium and dysprosium trichlorides. Russ Chem Bull 57(9):1831–1836
137. Vikram L, Sivasankar BN (2008) New nine coordinated hydrated heavier lanthanide ethyldiamine tetraacetates containing hydrazinium cation: Crystal structure of $N_2H_5[Dy(EDTA)(H_2O)_3(H_2O)_5$. Ind J Chem 47A:25–31
138. Juengsuwattananon K, Jaroenworaluck A, Panyathanmaporn T, Jinawath S, Supothina S (2007) Effect of water and hydrolysis catalyst on the crystal structure of nanocrystalline TiO_2 powders prepared by sol-gel method. Physica Status Solidi A 204(6):1751–1756
139. Baolong Z, Baishun C, Keyu S, Shangjin H, Xiaodong L, Zongjie D, Kelian Y (2003) Preparation and characterization of nanocrystal grain TiO_2 porous microspheres. Appl Catal B 40(4):253–258
140. Citron IM, Allan M (1964) Spectrophotometric determination of primary amines in aqueous solution with copper-(ethylenedinitrilo) tetraacetic acid. Anal Chem 36(1):208–210
141. Norwitz G, Kellher PN (1981) Spectrophotometric determination of aniline by the diazotation-coupling method with N-(1-napthyl) ethylenediamine as the coupling agent. Anal Chem 53:1238–1240

142. DeAtley WW (1970) Spectrophotometric determination of diphenylamine, 2-nitrodiphenylamine, and 4-nitrodipheylamine by oxidation with ferric ion. Anal Chem 42(6):662–664
143. Pandey OP, Bajpai DN, Giri S (2006) Practical chemistry, S. Chand & Company Ltd, New Delhi
144. Handrickson JB, Cram DJ, Hammond GS (1970) Organic chemistry, Chap. 4: A survey of the class of compounds, Tosho Printing Company Ltd, Tokyo, Japan, pp 107–109
145. Finar IL (2000) Organic chemistry, vol. 1: The fundamental principles, 6th edn, Chap. 23: Aromatic amino-compounds. Wiley, New York, p 664
146. Finar IL (2000) Organic chemistry vol. 1: The fundamental principles, 6th ed., Chap 23; Aromatic amino-compounds, Wiley, New York, p 655
147. Premchandran RS, Banerjee S, Wu XK, John VT, McPherson GL, Ayyagari M, Kaplan D (1996) Enzymatic synthesis of fluorescent naphthol-based polymers. Macromolecules 29(20):6452–6460
148. Cox PA (1992) Transition metal oxides: An introduction to their electronic structure and properties. Clarendon Press, Oxford
149. Yuhong Zhang,' Huaxing Zhang, Yongxi Xu, Yanguang Wang l (2004) Significant effect of lanthanide doping on the texture and properties of nanocrystalline mesoporous TiO_2. J Solid State Chem 177(10):3490–3495

Chapter 13
Sonoluminescence of Inorganic Ions in Aqueous Solutions

Pak-Kon Choi

Abstract Sonoluminescence from alkali-metal salt solutions reveals excited state alkali – metal atom emission which exhibits asymmetrically-broadened lines. The location of the emission site is of interest as well as how nonvolatile ions are reduced and electronically excited. This chapter reviews sonoluminescence studies on alkali-metal atom emission in various environments. We focus on the emission mechanism: does the emission occur in the gas phase within bubbles or in heated fluid at the bubble/liquid interface? Many studies support the gas phase origin. The transfer of nonvolatile ions into bubbles is suggested to occur by means of liquid droplets, which are injected into bubbles during nonspherical bubble oscillation, bubble coalescence and/or bubble fragmentation. The line width of the alkali-metal atom emission may provide the relative density of gas at bubble collapse under the assumption of the gas phase origin.

13.1 Introduction

Sonoluminescence (SL) can be used as a spectroscopic probe of species produced at the collapse of cavitating bubbles. The spectrum of multibubble sonoluminescence (MBSL) in argon-saturated water is known to consist of a broad continuum and OH radical bands [1–4]. A typical MBSL spectrum from argon-saturated water is shown in Fig. 13.1. The ultrasonic frequency used is 108 kHz and the ultrasonic power is 7.9 W. The continuum is suggested to originate from bremsstrahlung, blackbody radiation and/or excited-state molecular emission. The three peaks at 290, 310, and 340 nm are due to the electronic transitions of OH radicals including

P.-K. Choi (✉)
Department of Physics, School of Science and Technology, Meiji University, 1-1-1, Higashimita, Tama-ku, Kawasaki 214-8571, Japan
e-mail: pkchoi@isc.meiji.ac.jp

Pankaj and M. Ashokkumar (eds.), *Theoretical and Experimental Sonochemistry Involving Inorganic Systems*, DOI 10.1007/978-90-481-3887-6_13,

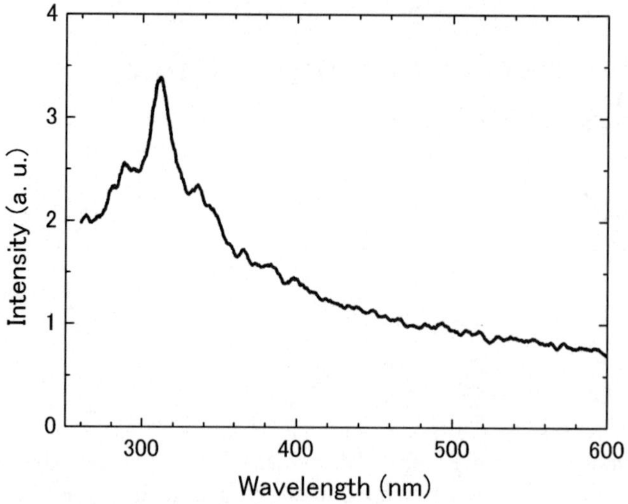

Fig. 13.1 Multibubble sonoluminescence spectrum obtained from argon-saturated water sonicated at a frequency of 108 kHz with an ultrasonic power of 7.9 W. The spectrum consists of a broad continuum and OH radical bands

transitions between different vibration states [1–3]. The SL of metal ions in aqueous solutions reveals the emission lines of metal atoms, which provide information on the cavitation conditions surrounding the emission species. It is important to clarify the mechanism for the SL of nonvolatile metal ions, which will be very useful for understanding the sonochemistry of nonvolatile species in aqueous solutions. The points that should be identified are how metal ions are reduced and metal atoms are excited, and the location of the atom emission site.

Günther et al. [5] were the first to investigate the SL spectra of alkali and alkaline-earth metal atoms. They used photographic techniques to obtain spectra from a number of solutions and found that the strong sodium D lines were exhibited although the resolution was poor. The SL spectrum of sodium atom emission was first reported by Heim [6]. He obtained a well-resolved spectrum of sodium D lines from xenon-saturated NaCl aqueous solutions with a concentration of 2 M at 294 kHz, with an effort requiring 3 weeks of exposure in a grating spectrograph.

Taylor and Jarman [1] observed SL spectra in the range of 280–740 nm from 2 M NaCl solutions saturated with argon, krypton and xenon sonicated at frequencies of 16 and 500 kHz. The spectra showed a continuum background with bands at about 310 nm and a peak of sodium D line, which exhibited appreciable asymmetric broadening, as shown in Fig. 13.2. The bands around 310 nm result from the $A^2\Sigma^+ - X^2\Pi$ transition of OH radicals. The OH bands are quenched in salt solutions compared with those in water, which suggests the energy transfer reaction

$$OH^* + Na \rightarrow OH + Na^*, \tag{13.1}$$

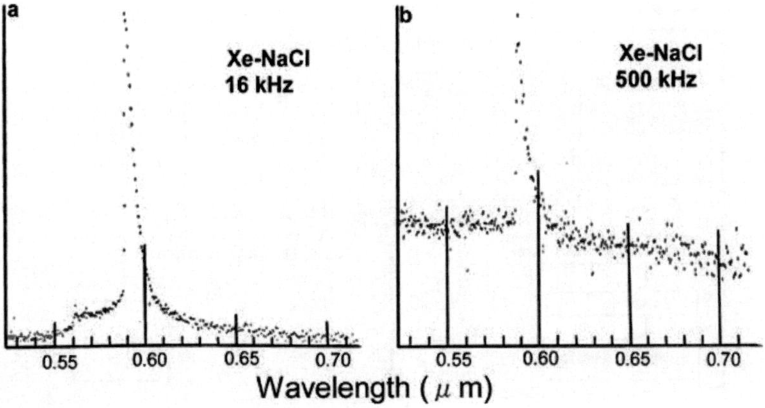

Fig. 13.2 MBSL spectra from 2M-NaCl aqueous solution saturated with xenon at frequencies of 16 kHz (**a**) and 500 kHz (**b**). The sodium D lines exhibit asymmetric broadening [1] (Reprinted from the CSIRO Publishing. With permission)

where the asterisk indicates an electronically excited state. Taylor and Jarman asserted that sodium ions enter the bubbles and that the sodium emission is chemiluminescence. Therefore, the temperature within bubbles cannot be measured from the relative intensities of different lines. However, they discussed the possibility of determining the relative density within bubbles from the amount of asymmetric line broadening in comparison with the spectroscopic data of the broadening. They also reported that no sodium line was observed in an air-saturated solution. This finding may be explained by the quenching of sodium emission due to interactions with nitrogen molecules, which was studied by Norrish and Smith [7].

The mechanism of alkali-metal atom emission has been discussed by various authors since Taylor and Jarman proposed the model of emission in the gas phase inside bubbles. This mechanism is still under debate and will be discussed in the following section.

13.2 Experimental System

The experimental system for measuring the sonoluminescence spectrum of alkali-metal atom emission from an aqueous solution is similar to that for measuring the MBSL spectrum from water. Degassing the solution is an important procedure because the presence of dissolved air affects the emission intensity. In an air-saturated solution, no observation of alkali-metal atom emission has been reported, whereas continuum emission can be observed. A typical experimental apparatus using ultrasonic standing waves is shown in Fig. 13.3 [8]. The cylindrical sample container is made of stainless steel, and its size is 46 mm in diameter and 150 mm in

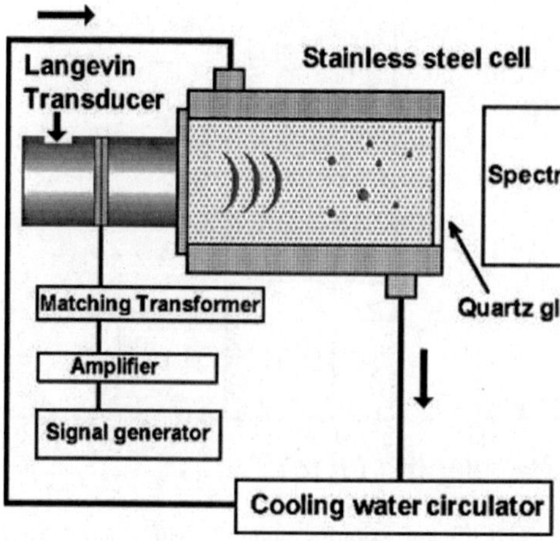

Fig. 13.3 Experimental apparatus for measuring MBSL spectrum using ultrasonic standing waves

length. The volume of solution required is 250 mL. The temperature of the solution can be controlled by circulating thermostated water around the container. The side faces of the container are equipped with a quartz glass window and a sandwich-type Langevin transducer, with a fundamental frequency of 28 kHz. The fundamental frequency and harmonic frequencies such as 138 kHz can be used as resonant frequencies of the container. The transducer can be exchanged with a ceramic transducer whose fundamental frequency is 1 MHz or higher. A signal from a function generator is amplified using a power amplifier and its impedance is matched to the transducer using a ferrite transformer. The transformer is particularly required for multi-frequency operation. The emitted light is analyzed using a spectrometer and a cooled CCD detector. The spectrometer is equipped with a grating of 1,200 grooves/mm blazed at 500 nm and a slit of 0.1 mm. The instrumental bandwidth of this system is experimentally estimated to be about 0.3 nm. This value is satisfactory for evaluating the line width of alkali-metal atom emission. An ultrasonic power of up to 20 W is applied, which can be determined by calorimetry using a thermocouple.

Another transducer system frequently used is a commercially available ultrasonic horn with a typical frequency of 20 kHz [9, 10]. High-power ultrasound is radiated from the horn tip immersed in a solution. A typical electric power in this case is in the range of 50–400 W. The power per unit area may be one order of magnitude higher than that in the standing-wave system described above. According to Yasui's theory [11, 12], a large acoustic amplitude at 20 kHz decreases the temperature inside bubbles. The amount of water vapor trapped inside collapsing bubbles is larger at a larger acoustic amplitude, resulting in a temperature decrease due to the endothermic dissociation of water vapor in heated bubbles and to the larger molar heat of water vapor. It has also been pointed out that the difference in cavitation field is large between the standing-wave type and horn-type system. Petrie et al. [13] compared the rate of hydrogen peroxide formation at 20 and

487 kHz at the same ultrasonic power and showed that the rate is six times larger at 487 kHz than that at 20 kHz. This suggests that the frequency and cavitation field may affect alkali-metal emission, because OH radicals are considered to participate in the reduction process of alkali-metal ions.

13.3 The Site of Emission

Sehgal et al. [14] reported the first experimental estimates of cavity temperature and pressure from the line broadening and shift of sodium and potassium atom emissions in argon-saturated alkali-metal salt solutions sonicated at 460 kHz. The resonance line of emitting species undergoes pressure-induced broadening and a shift due to the perturbation of its energy levels by neighboring atoms. The dependences of line broadening and the shift of alkali-metal atom emission on relative density are well documented in the literature [15]. The comparison of their results with values in the literature provides the relative density within a bubble at the time of alkali-metal atom emission. Sehgal et al. obtained a relative density of 36, from which a temperature of 3,400 K and a pressure of 310 atm were deduced under the assumption of adiabatic conditions. They also observed a blue satellite peak associated with the alkali-metal lines. On the basis of the previous observation of a similar satellite peak in a rapidly compressed mixture of alkali-metal vapor in argon, they claimed that the satellite peak lends support to the mechanism that alkali-metal atom emission occurs in highly gaseous phase conditions.

The blue satellite peak associated with resonance line of rubidium (Rb) saturated with a noble gas was closely examined by Lepoint-Mullie et al. [10] They observed SL from RbCl aqueous solution and from a 1-octanol solution of rubidium 1-octanolate saturated with argon or krypton at a frequency of 20 kHz. Figure 13.4 shows the comparison of the SL spectra of the satellite peaks of Rb-Ar and Rb-Kr in water (Fig. 13.4b) and in 1-octanol (Fig. 13.4c) with the gas-phase fluorescence spectra (Fig. 13.4a) associated with the B $\rightarrow$ X transition of Rb-Ar and Rb-Kr van der Waals molecules. The positions of the blue satellite peaks obtained in SL experiments, as indicated by arrows, exactly correspond to those obtained in the gas-phase fluorescence experiments. Lepoint-Mullie et al. attributed the blue satellites to B $\rightarrow$ X transitions of alkali-metal/rare-gas van der Waals species, which suggested that alkali-metal atom emission occurs inside cavitating bubbles. They estimated the intracavity relative density to be 18 from the shift of the resonance line by a similar procedure to that adopted by Sehgal et al. [14].

They also discussed the excitation mechanism of alkali-metal atoms as follows. The addition of a metal species from a liquid solution into cavitating bubbles is through the ablation of the bubble-liquid interface, the ablation of liquid jet or the evaporation of droplets, since the evaporation of salt is negligible. The salt molecules are released and decomposed into atoms via homolysis, analogous with the projection into a flame of metal species from salt solutions. The metal atoms are

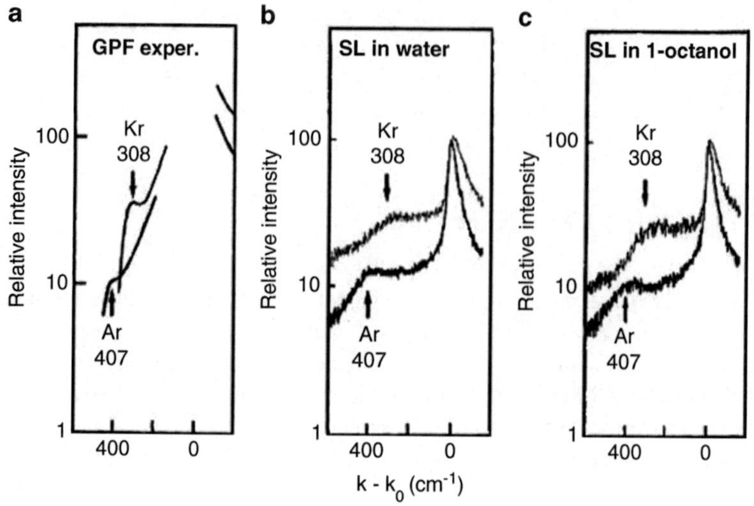

Fig. 13.4 Rubidium line spectra obtained by gas-phase fluorescence (**a**). Sonoluminescence spectra from RbCl aqueous solution saturated with Ar and Kr (**b**) and from 1-octanol solution of rubidium 1-octanolate (**c**). The numbers indicated in the figures correspond to the difference in wave number k in gas-phase fluorescence experiments, where k_0 is the reference attributed to the maximum intensity of the blue line [10] (Reprinted from Elsevior. With permission)

electronically excited through reactions including those with OH and H radicals such as

$$OH + H + Na \rightarrow H_2O + Na^* \tag{13.2}$$

$$H + H + Na \rightarrow H_2 + Na^* \tag{13.3}$$

Then, the collisional interactions with rare-gas atoms result in the formation of electronically excited alkali-metal/rare-gas molecules. These van der Waals molecules emit light. Lepoint-Mullie et al. claimed that the site of SL from alkali-metal atoms is in the gas phase inside bubbles and that the SL is chemiluminescence.

The proposition that alkali-metal atom emission occurs in the gas phase inside bubbles was challenged by Flint and Suslick [16]. They investigated the effects of solvent vapor pressure and dissolved-gas thermal conductivity on the line width and shift of potassium atom emission in primary alcohols and in water. They used high-intensity ultrasound generated by a horn-type transducer at a frequency of 20 kHz and a power of 60 W/cm^2. The potassium line intensity under argon decreased exponentially with increasing vapor pressure of solvents from 1-octanol to 1-propanol, as shown in Fig. 13.5. The increase in vapor pressure causes an increase in the molar heat of gases inside a bubble, leading to a decrease in the peak temperature at bubble collapse. The endothermal heat of the dissociation of the alcohols inside the bubble also decreases the peak temperature.

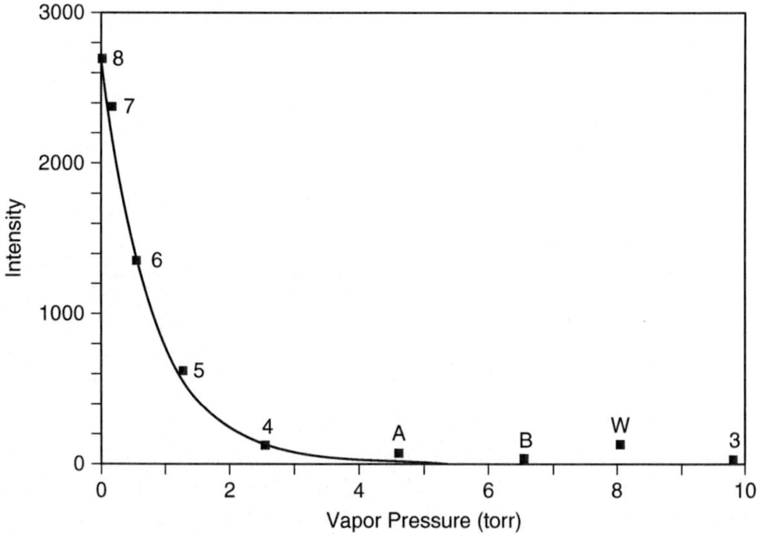

Fig. 13.5 Vapor pressure dependence of potassium line intensity in various 1-alcohols and water under argon. The numbers in the figure indicate the number of carbons in each alcohol. The symbols W, A and B denote in water, in 80/20 v/v 1-butanol/1-propanol and in 50/50 v/v 1-butanol/1-propanol, respectively [16] (Reprinted from American Chemical Society. With permission)

The line width and peak position of the potassium emission, on the other hand, showed no dependence on vapor pressure. The effect of gas thermal conductivity on the line width and peak position of the potassium emission was also examined in 1-octanol by changing the dissolved gas from argon to helium and varying the mixture ratio. The line width and peak position were unaffected by the change of the dissolved gas, although the cavitation condition was affected. Under high pressures of helium, the potassium lines should be slightly blue-shifted [15] if the emission occurs in the gas phase. On the basis of their experimental results, Flint and Suslick proposed that alkali-metal emission originates from the rapidly heated fluid immediately surrounding the collapsing bubble, and that radicals that are formed in the hot spot and diffuse out into the solution are responsible for the excitation of alkali-metal atoms. According to their explanation, the line profile of alkali-metal atom emission does not give any information on the cavitation condition at bubble collapse.

Grieser and Ashokkumar [17], and Ashokkumar et al. [18] pointed out that the bubble/liquid interface region may play an important role in the reduction process of alkali-metal ions. They observed sodium atom emission from an argon-saturated aqueous solution containing 10 mM sodium pentylsulfonate ($C_5H_{11}SO_3Na$; SPS). Figure 13.6 shows the SL spectrum obtained from SPS solution sonicated at 515 kHz. The spectrum exhibits the sodium line with a high intensity, higher than that observed from 1 M NaCl solution. SPS surfactant molecules are adsorbed onto the bubble interface and the bubble becomes negatively charged. This leads to an

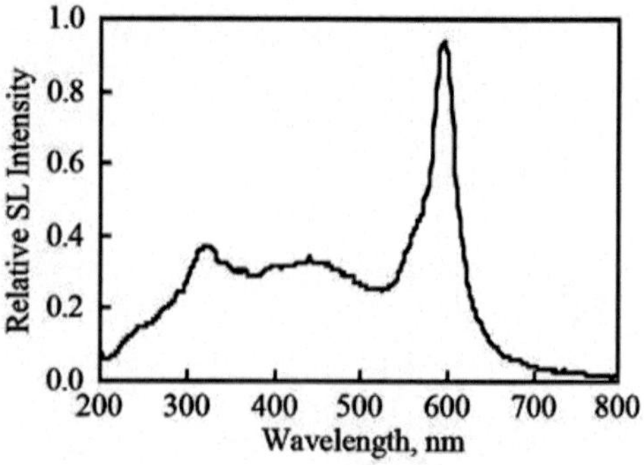

Fig. 13.6 Sonoluminescence spectrum obtained from an aqueous solution of 10 mM sodium pentylsulfonate sonicated at 515 kHz [18] (Reprinted from the authors. With permission)

enhanced Na^+ concentration in the vicinity of the bubble/liquid interface. The comparison of the sodium line intensity obtained from two different solutions suggests that the local Na^+ concentration at the surface of a surfactant-coated bubble is more than 1 M. This indicates the significance of Na^+ concentration at the bubble/liquid interface. Ashokkumar and coworkers suggested that the reduction of metal ions to form metal atoms occurs either at the bubble/liquid interface.

Choi and Funayama [19] also measured sodium atom emission from sodium dodecylsulfate (SDS) solutions in the concentration range of 0.1–100 mM at frequencies of 108 kHz and 1.0 MHz. The sodium line intensity observed at 1 MHz was nearly constant in the concentration range from 3 to 100 mM and was considerably higher than that at 108 kHz. This frequency dependence of the intensity is opposite that for NaCl aqueous solution. The dynamical behavior of the absorption and desorption of surfactant molecules onto the bubble surface may affect the reduction and excitation processes of sodium atom emission. This point should be clarified in the future.

The hydrostatic pressure and temperature dependences of the atom emission of alkali and alkali-earth metals (Na, Li, Mg and Ca) sonicated at 22 kHz were investigated by Gordeychuk and Kazachek [20]. When the pressure was increased from 1 to 1.8 atm, the intensity of sodium atom emission increased by one order of magnitude, whereas the continuum and OH* emission intensities increased by a factor of approximately 3. Decreasing temperature from 20°C to 1°C significantly enhanced the sodium atom emission. The pressure increase causes the more intense collapse of bubbles, and the temperature drop causes a decrease in the amount of water vapor in cavitation bubbles. Both effects lead to a higher temperature of the hot bubble core and therefore the surrounding interface layer. In the layer volume increased due to the higher temperature, a larger number of Na^+ ions are present. Gordeychuk and Kazachek explained that the reduction and excitation of the sodium atoms occur in the interface layer, where the temperature and pressure are close to or exceed the critical values (647 K, 218 atm). Na^+ ions can be reduced and excited by the capture of an electron from Cl^- ions.

The effect of dissolved gas or an impurity on alkali-metal atom emission provides important information about the site of the emission. Choi et al. [21]

investigated the effect of a small amount of ethanol on the line width and intensity of sodium-atom emission. It is well known that adding ethanol decreases the intensity of SL in water because ethanol molecules evaporate into bubbles [22, 23]. They measured the SL spectra from argon-saturated NaCl solutions at a frequency of 138 kHz at various acoustic powers from 1.8 to 16 W in the ethanol concentration range of 0–2 mM. Figure 13.7 shows the ethanol concentration dependence of the spectra of sodium atom emission from 2 M NaCl solutions doped with ethanol at a power of 7.3 W. The spectrum from NaCl solution with no ethanol, indicated as 'a' in Fig. 13.7, exhibits the sodium atom D line doublet. The peaks are not shifted from the original positions of the D line doublet of 589.0 nm ($^2P_{3/2} \rightarrow {}^2S_{1/2}$) and 589.6 nm ($^2P_{1/2} \rightarrow {}^2S_{1/2}$) and are broadened asymmetrically to the red (high-wavelength) side. With increasing ethanol concentration the line intensity is quenched rapidly and the line width increases, as shown by spectra 'b' to 'e' in Fig. 13.7. The acoustic power dependences of the sodium line width from 2 M NaCl solutions with ethanol concentrations of 0 mM (open circles), 0.5 mM (asterisks), 1 mM (closed triangles) and 2 mM (closed circles) are represented in Fig. 13.8. Since the doublet was too overbroadened and could not be resolved into two lines, the apparent values of total line width are plotted. The line width increases almost linearly up to a power of 16 W. This is in contrast to the case of no ethanol addition (open circles), where the line width takes a maximum at 4.8 W. Previous studies [8, 22] showed that the addition of ethanol to argon-saturated water caused SL quenching and a decrease in the intensity of the high-energy part of the SL spectrum, which suggests a temperature decrease in the bubble core. Several different molecules are produced by the decomposition of ethanol due to the high temperature within the bubbles. Yasui [23] simulated bubble oscillations in aqueous

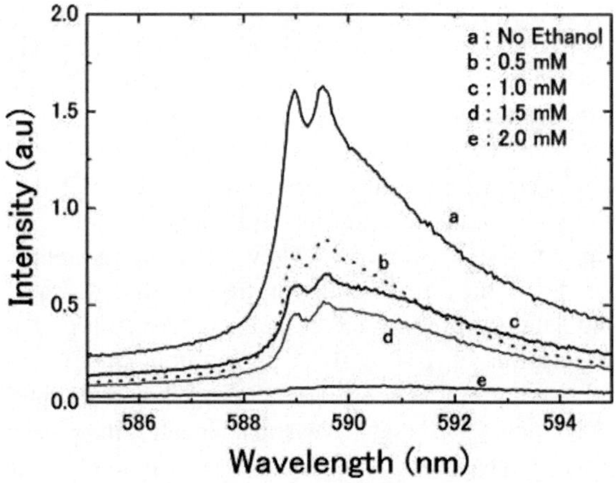

Fig. 13.7 Spectra of Na atom emission from 2 M NaCl solutions after adding ethanol with concentrations of 0.5 mM (b), 1 mM (c), 1.5 mM (d) and 2 mM (e). Line 'a' denotes the spectrum obtained without adding ethanol. The ultrasonic frequency is 138 kHz and the power is 7.3 W [21] (Reprinted from American Chemical Society. With permission)

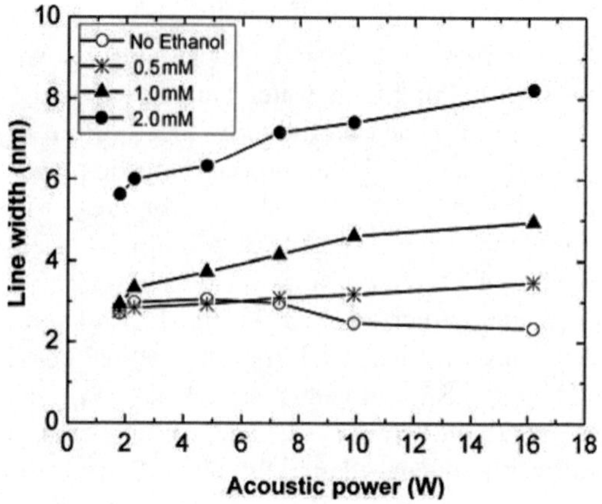

Fig. 13.8 Dependences of the line width of Na atom emission on acoustic power for 2 M NaCl solutions with ethanol concentrations of 0.5 mM (*asterisks*), 1 mM (*triangles*) and 2 mM (*closed circles*). The Na line width without the addition of ethanol is denoted by open circles and has a maximum at 4.8 W

methanol solution by considering various chemical reactions in bubbles, and showed that many gaseous products including H_2 and CO molecules contributed to the decrease in bubble temperature, which resulted in SL quenching. The enhancement of line broadening in Fig. 13.8 is explained by collisional interactions of excited sodium atoms with the hydrocarbon products of ethanol decomposition in addition to those with argon. Hooymayers and Alkemade [24] reported the strong quenching effect of CO, O_2, H_2 and N_2 on sodium emission in a flame. They inferred that this effect is associated with the small energy difference between the alkali-metal excitation energy and the nearest molecular vibrational energy. The strong quenching upon the addition of ethanol is thus accounted for by the interactions of sodium atoms with unsaturated hydrocarbons and H_2, which are produced during the decomposition of ethanol. Another reason for the strong quenching may be the decrease in the concentration of OH radicals due to reactions with ethanol, which is known to be a typical scavenger of OH radicals. The decrease in OH radical concentration prevents the excitation of sodium atoms via reaction (13.2). The intensity of sodium atom emission is well correlated with the quantity of OH radicals, which can be estimated from KI dosimetry [21]. Choi et al. [21] concluded that the effect of ethanol on the sodium line width and intensity strongly indicates the gas phase origin of sodium emission. They estimated the relative gas density and maximum pressure at the time of emission from the line width by comparing it with spectroscopic data. Comparing the collisional broadening of the sodium D_2 line for the argon perturber of 3.31×10^{-20} cm^{-1} (atom $cm^{-3})^{-1}$ at 2,000 K [25] with the measured maximum line width of 1.89 nm, Choi at al. obtained a relative density of 59.5. Assuming a perfect gas law and a temperature at bubble collapse of 4,300 K which was estimated from the C_2 molecular emission [26], they obtained a pressure of 873 atm. The value of relative density is comparable to the results of Sehgal et al. [14], who obtained a range from 36 to 50 for sodium emission at a frequency of 460 kHz. Lepoint-Mullie et al. [10] reported a relative density of 18 from rubidium emission at 20 kHz. McNamara et al. [27] obtained a relative density of 19, corresponding to a pressure of 300 atm from the peak shift of Cr atom emission in argon-saturated silicone oil at 20 kHz. These values may, of course, depend on experimental conditions and frequency. The results of the ethanol-doping effect obtained by Choi et al. imply that residual air in a solution may affect the line width and intensity of alkali-metal emission.

Potassium atom emission is favorable for estimating the line-broadening effect since the two lines of potassium doublet at 766.6 and 770.0 nm are more widely separated than those of the sodium doublet. The effect of ethanol on potassium atom emission from argon-saturated 2 M KCl aqueous solutions was studied by Abe and Choi [28]. The line width and intensity of potassium atom emission upon adding ethanol exhibited a similar tendency to that for sodium atom emission. The intensity was quenched by about half, and the line width increased twofold upon adding 1 mM ethanol. They also investigated the frequency dependence of the line width of potassium atom emission. Figure 13.9 shows the normalized spectra of potassium atom emission obtained at frequencies of 28, 115, 325 kHz and 1.0 MHz at an input electrical power of about 7 W. All the spectra are asymmetrically broadened to the red side, whereas only the spectrum at 28 kHz appears to be red shifted. The line width decreases with increasing frequency, and the intensity increases with frequency up to 115 kHz and then decreases at higher frequencies. This frequency dependence of the line width provides information on factors that determine the line width. It is possible to conclude that the relative density at the time of alkali-metal emission within a bubble is smaller at a higher ultrasonic frequency on the basis of the assumption that the line broadening is caused by collisional interactions with only argon gas. On the other hand, the results of the ethanol-doping effect described above indicate that a small amount of foreign gases may interact with alkali-metal atoms, giving rise to further line broadening. In this case the frequency dependence of the line width in Fig. 13.9 indicates the possibility that some of the gaseous content within the bubble contributes to line broadening. The gaseous content whose quantity is most likely to change with frequency is water vapor. H_2 gas is generated from the decomposition of water vapor due to the high temperature at

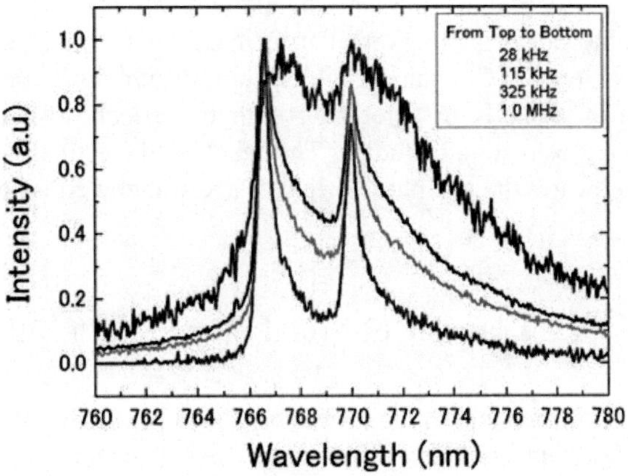

Fig. 13.9 Frequency dependence of potassium atom emission from argon-saturated 2 M KCl aqueous solutions. Shown are normalized spectra measured at frequencies of 28, 115, 325 kHz and 1.0 MHz at an input power of about 7 W. All spectra exhibit asymmetric broadening toward red side [42] (Reprinted from the Institute of Electronics, Information and Communication Engineers. With permission)

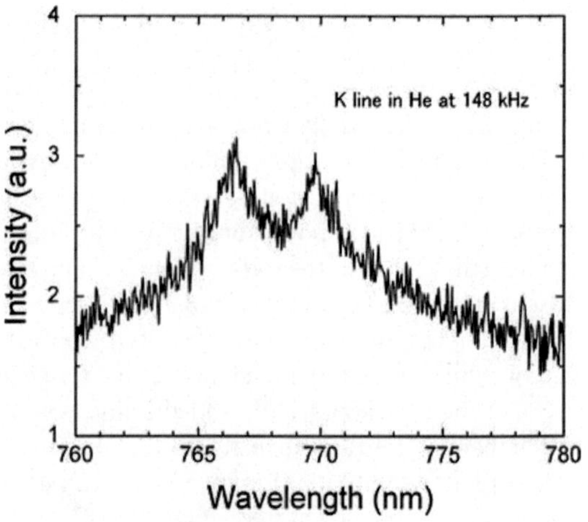

Fig. 13.10 Sonoluminescence spectrum of potassium-atom emission from helium-saturated KCl aqueous solution at 148 kHz. The spectrum shows slightly asymmetric broadening toward blue side, which is in contrast with the potassium line in argon-saturated solution

bubble collapse. Water vapor and H_2 are thus the most probable contributants to line broadening. Water vapor can enter bubbles by evaporation during bubble expansion more easily at a lower frequency than at a higher frequency because the bubble radius is larger at a lower frequency. Yasui et al. [29] calculated the amount of water molecules at bubble collapse at various frequencies from 20 kHz to 1 MHz, and showed that the water vapor fraction at the end of bubble collapse decreases with increasing frequency. Further studies including those on bubble dynamics are required to clarify the frequency effect.

Spectroscopic studies [30, 31] have shown that the effect of helium perturbers on the collisional broadening of alkali-metal atom emission is markedly different from that of argon perturbers. Helium perturbers cause slightly asymmetric broadening toward blue side whereas argon perturbers cause asymmetric broadening toward red side. The SL spectrum of potassium emission from helium-saturated KCl aqueous solution at 148 kHz was measured by Hayashi and Choi [32]. As shown in Fig. 13.10, the potassium doublet with slightly asymmetric line broadening toward blue side was observed. This is in contrast with the spectrum obtained from argon-saturated KCl solution, which shows asymmetric line broadening toward red side. The result is in agreement with the effects of foreign gas perturbers obtained in spectroscopic studies. The result of investigating the disolved-gas effect also indicates the gas-phase origin of alkali-metal emission inside bubbles.

13.4 Transfer of Metal Species into Bubbles

Two mechanisms have been proposed to explain how nonvolatile species become heated in a collapsing bubble [33, 34]: (1) an interfacial liquid layer around the hot cavitation bubble is heated. The metal ions in the initially liquid layer are reduced

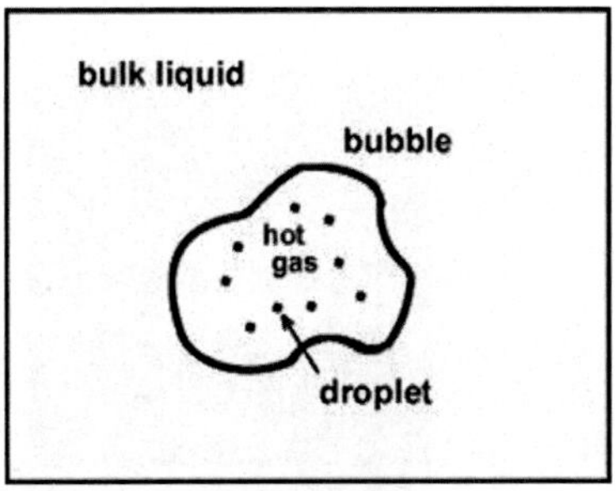

Fig. 13.11 Injected droplet model proposed to explain how nonvolatile species become heated in a collapsing bubble. Nanosized liquid droplets enter the bubble as a result of interfacial instabilities

and excited by radicals formed in the gas phase within the bubble. (2) Nanosized droplets of the solution are injected into the bubble and heated in the hot core of the collapsing bubble, as illustrated in Fig. 13.11. The droplets enter the bubble as a result of interfacial instabilities such as capillary surface waves and microjet formation. The coalescence and fragmentation of bubbles are also likely to induce droplet formation.

In explaining the difference in MBSL and single-bubble (SB) SL spectra from NaCl solution, Matula et al. [35] indicated that the injection of liquid droplets is responsible for the sodium emission in MBSL. The SBSL spectrum exhibited no sodium line because of the highly symmetric bubble collapse and the absence of surface deformation. After droplets enter a bubble, water molecules are removed from the droplets because of the high temperature during the bubble collapse phase, which leads to the formation of salt molecules followed by homolysis, thus generating alkali-metal atoms. The above processes are similar to those observed upon spraying aqueous NaCl solution into a flame [36]. The alkali-metal atoms are then electronically excited by OH and H radicals, as discussed previously, by reactions (13.2) and (13.3). Reaction (13.3) may plausibly occur since OH and H radicals are produced from water vapor within the high-temperature region of the bubble core. The H radicals are, however, known to be converted to H_2 in the gas phase or to OH radicals in a reaction with H_2O [37]. Thus, the dominant mechanism of the excitation of alkali-metal atoms is reaction (13.2).

13.5 Alkali-Metal Atom Emission and Continuum Emission

Recent observations of MBSL from a sulfuric acid solution of Na_2SO_4 [38, 39] revealed the spatial separation of sodium atom emission (orange) from continuum emission (blue-white). These results strongly support the injected droplet model regarding the emission site of nonvolatile species. Xu et al. [38] captured a photograph of SL, as shown in Fig. 13.12a, radiating under a horn-type transducer at 20 kHz. The orange color is spatially separated from the blue-white color. They asserted that if sodium emission occurs in an interfacial layer of a cavitating bubble, the sodium emission should always be present whenever conditions are suitable to

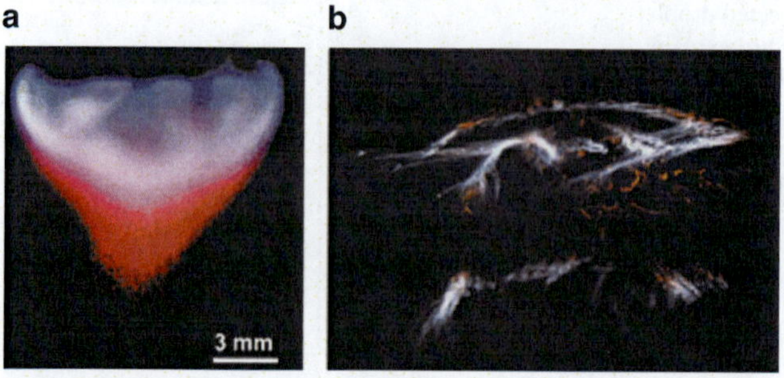

Fig. 13.12 Photographs of sonoluminescence from sulfuric acid solution of Na_2SO_4, illustrating the spatial separation of sodium (*orange*) and continuum (*blue-white*) emissions using a horn-type transducer at 20 kHz (**a**) [38] (Reprinted from American Chemical Society. With permission) and using standing-wave fields at 28 kHz in a cylindrical beaker (**b**) [39]

give rise to continuum emission inside the bubble. Hatanaka et al. [39] observed the separation of sodium emission in sulfuric solution in standing-wave fields at 28 kHz. Figure 13.12b shows a photograph of the SL of sodium and continuum emissions in a cylindrical beaker, from the bottom of which ultrasonic waves are irradiated and standing-wave fields are established. The region of blue-white emission corresponds to the antinode of the standing waves, and the orange emission propagates towards the blue-white region from the node of the standing waves. This suggests that the populations of bubbles producing sodium and continuum emissions are clearly different.

Similar studies on the separation of sodium atom emission from continuum emission have been made using aqueous solutions [40, 41]. Sunartio et al. [40] captured photographs of sonoluminescence and sonochemical luminescence images from water, 0.1 M NaCl, 10 mM SDS and luminol solutions sonicated at 139 kHz, and compared the regions of sodium atom emission and sonochemical emission in a rectangular container. The regions emitting orange light (sodium atom emission) closely correspond to those emitting blue light (luminol sonochemical emission), and do not correspond to those of continuum emission. They concluded that the sodium emission originates from sonochemically active bubbles, which have different populations from SL bubbles giving rise to the continuum emission. In SDS solution, the intensity of sodium emission is much higher than that in 0.1 M NaCl solution, and the intensity of the continuum emission is somewhat lower than that in water. The former result is due to the higher excess of SDS molecules at the bubble/solution interface. The latter result, the partial quenching of the continuum emission is due to the decomposition of the adsorbed surfactant molecules in the high-temperature environment in the bubble core, leading to a lowering of the bubble temperature. The different tendency of the sodium atom and continuum emissions in SDS solution suggests the existence of two different populations of bubbles.

Abe and Choi [41] succeeded in the spatiotemporal separation of sodium atom emission from continuum emission in argon-saturated NaCl aqueous solution

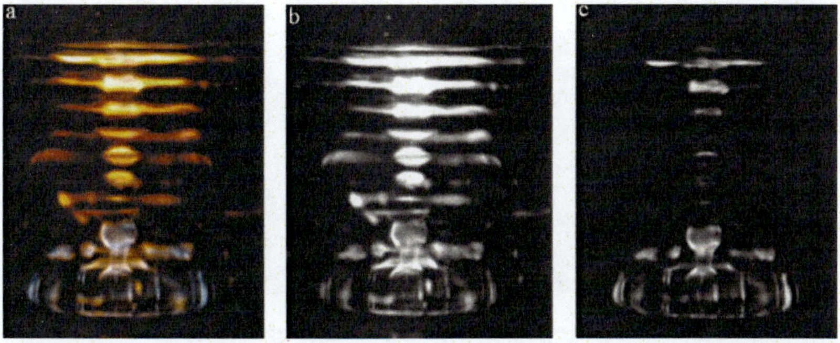

Fig. 13.13 Photographs of sonoluminescence from NaCl aqueous solution sonicated at 135 kHz in a cylindrical beaker (**a**). Image (a) was digitally processed to obtain the red component (**b**), which corresponds to Na atom emission, and the blue component (c), which corresponds to continuum emission [41] (Reprinted from the Japan Society of Applied Physics. With permission)

sonicated at 137 kHz. Figure 13.13a shows a photograph of SL viewed from the cylindrical side of a beaker. Both orange and blue-white regions can be seen in the image. The orange region expands outward as the applied voltage increases. The red and blue components of Fig. 13.13a were digitally processed and are illustrated in Figs. 13.13b and 13.13c, respectively. The red component corresponding to Na-atom emission spreads over a wide region compared with the blue component corresponding to the continuum emission. The regions of continuum emission are near the bottom of the cell and below the surface of the solution. The sound pressure may be high in these regions. At even higher applied voltages, the 'continuum region' expands and exhibits a similar distribution to that in Fig. 13.13b. These results indicate that sodium-atom emission originates from bubbles that are subjected to low pressure, and continuum emission originates from bubbles that are subjected to high pressure. In other words, sodium-atom emission originates from low-temperature bubbles and continuum emission originates from high-temperature bubbles.

To achieve the temporal separation of sodium atom emission from continuum emission, Abe and Choi [41] measured SL pulses spectrally isolated above and below 500 nm using a dichroic mirror and a set of two photomultipliers. The oscilloscope traces of SL pulses are shown in Figs. 13.14a–c, which were obtained at applied voltages of 16.5, 23.5 and 35 V_{pp}, respectively. Upper and lower oscilloscope traces correspond to sodium atom emission and continuum emission, respectively. The traces of SL pulses were triggered by an input ultrasound signal, and the SL pulses were accumulated for about 1 s. In Fig. 13.14a, only the sodium atom emission pulses were obtained. At higher voltages, as shown in Fig. 13.14b, the SL pulses are almost synchronous with the ultrasound period, and the timing of SL pulse emission has a wide distribution. The timing of the sodium atom emission has a wider distribution than that of the continuum emission. In Fig. 13.14c, the pulses due to sodium atom emission exhibit chaotic behavior, whereas the continuum emission pulse maintains a synchronous property. A simulation of bubble

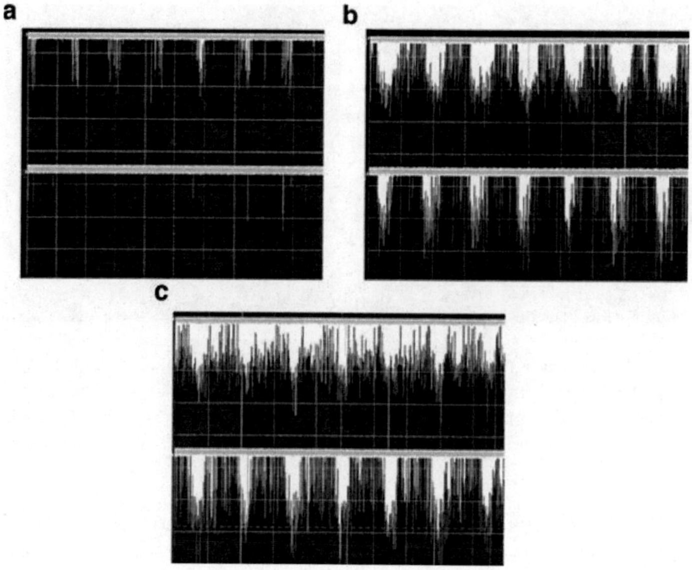

Fig. 13.14 Oscilloscope traces of sonoluminescence pulses from NaCl solutions for applied voltages of 16.5 V_{pp} (**a**), 23.5 V_{pp} (**b**) and 35 V_{pp} (**c**). The upper trace represents Na atom emission, and the lower trace represents continuum emission [41] (Reprinted from the Japan Society of Applied Physics. With permission)

radius and bubble temperature using the Keller-Mixis equation revealed that the distribution of initial bubble radius is responsible for the distribution of the timing of SL pulses. This suggests that sodium atom emission results from larger bubbles with a wider radius distribution than those producing continuum emission. The peak temperature in such bubbles is relatively low compared with that in bubbles that produce continuum emission.

The dependence of the broad-band SL spectrum from argon-saturated KCl solutions on ultrasonic frequency was obtained by Abe and Choi [42], as shown in Fig. 13.15. For potassium atom emission, the intensity is low at 28 kHz even though it is prominent in the spectrum. Intensity increases up to 115 kHz and then decreases. The intensity may be mainly determined by the degree of transfer of the alkali-metal species into bubbles, and also by the quenching rate due to interactions with gas molecules. Water droplets can easily enter bubbles at lower frequencies, because the probabilities of nonspherical oscillation and bubble coalescence are larger due to the larger bubble size. However the hydrogen or oxygen molecules produced by the decomposition of water vapor may quench alkali-metal emission, resulting in the small intensity. Another factor that affects intensity is the quantity of OH radicals produced in a bubble. OH radicals may participate in the excitation of alkali-metal atoms, and the production of OH radicals has been reported to be most efficient at about 200 kHz [43].

For continuum emission, on the other hand, the intensity is extremely low at 28 kHz and increases with increasing frequency. The intensity is highest at 1 MHz.

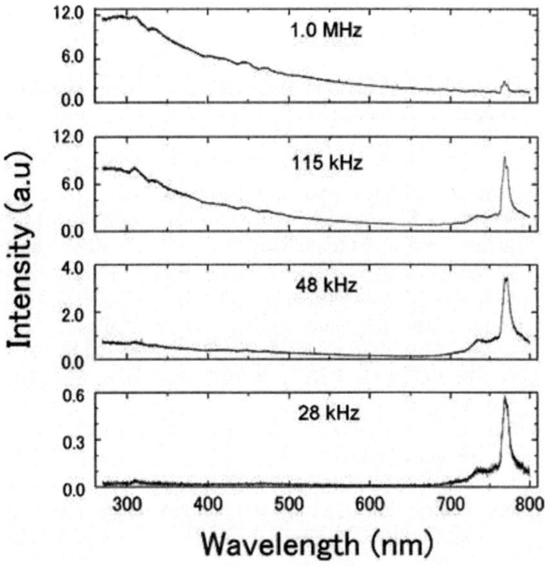

Fig. 13.15 Frequency dependence of broad band SL spectra from argon-saturated KCl solutions. The ultrasonic frequencies used are 28, 48, 115 kHz and 1.0 MHz (from *bottom* to *top*) [42] (Reprinted from the Institute of Electronics, Information and Communication Engineers. With permission)

The almost featureless spectrum at 1.0 MHz is reminiscent of the result by Matula et al. [35], i.e., the SBSL spectrum from NaCl solution indicated only continuum emission with no atomic lines. Sonoluminescing bubbles at higher frequencies are smaller and interact less with surrounding bubbles. These factors may explain why the MBSL spectrum at 1 MHz is similar to that of SBSL.

13.6 Conclusions

Fifty years have passed since SL from alkali-metal aqueous solutions was first observed, and since then many experimental studies have been performed on SL. Although the emission mechanism has not been fully clarified, many studies suggest that alkali-metal atom emission occurs in the gas phase inside the bubble. Alkali-metal ions may be injected into a bubble as liquid droplets when the bubble oscillates asymmetrically during surface oscillation and/or at the occasion of bubble coalescence or fragmentation. An unresolved problem is the reason for the strong sodium-atom emission in a solution containing surface active molecules, in which asymmetric bubble oscillations are suppressed. The direct observation of bubble dynamics using a high-speed camera will help to solve this problem.

Acknowledgments The author thanks Shogo Abe and Yuichi Hayashi for their experimental studies, and Dr. Shin-ichi Hatanaka for his valuable discussion.

References

1. Taylor KJ, Jarman PD (1970) The spectra of sonoluminescence. Aust J Phys 23:319–334
2. Sehgal C, Sutherland RG, Verrall RE (1980) Optical spectra of sonoluminescence from transient and stable cavitation in water saturated with various gases. J Phys Chem 84:388–395
3. Didenko YT, Gorrdeychuk TV, Koretz VL (1991) The effect of ultrasound power on water sonoluminescence. J Sound Vib 147:409–416
4. Weninger KR, Camara CG, Putterman SJ (2000) Observation of bubble dynamics within luminescent cavitation clouds: Sonoluminescence at the nano-scale. Phys Rev E 63:016310
5. Günther P, Zeil W, Grisar U, Heim E (1957) Versuche über die Sonolumineszenz wäßriger Lösungen. Z Electrochem 61:188–201
6. Heim E (1960) Asymmetrisch verbreiterte Emissionslinien in den Sonolumineszenzspektren wäßriger Salzlösungen. Z Angew Phys 12:423–424
7. Norrish RGW, MacF SW (1940) The quenching of the resonance radiation of sodium. Proc R Soc A176:295–312
8. Hayashi Y, Choi P-K (2006) Effects of alcohols on multi-bubble sonoluminescence spectra. Ultrasonics 44:e421–e423
9. Didenko YT, McNamara WB III, Suslick KS (2000) Effect of noble gases on sonoluminescence temperatures during multibubble cavitation. Phys Rev Lett 84:777–780
10. Lepoint-Mullie F, Voglet N, Lepoint T, Avni R (2001) Evidence for the emission of 'alkali-metal-noble-gas' van der Waals molecules from cavitation bubbles. Ultrason Sonochem 8:151–158
11. Yasui K (2001) Temperature in multibubble sonoluminescence. J Chem Phys 115:2893–2896
12. Yasui K, Tuziuchi T, Sivakumar M, Iida Y (2004) Sonoluminescence. Appl Spectrosc Rev 39:399–436
13. Petrie C, Lamy M-F, Francony A, Benahcene A, David B (1994) Sonochemical degradation of phenol in dilute aqueous solutions: Comparison of the reaction rates at 20 and 487 kHz. J Phys Chem 98:10514–10520
14. Sehgal C, Steer RP, Sutherland RG, Verrall RE (1979) Sonoluminescence of argon saturated alkali metal salt solutions as a probe of acoustic cavitation. J Chem Phys 70:2242–2248
15. Ch'en SY, Takeo M (1957) Broadening and shift of spectral lines due to the presence of foreign gases. Rev Mod Phys 29:20–99
16. Flint EB, Suslick KS (1991) Sonoluminescence from alkali-metal salt solutions. J Phys Chem 95:1484–1488
17. Grieser F, Ashokkumar M (2001) The effect of surface active solutes on bubbles exposed to ultrasound. Adv Colloid Interface Sci 89–90:423–438
18. Ashokkumar M, Vu T, Grieser F (2004) A quest to find the mechanism for the formation of excited state metal atoms during acoustic cavitation. Proc 18th Int Congr Acoust 4:2935–2936
19. Choi PK, Funayama K (2007) Multibubble sonoluminescence and Na atom emission in sodium dodecyl sulfate surfactant solutions. Jpn J Appl Phys 46:4768–4770
20. Gordeychuk TV, Kazachek MV (2009) Experimental observation of the intense enhancement of metal sonoluminescence under pressure and temperature. Opt Spectro 106:238–241
21. Choi PK, Abe S, Hayashi Y (2008) Sonoluminescence of Na atom from NaCl solutions doped with ethanol. J Phys Chem B 112:918–922
22. Ashokkumar M, Crum LA, Frenley CA, Grieser F, Matula TJ, McNamara WB III, Suslick KS (2000) Effect of solutes on single-bubble sonoluminescence in water. J Phys Chem A 104:8462–8465
23. Yasui K (2002) Effect of volatile solutes on sonoluminescence. J Chem Phys 116:2945–2954
24. Hooymayers HP, Alkemade CTJ (1966) Quenching of excited alkali atoms and related effects in flames: Part II. Measurements and discussion. J Quant Spectrosc Radiat Transfer 6:847–874

25. Jongerius MJ, Van Bergen ARD, Hollander T, Alkemade CTH (1981) An experimental study of the collisional broadening of the Na-D lines by Ar, N_2 and H_2 perturbers in flames and vapor cells—I. The line core. J Quant Spectrosc Radiat Transfer 25:1–18
26. Didenko YT, McNamara WB III, Suslick KS (2000) Effect of noble gases on sonoluminescence temperatures during multibubble cavitation. Phys Rev Lett 84:777–780
27. McNamara WB III, Didenko YT, Suslick KS (2003) Pressure during sonoluminescence. J Phys Chem B 107:7303–7306
28. Abe S, Choi P- K (2008) Effect of frequency on sonoluminescence spectrum from alkali-metal solutions. Nonlinear acoustics- Fundamentals and applications. AIP Confer Proc 22:189–192
29. Yasui K, Tujiuti T, Kozuka T, Towata A, Iida Y (2007) Relationship between the bubble temperature and main oxidant created inside an air bubble under ultrasound. J Chem Phys 127:154502
30. Lwin N, McCartan DG (1978) Collision broadening of the potassium resonance lines by noble gases. J Phys B 11:3841–3849
31. Mullamphy DFT, Peach G, Venturi V, Whittingham IB, Gibson SJ (2007) Collisional broadening of alkali doublets by helium perturbers. J Phys B 40:1141–1152
32. Hayashi Y, Choi P-K (2010) Effects of rare gases on MBSL spectrum of K atom emission. Proc 20th Inter Congr Acoust:248–251
33. Suslick KS, Hammerton DA, Cline RE Jr (1986) The sonochemical hot spot. J Am Chem Soc 108:5641–5642
34. Henglein A (1995) Chemical effects of continuous and pulsed ultrasound in aqueous solutions. Ultason Sonochem 2:S115–S121
35. Matula TJ, Roy RA, Mourad PD (1995) Comparison of multibubble and single-bubble sonoluminescence spectra. Phys Rev Lett 75:2602–2605
36. Alkemade C, Th J, Hermann R (1979) Fundamentals of analytical flame spectroscopy. Adam Hilger, Bristol
37. Tauber A, Mark G, Schuchmann H-P, Von Sonntag C (1999) Sonolysis of tert-butyl alcohol in aqueous solution. J Chem Soc, Perkin Trans 2:1129–1136
38. Xu H, Eddinsaas NC, Suslick KS (2009) Spatial separation of cavitatin bubble populations: The nanodroplet injection model. J Am Chem Soc 131:6060–6061
39. Hatanaka S, Hayashi S, Abe S, Choi P-K (2008) High-intensity Na* emission during multibubble sonoluminescence in surfuric acid. Nonlinear Acoustics-Fundamentals and Applications. AIP Confer Proc 1022:205–208
40. Sunartio D, Yasui K, Tuziuti T, Kozuka T, Iida Y, Ashokkumar M, Grieser F (2007) Correlation between Na* emission and "chemically active" acoustic cavitation bubbles. Chemphyschem 8:2331–2335
41. Abe S, Choi P-K (2009) Spatiotemporal separation of Na-atom emission from continuum emission in sonoluminescence. Jpn J Appl Phys 48:07GH02
42. Abe S, Choi P-K (2008) Effect of frequency on sonoluminescence from alkali-metal atoms. IEICE Tech Rep US2007-95:1-4
43. Koda S, Kondo T, Kimura T, Mitome H (2003) A standard method to calibrate sonochemical efficiency of an individual reaction system. Ultrason Sonochem 10:149–156

Chapter 14
The Role of Salts in Acoustic Cavitation and the Use of Inorganic Complexes as Cavitation Probes

Adam Brotchie, Franz Grieser and Muthupandian Ashokkumar

Abstract Inorganic complexes have been employed as cavitation probes to ascertain invaluable quantitative information pertaining to bubble collapse temperatures and pressures, and provide qualitative insight into other fundamental aspects of cavitation. In addition to serving as cavitation probes, simple salts exert a marked influence on critical facets of acoustic cavitation ranging from nucleation, to inter-bubble interactions and the bubble size distribution. Multi-bubble sonoluminescence intensities can be elevated by almost an order of magnitude at high electrolyte concentration and the coalescence behaviour exhibited between bubbles scales with the 'salting-out' effect of a range of solutes, adding an interesting insight in the context of ion-specific electrolyte coalescence inhibition. This chapter provides a review of the literature available in these areas.

14.1 The Use of Inorganic Complexes to Probe the Conditions of Cavitation

One of the most important properties of any cavitation event is the collapse temperature. Noltingk and Neppiras developed the so-called *hot spot* model in 1950 and demonstrated that adiabatic heating during bubble collapse would generate temperatures in the range 5,000–10,000 K [1]. Some of the earliest experimental temperature estimations were obtained through the blackbody treatment of the broadband sonoluminescence (SL) from the emitting bubbles [2, 3]. However, Gompf et al. have shown quite conclusively that a blackbody radiation mechanism is untenable and, in light of this, the values obtained from blackbody fitting are not reliable estimates of the bubble temperature [4].

A. Brotchie (✉)
School of Chemistry, The University of Melbourne, Parkville, Victoria, Australia
e-mail: abrotchie@gmail.com

Pankaj and M. Ashokkumar (eds.), *Theoretical and Experimental Sonochemistry Involving Inorganic Systems*, DOI 10.1007/978-90-481-3887-6_14,

Important experimental determinations of both bubble collapse temperature and pressure have been conducted using inorganic salts and complexes. As detailed in the preceding chapter, the SL spectrum, which exhibits a predominantly continuum profile, is punctuated in alkali salt solutions by intense emission lines from excited metal atoms. Sehgal et al., through analysis of the SL spectra in metal halide salts (NaI and KI) were able to deduce both temperature and pressure values [5]. The observation of line shifts and asymmetry towards the red was used as evidence for SL emission from the compressed gas phase during the final stages of compression, thus providing firm evidence in favour of the hot-spot model. The pressure broadening (determined from the emission line half widths) was used to calculate a radial compression ratio, R_F/R_C, which, in turn was used to calculate the final state of both temperature and pressure within the gas core. The values they arrived at were 310 ± 50 atm and 3,360 ± 330 K, which are quite consistent with subsequent determinations employing different, independent quantitative techniques. Flint and Suslick however, found no evidence of broadening or wavelength shifts in alkali metal emission lines, leading them to the opposite conclusion of Sehgal et al., that the site of metal emission is from a heated liquid shell surrounding the bubble rather than from the gas phase [6]. It should be noted that subsequent work from the same group strongly supports a gas phase emission mechanism (through a liquid droplet injection model, as discussed in the previous chapter) [7–9].

Suslick and co-workers found, in their measurements of Cr* emission in the multi-bubble SL (MBSL) from $Cr(CO)_6$ solutions in silicon oil, that a significant proportion (ca. 50%) of the metal line width was attributed to scattering from the bubble cloud, rendering this method unreliable in the acoustic system [10]. Alternatively, line shifts (independent of light scattering) of the triplet centred at ~27,820 cm^{-1} were used allowing the pressure to be determined through comparison with line shift data present in the literature for ballistic compressors. Figure 14.1a shows the dependence of the red shift for the three peaks of the triplet as a function of pressure and Fig. 14.1b contains a comparison of low pressure Cr* emission from a cathode lamp (light blue) to that of the MBSL (orange). The wavenumber shift is indicated above each peak, for which pressures of 240, 320 and 310 bar are obtained.

In the mid 1980s, Suslick's group made the first of its many experimental contributions in this area [12]. Keeping the total vapour pressure constant (through using different mixtures of alkanes), the first order rates of metal (Fe, Cr, Mo, W) carbonyl ligand substitution were studied in the presence of phenylphosphine, as a function of carbonyl vapour pressure. The data revealed two sites of reaction: one dependent upon the vapour pressure (i.e., the gas phase) and one independent of vapour pressure (i.e., the liquid phase). Employing the technique of comparative rate thermometry, originally developed for use in a shock tube, and known Arrhenius parameters, they determined the temperature of the two distinct domains of sonochemical activity to be 5,200 ± 600 and 1,900 K. The former temperature corresponds to the gas phase; the latter to the liquid shell, calculated using a heat transport model to be approximately 200 nm thick surrounding the bubble. The temporal evolution of the reaction rates within

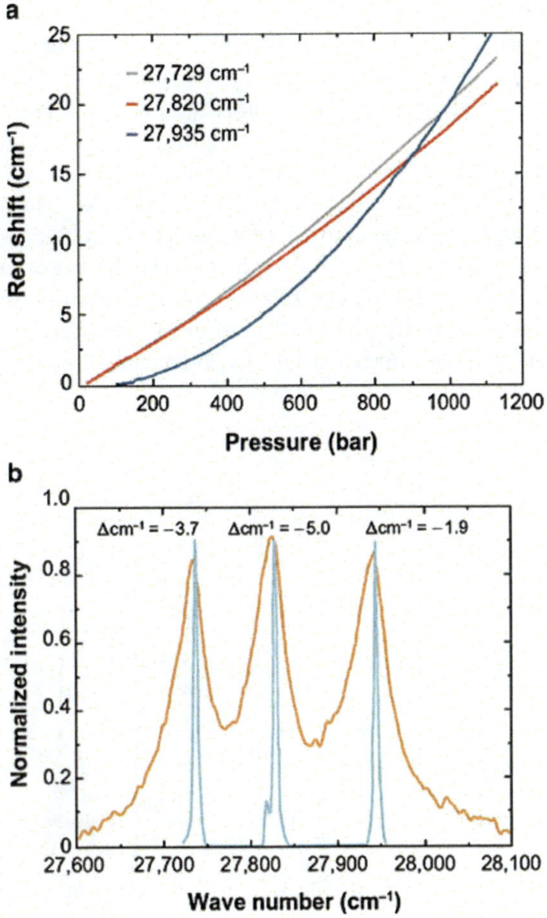

Fig. 14.1 (**a**) Red shift of Cr* emission line peaks as a function of Ar bath gas pressure at 3,230 K. The three curves correspond to the three peaks of the triplet centred on 27,820 cm^{-1}. (**b**) Corrected MBSL spectra (*orange*) and Cr* emission from a hollow cathode lamp at low pressure (*blue*). Relative red shifts for each peak are indicated [11] (reprinted with permission from Annual Reviews)

this shell were also determined, with the transient shell estimated to exist for less than 2 μs.

Importantly, Suslick developed additional spectroscopic techniques which have proved to be immensely powerful tools in probing the temperature conditions of bubble collapse. Using the same metal carbonyls used in the aforementioned comparative rate thermometry study, they found that the MBSL spectra in dodecane and silicone oil contained atomic emission lines for the thermally excited metals [13]. With greater spectral resolution, they later analysed SL spectra, calculating the bubble temperature from the ratio of two integrated emission line intensities (I_1/I_2) within a given multiplet [14]. This ratio relates to the temperature (T) through the following equation:

$$\frac{I_1}{I_2} = \frac{g_1 A_1 \lambda_2}{g_2 A_2 \lambda_1} \exp\left[\frac{(E_1 - E_2)}{kT}\right]$$

where g is the degeneracy of the electronic state, A is the Einstein transition probability, λ is the wavelength, E is the energy of the electronic state and k is the Boltzmann constant. Synthetic spectra were generated for different temperatures and the best fit of the experimental spectra reported. The temperatures determined (in silicone oil) using the Cr and Mo emission lines were 4,700 $\pm$ 300 and 4,800 $\pm$ 300 K, respectively, and are comparable to those determined through comparative rate thermometry. The addition of a volatile hydrocarbon additive, propane, was found to severely reduce the bubble temperature. Further details and discussion on the possible explanations for temperature reduction and SL quenching by volatile organic species can be found elsewhere in the literature [15–17]. Figure 14.2a shows the effect of gas temperature on the relative intensities of two Cr* emission lines and Fig. 14.2b contains the experimental SL spectrum and the best fit simulation at 4,700 K.

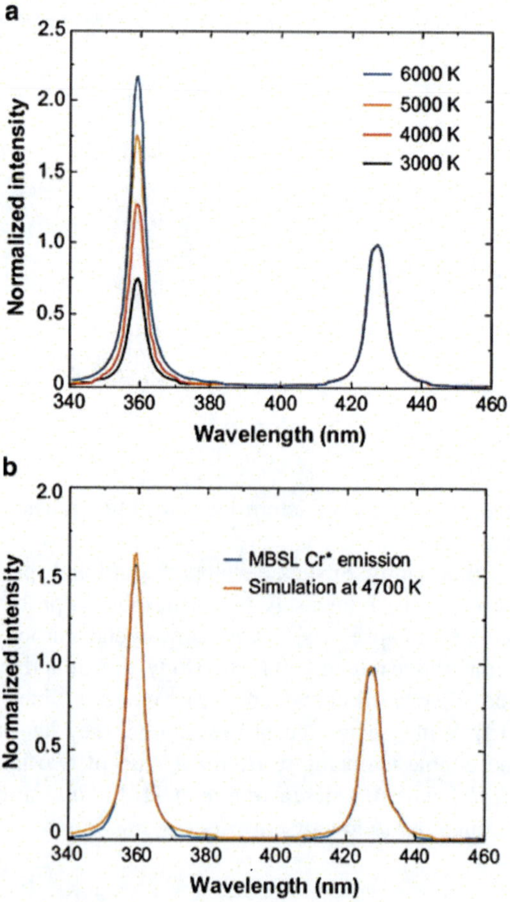

Fig. 14.2 (**a**) Emission spectra calculated for excited Cr at different temperatures, normalised with respect to the peak at 424 nm. (**b**) Experimental MBSL spectrum from 2.5 mM $Cr(CO)_6$ in silicon oil and the best fit simulated spectrum at 4,700 K. Sonication was performed at 20 kHz and 90 W cm^{-2} [11] (reprinted with permission from Annual Reviews)

It should be noted that these results are consistent with those of the same group when using the SL spectra of both organic and gas probes. For example, Flint and Suslick calculated a temperature of 5,075 ± 156 K by analysing the C_2 swan bands ($d3_g - a3_u$) in silicone oil, with comparison to synthesised spectra [18]. Didenko et al. extended this work to aqueous systems using the C_2 swan bands from benzene-water solutions, calculating a temperature of 4,300 ± 200 K [19, 20]. The lower temperature obtained in the aqueous system possibly owes to the relatively high vapour pressure compared with the silicone oil and long chain length alkanes used in the aforementioned studies.

All of the mentioned studies have concerned multi-bubble cavitation systems. Flannigan and Suslick [8] have also extended their method to analyse single bubble SL (SBSL) spectra from a moving single bubble in 85% H_2SO_4 in varying mixtures of argon and neon. In pure argon, they found the temperature (determined through the Ar* emission lines) to increase with increasing acoustic pressure from 8,000 K at 2.3 bar to about 15,000 K at 2.8 bar. They also found that as the neon content of the bubble was increased, the temperature decreased from 15,000 K in 100% Ar to about 2,000 K in 100% Ne (temperature determined using SO emission lines), owing to the increased thermal conductivity of neon. From the pressure broadening of an Ar* emission line (deconvoluted to account for instrumental broadening) a pressure of 1,400 K was calculated, which correlated well with that theoretically predicted from a Rayleigh–Plesset treatment of the cavity dynamics. The SL spectrum containing the Ar* emission lines and the best fit synthetic spectra is shown in Fig. 14.3a and the SL spectrum is compared with the low pressure Hg(Ar) lamp emission in Fig. 14.3b.

Several other methods have been employed to access the conditions of bubble collapse. Misik et al. studied H_2O-D_2O mixtures and through measurements with the use of spin traps, were able to determine the temperature from the relative rates of O–H and O–D cleavage [21]. They reported temperatures ranging from 2,000 to 4,000 K. Hart et al. developed a method based on the gas phase recombination of methyl radicals (MRR method), formed from the decomposition of methane [22]. They calculated temperatures of 2,000–2,800 K depending on the methane concentration.

Tauber et al. [23] following the same method as Hart et al. but using *tert*-butanol as the methyl radical source, obtained a temperature of 3,600 K in 10^{-3} M *tert*-butanol and reported, similar to Hart et al. that this temperature decreased with increasing *tert*-butanol concentration. More recently, this method was adopted by Rae et al. [24] and Ciawi et al. [25, 26] in aqueous solutions. Rae et al. examined the effect of concentration of a series of aliphatic alcohols, extrapolating a maximum temperature of about 4,600 K at zero alcohol concentration [24]. They also observed a decrease in temperature with increasing alcohol concentration, which correlated well with the alcohol surface-excess and SL measurements obtained in the same system. Ciawi et al. investigated the effects of ultrasound frequency, solution temperature and dissolved gas on bubble temperature [26].

It should be noted however, that *all* of these methods described are unable to provide information on the absolute *peak* temperature; rather they yield values that

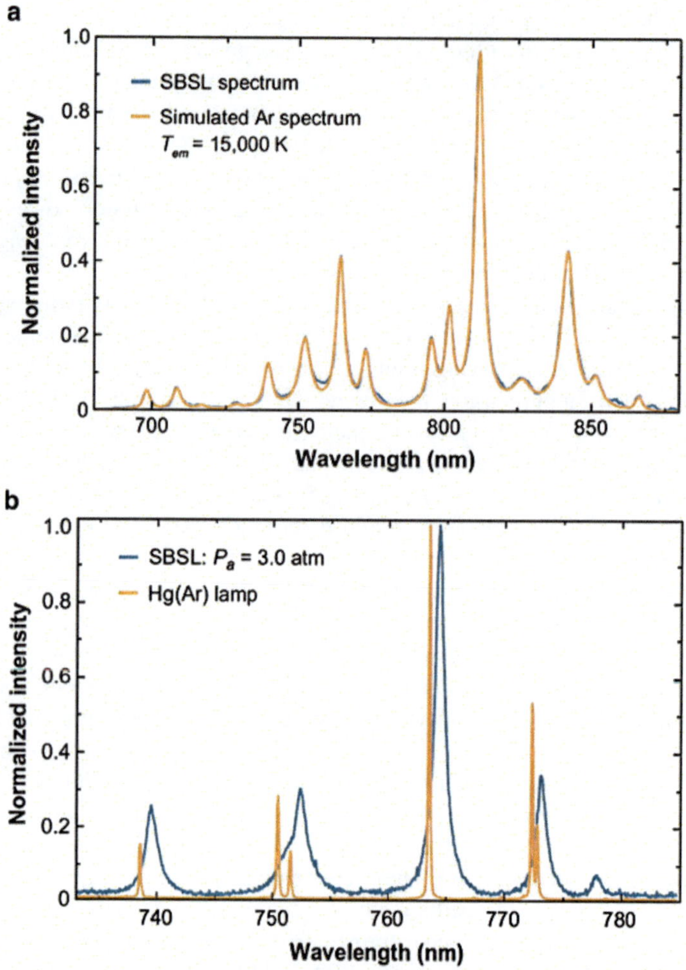

Fig. 14.3 (**a**) SBSL spectra containing Ar* emission lines from 85% H_2SO_4 and the best fit synthetic spectra at 15,000 K. (**b**) SBSL spectra and the low pressure spectra from a Hg(Ar) calibration lamp [11] (reprinted with permission from Annual Reviews)

are averaged both spatially and temporally. The spectroscopic techniques better approximate this peak value due to the fact that SL emission is relatively more confined both spatially and temporally, and occurs towards the final stages of collapse when the pressure and temperature are close to maximum. The chemical methods (MRR method, EPR spin trapping, comparative rate thermometry, etc.) report values that can be considered to be effective 'chemical' values representing the temperature throughout the bubble over a time period significantly longer than the SL flash width. Moreover, chemical methods can probe the sonochemically active but non-sonoluminescing bubble population.

In addition to permitting quantitative temperature and pressure studies to be conducted, inorganic probes, particularly alkali metal salts, have been used qualitatively to reveal important characteristics of acoustic bubbles and acoustic fields. Much of this area has been addressed in the previous chapter. In a standing-wave reactor, Sunartio et al. investigated photographically, the spatial characteristics of SL emission from pure water, SL emission from water containing NaCl and surfactant, sodium dodecyl sulphate (SDS), and secondary sonochemical luminescence (SCL; from alkaline luminol solution) [27]. They reported good spatial correlation between Na* emission and luminol SCL. The SL emitting region however, was found to be spatially distinct from the two forms of chemical luminescence, as is evident in Fig. 14.4a. In this figure, the SL can be seen to be predominantly emanating from the central column of the reactor, whereas the red-orange Na* emission is marginalised to the sides, similar to the luminol emission in Fig. 14.4b. Abe and Choi have subsequently reported similar findings observing both spatial and temporal separation of Na* emission from the continuum SL emission at 137 kHz [28].

It should be noted that it has recently been found that not only are SL and luminol SCL (and Na* emission) produced from bubbles separated spatially and temporally, and presumably having different temperature and pressure profiles during collapse, but they are emitted from bubbles of different size distributions. Brotchie et al. [29] found significant differences in the size distributions of the two populations, measured using the decay of SL and luminol SCL as a function of ultrasound pulse separation, (technique described in detail in reference [30]), with the SCL bubble population being of smaller size with a very broad distribution and the SL emitting bubbles being significantly larger and having a very narrow distribution. This is consistent with the results of Abe and Choi [28] with respect to the relative narrowness of the size distributions. However, in their study they deduced indirectly that the SL bubble size was, in fact, *smaller* than that of the SCL emitting bubbles.

These observations have been corroborated by those of Suslick and co-workers who compared the MBSL spectra of 0.1 M Na_2SO_4 in 95% H_2SO_4 from two spatially distinct regions of the cone-shaped cavitation zone under a 20 kHz sonotrode [7]. The spectrum recorded close to the horn tip was a featureless continuum, whereas the spectrum recorded far from the tip exhibited pronounced

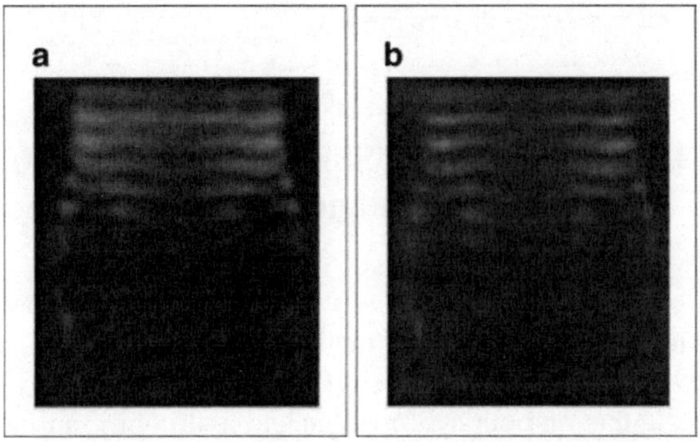

Fig. 14.4 Photographs of (**a**) SL emission and (**b**) luminol SCL from argon saturated 1 mM SDS solutions. Sonication was performed at 159 kHz and 0.06 W cm^{-3} [27] (reprinted with permission from Wiley Interscience)

Na* and Na-Ar exciplex peaks. There are two possible mechanisms for metal ion reduction and subsequent metal atom emission: (i) injection or nebulisation of liquid droplets containing the metal ions into the core of the collapsing bubble and (ii) the heating of the metal in a thin liquid shell surrounding the bubble [6, 31, 32]. This observation was interpreted as evidence for the droplet injection model and the difference between the two spectra explained in terms of the bubbles' translational velocities in the medium. Bubbles far from the horn were argued to move at much higher velocities leading to greater asymmetry upon collapse resulting in the injection of liquid droplets into the bubble. Conversely, bubbles near to the horn tip, which travel at relatively slower velocities, were suggested to collapse with a greater degree of symmetry, precluding droplet injection. The droplet injection model is also strongly supported by spectral differences between the MBSL and SBSL spectra. Matula et al. [32] observed Na* and K* emission in the MBSL but not in the SBSL spectra. This was attributed to the absence of surface waves and microjetting for the spherical single bubble preventing droplet injection. It is also noteworthy that Didenko et al. [9] found that a stable, *stationary* single bubble (highly symmetric) emitted a featureless continuum spectrum whereas a *moving* single bubble (less symmetric) exhibited pronounced molecular emission [CN emission ($B^2\Sigma - X^2\Sigma$) from adiponitrile solution].

Thus, a picture has been developed in which spatially separated populations of bubbles with intrinsically different properties lead to stark differences in emission spectra. Na* and luminol SCL can be attributed to relatively cooler bubbles, presumably undergoing a more asymmetric collapse, whereas the SL emitting population can be attributed to bubbles, which attain hotter conditions through more symmetric collapse. Suslick and co-workers have provided quantitative experimental verification of this hypothesis [33]. They used two independent spectroscopic temperature probes, •OH (formed from water vapour hydrolysis) and PO• radicals (formed by the decomposition of non-volatile H_3PO_4). Emission from •OH came predominantly from the cavitation region close to the horn where bubble collapse was assumed to be relatively symmetric; PO• emission was mainly from the region far from the horn where a high degree of asymmetry exists, analogous to alkali metal emission. The temperature determined of the former region was significantly higher (~9,500 K) than the latter (~4,000 K), validating the interpretation of spatially separated cavitation regions in which the symmetry of bubble collapse differs greatly.

14.2 The Effect of Simple Electrolytes and Gas Type on Cavitation and Sonoluminescence

Electrolytes influence almost every facet of cavitation from nucleation to coalescence, to the conditions of the bubble interior which determine the nature and severity of inertial collapse. Bunkin et al. [34] investigated optical cavitation and found that on both hydrophobic and hydrophilic surfaces the cavitation probability increased with concentration for several 1:1 electrolytes. Higher cavitation probability on the hydrophobic surface was attributed to the greater presence of surface nanobubbles and the electrolyte effect, which exhibited ion-specificity, was tentatively attributed to ion-specific adsorption influencing the geometry of such gas entities. The implication of submicroscopic gas bubbles is well established in phenomena such as emulsion and colloid stability, slippage at liquid-solid boundaries and the highly contentious long-range hydrophobic attractive force [35]. However, the seemingly high stability of surface nanobubbles with respect to high acoustic pressure seemingly precludes their direct involvement in nucleation [36, 37]. Ceccio et al. [38] found that although sea-water exhibited a lower (hydrodynamic) cavitation threshold pressure compared with fresh water, both the bubble size and number of cavitation events were reduced. This was attributed to a lower number of relatively smaller sized freestream nuclei in the salt water producing less, and smaller, macroscopic bubbles.

MBSL studies have found that the presence of various electrolytes can enhance the emission intensity by almost an order of magnitude compared to pure water [39, 40]. This can be seen in Fig. 14.5a, which contains the SL intensity, normalised with respect to pure water, as a function of NaCl concentration for sonication at 20 and 355 kHz. Wall et al. [39] reported ion-specific MBSL enhancement and found good correlation between the enhancement and the dissolved O_2 concentration (Fig. 14.5b), which scales with salt concentration with different proportionality constants for the different electrolytes. The physical mechanism proposed [39, 40] can be summarised as follows: although the gas pressure in a stable bubble is defined by its surface tension and radius through the Laplace equation, when a bubble undergoes a rapid and explosive size excursion prior to collapse, typical of transient cavitation, the internal gas pressure is far less than the equilibrium pressure. For high salt concentrations, which reduce the dissolved gas concentration, the diffusion of gas into the bubble will be reduced during this expansion phase. The lowered internal pressure prior to collapse will produce a higher collapse temperature and therefore a greater SL intensity. Although plausible, this explanation neglects the effect of the salts on nucleation and inter-bubble processes.

Similar to the trends in MBSL, sonochemical activity has been found to exhibit an inverse dependence on the dissolved gas concentration, as shown in Fig. 14.6. Interestingly, Okitsu et al. have demonstrated a strong positive linear dependence of sonochemical activity on gas concentration for a range of different gases [42]. This point will be discussed later in this section.

An issue as interesting as it is contentious is that of electrolyte inhibition of bubble coalescence. Recently, a number of studies have reported the ion-specific nature of electrolyte inhibition of bubble coalescence, albeit in static (non-acoustic) fields [43–49]. Some electrolytes appear to be highly efficacious whereas others almost completely ineffectual in inhibiting coalescence and ion combination rules have been devised to predict the behavior of various ion pairs. Various explanations have been proposed, most implying a gas-liquid interfacial mechanism. Christenson and Yaminsky [44] have reported a correlation between the inverse Marangoni factor, $(d_\gamma/[dc])^{-2}$, and coalescence inhibition ability for several different

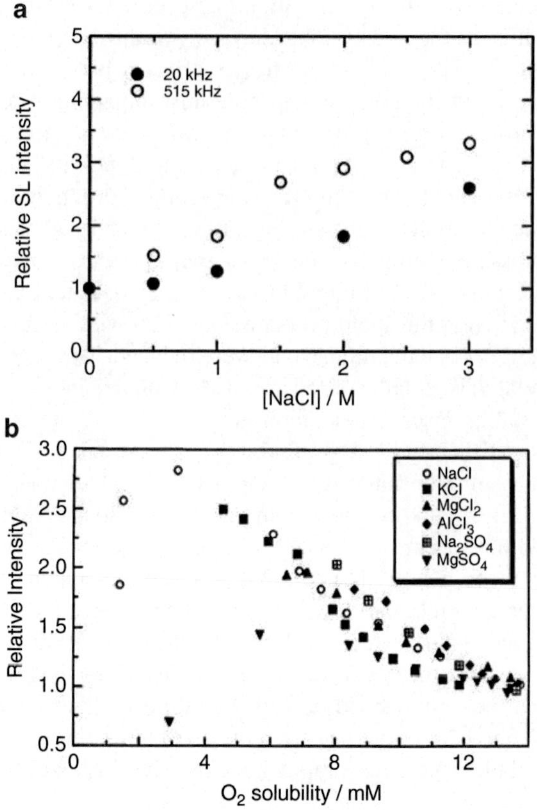

Fig. 14.5 (**a**) Normalised MBSL intensity as a function of NaCl concentration under sonication at 20 kHz and 515 kHz (reprinted with permission from the American Chemical Society). (**b**) Normalised MBSL intensity as a function of O_2 solubility for various salts under 515 kHz sonication [39] (reprinted with permission from Elsevier)

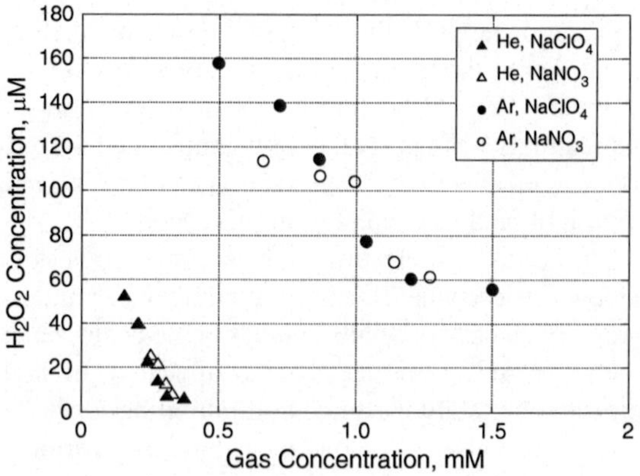

Fig. 14.6 (**a**) Concentration of hydrogen peroxide formed after 30 min sonication as a function of dissolved gas concentration for helium and argon saturated $NaNO_3$ and $NaClO_4$ solutions (data adapted from reference [41])

electrolyte species. However, some solutes which increase surface tension are also observed to inhibit coalescence. Marčelja [47] has argued that the tendency of certain ions to concentrate and others to avoid the bubble surface results in ion separation resulting in electric double-layer repulsions. Craig et al. [43] proposed a mechanism based on the reduction in the range of the hydrophobic attraction between bubbles through their influence on the water structure. Henri et al. [50] have provided evidence in mixed electrolyte systems that surface tension gradients and Gibbs elasticity do not correlate well with coalescence inhibition and instead, propose a mechanism based on ion separation in the interfacial region leading to electric double layer repulsion, and reduction in the magnitude of the hydrophobic attraction. Weissenborn and Pugh [45] calculated the film rupture thickness to be much greater than the distance over which short-range van der Waals and electrostatic forces are significant leaving the long-range hydrophobic force as the only remaining factor in explaining their observed electrolyte coalescence inhibition. They have proposed that the hydrophobic force is affected by the reduction of dissolved gas, which decreases the microscopic bubble concentration therefore weakening a bridging attraction between macroscopic bubbles.

In their original MBSL study, Wall et al. [39] compared their data with the coalescence data from the non-acoustic systems and found little correlation leading them to conclude that coalescence effects were not important in the SL enhancement effect. In subsequent years, a relatively simple method was developed to quantify acoustic bubble coalescence through measuring the volume change (ΔV_T) in a capillary tube during sonication [51]. Using this method, Brotchie et al. [41] investigated coalescence in a range of electrolytes including those found to have a pronounced effect on coalescence in quasi-static systems (e.g., NaCl, KCl) and those found to have little or no discernable effect (e.g., HCl, $NaClO_4$). This investigation revealed several fundamental differences between the results from the quasi-static (non-acoustic) and the acoustic systems. Firstly, the concentrations required to induce a discernable drop in the extent of coalescence in the acoustic system far exceed those reported in the literature. For example, Craig et al. [52] reported a transition concentration for NaCl of the order 10^{-2} M, with effectively 100 % coalescence inhibition at 10^{-1} M. Although an acoustic transition concentration was not determined, effectively no inhibition was observed at concentrations below 0.5 M and as can be seen in Fig. 14.7a, at 1 M there is only a 10–20% decrease for all salts studied with the exception of HCl. The salting out parameters of the different salts permit the dissolved gas concentration to be easily calculated [53, 54]. It can be seen in Fig. 14.7b that there is a very good correlation between the dissolved argon concentration and the extent of coalescence for all salts studied. The same trend was found in other gases, air and helium. Thus, it is clear that coalescence in an acoustic field is a function of gas concentration and that differences in the inhibition capacity of the different salts can be rationalised in terms of their differing 'salting out' capacities. For example, in argon the 'salting out' constant (Sechenov constant) of HCl is 3.55×10^{-2} m^3 $kmol^{-1}$, whereas for NaCl it is 1.43×10^{-1} m^3 $kmol^{-1}$.

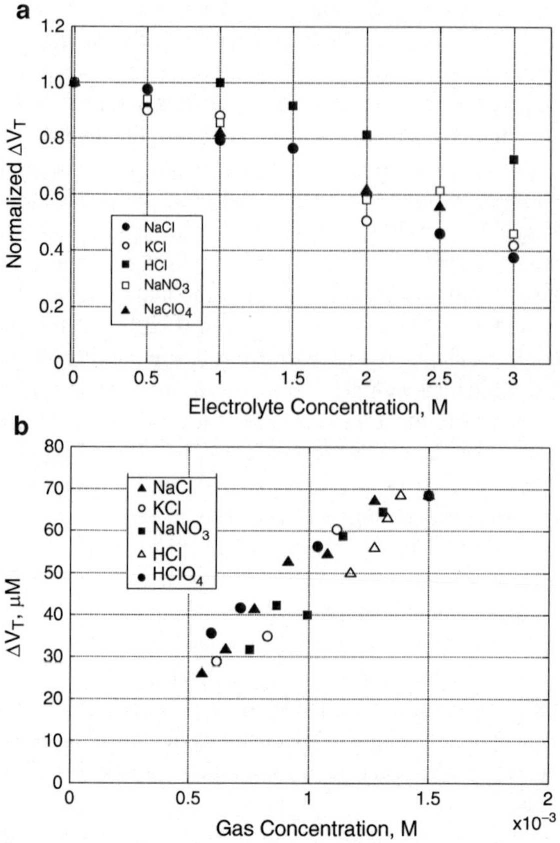

Fig. 14.7 (**a**) Normalised extent of bubble coalescence (ΔV_T) as a function of bulk electrolyte concentration. (**b**) Absolute ΔV_T as a function of dissolved gas concentration. Solutions were argon-saturated and sonicated at 355 kHz (figures adapted from reference [41])

The acoustic bubble size, determined through a pulsed MBSL method developed by Lee at al. [30], was also found to obey a similar dependence on gas concentration as did the coalescence in the same electrolyte solutions [41], as can be seen in Fig. 14.8. It can be inferred from these results that gas concentration controls the extent of coalescence, which itself is the main determinant of the bubble size in an acoustic field.

The interpretation of trends in MBSL and sonochemical yield with electrolyte concentration needs to be revised in light of the aforementioned finding as changes in bubble size distribution and number population not only determine the number of cavitation events occurring but will have a marked effect on sound wave transmission and the local environment surrounding bubbles, influencing collapse symmetry.

Whereas the amount of gas in the system influences the extent of bubble–bubble coalescence and, in turn, the bubble radius, the composition of the gas atmosphere within the bubble plays a crucial role in determining the conditions of collapse.

Fig. 14.8 Bubble radius as a function of the dissolved gas concentration for a solution of argon-saturated NaCl under sonication at 515 kHz (data adapted from reference [41])

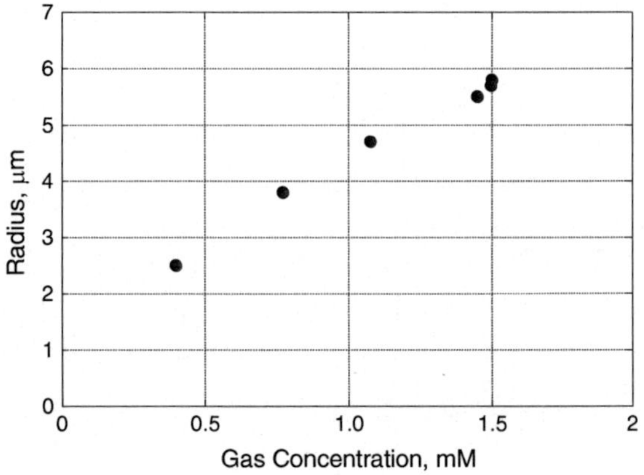

Simple thermodynamic treatment of gas compression stipulates that the maximum temperature ($T_{\max}$) attained is dependent on the adiabatic index (γ) of the gas:

$$T_{max} = T_0\left\{\frac{P_m(\gamma - 1)}{P_v}\right\}$$

where T_0 is the ambient temperature, P_m is the pressure in the liquid (sum of the hydrostatic and acoustic pressures) and P_v is the pressure in the bubble at its maximum size, usually assumed to be equal to the vapour pressure of the liquid.

If the condition of adiabaticity were to be true, all monoatomic gas (i.e., rare gas) bubbles would produce the same temperature upon collapse according to this model. It was early recognised however [55, 56], that the MBSL intensity varied markedly down the noble gas group and, as shown in Fig. 14.9, exhibits a strong dependence on the thermal conductivity of the gas. Briefly, it was proposed that as the thermal conductivity increases, a greater amount of energy will be lost from the bubble, reducing the collapse temperature, and in turn lowering the SL luminosity. In other words, greater thermal conductivity decreases the adiabaticity of collapse.

A more advanced theoretical temperature model taking into consideration thermal conductivity losses proposed by Young [55], yields quite good correlation between the observed MBSL intensity and the gas temperature, as can be seen in Fig. 14.10a. The near linearity of the log-log plot is consistent with the $L \alpha T^4$ dependence for blackbody radiation (although for reasons outlined earlier, it has since been shown that a blackbody mechanism is untenable) and this was interpreted as very strong support for the hot-spot model. Although solubility effects also play a very large role in determining the SL luminosity, SBSL experiments substantiate the MBSL data. Figure 14.10b contains SBSL spectra for the rare gases; the trend of luminosity with thermal conductivity is apparent.

Using the spectroscopic technique previously described, Didenko et al., investigated the collapse temperature for rare gases in organic solvents (octanol and dodecane) with $Cr(CO)_6$ as the spectroscopic probe [58]. They observed a trend in temperature consistent with that predicted based on differences in thermal

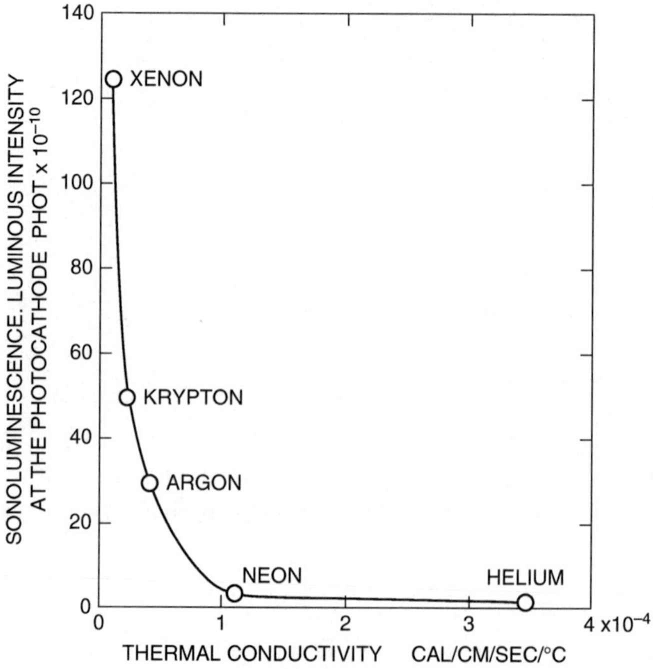

Fig. 14.9 SL intensity as a function of thermal conductivity for the rare gases [55] (reprinted with permission from the American Institute of Physics)

conductivity, with the temperature increasing in the order: He < Ne < Ar < Kr < Xe. In octanol, temperatures ranged from between 3,800 K for He and 5,100 K for Xe; in dodecane the range was between 2,600 K and 3,800 K.

Although the single bubble experiment in Fig. 14.10b and the aforementioned multi-bubble work of Didenko et al. does support the hypothesis that thermal conductivity is a defining parameter of SL emission intensity, an alternative explanation attributes the trend in multi-bubble systems to the gas solubility, rather than the thermal conductivity. If the SL data from Fig. 14.9 is re-plotted as a function of the gas solubility, as shown below in Fig. 14.11, a very good correlation is found. This explanation is supported by several studies by Okitsu et al. [42, 59]. They found sonochemical activity to obey the same trend for the rare gases as for thermal conductivity, SL luminosity and temperature, as described above. This is evident in Fig. 14.12, which shows the sonochemical reduction of Au(III) to colloidal gold as a function of sonication time for different gas atmospheres.

Okitsu et al. however, found that the effective sonochemical bubble temperature (using the MRR method) was the same for all rare gases leading them to suggest that sonochemical trends for the gases are due not to differences in thermal conductivity, but solubility in water. Figure 14.13a contains bubble temperatures obtained for the different gases studied, which all yield a value of about 3,900 K. It was proposed that despite the SL intensity and temperature (as measured using

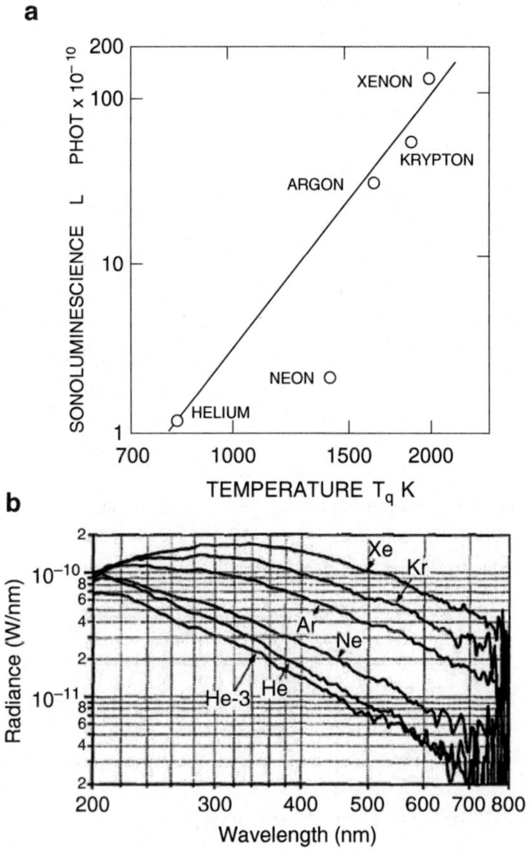

Fig. 14.10 (**a**) MBSL intensity as a function of the theoretical collapse temperature for the rare gases (reprinted with permission from reference [55], copyright 1976, Acoustical Society of America). (**b**) SBSL spectra from different rare gases [57] (reprinted with permission from Elsevier)

SL thermometry) exhibiting a strong dependence on thermal conductivity, the more spatially and temporally averaged temperature (from the MRR method), which is more representative for sonochemical processes, does not show such dependence. It is probable that it is actually the water vapour content within the bubble not the gas itself that dictates the bubble temperature.

When they examined the sonochemical production of hydrogen peroxide in water saturated with different rare gases and their mixtures, a very strong correlation between sonochemical yield and gas solubility became apparent, as shown in Fig. 14.13b. This was attributed to a correlation between the gas concentration and the number of active cavities. Although a sound physical explanation for this relationship is lacking, it is likely that at higher gas concentration, rates of bubble nucleation and growth (through rectified diffusion) are increased, tending to increase the bubble population. Irrespective of the exact mechanism, support for this argument can be found in the results of Brotchie et al. described earlier, in which the extent of bubble coalescence (indicative of the total bubble population) correlated strongly with gas concentration [41].

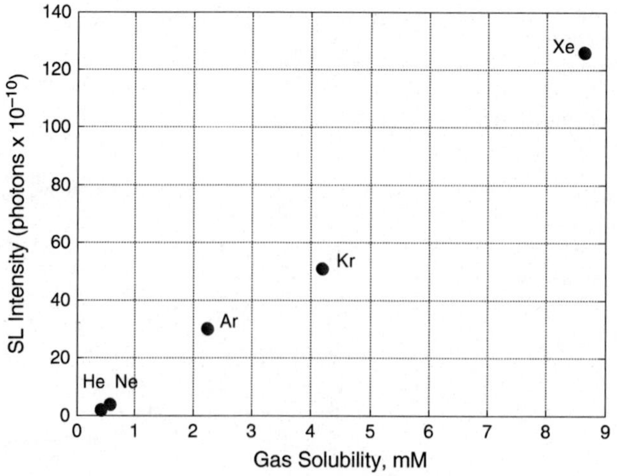

Fig. 14.11 Sonoluminescence intensity (estimated from Fig. 14.9) plotted as a function of the dissolved rare gas concentration at 273 K and 1 atm.

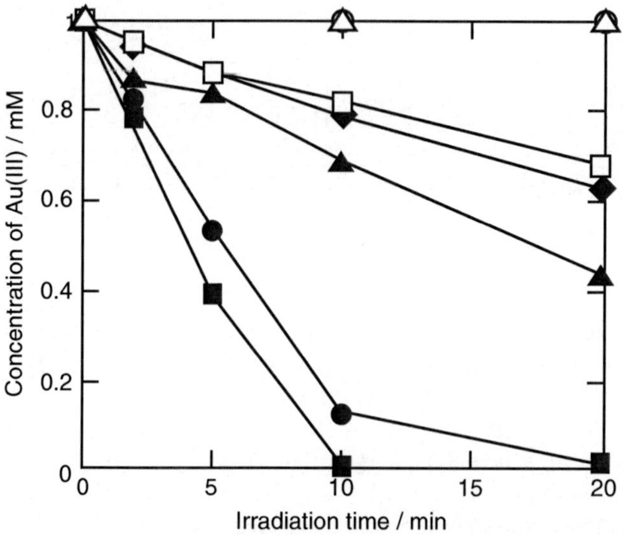

Fig. 14.12 Concentration of Au(III) as a function of sonication time in different gas atmospheres: (■) Kr, (●) Ar, (▲) Ne, (◆) He, (□) N_2, (○) CO_2, (Δ) CH_4 [59] (reprinted with permission from the Chemical Society of Japan)

Vu [60] also used the MRR method described earlier to investigate differences between two rare gases, helium and argon, and the effect of various organic SL quenchers. Figure 14.14 shows the MBSL intensity for both gases as functions of ethanol bulk concentration.

It can be noticed that the extent of SL quenching (and the extent of temperature lowering, as shown later in Fig. 14.15) is significantly lower in helium saturated

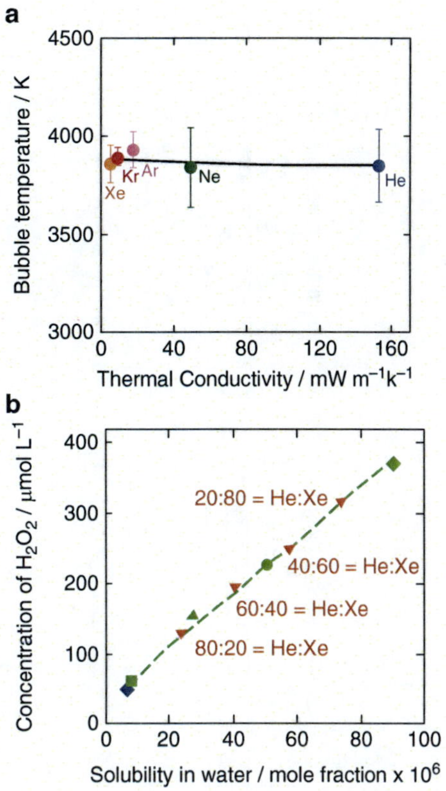

Fig. 14.13 (**a**) Bubble temperatures estimated using the MRR method as a function of thermal conductivity for the rare gases. (**b**) Hydrogen peroxide concentration following sonication of pure water as a function of gas solubility in different rare gases: (◆) He; (■) Ne; (▲) Ar; (●) Kr; (◆) Xe; (▼) He/Xe mixture [42] (reprinted with permission from the American Chemical Society)

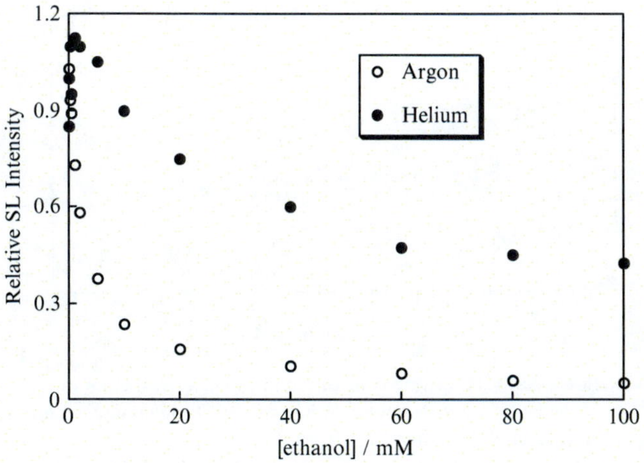

Fig. 14.14 The effect of ethanol concentration on the relative SL intensity in argon and helium saturated water normalised with respect to pure water) at 363 kHz [60]

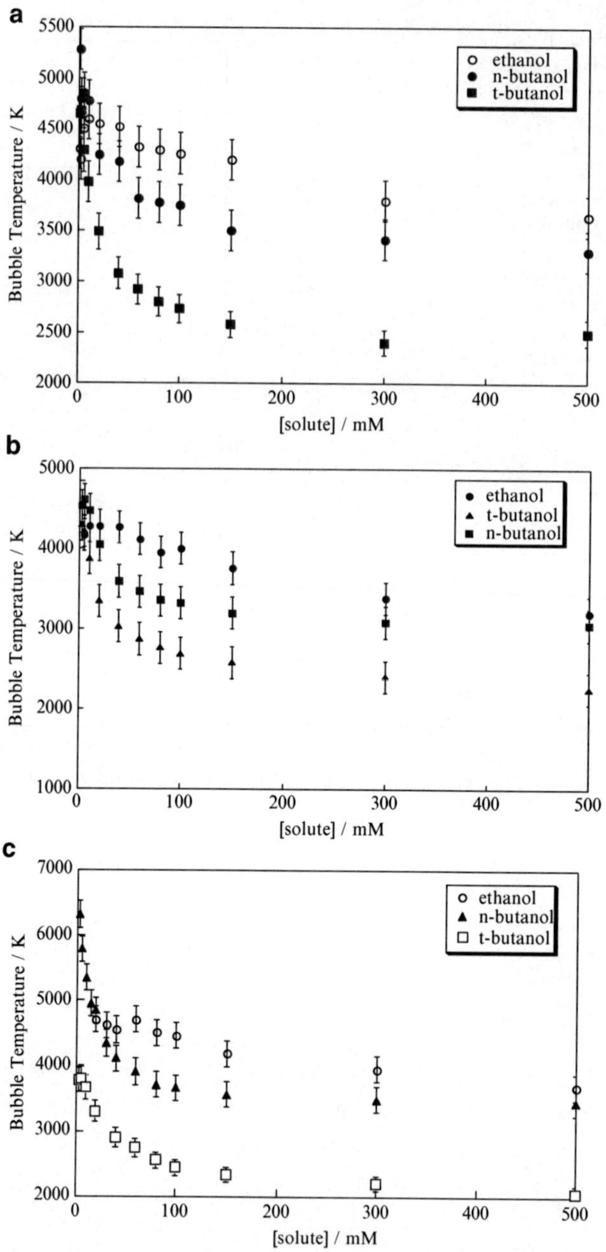

Fig. 14.15 Cavitation bubble temperature at different alcohol concentrations at 363 kHz in (**a**) helium saturated and (**b**) argon saturated solutions. (**c**) Temperature at different alcohol concentrations in argon saturated solutions at 1,056 kHz [60]

alcohol solutions compared to the argon saturated solutions. As has been discussed, the 'peak' temperature (or temperature during the emission of SL) of a helium bubble would be expected to be lower than that of an argon bubble. The extent of SL quenching is related to the amount of hydrocarbon products formed within the bubble. The greater extent of SL quenching in argon therefore suggests that a greater extent of hydrocarbon decomposition occurs for argon bubbles. The amount of hydrocarbon products generated in helium saturated solutions was found to be significantly lower than that in argon saturated solutions. In some cases, the product amounts were insufficient to measure the bubble temperatures.

In a multi-bubble system it is difficult to distinguish between population and temperature effects, and both factors will acutely influence sonochemical yields. The higher solubility of argon (~60 mg kg^{-1}) compared with helium (~1.5 mg kg^{-1}) leads to a greater bubble population (presumably through a faster rate of nucleation and bubble growth). Coalescence studies have shown that although a greater extent of coalescence is observed in argon saturated solutions, this is only about double that for helium. In contrast, the chemical activity measured in the same system (see Fig. 14.6) is about an order of magnitude different. Therefore, although solubility effects contribute to the observed differences in sonochemical activity, this factor alone would seem insufficient to account for this behaviour and consequently a higher temperature for the argon bubbles can be deduced. Thus, this observation is in contrast with the work of Okitsu et al.

Assuming a cooler temperature for helium, this means the amount of heat energy available for pyrolysis is less, which would produce less hydrocarbon products. This ultimately would lead to a relatively smaller SL quenching compared to that observed in argon saturated solutions. The relative change in bubble temperature would also be less in helium saturated solutions due to the lower amount of hydrocarbon products generated. Solubility differences would re-enforce this.

Experimental temperature determination reveals little difference between the gases at 355 kHz, as is evident in Figs. 14.15a and b, similar to the results of Okitsu et al. The likely explanation for this seemingly contradictory result is that the MRR method, which requires the presence of a significant amount of hydrocarbon within the bubble core, yields a temperature value that is dictated by the hydrocarbon itself. In other words, the probe itself determines the temperature measured. This becomes apparent at higher frequency (1,056 kHz; Fig. 14.15c) where it is clear that the temperature that is obtained when one extrapolates to zero hydrocarbon concentration varies depending on the probe molecule. The lower temperature for *tert*-butanol is attributed to a greater rate of hydrocarbon product formation.

The use of ultrasound in both the synthesis and crystallisation of a broad array of both organic and inorganic materials has been intensively researched and is well documented [61–64]. An application of ultrasound that has received relatively less attention however, is in the dissolution of colloidal particles. Prakash and Ghosh [65] reported on the dissolution of silver colloids under 1 MHz ultrasound irradiation, proposing that the silver is oxidised by sonochemically produced hydroxyl radicals. Sostaric et al. [66] investigated the dissolution of MnO_2 colloids in the presence of aliphatic alcohols at a lower frequency of 20 kHz. They found that the dissolution rate correlated strongly with the alcohol Gibbs surface-excess (at the air–water interface). A reductive dissolution mechanism was proposed as follows:

$$H_2O \rightarrow H^\bullet + {}^\bullet OH$$

$${}^\bullet OH + {}^\bullet OH \rightarrow H_2O_2$$

$$H_2O_2 + 2H^+ + MnO_2 \rightarrow Mn^{2+} + 2H_2O + O_2$$

The alcohol serves as a radical scavenger and accelerates the dissolution through the following reactions:

$$H^\bullet({}^\bullet OH) + RHOH \rightarrow {}^\bullet ROH + H_2(H_2O)$$

$$2^\bullet ROH + 2H^+ + MnO_2 \rightarrow Mn^{2+}(aq) + 2H_2O + 2RO$$

This mechanism is the same as that implicated in the sonochemical reduction of metal salts to metal colloids.

Sostaric et al. [67] also found that dissolution of CdS could be achieved through sonochemical reduction of the sulphur by hydroxyl radicals, hydrogen peroxide and superoxide:

$$Cds + 2^\bullet OH \rightarrow Cd^{2+}(aq) + S^0 + 2OH^-$$

$$Cds + H_2O_2 \rightarrow Cd^{2+}(aq) + S^0 + 2OH^-$$

$$Cds + 2O_{2^-} + 4H_2O \rightarrow Cd^{2+}(aq) + S^0 + 2H_2O_2 + 4OH^-$$

These dissolution reactions were found to be inhibited by the presence of Na_2S or alcohol, which scavenge the active oxidising radicals.

14.3 Conclusions

The uses of inorganic metal compounds and rare gases to probe the conditions of cavitation collapse have become some of the most important methods available in fundamental ultrasonics. Quantitative determination of collapse temperatures and pressures, and qualitative determination of fundamental aspects of the nature of the cavitation field have been achieved, largely through SL spectroscopic methods. The presence of salts has a marked influence on properties on the acoustic systems, such as the extent of coalescence and bubble size, and the sonochemical activity and SL intensity.

References

1. Neppiras EA, Noltingk BE (1951) Cavitation produced by ultrasonics: theoretical conditions of the onset of cavitation. Proc Phys Soc B 64:1032–1038
2. Srinivasan D, Holroyd LV (1961) Optical spectrum of the sonoluminescence emitted by cavitated water. J Appl Phys 32:446–449
3. Gunther P, Heim E, Borgstedt HV (1959) Uber die Kontinuierichen Sonolumineszenzspektren Wassriger Losungen. Z Elektrochem 63:43–47
4. Gompf B, Gunther R, Nick G, Pecha R, Eisenmenger W (1997) Resolving sonoluminescence pulse width with time-correlated single photon counting. Phys Rev Lett 79:1405–1408
5. Sehgal C, Steer RP, Sutherland RG, Verrall RE (1979) Sonoluminescence of argon saturated alkali-metal salt-solutions as a probe of acoustic cavitation. J Chem Phys 70:2242–2248
6. Flint EB, Suslick KS (1991) Sonoluminescence from alkali-metal salt-solutions. J Phys Chem 95:1484–1488
7. Xu HX, Eddingsaas NC, Suslick KS (2009) Spatial separation of cavitating bubble populations: the nanodroplet injection model. J Am Chem Soc 131:6060–6061
8. Flannigan DJ, Suslick KS (2007) Emission from electronically excited metal atoms during single-bubble sonoluminescence. Phys Rev Lett 99:134301
9. Didenko YT, McNamara WB, Suslick KS (2000) Molecular emission from single-bubble sonoluminescence. Nature 407:877–879
10. McNamara WB, Didenko YT, Suslick KS (2003) Pressure during sonoluminescence. J Phys Chem B 107:7303–7306
11. Suslick KS, Flannigan DJ (2008) Inside a collapsing bubble: sonoluminescence and the conditions during cavitation. Annu Rev Phys Chem 59:659–683
12. Suslick KS, Hammerton DA, Cline RE Jr (1986) The sonochemical hot spot. J Am Chem Soc 108:5641–5642
13. Suslick KS, Flint EB, Grinstaff MW, Kemper KA (1993) Sonoluminescence from metal carbonyls. J Phys Chem 97:3098–3099
14. McNamara WBI, Didenko YT, Suslick KS (1999) Sonoluminescence temperatures during multi-bubble cavitation. Nature 401:772–775
15. Guan JF, Matula TJ (2003) Time scales for quenching single-bubble sonoluminescence in the presence of alcohols. J Phys Chem B 107:8917–8921
16. Ashokkumar M, Mulvaney P, Grieser F (1999) The effect of pH on multibubble sonoluminescence from aqueous solutions containing simple organic weak acids and bases. J Am Chem Soc 121:7355–7359
17. Ashokkumar M, Crum LA, Frensley CA, Grieser F, Matula TJ, McNamara WB, Suslick KS (2000) Effect of solutes on single-bubble sonoluminescence in water. J Phys Chem A 104:8462–8465
18. Flint EB, Suslick KS (1991) The temperature of cavitation. Science 253:1397–1399
19. Didenko YT, McNamara WB, Suslick KS (1999) Hot spot conditions during cavitation in water. J Am Chem Soc 121:5817–5818
20. Didenko YT, McNamara WB, Suslick KS (1999) Temperature of multibubble sonoluminescence in water. J Phys Chem A 103:10783–10788
21. Misik V, Miyoshi N, Riesz P (1995) EPR spin-trapping studies of the sonolysis of H_2O/D_2O mixtures: probing the temperatures of cavitation regions. J Phys Chem 99:3605–3611
22. Hart EJ, Fischer CH, Henglein A (1990) Sonolysis of hydrocarbons in aqueous solution. Radiat Phys Chem 36:511–516
23. Tauber A, Mark G, Schuchmann H-P, von Sonntag C (1999) Sonolysis of tert-butyl alcohol in aqueous solution. J Chem Soc Perkin Trans 2:1129–1136
24. Rae J, Ashokkumar M, Eulaerts O, von Sonntag C, Reisse J, Grieser F (2005) Estimation of ultrasound induced cavitation bubble temperatures in aqueous solutions. Ultrason Sonochem 12:325–329

25. Ciawi E, Ashokkumar M, Grieser F (2006) Limitations of the methyl radical recombination method for acoustic cavitation bubble temperature measurements in aqueous solutions. J Phys Chem B 110:9779–9781
26. Ciawi E, Rae J, Ashokkumar M, Grieser F (2006) Determination of temperatures within acoustically generated bubbles in aqueous solutions at different ultrasound frequencies. J Phys Chem B 110:13656–13660
27. Sunartio D, Ashokkumar M, Grieser F (2007) Study of the coalescence of acoustic bubbles as a function of frequency, power, and water-soluble additives. J Am Chem Soc 129:6031–6036
28. Abe S, Choi P-K (2009) Spatiotemporal separation of Na-atom emission from continuum emission in sonoluminescence. Jpn J Appl Phys 48:07GH02
29. Brotchie A, Grieser F, Ashokkumar M (2009) The effect of power and frequency on acoustic cavitation bubble size distributions. Phys Rev Lett 102:084302
30. Lee J, Ashokkumar M, Kentish S, Grieser F (2005) Determination of the size distribution of sonoluminescence bubbles in a pulsed acoustic field. J Am Chem Soc 127:16810–16811
31. Lepoint-Mullie F, Voglet N, Lepoint T, Avni R (2001) Evidence for the emission of 'alkali-metal-noble-gas' van der Waals molecules from cavitation bubbles. Ultrason Sonochem 8:151–158
32. Matula TJ, Roy RA, Mourad PD, McNamara WB, Suslick KS (1995) Comparison of multi-bubble and single-bubble sonoluminescence spectra. Phys Rev Lett 75:2602–2605
33. Xu H, Glumac NG, Suslick KS (2010) Temperature inhomogeneity during multibubble sonoluminescence. Angew Chem 122:1097–1100
34. Bunkin NF, Kiseleva OA, Lobeyev AV, Movchan TG, Ninham BW, Vinogradova OI (1997) Effect of salts and dissolved gas on optical cavitation near hydrophobic and hydrophilic surfaces. Langmuir 13:3024–3028
35. Attard P (2003) Nanobubbles and the hydrophobic attraction. Adv Colloid Interface Sci 104:75–91
36. Borkent BM, Dammer SM, Schonherr H, Vancso GJ, Lohse D (2007) Superstability of surface nanobubbles. Phys Rev Lett 98:204502
37. Brotchie A, Ashokkumar M, Zhang XH (2010) Behaviour of interfacial nanofluids under ultrasound irradiation. Manuscript under review
38. Ceccio SL, Gowing S, Shen Y (1997) The effects of salt water on bubble cavitation. J Fluid Eng 119:155–163
39. Wall M, Ashokkumar M, Tronson R, Grieser F (1999) Multibubble sonoluminescence in aqueous salt solutions. Ultrason Sonochem 6:7–14
40. Tronson R, Ashokkumar M, Grieser F (2002) Comparison of the effects of water-soluble solutes on multibubble sonoluminescence generated in aqueous solutions by 20- and 515-kHz pulsed ultrasound. J Phys Chem B 106:11064–11068
41. Brotchie A, Statham T, Zhou M, Devandra L, Grieser F, Ashokkumar M (2010) Acoustic bubble sizes, coalescence and sonochemical activity in aqueous electrolyte solutions saturated with different gases. Langmuir 26:12690–12695
42. Okitsu K, Suzuki T, Takenaka N, Bandow H, Nishimura R, Maeda Y (2006) Acoustic multibubble cavitation in water: a new aspect of the effect of a rare gas atmosphere on bubble temperature and its relevance to sonochemistry. J Phys Chem B 110:20081–20084
43. Craig VSJ, Ninham BW, Pashley RM (1993) Effect of electrolytes in bubble coalescence. Nature 364:317–319
44. Christenson HK, Yaminsky VV (1995) Solute effects on bubble coalescence. J Phys Chem 99:10420
45. Weissenborn PK, Pugh RJ (1996) Surface tension of aqueous solutions of electrolytes: relationship with hydration, oxygen solubility, and bubble coalescence. J Colloid Interface Sci 184:550–553
46. Deschenes LA, Barret J, Muller LJ, Fourkas JT, Mohanty U (1998) Inhibition of bubble coalescence in aqueous solutions. 1. Electrolytes. J Phys Chem B 102:5115–5119

47. Marcelja S (2006) Selective coalescence of bubbles in simple electrolytes. J Phys Chem B 110:13062–13067
48. Henri CL, Parkinson L, Ralston JR, Craig VSJ (2008) A mobile gas-water interface in electrolyte solutions. J Phys Chem C 112:15094–15097
49. Christenson HK, Bowen RE, Carlton JA, Denne JRM, Lu Y (2008) Electrolytes that show a transition to bubble coalescence inhibition at high concentrations. J Phys Chem C 112:794–796
50. Henri CL, Dalton CN, Scruton L, Craig VSJ (2007) Ion-specific coalescence on bubbles in mixed electrolyte solutions. J Phys Chem C 2007:1015–1023
51. Lee J, Kentish SE, Ashokkumar M (2005) The effect of surface active solutes on bubble coalescence in the presence of ultrasound. J Phys Chem B 109:5095–5099
52. Craig VSJ, Ninham BW, Pashley RM (1993) The effect of electrolytes on bubble coalescence in water. J Phys Chem 97:10192–10197
53. Lang W, Zander R (1986) Salting-out of oxygen from aqueous electrolyte solutions: prediction and measurement. Ind Eng Chem Fundam 25:775–782
54. Hermann C, Dewes I, Schumpe A (1995) The estimation of gas solubilities in salt solutions. Chem Eng Sci 50:1673–1675
55. Young RF (1976) Sonoluminescence from water containing dissolved gases. J Acoust Soc Am 60:100–104
56. Prudhomme RO, Guilmart T (1957) photogenese ultraviolette par irradiation ultrasonore de l'eau en presence des gaz rares. J Chim Phys 54:336
57. Barber BP, Hiller RA, Lofstedt R, Putterman SJ, Weninger KR (1997) Defining the unknowns of sonoluminescence. Phys Rep 281:65–143
58. Didenko YT, McNamara WB, Suslick KS (2000) Effect of noble gases on sonoluminescence temperatures during multibubble cavitation. Phys Rev Lett 84:777–780
59. Okitsu K, Yue A, Tanabe S, Matsumoto H, Yobiko Y, Yoo Y (2002) Sonolytic control of rate of gold (III) reduction and size of formed gold nanoparticles: relation between reduction rates and sizes of formed nanoparticles. Bull Chem Soc Jpn 75:2289–2296
60. Vu T (2004) MSc. Thesis, Effects of organic and inorganic solutes on multibubble sonoluminesence in aqueous solutions. The University of Melbourne
61. Ashokkumar M, Grieser F (1999) Ultrasound assisted chemical processes. Rev Chem Eng 15:41–83
62. Suslick KS, Price GJ (1999) Applications of ultrasound to materials chemistry. Annu Rev Mater Sci 29:295–326
63. Gedanken A (2004) Using sonochemistry for the fabrication of nanomaterials. Ultrason Sonochem 11:47–55
64. Ruecroft G, Hipkiss D, Ly T, Maxted N, Cains PW (2005) Sonocrystallization: the use of ultrasound for improved industrial crystallization. Org Process Res Dev 9:923–932
65. Prakash P, Ghosh AK (1959) The influence of ultrasonic waves on silver hydrosol. J Colloid Sci 14:338–342
66. Sostaric JZ, Mulvaney P, Grieser F (1995) Sonochemical dissolution of MnO_2 colloids. J Chem Soc Faraday Trans 91:2843–2846
67. Sostaric JZ, Caruso-Hobson RA, Mulvaney P, Grieser F (1997) Ultrasound-induced formation and dissolution of colloidal CdS. J Chem Soc Faraday Trans 93:1791–1795

Chapter 15
Introductory Experiments in Sonochemistry and Sonoluminescence

Pankaj, Mayank Verma, Shikha Goyal, and Adam Brotchie

Abstract Having discussed many aspects of sonochemistry and its application in the previous chapters, a few introductory experiments in sonochemistry and sonoluminescence are presented in this chapter. These physical demonstrations are especially aimed at beginners in the field of sonochemistry making them e.g. aware of the power of ultrasound.

15.1 Introduction

Fascination with a subject increases when one sees its physical demonstration too. One of the authors recalls the very first lecture on Sonochemistry by Prof. T.J. Mason at Coventry Polytechnic, Coventry, UK and then his another lecture at Ultrasonics International, '91, at Le Tuqoute, Paris, France during 1–4 July, 1991, with few basic experiments and their results being shown to demonstrate the power of ultrasound, such as;

1. How a piece of kitchen aluminium foil in water could be eroded in an ultrasonic field due to acoustic cavitation. The foil, after a few minutes of sonication, had several holes.
2. How the reactivity of Zn metal pieces with 5N HCl increased many fold in an ultrasonic bath and the H_2 gas was simultaneously removed from the test solution as a result of the degassing effect of ultrasound.
3. How the oxidation of iodide ions to form molecular iodine by the sonochemically generated radicals can be monitored by the blue colour generated in the presence of freshly prepared starch solution.

Pankaj (✉)
Department of Chemistry, Dayalbagh Educational Institute, AGRA 282 110, India
e-mail: pankaj2@sancharnet.in

Pankaj and M. Ashokkumar (eds.), *Theoretical and Experimental Sonochemistry Involving Inorganic Systems*, DOI 10.1007/978-90-481-3887-6_15,

In the preceding chapters many aspects of sonochemistry and its application have already been discussed in details and now to conclude, few experiments are being discussed here to make the beginners in the field of sonochemistry, especially the undergraduate students, to ride on the sound wave and begin their journey of sonochemistry with some of these experiments, which can be conveniently carried out with an ultrasonic cleaning bath (Fig. 15.1) or an ultrasonic probe (Fig. 15.2) of 20 kHz, available commercially abundantly.

Following are some of the introductory experiments in sonochemistry and sonoluminescence.

1. Erosion and pitting on the metal surface – experiment with aluminium foil.
2. Experiment of Zn metal and dil. HCl under normal and sonicated conditions.
3. Crystallisation of inorganic materials under ultrasound.
4. Effect of ultrasound in the synthesis of simple organic compounds.
5. Effect of ultrasound on the degradation of indicators.
6. Effect of ultrasound on the degradation of phenols.
7. Effect of ultrasound on crude oil.

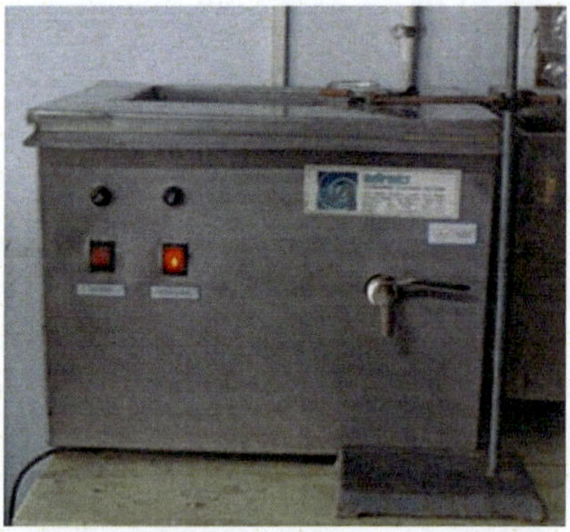

Fig. 15.1 Meltronics ultrasonic bath, 20 kHz, 250 W

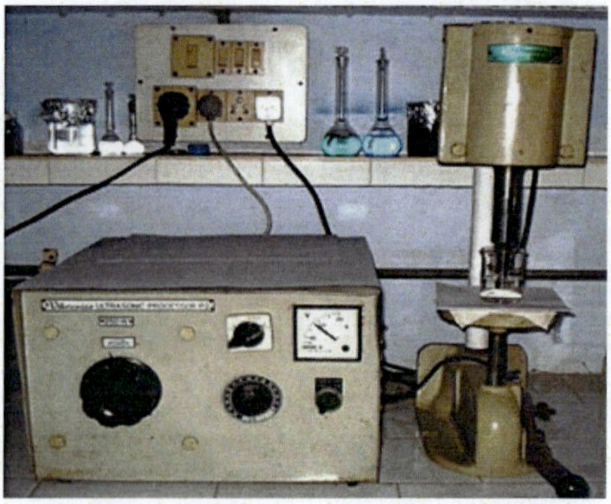

Fig. 15.2 Vibronics processor, 20 kHz, 250 W

8. Effect of ultrasound on the photocatalytic decomposition of KI using TiO_2 doped with rare earths.
9. Sonoluminescence and Sonochemical Luminescence.

15.1.1 Experiment

Objective: To examine the cavitation effect of ultrasound

Procedure: Switch on the ultrasonic cleaning tank filled with water and watch for the position of cavitation activity through the circular rings formed on the surface of water. Hold the Al-foil on the circular ring for 2–3 min and remove. Watch the foil against the light or through an overhead projector for the holes created on it as a result of cavitation.

Result: Since aluminium is a soft metal and the thin foil gets eroded easily due to the bursting of bubbles on its surface, holes are created.

Abrasion and holes created by ultrasound on aluminum foil

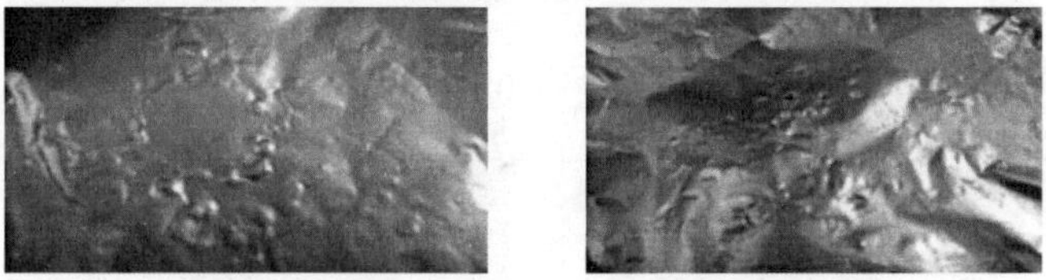

15.1.2 Experiment

Objective: To monitor the reaction of Zn metal pieces with dil. HCl

Reagents: 50 mL of 5 N HCl solution in a 100 mL conical flask and 15–20 g of Zn granules

Procedure: Add a few pieces of Zn metal granules to the 5 N HCl solution and watch for the bubble formation due to the reaction between Zn metal and the acid, producing H_2 gas, the intensity of which, however, decreases after some time. Now, place the flask in the ultrasonic bath and watch the change.

Result: The diminishing chemical reaction between Zn metal and the acid solution suddenly is accelerated and bubble formation intensifies many fold. However, these bubbles under the effect of ultrasound are quickly removed to the top layer of the reaction mixture.

Reason: Initial fast reaction of Zn metal with acid decreases due to a thin oxide coating on the surface of the metal, hindering the further intimate contact of metal with acid. However, when the solution flask was immersed in the ultrasonic cleaning bath, the surface of the metal is cleaned by the agitation generated due to mechanical vibration and acoustic cavitation, exposing the fresh metal surface for reaction with the acid. As a secondary effect of ultrasound, the H_2 gas bubbles are quickly removed to the surface of the reaction solution and thus in spite of a much faster reaction, the solution looks cleaner.

15.1.3 Experiment

Objective: Crystallisation of inorganic materials under normal and ultrasonic conditions

Procedure: Prepare a saturated solution of inorganic salts in water and crystallize by keeping over ultrasonic bath for an hour or until the crystals just appear to form. Differentiate the morphology of crystals formed under normal and ultrasonic conditions under microscope. SEM pictures of some of such materials are being given here for comparison.

Observation: As can be seen from the following SEM pictures that there is growth and breaking of crystals in $PbCl_2$ and PbI_2, blistering in Au_2S, surface cleaning in PtS_2, crystal formation in $PdO.nH_2O$, doping in $CrCl_3$ – $CeCl_3$ and elongation in $CuAc_2 - CeCl_3$ in the materials synthesized under sonicated condition compared to the synthesis of same material under normal condition.

Crystals of inorganic materials as seen through digital camera

Inorganic materials	Unsonicated	Sonicated	Changes
$MnCl_2$ –YCl_3			Crystal geometry
$CoCl_2$ – YCl_3			Colour
$CuSO_4$ – YCl_3	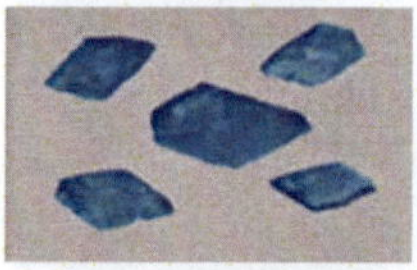	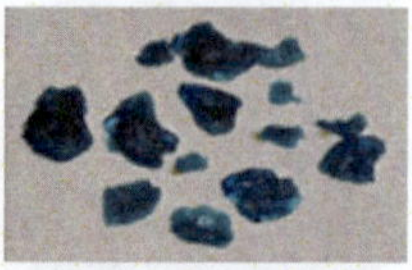	Colour
$CuSO_4$ – $DyCl_3$			Crystal geometry
$MnCl_2$ – $CeCl_3$			Crystal geometry

Crystals of inorganic materials as seen through SEM

Inorganic materials	Unsonicated	Sonicated	Changes
$CuSO_4 - DyCl_3$	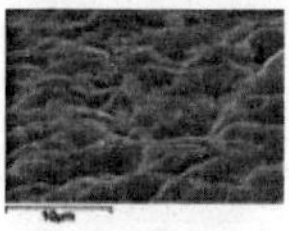		Crystallinity
$CuAc_2 - DyCl_3$			Fragmentation
$CeCl_3 - CuAc_2$			Rod-shape crystals
$CeCl_3 - CrCl_3$			Crystallinity

15.1.4 Experiment

Objective: Effect of ultrasound on simple one step organic syntheses.

Theory: Cavitational properties of ultrasound accelerate those organic reactions which involving free radical mechanism, hydrolysis, protonation, etc. However the sonochemical effects are negative for those reactions in which one of the reactants is volatile.

Procedure: Effects of ultrasound on the following reaction can be seen clearly, when the reactions are carried out both in the ultrasonic field as well as under normal conditions. The yield in most of the reactions increased due to cavitation and mass flow but a decrease could also be sometimes expected due to volatility of one of the reagents. The average increase/decrease in the yield of product, reported here, is an average of two sets of reactions carried out separately under sonicated and mechanical stirring conditions.

15.1.4.1 Synthesis of Benzanilide

Method of Preparation: Dissolve 1 mL of benzoyl chloride in 5 mL of pure benzene and treat slowly with 1 mL of aniline dissolved in benzene until the odour of the benzoyl chloride was no longer detectable. After vigorous stirring, treat the mixture with dilute HCl. Evaporate to remove excess of amine and the benzene from the solution and recrystallise the product in aqueous alcohol.

$$PhCOCl + H_2NPh \rightarrow PhCONHPh + HCl$$

Preparation of benzanilide (yield in grams)			
	Normal condition	Under sonication	% Variation
First set of experiment	0.64	0.80	(+) 20
Second set of experiment	0.62	0.79	(+) 17
Average	0.63	0.795	(+) 18.5

15.1.4.2 Synthesis of Phenylbenzoate

Method of Preparation: Add 20 mL of 5% sodium hydroxide to about 1 g of phenol in a conical flask followed by the drop-wise addition of 2 mL of benzoyl chloride with vigorous shaking / sonication, after stoppering the flask securely in between the each addition of the reagent. Cool the mixture throughout the addition and shake further for 5–10 min after the final addition of benzoyl chloride. Filter the product, wash with water and recrystallise in alcohol.

$$PhOH + C_6H_5COCl \rightarrow C_6H_5COOPh + HCl$$

Preparation of phenylbenzoate (yield in grams)			
	Normal condition	Under sonication	% Variation
First set of experiment	1.10	1.56	(+)46
Second set of experiment	1.05	1.40	(+)35
Average	1.075	1.48	(+)40.5

15.1.4.3 Synthesis of Bromoderative of Phenol

Method of Preparation: Dissolve 1 g of phenol and 4 mL of liquid Br_2 separately in 10–15 mL of glacial acetic acid. Add Br_2 solution to phenol solution until the decolourisation of bromine ceases. Allow the mixture to stand for 20 min and add more bromine solution, if the colour faded. Pour the liquor into 70 mL of water and filter the product. Wash with water and recrystallise in ethanol.

$$PhOH + Br_2 \rightarrow BrPhOH + HBr$$

Preparation of p-bromophenol (yield in grams)			
	Normal condition (g)	Under sonication (g)	% Variation
First set of experiment	2.20	2.41	(+)19
Second set of experiment	2.15	2.46	(+)31
Average	2.18	2.44	(+)25

15.1.4.4 Synthesis of Acetanilide

Method of Preparation: Add 5 mL acetic anhydride and 5mL glacial acetic acid to 5 mL of aniline in a 150 mL conical flask and boil gently for about 15 min in a round bottom flask fitted with an air condenser. Pour the hot mixture in 200 mL of ice-cold water with continuous stirring. The acetanilide would rapidly crystallize. Wash the preparation with cold water and recrystallise in dil. acetic acid.

$$C_6H_5NH_2 + (CH_3CO)_2O \rightarrow C_6H_5NHCOCH_3 + CH_3COOH$$

Preparation of acetanilide (yield in grams)

	Normal condition (g)	Under sonication (g)	% Variation
First set of experiment	2.850	3.945	(+)27.7
Second set of experiment	2.700	3.900	(+)30.7
Average	2.775	3.9225	(+)29.2

15.1.4.5 Synthesis of Aspirin

Method of Preparation: Add 0.5 mL of H_2SO_4 to a mixture of 5 g of salicylic acid and 5 mL acetic anhydride in a 100 mL flask. Continue stirring and heat on water bath for 15 min, keeping the temperature between 50°C and 60°C. Pour the mixture in about 80 mL of ice cold water, filter and crystallize with a mixture of equal volumes of acetic acid and water.

$$OHC_6H_4COOH + (CH_3CO)_2O \rightarrow CH_3COOC_6H_4COOH + CH_3COOH$$

Preparation of aspirin (yield in gram)

	Normal condition (g)	Under sonication (g)	% Variation
First set of experiment	2.700	3.310	(+)18.4
Second set of experiment	2.600	3.010	(+)13.6
Average	2.650	3.160	(+)16

15.1.4.6 Synthesis of Anthranilic Acid

Method of Preparation: Add 1.8 mL of liquid Br_2, with continuous stirring, to a solution of 6.5 g of NaOH dissolved in 25 mL of ice cooled water. Cool the contents to 0°C and add 5 g of phthalimide, followed by the addition of a solution of 5 g of NaOH, dissolved in 20 mL of water. Heat the mixture to about 80°C, for 3–4 min and filter hot. Cool the filtrate in ice and add conc. HCl slowly to neutralize the solution. Add about 5 mL of glacial acetic acid to precipitate anthranilic acid completely. Filter the product, wash with little cold water and recrystallise in boiling water.

Preparation of anthranilic acid (yield in gram)

	Normal condition (g)	Under sonication (g)	% Variation
First set of experiment	1.260	2.300	(+)45.4
Second set of experiment	1.300	2.320	(+)44.0
Average	1.280	2.310	(+)44.7

15.1.4.7 Synthesis of Benzamide

Method of Preparation: Treat 1 mL of benzoyl chloride with 10 mL of liquor ammonia solution and stir the mixture for 30 min. Recrystallise the amide in aqueous ethanol.

$$PhCOCl + NH_3 \rightarrow PhCONH_2 + HCl(NH_4Cl)$$

Preparation of benzamide (yield in grams)

	Normal condition (g)	Under sonication (g)	% Variation
First set of experiment	0.87	0.24	(−)63
Second set of experiment	0.64	0.23	(−)41
Average	0.75	0.235	(−)52

Result: The decrease in yield of benzamide could be attributed to the degassing effect of ultrasound which removes NH_3 from the solution due to its volatile nature in water.

15.1.5 Experiment

Objective: Examination of the effect of ultrasound on the degradation of Indicators (Murexide, Erichrome black T, Xylenol orange and Fast sulphone black F)

Methodology: The ultrasonic probe is used for the experiment which allows relatively more acoustic energy directly into the system. Indicators were sonicated for different durations and the absorbance was recorded using UV-vis spectrophotometer. The values, however, given in the table below give qualitative idea of the degradation of indicators.

	Absorbance			
	Unsonicated	30 min	60 min	90 min
Murexide	1.09	1.00	0.91	0.82
Erichrome black T	0.23	0.20	0.19	0.18
Xylenol orange	1.38	1.35	1.33	1.31
Fast sulphon black F	0.19	0.19	0.20	0.19

Result: The degradation increased with an increase in sonication time. The ultrasound was effective in degradation of murexide, erichrome black T and xylenol orange indicators due to breaking of some of weak bonds in the molecule, however, the degradation of fast sulphon black F was not as effective due to higher stability of bonds therefore the energy of ultrasound was not sufficient to degrade this indicator. A slight change in the values of absorption could be due to the formation of some precipitate and later its de-agglomeration.

15.1.6 Experiment

Objective: To examine the effect of ultrasound on the degradation of phenol, p-aminophenol, and p-nitrophenol.

Procedure: The percentage degradation of phenol, p-aminophenol, p-nitrophenol was determined volumetrically using 0.1M thiosulphate. The free radical, produced during cavitation, had marked effect on the degradation.

	Time (min)	Volume of hypo consumed (mL)	% Degradation
Phenol	0	00.0	00.0
	60	18.0	28.0
	120	22.4	67.5
p-Aminophenol	0	00.0	00.0
	60	25.3	59.1
	120	26.0	75.0
p-Nitrophenol	0	00.0	00.0
	60	24.0	53.0
	120	26.1	73.0

Result: The sonochemical degradation of phenol is less than either p-aminophenol or p-nitrophenol. However, the sonochemical degradation of p-aminophenol is more than that of p-nitrophenol. This difference in the degradability of individual organic moieties can be understood / explained as below.

In case of p-aminophenol both amino as well as hydroxy groups release electrons to the ring, which cancel the mesomeric / resonance effect of each other but at the same time cause greater localisation of electrons in the ring structure. Whereas, in case of p-nitrophenol, the nitro group only partially localises the electrons of the ring due to its electron withdrawing nature. Needless to mention here that amongst the three moieties, only phenol has a delocalised system of electrons contrary to localised systems in the other two moieties, hence least degradable.

15.1.7 Experiment

Object: Effect of ultrasound on various fractions of distilled crude oil.

Methodology: Various fraction of crude oil was distilled under different condition after being sonicated by an ultrasonic processor for different intervals of time. Density of these fractions was determined by dialatometer. Viscosities of these fractions were measured using Ubbelohde viscometer and the results given as under;

Fraction	Boiling range (°C)	Volume of distillate under atmospheric pressure (mL)	Volume of distillate under reduced pressure (mL)	Volume of distillate under atmospheric and reduced pressure (mL)
1	50–90	5	8	5
2	90–120	3	12	7
3	120–200	15	17	15

(*continued*)

Fraction	Boiling range (°C)	Volume of distillate under atmospheric pressure (mL)	Volume of distillate under reduced pressure (mL)	Volume of distillate under atmospheric and reduced pressure (mL)
4	200–240	8	14	14
5	240–300	39	8	18
Total recovery %		70	45	59

Fraction	Boiling range (°C)	Volume of distillates after sonication (mL)			
		15 min	30 min	45 min	60 min
1	50–90	9	7	4	4
2	90–120	8	6	4	5
3	120–200	8	9	15	11
4	200–240	9	8	9	12
5	240–300	40	40	49	49
6 (last fraction distilled under reduced pressure)	260	13	12	6	4
Total recovery %		87	82	87	85

Fraction	Density of unsonicated sample (g/mL)	Density of sonicatede sample (g/mL)
3	0.7586	0.7423
4	0.8016	0.7812
5	0.8489	0.8428

Sonication time (min)	Viscosity (milli poise)
0	23.7869
15	27.9016
30	27.342
45	45.0292
60	37.3712

Result: The recovery of different fractions of petroleum distillate under atmospheric pressure was more than under reduced pressure because at lower pressure the vapour pressure of lighter molecule of crude oil increased so that they were siphoned out from the system without being condensed. Whereas a combination of distillation of lighter fraction under normal atmospheric pressure followed by the distillation of heavier contents under reduced pressure showed an improvement in the recovery of petroleum products. Recovery of distillates was still more when crude oil was first sonicated and then distilled under normal and reduced pressures. The viscosity of distillate increased with sonication whereas there was a decrease in value of density.

15.1.8 Experiment

Object: Effect of ultrasound on the photocatalytic decomposition of KI using TiO_2 doped with Eu, Gd, Ce, Dy.

Theory: TiO_2 acts as photocatalyst due to generation of photoexcited electrons and holes which were involved in decomposition of KI. Under ultrasonic irradiation efficiency of iodine release increased almost linearly with irradiation time.

Procedure: 10% aqueous solution of potassium iodide, KI, when exposed to sunlight, liberated I_2 due to the photolytic decomposition and gave blue colour with freshly prepared starch solution. The intensity of blue coloured complex with the starch increased many fold when the same solution was kept in the ultrasonic cleaning bath. As an extension of the experiment, the photochemical decomposition of KI could be seen to be increasing in the presence of a photocatalyst, TiO_2, showing an additive effect of sonication and photocatalysis (sono-photocatalysis) However, the addition of different rare earth ions affect the process differently due to the different number of electrons in their valence shells.

Absorbance of iodine liberated from KI in different condition.

Condition/ photocatalysts	Absorbance of the liberated iodine-starch complex for different durations of sonication				
	5 min	10 min	15 min	20 min	25 min
Ultrasound,)))	0.099	0.143	0.145	0.149	0.152
TiO_2 +)))	0.210	0.269	0.403	0.462	0.547
Ce + TiO_2	0.232	0.237	0.225	0.221	0.217
Ce +TiO_2 +)))	0.298	0.322	0.299	0.295	0.289
Eu +TiO_2	0.215	0.379	0.504	0.576	0.648
Eu +TiO_2 +)))	0.248	0.391	0.525	0.591	0.667
Gd +TiO_2	0.312	0.407	0.475	0.547	0.618
Gd +TiO_2 +)))	0.412	0.528	0.576	0.638	0.715
Dy +TiO_2	0.258	0.267	0.249	0.247	0.262
Dy +TiO_2 +)))	0.295	0.309	0.299	0.262	0.235

Result: The decomposition of KI was faster in TiO_2 used along with ultrasound than in TiO_2 alone due to chemical effect produced by ultrasonic cavitation. The cerium doped TiO_2 lowered the photocatalytic effect and dysprosium doping did not affect the photocatalytic effect.The negative effect was due to anticipated transformation of anatase phase of TiO_2 to its rutile phase at 600°C during the thermolytic recombination of titanium with cerium. This blocked the surface sites required for photocatalysis. Eu and Gd doped TiO_2 increased the photocatalytic effect due to the faster rate of charge transfer of photogenerated carriers.

Similar small experiments, demonstrating the sterilization of potable water, reduction in the hardness of water, degradation of phenol, amines, potassium iodide and indicators, degradation of complexes, formation of complexes may still be added as found in the preceding chapters of this book.

15.2.1 Experiment

Objective: To produce, visualise and quantify different forms of luminescence produced by ultrasound: sonoluminescence and luminol sonochemical luminescence, and perform simple experiments to examine fundamental principles of acoustic cavitation.

Equipment and reagents: Ultrasound generator, pulse generator, photomultiplier tube, oscilloscope, light-insulated cabinet, rare gas source (e.g. argon), 3-aminophthalhydrazide (luminol), sodium hydroxide, alcohol or other volatile organic solutes.

Theory: Collapse of gas/vapour cavities in an acoustic field produces extremely high pressures and temperatures capable of causing the emission of light from the core of the collapsing cavity (sonoluminescence) and also the formation of oxidising radical species that can react in the solution with molecules, such as luminol, to produce a secondary, chemical luminescence.

Procedure: Set up an acoustic reactor in a light-proof cabinet with a photomultiplier (PM) tube positioned facing the cell as shown in Fig. 15.3a and b. Fill the cell with distilled water and close the cabinet. A potential should now be applied to the PM tube, the output (spectrally integrated) of which is produced on an oscilloscope (note that the ultrasound cell can easily be placed inside a commercial spectrometer in order to record the emission spectrum). Switch on the ultrasound and you should observe on the oscilloscope a change in voltage, directly proportional to the intensity of sonoluminescence emission. The following experiments can be performed to explore the different types of light emission and some of the factors that influence these emission processes.

Experiment 1: A pulse generator can be connected to the function generator in order to produce a pulsed waveform. Monitor the initial (*ca.* 100) pulses and explain the initial growth in the signal and the shape of the pulses. A typical initial growth profile is shown in Fig. 15.4.

a b

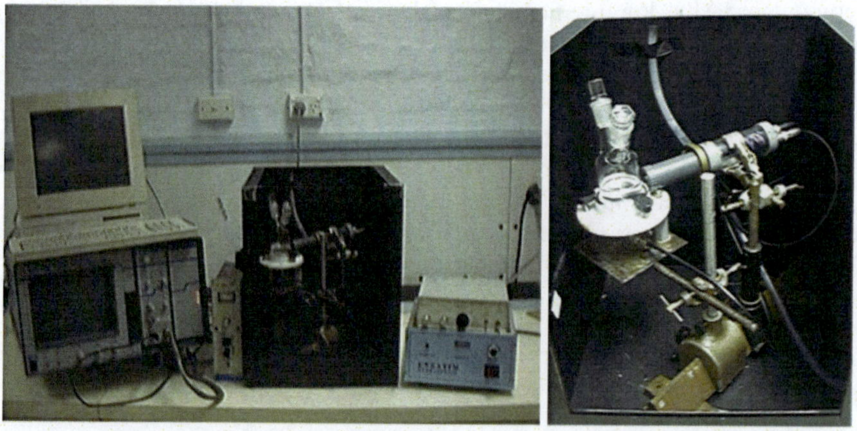

Fig. 15.3 Typical experimental arrangement for the study of multi-bubble sonoluminescence. The ultrasound transducer used here is 515 kHz and produces a standing wave pattern in the reaction cell. A horn-type sonifier (usually 20 kHz) can also be used in such an arrangement

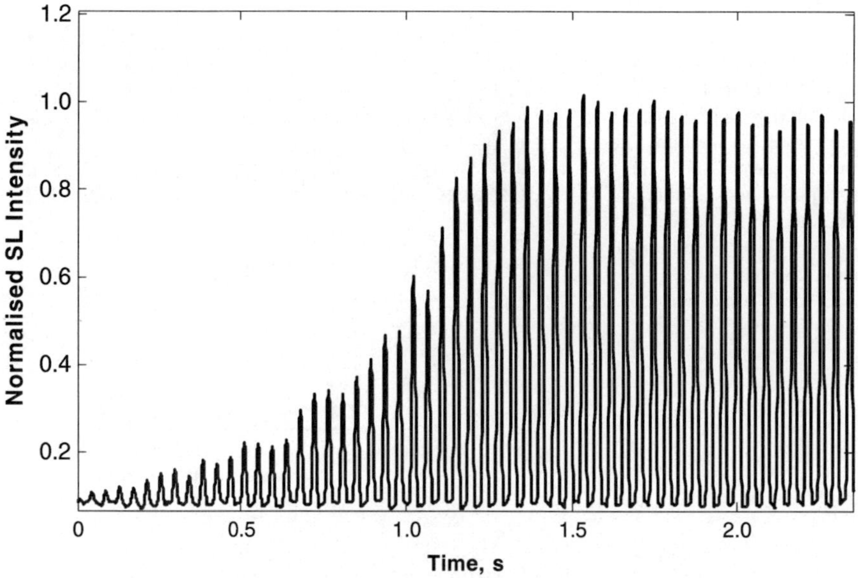

Fig. 15.4 Initial growth in sonoluminescence intensity from a pulsed sound field

Experiment 2: Saturate distilled water with a rare gas and compare the intensity of the signal with that from air. The luminosity will be enhanced in the rare gas saturated solutions. For any gas atmosphere, add small amounts of volatile water-soluble solutes (e.g. alkyl series alcohols) and quantify the quenching of sonoluminescence as a function of both bulk quencher concentration and surface excess. Good correlation between the extent of quenching and the Gibbs surface excess should be observed. Explain the changes in sonoluminescence intensity when a rare gas atmosphere is used and the quenching of volatile solutes, in terms of simple thermodynamics.

Experiment 3: In general, sonoluminescence emission is not discernable with the naked eye. The luminosity of the secondary emission from luminol (oxidised by sonochemically produced OH radicals) however, is several orders of magnitude brighter and is easily seen in a dark room. Prepare a 0.1 mM aqueous luminol solution in 0.1 M NaOH. Sonicate this solution and observe the emission pattern. This will appear as bands of light and dark if a standing wave reactor is used or in more elaborate forms in different reactors. If a 20 kHz horn is used, a cone shaped zone of luminescence will be observed. Explain the emission pattern.

The effect of gas type on this form of emission can also be tested either qualitatively with the naked eye or quantitatively using the PM tube and oscilloscope. Compare and explain the effect of oxygen, argon and air with recourse to the mechanism of chemiluminescence as well as thermodynamics.

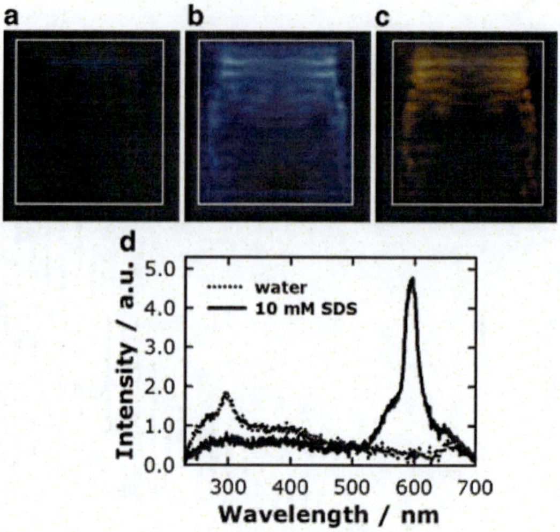

Fig. 15.5 Long-exposure photographs recorded for argon-saturated water (**a**) and luminol solution (**b**) and (**c**) 10 mM sodium dodecyl sulphate (SDS) solution. The emission spectra in (**d**) are from pure argon-saturated water and 10 mM SDS solution. Note the sodium D line at 589 nm. In this experiment, sonication was performed at 159 kHz

In the presence of an alkali salt, strong metal atom emission can be seen both in the emission spectrum and visually. This form of emission is described in detail in Chapter 13. Long-time exposure photographs comparing sonoluminescence and luminol and Na* sonochemical luminescence are shown in Fig. 15.5a–c.

Index